Finsler geometry generalizes Riemannian geometry in the same sense that Banach spaces generalize Hilbert spaces. This book presents an expository account of seven important topics in Riemann–Finsler geometry, which have recently undergone significant development but have not had a detailed pedagogical treatment elsewhere. Each article will open the door to an active area of research and is suitable for a special topics course in graduate-level differential geometry.

Álvarez and Thompson discuss the theory of volumes for normed spaces and Finsler spaces and show how it unifies a wide range of geometric inequalities. Bellettini studies the evolution of crystals, where the driving agent is the mean curvature of the facets. Aikou reviews the essential role played by Finsler metrics in complex differential geometry. Chandler and Wong explain why parametrized jet bundles admit only Finsler metrics and develop machinery which they use to prove the Kobayashi conjecture (1960) and a special case of the Green–Griffiths (1979) conjecture. Bao and Robles focus on the flag and Ricci curvatures of Finsler manifolds, with an emphasis on Einstein metrics of Randers type. Rademacher gives a detailed and new account of his Sphere Theorem for nonreversible Finsler metrics. Shen's article explains why Finsler manifolds are colorful objects and examines the interplay among the flag, S-, and Landsberg curvatures in Finsler geometry.

Mathematical Sciences Research Institute
Publications

50

A Sampler of Riemann–Finsler Geometry

Mathematical Sciences Research Institute Publications

Volumes 1–4 and 6–27 are published by Springer-Verlag

A Sampler of
Riemann–Finsler Geometry

Edited by

David Bao

University of Houston

Robert L. Bryant

Duke University

Shiing-Shen Chern

UC Berkeley and Nankai Institute of Mathematics

Zhongmin Shen

Indiana University – Purdue University Indianapolis

CAMBRIDGE
UNIVERSITY PRESS

CAMBRIDGE UNIVERSITY PRESS
Cambridge, New York, Melbourne, Madrid, Cape Town, Singapore,
São Paulo, Delhi, Dubai, Tokyo, Mexico City

Cambridge University Press
The Edinburgh Building, Cambridge CB2 8RU, UK

Published in the United States of America by Cambridge University Press, New York

www.cambridge.org
Information on this title: www.cambridge.org/9780521168731

First published 2004
First paperback edition 2010

A catalogue record for this publication is available from the British Library

Library of Congress Cataloguing in Publication Data

ISBN 978-0-521-83181-9 Hardback
ISBN 978-0-521-16873-1 Paperback

Riemann–Finsler Geometry
MSRI Publications
Volume **50**, 2004

Contents

Riemann–Finsler Geometry
MSRI Publications
Volume **50**, 2004

Preface

This volume contains seven expository articles and concerns three facets of Riemann–Finsler geometry that have undergone important recent developments:

1. The concept of volumes on normed spaces and Finsler manifolds, and crystalline motion by mean curvature in phase transitions.

2. The essential role played by Finsler metrics in complex manifold theory, together with the resolution of the Kobayashi conjecture and a special case of the Green–Griffiths conjecture.

3. The significance of the flag, Ricci, and S-curvatures of Finsler metrics, as well as the Sphere Theorem for nonreversible Finsler structures.

Conspicuously absent from the above are two highly geometrical areas: Bryant's use of exterior differential systems to understand Finsler metrics of constant flag curvature, and Foulon's dynamical systems approach to Finsler geometry. They are not included here because reasonable expositions already exist in a special Chern issue of the *Houston Journal of Mathematics* **28** (2002), 221–262 (Bryant) and 263–292 (Foulon).

Our goal is to render the aforementioned developments accessible to the differential geometry community at large. It is not our intention to present an encyclopedic picture of the field. What we do covet are concrete examples, instructive graphics, meaningful computations, and care in organizing technical arguments. The resulting articles appear to have met these criteria at an above-average level.

All the articles have been refereed. In fact, a total of 26 referee reports were obtained, some addressing the mathematics, others critiquing expository matters. After a few rounds of revision, each article was line-edited by at least one mathematician who is *not* familiar with the topic in question, in the hope that this would uncover most typographical mistakes. It is certain that, in spite of these attempts at quality control, layers of blemishes are still awaiting their turn to surface. We urge readers to bring those mistakes to the attention of one of us (Bao).

Thanks are due all reviewers who generously donated their time. We are also grateful to Dr. Silvio Levy, Editor of the MSRI Series, for his prompt responses to all our queries.

David Bao, Houston, USA
Robert L. Bryant, Durham, USA
Shiing-Shen Chern, Nankai, China
Zhongmin Shen, Indianapolis, USA

Riemann–Finsler Geometry
MSRI Publications
Volume **50**, 2004

Synopses

1. Part One

- **Álvarez–Thompson**: This paper is a concise introduction to the theory of volumes on normed and Finsler spaces. The definitions of Holmes–Thompson, Busemann, and Benson–Gromov are studied and their convexity (ellipticity) properties are discussed in detail. The authors show how the theory of volumes provides a unified context for a diverse range of geometric inequalities. The article is intended for students and researchers in differential, integral, and convex geometry.

- **Bellettini**: Crystalline motion driven by the mean curvature is an evolution process arising in material science and phase transitions. It is an anisotropic flow in an ambient space endowed with a piecewise linear norm. For three-dimensional crystals, the crystalline mean curvature of a facet is defined and identified with the initial velocity in the evolution process. Facets with constant crystalline mean curvature are important because they are expected not to break or curve under the evolution. The problem of characterizing such facets is discussed.

2. Part Two

- **Aikou**: This article highlights the essential role played by Finsler metrics in complex differential geometry. It describes a few situations for which techniques based solely on Hermitian metrics are hopelessly inadequate. These include Kobayashi's characterization of negative holomorphic vector bundles over compact complex manifolds, in terms of the existence of negatively curved pseudoconvex Finsler metrics.

- **Chandler–Wong**: The authors present the proof of the Kobayashi conjecture (1960) on the hyperbolicity of generic algebraic surfaces of degree $d \geqslant 5$ in $\mathbb{P}^3$. They also address the Green–Griffiths conjecture (1979) that every holomorphic map $f : \mathbb{C} \to X$ to a surface X of general type is algebraically degenerate. Their paper establishes the latter for the *special* case where X is minimal, $\mathrm{Pic}(X) \cong \mathbb{Z}$, and $p_g(X) > 0$. The main tool used is their generalization of the classical Schwarz lemma for complex curves, to varieties of every dimension. In this crucial step, algebraic geometric arguments are used to construct a Finsler metric of logarithmic type, thereby reducing the problem to one in which a certain estimate (the lemma of logarithmic derivatives) is applicable.

3. Part Three

- **Bao–Robles**: Many recent developments have advanced our understanding of the flag and Ricci curvatures of Finsler metrics. This paper is a uniform presentation of these results and their underlying techniques. Included is a geometric definition of Einstein–Finsler metrics. Einstein metrics of Randers type are studied via their representation as solutions to Zermelo navigation on Riemannian manifolds. This viewpoint leads to the classification of *all* constant flag curvature Randers metrics. It also yields a Schur lemma, and settles a question of rigidity in three dimensions, for Einstein–Randers metrics. The theory is illustrated with a diverse array of explicit examples.

- **Rademacher**: The author shows in detail how the classical Sphere Theorem in Riemannian geometry is extended to the case of nonreversible Finsler metrics. The proof hinges on a fruitful definition of the notion of reversibility, and how that can be used to effect some crucial estimates, such as the injectivity radius, the length of nonminimal geodesics between two fixed points, and the length of nonconstant geodesic loops. The proof also capitalizes on an idea of Klingenberg: that Morse theory of the energy functional allows us to circumvent Toponogov's comparison theorem for geodesic triangles. This idea renders irrelevant the "handicap" that, in Finsler geometry proper, there is no Toponogov's theorem.

- **Shen**: This paper is about the interaction between the generalized Riemann curvature and other non-Riemannian quantities in Finsler geometry. The latter include the S-curvature, whose vanishing is equivalent to having constant "distortion" along each geodesic. The S-curvature quantifies some aspect of the change in the Minkowski model ("infinitesimal color pattern") as one moves from one tangent space to another, along geodesics on a Finslerian landscape. The information it provides is complementary to that supplied by a certain contracted version of the Berwald curvature. These theoretical constructs are exemplified by Finsler metrics with a broad array of special curvature properties. The author also proves a number of local and global theorems using certain curvature equations along geodesics.

Riemann–Finsler Geometry
MSRI Publications
Volume **50**, 2004

Volumes on Normed and Finsler Spaces

J. C. ÁLVAREZ PAIVA AND A. C. THOMPSON

CONTENTS

1. Introduction

The study of volumes and areas on normed and Finsler spaces is a relatively new field that comprises and unifies large domains of convexity, geometric tomography, and integral geometry. It opens many classical unsolved problems in these fields to powerful techniques in global differential geometry, and suggests new challenging problems that are delightfully geometric and simple to state.

Keywords: Minkowski geometry, Hausdorff measure, Holmes–Thompson volume, Finsler manifold, isoperimetric inequality.

The theory starts with a simple question: How does one measure volume on a finite-dimensional normed space? At first sight, this question may seem either unmotivated or trivial: normed spaces are metric spaces and we can measure volume using the Hausdorff measure, period. However, if one starts asking simple, naive questions one discovers the depth of the problem. Even if one is unwilling to consider that definitions of volume other than the Hausdorff measure are not only possible but may even be better, one is faced with questions such as these: What is the $(n-1)$-dimensional Hausdorff measure of the unit sphere of an n-dimensional normed space? Do flat regions minimize area? For what normed spaces are metric balls also the solutions of the isoperimetric problem? These questions, first posed in convex-geometric language by Busemann and Petty [1956], are still open, at least in their full generality. However, one should not assume too quickly that the subject is impossible. Some beautiful results and striking connections have been found. For example, the fact that the $(n-1)$-Hausdorff measure in a normed space determines the norm is equivalent to the fact that the areas of the central sections determine a convex body that is symmetric with respect to the origin. This, in turn, follows from the study of the spherical Radon transform. The fact that regions in hyperplanes are area-minimizing is equivalent to the fact that the intersection body of a convex body that is symmetric with respect to the origin is also convex.

But the true interest of the theory can only be appreciated if one is willing to challenge Busemann's dictum that the natural volume in a normed or Finsler space is the Hausdorff measure. Indeed, thinking of a normed or Finsler space as an anisotropic medium where the speed of a light ray depends on its direction, we are led to consider a completely different notion of volume, which has become known as the *Holmes–Thompson volume.* This notion of volume, introduced in [Holmes and Thompson 1979], uncovers striking connections between integral geometry, convexity, and Hamiltonian systems. For example, in a recent series of papers, [Schneider and Wieacker 1997], [Alvarez and Fernandes 1998], [Alvarez and Fernandes 1999], [Schneider 2001], and [Schneider 2002], it was shown that the classical integral geometric formulas in Euclidean spaces can be extended to normed and even to projective Finsler spaces (the solutions of Hilbert's fourth problem) if the areas of submanifolds are measured with the Holmes–Thompson definition. That extensions of this kind are not possible with the Busemann definition was shown by Schneider [Schneider 2001].

Using Finsler techniques, Burago and Ivanov [2001] have proved that a flat two-dimensional disc in a finite-dimensional normed space minimizes area among all other immersed discs with the same boundary. Ivanov [2001] has shown, among other things, that Pu's isosystolic inequality for Riemannian metrics in the projective plane extends to Finsler metrics, and the Finslerian extension of Berger's infinitesimal isosystolic inequality for Riemannian metrics on real projective spaces of arbitrary dimension has been proved by Álvarez [2002].

Despite these and other recent interdisciplinary results, we believe that the most surprising feature of the Holmes–Thompson definition is the way in which it organizes a large portion of convexity into a coherent theory. For example, the sharp upper bound for the volume of the unit ball of a normed space is given by the Blaschke–Santaló inequality; the conjectured sharp lower bound is Mahler's conjecture; and the reconstruction of the norm from the area functional is equivalent to the famous Minkowski's problem of reconstructing a convex body from the knowledge of its curvature as a function of its unit normals.

In this paper, we have attempted to provide students and researchers in Finsler and global differential geometry with a clear and concise introduction to the theory of volumes on normed and Finsler spaces. To do this, we have avoided the temptation to use auxiliary Euclidean structures to describe the various concepts and constructions. While these auxiliary structures may render some of the proofs simpler, they hinder the understanding of the subject and make the application of the ideas and techniques to Finsler spaces much more cumbersome. We also believe that by presenting the results and techniques in intrinsic terms we can make some of the beautiful results of convexity more accessible and enticing to differential geometers.

In the course of our writing we had to make some tough choices as to what material should be left out as either too advanced or too specialized. At the end we decided that we would concentrate on the basic questions and techniques of the theory, while doing our best to exhibit the general abstract framework that makes the theory of volumes on normed spaces into a sort of Grand Unified Theory for many problems in convexity and Finsler geometry. As a result there is little Finsler geometry *per se* in the pages that follow. However, just as tensors, forms, spinors, and Clifford algebras developed in invariant form have immediate use in Riemannian geometry, the more geometric constructions with norms, convex bodies, and k-volume densities that make up the heart of this paper have immediate applications to Finsler geometry.

2. A Short Review of the Geometry of Normed Spaces

This section is a quick review of the geometry of finite-dimensional normed spaces adapted to the needs and language of Finsler geometry. Unless stated otherwise, *all vector spaces in this article are real and finite-dimensional.* We suggest that the reader merely browse through this section and come back to it if and when it becomes necessary.

DEFINITION 2.1. A *norm* on a vector space X is a function

$$\| \cdot \| : X \to [0, \infty)$$

satisfying the following properties of positivity, homogeneity, and convexity:

(1) If $\|\boldsymbol{x}\| = 0$, then $\boldsymbol{x} = \boldsymbol{0}$;

(2) If t is a real number, then $\|t\boldsymbol{x}\| = |t|\|\boldsymbol{x}\|$;

(3) For any two vectors $\boldsymbol{x}$ and $\boldsymbol{y}$ in X, $\|\boldsymbol{x} + \boldsymbol{y}\| \leq \|\boldsymbol{x}\| + \|\boldsymbol{y}\|$.

If $(X, \|\cdot\|)$ is a finite-dimensional normed space, the set

$$B_X := \{\boldsymbol{x} \in X : \|\boldsymbol{x}\| \leq 1\}$$

is the *unit ball* of X and the boundary of B_X, S_X, is its *unit sphere*. Notice that B_X is a compact, convex set with nonempty interior. In short, it is a *convex body* in X. Moreover, it is symmetric with respect to the origin. Conversely, if $B \subset X$ is a *centered convex body* (*i.e.*, a convex body symmetric with respect to the origin), it is the unit ball of the norm

$$\|\boldsymbol{x}\| := \inf \{t \geq 0 : t\boldsymbol{y} = \boldsymbol{x} \text{ for some } \boldsymbol{y} \in B\}.$$

We shall now describe various classes of normed spaces that will appear repeatedly throughout the paper.

Euclidean spaces. A Euclidean structure on a finite-dimensional vector space X is a choice of a symmetric, positive-definite quadratic form $\mathcal{Q} : X \to \mathbb{R}$. The normed space $(X, \mathcal{Q}^{1/2})$ will be called a Euclidean space. Note that a normed space is Euclidean if and only if its unit sphere is an ellipsoid, which is if and only if the norm satisfies the parallelogram identity:

$$\|\boldsymbol{x} + \boldsymbol{y}\|^2 + \|\boldsymbol{x} - \boldsymbol{y}\|^2 = 2\|\boldsymbol{x}\|^2 + 2\|\boldsymbol{y}\|^2.$$

EXERCISE 2.2. Let $B \subset \mathbb{R}^n$ be a convex body symmetric with respect to the origin. Show that if the intersection of B with every 2-dimensional plane passing through the origin is an ellipse, then B is an ellipsoid.

The ℓ_p spaces. If $p \geq 1$ is a real number, the function

$$\|\boldsymbol{x}\|_p := \left(|x_1|^p + \cdots + |x_n|^p\right)^{1/p}$$

is a norm on $\mathbb{R}^n$. When p tends to infinity, $\|\boldsymbol{x}\|_p$ converges to

$$\|\boldsymbol{x}\|_\infty := \max\{|x_1|, \ldots, |x_n|\}.$$

The normed space $(\mathbb{R}^n, \|\cdot\|_p)$, $1 \leq p \leq \infty$, is denoted by ℓ_p^n.

The unit ball of ℓ_∞^n is the n-dimensional cube of side length two, while the unit ball of ℓ_1^n is the n-dimensional *cross-polytope*. In general, norms whose unit balls are polytopes are called *crystalline norms*.

Subspaces of $L_1([0,1], dx)$. The space of measurable functions $f : [0,1] \to \mathbb{R}$ satisfying

$$\|f\| := \int_0^1 |f(x)|\, dx < \infty$$

is a normed space denoted by $L_1([0,1], dx)$. The geometry of finite-dimensional subspaces of $L_1([0,1], dx)$ is closely related to problems of volume, area, and integral geometry on normed and Finsler spaces. In the next few paragraphs, we will summarize the properties of these subspaces that will be used in the rest of the paper. For proofs, references, and to learn more about hypermetric spaces, we recommend the landmark paper [Bolker 1969], as well as the surveys [Schneider and Weil 1983] and [Goodey and Weil 1993].

First we begin with a beautiful metric characterization of the subspaces of $L_1([0,1], dx)$.

DEFINITION 2.3. A metric space (M, d) is said to be *hypermetric* if it satisfies the following stronger version of the triangle inequality: If $m_1, \ldots, m_k$ are elements of M and $b_1, \ldots, b_k$ are integers with $\sum_i b_i = 1$, then

$$\sum_{i,j=1}^{k} d(m_i, m_j) b_i b_j \leq 0.$$

THEOREM 2.4. *A finite-dimensional normed space is hypermetric if and only if it is isometric to a subspace of $L_1([0,1], dx)$.*

An important analytic characterization of a hypermetric normed space can be given through the Fourier transform of its norm:

THEOREM 2.5. *A norm on $\mathbb{R}^n$ is hypermetric if and only if its distributional Fourier transform is a nonnegative measure.*

The characterizations above, important as they are, are hard to grasp at first sight. A much more visual approach will be given after we review the duality of normed spaces.

Minkowski spaces. Minkowski spaces are normed spaces with strict smoothness and convexity properties. In precise terms, a norm $\| \cdot \|$ on a vector space X is said to be a *Minkowski norm* if it is smooth outside the origin and the Hessian of the function $\| \cdot \|^2$ at every nonzero point is a positive-definite quadratic form.

The unit sphere of a Minkowski space X is a smooth convex hypersurface S_X such that for any Euclidean structure on X the principal curvatures of S_X are positive.

2.1. Maps between normed spaces.
An important feature of the geometry of normed spaces is that the space of linear maps between two normed spaces carries a natural norm.

DEFINITION 2.6. If $T : X \to Y$ is a linear map between normed spaces $(X, \| \cdot \|_X)$ and $(Y, \| \cdot \|_Y)$, we define the *norm of T* as the supremum of $\|Tx\|_Y$ taken over all vectors $x \in X$ with $\|x\|_X \leq 1$.

A linear map $T : X \to Y$ is said to be *short* if its norm is less than or equal to one. In other words, a short linear map does not increase distances. Two important types of short linear maps between normed spaces are isometric embeddings and isometric submersions:

DEFINITION 2.7. An injective linear map $T : X \to Y$ between normed spaces $(X, \| \cdot \|_X)$ and $(Y, \| \cdot \|_Y)$ is said to be an *isometric embedding* if $\|T\boldsymbol{x}\|_Y = \|\boldsymbol{x}\|_X$ for all vectors $\boldsymbol{x} \in X$.

DEFINITION 2.8. A surjective linear map $T : X \to Y$ between normed spaces $(X, \| \cdot \|_X)$ and $(Y, \| \cdot \|_Y)$ is said to be an *isometric submersion* if

$$\|T\boldsymbol{x}\|_Y = \inf \{\|\boldsymbol{v}\|_X : \boldsymbol{v} \in X \text{ and } T\boldsymbol{v} = T\boldsymbol{x}\}$$

for all vectors $x \in X$.

In terms of the unit balls, $T : X \to Y$ is an isometric embedding if and only if $T(B_X) = T(X) \cap B_Y$, and T is an isometric submersion if and only if $T(B_X) = B_Y$.

2.2. Dual spaces and polar bodies. From the previous paragraph, we see that if $(X, \| \cdot \|)$ is a normed space, then the set of all linear maps onto the one-dimensional normed space $(\mathbb{R}, | \cdot |)$ carries a natural norm. The resulting normed space is called the *dual* of $(X, \| \cdot \|)$ and is denoted by $(X^*, \| \cdot \|^*)$. It is easy to see that the double dual (*i.e.*, the dual of the dual) of a finite-dimensional normed space can be naturally identified with the space itself. The unit ball of $(X^*, \| \cdot \|^*)$ is said to be the *polar* of the unit ball of $(X, \| \cdot \|)$.

Example. Hölder's inequality implies that if $p > 1$, the dual of ℓ_p^n is ℓ_q^n, where $1/p + 1/q = 1$. Likewise, it is easy to see that the dual of ℓ_1^n is ℓ_∞^n.

If $T : X \mapsto Y$ is a linear map then the *dual map* $T^* : Y^* \mapsto X^*$ is defined by

$$(T^*\boldsymbol{\xi})(\boldsymbol{x}) = \boldsymbol{\xi}(T\boldsymbol{x}).$$

Note that $\|T^*\| = \|T\|$.

EXERCISE 2.9. Show that if $T : X \to Y$ is an isometric embedding between normed spaces X and Y, the dual map $T^* : Y^* \to X^*$ is an isometric submersion.

Many of the geometric constructions in convex geometry and the geometry of normed spaces are functorial. More precisely, if we denote by $\mathcal{N}$ the category whose objects are finite-dimensional normed spaces and whose morphisms are short linear maps, many classical constructions define functors from $\mathcal{N}$ to itself.

EXERCISE 2.10. Show that the assignment $(X, \| \cdot \|) \mapsto (X^*, \| \cdot \|^*)$ is a contravariant functor from $\mathcal{N}$ to $\mathcal{N}$.

Duals of hypermetric normed spaces. As advertised earlier in this section, the notion of duality allows us to give a more geometric characterization of hypermetric spaces.

DEFINITION 2.11. A polytope in a vector space X is said to be a *zonotope* if all of its faces are symmetric. A convex body is said to be a *zonoid* if it is the limit (in the Hausdorff topology) of zonotopes.

Notice that an n-dimensional cube, as well as all its linear projections, are zonotopes. In fact, it can be shown that any zonotope is the linear projection of a cube (see, for example, Theorem 3.3 in [Bolker 1969]).

THEOREM 2.12. *Let X be a finite-dimensional normed space with unit ball B_X. The dual of X is hypermetric if and only if B_X is a zonoid.*

Notice that this immediately implies that the space ℓ_1^n, $n \geq 1$, is hypermetric.

Duality in Minkowski spaces. If $(X, \|\cdot\|)$ is a Minkowski space, the differential of the function $L := \|\cdot\|^2/2$,

$$dL(\boldsymbol{x})(\boldsymbol{y}) := \frac{1}{2}\frac{d}{dt}\|\boldsymbol{x} + t\boldsymbol{y}\|_{t=0}^2,$$

is a continuous linear map from X to X^* that is smooth outside the origin and homogeneous of degree one. This map is usually called the *Legendre transform*, although that term is also used to describe some related concepts (see, for example, §2.2 in [Hörmander 1994]). The following exercise describes the most important properties of the Legendre transform.

EXERCISE 2.13. Let $(X, \|\cdot\|)$ be a Minkowski space and let

$$\mathcal{L} : X \setminus \mathbf{0} \to X^* \setminus \mathbf{0}$$

be its Legendre transform.

(1) Show that if $\boldsymbol{x} \in X$ is a unit vector, then $\mathcal{L}(\boldsymbol{x})$ is the unique covector $\boldsymbol{\xi} \in X^*$ such that the equation $\boldsymbol{\xi} \cdot \boldsymbol{y} = 1$ describes the tangent plane to the unit sphere S_X at the point $\boldsymbol{x}$.
(2) Show that the Legendre transform defines a diffeomorphism between the unit sphere and its polar.
(3) Show that the inverse of the Legendre transform from $X \setminus \mathbf{0}$ to $X^* \setminus \mathbf{0}$ is just the Legendre transform from $X^* \setminus \mathbf{0}$ to $X \setminus \mathbf{0}$.
(4) Show that the Legendre transform is linear if and only if X is a Euclidean space.

EXERCISE 2.14. Show that a normed space is a Minkowski space if its unit sphere and the unit sphere of its dual are smooth.

2.3. Sociology of normed spaces. If $\|\cdot\|_1$ and $\|\cdot\|_2$ are two norms on a finite-dimensional vector space X, it is easy to see that there are positive numbers m and M such that

$$m\|\cdot\|_2 \leq \|\cdot\|_1 \leq M\|\cdot\|_2.$$

If we take the numbers m and M such that the inequalities are sharp, then $\log(M/m)$ is a good measure of how far away one norm is from the other.

For example, the following well-known result states that we can always approximate a norm by one whose unit sphere is a polytope or by one such that its unit sphere and the unit sphere of its dual are smooth.

PROPOSITION 2.15. *Let* $\| \cdot \|$ *be a norm on the finite-dimensional vector space* X. *Given* $\varepsilon > 0$, *there exist a crystalline norm* $\| \cdot \|_1$ *and a Minkowski norm* $\| \cdot \|_2$ *on* X *such that*

$$\| \cdot \|_1 \leq \| \cdot \| \leq (1+\varepsilon)\| \cdot \|_1,$$
$$\| \cdot \|_2 \leq \| \cdot \| \leq (1+\varepsilon)\| \cdot \|_2.$$

For a short proof see Lemma 2.3.2 in [Hörmander 1994] .

In many circumstances, one wants to measure how far is one normed space from being isometric to another. The straightforward adaptation of the previous idea leads us to the following notion:

DEFINITION 2.16. *The Banach–Mazur distance* between n-dimensional normed spaces X and Y, is the infimum of the numbers $\log(\|T\|\|T^{-1}\|)$, where T ranges over all invertible linear maps from X to Y.

Notice that the Banach–Mazur distance is a distance on the set of isometry classes of n-dimensional normed spaces: two such spaces are at distance zero if and only if they are isometric.

An important question is to determine how far a general n-dimensional normed space is from being Euclidean. The answer rests on two results of independent interest:

THEOREM 2.17 (LOEWNER). *If B is a convex body in an n-dimensional vector space X, there exists a unique n-dimensional ellipsoid $E \subset B$ such that for any Lebesgue measure on X, the ratio* $\mathrm{vol}(B)/\mathrm{vol}(E)$ *is minimal.*

THEOREM 2.18 [John 1948]. *Let X be an n-dimensional normed space with unit ball B. If $E \subset B$ is the Loewner ellipsoid of B, then*

$$E \subset B \subset \sqrt{n}E.$$

EXERCISE 2.19. Show that the Banach–Mazur distance from an n-dimensional normed space to a Euclidean space is at most $\log(n)/2$.

The structure of the set of isometry classes of n-dimensional normed spaces is given by the following theorem (see [Thompson 1996, page 73] for references and some of the history on the subject):

THEOREM 2.20. *The set of isometry classes of n-dimensional normed spaces, $\mathcal{M}_n$, provided with the Banach–Mazur distance is a compact, connected metric space.*

The *Banach–Mazur compactum*, $\mathcal{M}_n$, enters naturally into Finsler geometry by the following construction: Let $\pi : \zeta \to M$ be a vector bundle with n-dimensional fibers such that every fiber $\zeta_m = \pi^{-1}(m)$ carries a norm that varies continuously with the base point (a *Finsler bundle*). The (continuous) map

$$\mathfrak{J} : M \longrightarrow \mathcal{M}_n$$

that assigns to each point $m \in M$ the isometry class of ζ_m measures how the norms vary from point to point.

Currently, there are not many results that describe the map $\mathfrak{I}$ under different geometric and/or topological hypotheses on the bundle. However the following exercise (and its extension in [Gromov 1967]) shows that such results are possible.

EXERCISE 2.21. Let $\pi : \zeta \to S^2$ be a Finsler bundle whose fibers are 2-dimensional. Show that if the bundle is nontrivial and the map $\mathfrak{I}$ is constant, then the image of S^2 under $\mathfrak{I}$ is the isometry class of 2-dimensional Euclidean spaces.

A corollary of this exercise is that if X is a three-dimensional normed space such that all its two-dimensional subspaces are isometric, then X is Euclidean. Another interesting corollary is that a Berwald (Finsler) metric on S^2 must be Riemannian.

3. Volumes on Normed Spaces

In defining the notion of volume on normed spaces, it is best to adopt an axiomatic approach. We shall impose some minimal set of conditions that are reasonable and then try to find out to what extent they can be satisfied, and to what point they determine our choices.

In a normed space, all translations are isometries. This suggests that we require the volume of a set to be invariant under translations. Since any finite-dimensional normed space is a locally compact, commutative group, we can apply the following theorem of Haar:

THEOREM 3.1. *If μ is a translation-invariant measure on $\mathbb{R}^n$ for which all compact sets have finite measure and all open sets have positive measure, then μ is a constant multiple of the Lebesgue measure.*

Proofs of this theorem can be found in many places. A full account is given in [Cohn 1980] and an abbreviated version in [Thompson 1996].

In the light of Haar's theorem, in order to give a definition of volume in *every* normed space, we must assign to every normed space X a multiple of the Lebesgue measure. Since the Lebesgue measure is not intrinsically defined (it depends on a choice of basis for X), it is best to describe this assignment as a choice of a norm μ in the 1-dimensional vector space $\Lambda^n X$, where n is the dimension of X; if $\boldsymbol{x}_1, \ldots, \boldsymbol{x}_n \in X$, we define $\mu(\boldsymbol{x}_1 \wedge \boldsymbol{x}_2 \wedge \cdots \wedge \boldsymbol{x}_n)$ as the volume of the parallelotope formed by these vectors.

Another natural requirement is *monotonicity*: if X and Y are n-dimensional normed spaces and $T : X \to Y$ is a short linear map (*i.e.*, a linear map that does not increase distances), we require that T does not increase volumes. Notice that this implies that isometries between normed spaces are volume-preserving.

The monotonicity requirement makes a definition of volume on normed spaces into a functor from $\mathcal{N}$ to itself that takes the n-dimensional normed space $(X, \|\cdot\|)$ to the 1-dimensional normed space $(\Lambda^n X, \mu)$. While we shall often abandon this viewpoint, it is a guiding principle throughout the paper with which we would like to acquaint the reader early on.

DEFINITION 3.2. A *definition of volume* on normed spaces assigns to every n-dimensional, $n \geq 1$, normed space X a normed space $(\Lambda^n X, \mu_X)$ with the following properties:

(1) If X and Y are n-dimensional normed spaces and $T : X \to Y$ is a short linear map , then the induced linear map $T_* : \Lambda^n X \to \Lambda^n Y$ is also short.
(2) The map $X \mapsto (\Lambda^n X, \mu_X)$ is continuous with respect to the topology induced by the Banach–Mazur distance.
(3) If X is Euclidean, then μ_X is the standard Euclidean volume on X.

Before presenting the principal definitions of volume in normed spaces, let us make the first link between these concepts and the affine geometry of convex bodies.

EXERCISE 3.3. Assume we have a definition of volume in normed spaces and use it to assign a number to any centrally symmetric convex body $B \subset \mathbb{R}^n$ by the following procedure: Consider $\mathbb{R}^n$ as the normed space X whose unit ball is B and compute

$$\mathcal{V}(B) := \mu_X(B) = \int_B \mu_X.$$

Show that if $T : \mathbb{R}^n \to \mathbb{R}^n$ is an invertible linear map, then $\mathcal{V}(B) = \mathcal{V}(T(B))$, and write the monotonicity condition in terms of the affine invariant $\mathcal{V}$.

Notice that we can turn the tables and start by considering a suitable affine invariant $\mathcal{V}$ of centered convex bodies and give a definition of volume in normed spaces by prescribing that the volume of the unit ball B of a normed space X be given by $\mathcal{V}(B)$.

EXERCISE 3.4. Let μ be a definition of volume for 2-dimensional normed spaces. Use John's theorem to show that if B is the unit disc of a two-dimensional normed space X, then $\pi/2 \leq \mu_X(B) \leq 2\pi$.

3.1. Examples of definitions of volume in normed spaces. The study of the four definitions of volume we shall describe below makes up the most important part of the theory of volumes on normed and Finsler spaces.

The Busemann definition. The Busemann volume of an n-dimensional normed space is that multiple of the Lebesgue measure for which the volume of the unit ball equals the volume of the Euclidean unit ball in dimension n, ε_n, . In other words, we have chosen as our affine invariant the constant ε_n, where n is the dimension of the space.

Another way to define the Busemann volume of a normed space X is by setting

$$\mu^b(\boldsymbol{x}_1 \wedge \boldsymbol{x}_2 \wedge \cdots \wedge \boldsymbol{x}_n) = \frac{\varepsilon_n}{\mathrm{vol}(B; \boldsymbol{x}_1 \wedge \boldsymbol{x}_2 \wedge \cdots \wedge \boldsymbol{x}_n)},$$

where the notation $\mathrm{vol}(B; \boldsymbol{x}_1 \wedge \boldsymbol{x}_2 \wedge \cdots \wedge \boldsymbol{x}_n)$ indicates the volume of B in the Lebesgue measure determined by the basis $\boldsymbol{x}_1, \ldots, \boldsymbol{x}_n$.

Using Brunn–Minkowski theory, Busemann showed in [1947] that the Busemann volume of an n-dimensional normed space equals its n-dimensional Hausdorff measure. Hence, from the viewpoint of metric geometry, this is a very natural definition.

EXERCISE 3.5. Show that the Busemann definition of volume satisfies the axioms in Definition 3.2.

The Holmes–Thompson definition. Let X be an n-dimensional normed space and let $B^* \subset X^*$ be the dual unit ball. If $\boldsymbol{x}_1, \ldots, \boldsymbol{x}_n$ is a basis of X and $\boldsymbol{\xi}_1, \ldots, \boldsymbol{\xi}_n$ is its dual basis, define

$$\mu^{\mathrm{ht}}(\boldsymbol{x}_1 \wedge \boldsymbol{x}_2 \wedge \cdots \wedge \boldsymbol{x}_n) := \varepsilon_n^{-1} \, \mathrm{vol}(B^*; \boldsymbol{\xi}_1 \wedge \boldsymbol{\xi}_2 \wedge \cdots \wedge \boldsymbol{\xi}_n).$$

Another way of defining the Holmes–Thompson volume is by considering the set $B \times B^*$ in the product space $X \times X^*$. Since $X \times X^*$ has a natural symplectic structure defined by

$$\omega((\boldsymbol{x}_1, \boldsymbol{\xi}_1), (\boldsymbol{x}_2, \boldsymbol{\xi}_2)) := \boldsymbol{\xi}_2(\boldsymbol{x}_1) - \boldsymbol{\xi}_1(\boldsymbol{x}_2),$$

it has a canonical volume (the *symplectic* or *Liouville* volume) defined by the n-th exterior power ω^n of ω, divided by $n!$. The Holmes–Thompson volume of the n-dimensional normed space X is the multiple of the Lebesgue measure for which the volume of the unit ball equals the Liouville volume of $B \times B^*$ divided by the volume of the Euclidean unit ball of dimension n. We mention in passing that in convex geometry it is usual to denote the Liouville volume of $B \times B^*$ as the *volume product* of B, $\mathrm{vp}(B)$.

The Holmes–Thompson definition — introduced in [Holmes and Thompson 1979] — was originally motivated by purely geometric considerations. However, from the physical point of view it is the natural definition of volume if we think of normed spaces as homogeneous, anisotropic media: media in which the speed of light varies with the direction of the light ray, but not with the point at which the propagation of light originates.

It is interesting to remark that the Busemann definition and the Holmes–Thompson definition are *dual functors:* to obtain the Holmes–Thompson volume of an n-dimensional normed space X we pass to the dual normed space X^*, we apply the "Busemann functor" to obtain $(\Lambda^n X^*, \mu^b_{X^*})$ and then pass to the dual of the normed space $(\Lambda^n X^*, \mu^b_{X^*})$.

EXERCISE 3.6. Consider a definition of volume $(X, \|\cdot\|) \mapsto (\Lambda^n X, \mu_X)$, where n is the dimension of X, and define its *dual definition* by the map $(X, \|\cdot\|) \mapsto$

$(\Lambda^n X, \mu_X^*) := (\Lambda^n X^*, \mu_{X^*})^*$. Show that μ^* also satisfies the axioms in Definition 3.2.

The notion of duality is somewhat mysterious and is closely related to the duality between intersections and projections proposed in [Lutwak 1988], and which led to the development of the dual Brunn–Minkowski theory. We shall have a little more to say about this duality after presenting a second dual pair of volume definitions.

Gromov's mass. If X is an n-dimensional normed space, define $\mu^m : \Lambda^n X \to [0, \infty)$ by the formula

$$\mu^m(a) := \inf\left\{ \prod_{i=1}^{n} \|\boldsymbol{x}_i\| : \boldsymbol{x}_1 \wedge \boldsymbol{x}_2 \wedge \cdots \wedge \boldsymbol{x}_n = a \right\}.$$

Another way to define the mass of an n-dimensional normed space X is as the multiple of the Lebesgue measure for which the volume of the maximal cross-polytope inscribed to the unit ball is $2^n/n!$.

EXERCISE 3.7. Consider the 2-dimensional normed space whose unit disc D is a regular hexagon. What is $\mu^m(D)$?

The Benson definition or Gromov's mass∗. One way to make the Benson definition is as the dual of mass: given an n-dimensional normed space X together with a basis $\boldsymbol{x}_1, \ldots, \boldsymbol{x}_n$, we take the dual basis $\boldsymbol{\xi}_1, \ldots, \boldsymbol{\xi}_n$ in X^* and define

$$\mu_X^{m*}(\boldsymbol{x}_1 \wedge \boldsymbol{x}_2 \wedge \cdots \wedge \boldsymbol{x}_n) := \frac{1}{\mu_{X^*}^m(\boldsymbol{\xi}_1 \wedge \boldsymbol{\xi}_2 \wedge \cdots \wedge \boldsymbol{\xi}_n)}.$$

This is Gromov's definition [1983]. Benson [1962] originally defined the mass∗ of an n-dimensional normed space as the multiple of the Lebesgue measure for which the volume of a minimal parallelotope circumscribed to the unit ball is 2^n.

EXERCISE 3.8. Consider the 2-dimensional normed space whose unit disc D is a regular hexagon. What is $\mu^{m*}(D)$?

The following exercise gives a third characterization of mass∗.

EXERCISE 3.9. Let X be an n-dimensional normed space and let B be its unit ball. A basis $\boldsymbol{\xi}_1, \ldots, \boldsymbol{\xi}_n$ of X^* is said to be *short* if $|\boldsymbol{\xi}_i(\boldsymbol{x})| \leq 1$ for all $\boldsymbol{x} \in B$ and all i, $1 \leq i \leq n$ (*i.e.*, if all the vectors in the basis are in the dual unit ball). Show that for any n-vector $a \in \Lambda^n X$

$$\mu^{m*}(a) = \sup\{|\boldsymbol{\xi}_1 \wedge \boldsymbol{\xi}_2 \wedge \cdots \wedge \boldsymbol{\xi}_n (a)| : \boldsymbol{\xi}_1, \ldots, \boldsymbol{\xi}_n \text{ is a short basis of } X^*\}$$

It is not hard to come up with other definitions of volume. For example, instead of considering inscribed cross-polytopes and circumscribed parallelotopes one might consider maximal inscribed or minimal circumscribed ellipsoids (as in Loewner's theorem cited above) and then specify the volume of either to be

ε_n. However, as we shall see in the next two sections, a good definition of volume must satisfy some additional conditions that are very hard to verify. The examples given above are important mainly because their study provides a common context to many problems in convex, integral, and differential geometry.

3.2. The volume of the unit ball. If we are given a definition of volume and a normed space, we would like to compute the volume of the unit ball. This is, of course, trivial if we work with the Busemann definition, but for the other definitions it is a challenging problem. Let us start with some simple experiments.

EXAMPLE 3.10. In the table below we use 7 different norms in $\mathbb{R}^3$ whose unit balls are, in order, the Euclidean unit ball; the cube with vertices at $(\pm 1, \pm 1, \pm 1)$; the octahedron with vertices at $\pm(1, 0, 0)$, $\pm(0, 1, 0)$, $\pm(0, 0, 1)$; the right cylinder over the unit circle in the xy-plane and with $-1 \leq z \leq 1$; its dual, the double cone which is the convex hull of the unit circle in the xy-plane and the points $\pm(0, 0, 1)$; the affine image of the cuboctahedron that has vertices at $\pm(1, 0, 0)$, $\pm(0, 1, 0)$, $\pm(0, 0, 1)$, $\pm(1, -1, 0)$, $\pm(1, 0, -1)$, $\pm(0, 1, -1)$; and its dual, the affine image of the rhombic dodecahedron, that has vertices at $\pm(1, 1, 1)$, $\pm(0, 1, 1)$, $\pm(1, 0, 1)$, $\pm(1, 1, 0)$, $\pm(0, 0, 1)$, $\pm(0, 1, 0)$, $\pm(1, 0, 0)$. These are listed in the first column. In the subsequent columns are the volumes of each unit ball using the different definitions of volume.

The ball B	$\mu^b(B)$	$\mu^{ht}(B)$	$\mu^{m*}(B)$	$\mu^m(B)$
ball	$4\pi/3$	$4\pi/3$	$4\pi/3$	$4\pi/3$
cube	$4\pi/3$	$8/\pi$	8	2
octahedron	$4\pi/3$	$8/\pi$	$16/3$	$4/3$
cylinder	$4\pi/3$	π	2π	π
double cone	$4\pi/3$	π	$4\pi/3$	$2\pi/3$
cuboctahedron	$4\pi/3$	$10/\pi$	$20/3$	$10/3$
rhombic dodecahedron	$4\pi/3$	$10/\pi$	4	2

EXERCISE 3.11. Verify these numbers.

Given a definition of volume, an interesting problem is to determine sharp upper and lower bounds for the volume of the unit balls of n-dimensional normed spaces. In the case of the Holmes–Thompson definition, this question has a classical reformulation: give sharp upper and lower bounds for the volume product of an n-dimensional centrally-symmetric convex body.

THEOREM 3.12 (BLASCHKE–SANTALÓ INEQUALITY). *The Holmes–Thompson volume of the unit ball of an n-dimensional normed space is less than or equal to the volume of the Euclidean unit ball of dimension n. Moreover, equality holds if and only if the space is Euclidean.*

The sharp lower bound for the Holmes–Thompson volume of unit balls is a reformulation of a long-standing conjecture of Mahler [1939]:

CONJECTURE. *The Holmes–Thompson volume of the unit ball of an n-dimensional normed space is greater than or equal to $4^n/\varepsilon_n n!$. Moreover, equality holds if and only if the unit ball is a parallelotope or a cross-polytope.*

This conjecture has been verified by Mahler [1939] in the two-dimensional case and by Reisner [1985, 1986] in the case when either the normed space or its dual is hypermetric.

For $\mu^{m*}(B)$ the upper bound of 2^n is attained for a parallelotope and for $\mu^m(B)$ the equivalent lower bound of $2^n/n!$ is attained by a cross-polytope. One also has $\mu^{m*}(B) \geq 2^n/n!$ and $\mu^m(B) \leq 2^n$ but these are far from optimal; better bounds will be obtained after studying the relationship between the different definitions of volume.

3.3. Relationship between the definitions of volume.

There are several relationships between the various measures we are considering. For example, the Blaschke–Santaló inequality is clearly equivalent to the following theorem:

THEOREM 3.13. *If X is an n-dimensional normed space, then $\mu_X^{ht} \leq \mu_X^b$ with equality if and only if X is Euclidean.*

For mass and mass∗ we have the following inequality:

PROPOSITION 3.14. *If X is an n-dimensional normed space, then $\mu_X^m \leq \mu_X^{m*}$.*

PROOF. Let P be a minimal circumscribed parallelotope to the unit ball B. Then (see for example [Thompson 1999], but there are many other possible references) the midpoint of each facet of P is a point of contact with B. The convex hull of these midpoints is a cross-polytope C inscribed to B. Also, if P is given the volume 2^n, then C has volume $2^n/n!$. Hence, in this situation, a *maximal* inscribed cross-polytope will have volume greater than or equal to $2^n/n!$. □

THEOREM 3.15. *If X is an n-dimensional normed space, then $\mu_X^m \leq \mu_X^b$ with equality if and only if X is Euclidean.*

The proof depends on the following theorem of McKinney [1974]:

THEOREM 3.16. *Let $K \subset X$ be a convex set symmetric about the origin and let S be a maximal simplex contained in K with one vertex at the origin, then for any Lebesgue measure λ*

$$\lambda(S)/\lambda(K) \geq 1/n!\varepsilon_n$$

with equality if and only if K is an ellipsoid.

PROOF OF THEOREM 3.15. If B is the unit ball of X then $\mu^b(B) = \varepsilon_n$ and $\mu^m(B) = 2^n\lambda(B)/n!\lambda(C)$ where C is a maximal cross-polytope inscribed to B. Moreover, C is the convex hull of $S \cup -S$, where S is a maximal simplex inscribed to B with one vertex at the origin. It follows from the theorem that $\lambda(C)/\lambda(B) \geq 2^n/n!\varepsilon_n$ which, upon rearrangement, gives $\mu^b(B) \geq \mu^m(B)$. □

The relationship between mass* and the Holmes–Thompson volume follows from Theorem 3.15 and the following simple exercise:

EXERCISE 3.17. Let μ and ν be two definitions of volume, and let μ^* and ν^* be their dual definitions. Show that if for every normed space X

$$\mu_X \le \nu_X, \quad \text{then} \quad \nu_X^* \le \mu_X^*.$$

COROLLARY 3.18. *If X is an n-dimensional normed space, then $\mu_X^{\mathrm{ht}} \le \mu_X^{m*}$ with equality if and only if X is Euclidean.*

The previous inequalities are summarized by the diagram

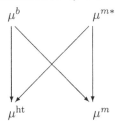

Notice that as a consequence of the Mahler–Reisner inequality we have the following lower bounds for the mass and mass* of unit balls in normed spaces and their duals.

COROLLARY 3.19. *For any unit ball B, we have $\mu^m(B) \le \varepsilon_n$ and, if B is either a zonoid or the dual of a zonoid, $\mu^{m*}(B) \ge 4^n/n!\varepsilon_n$.*

PROBLEM. *Is the mass* of the unit ball of an n-dimensional normed space at least $4^n/n!\varepsilon_n$? This is a weaker version of Mahler's conjecture.*

3.4. Extension to Finsler manifolds

DEFINITION 3.20. A *volume density* on an n-dimensional manifold M is a continuous function

$$\Phi : \Lambda^n TM \longrightarrow \mathbb{R}$$

such that for every point $m \in M$ the restriction of Φ to $\Lambda^n T_m M$ is a norm. A volume density is said to be smooth if the function Φ is smooth outside the zero section.

If M is an oriented manifold and Φ is a volume density on M, then we can define a volume form Ω on M whose value at a basis $\boldsymbol{x}_1, \dots, \boldsymbol{x}_n$ of $T_m M$ is $\Phi(m; \boldsymbol{x}_1 \wedge \boldsymbol{x}_2 \wedge \cdots \wedge \boldsymbol{x}_n)$ if the basis is positively oriented and $-\Phi(m; \boldsymbol{x}_1 \wedge \boldsymbol{x}_2 \wedge \cdots \wedge \boldsymbol{x}_n)$ if it is negatively oriented. For any positively oriented n-dimensional submanifold $U \subset M$ we have that

$$\int_U \Phi = \int_U \Omega.$$

However, the integral of a volume density does not depend on the orientation and volume densities can be defined in nonorientable manifolds like the projective plane where no volume form exists.

DEFINITION 3.21. A *continuous Finsler metric* F on a manifold M assigns a norm, $F(m, \cdot)$, to each tangent space $T_m M$ in such a way that the norm varies continuously with the base point. A *continuous Finsler manifold* is a pair (M, F), where M is a manifold and F is a continuous Finsler metric on M.

An important class of examples of Finsler manifolds are finite-dimensional sub-manifolds of normed spaces. If M is a submanifold of a finite-dimensional normed space X, at each point $m \in M$ the tangent space $T_m M$ can be thought of as a subspace of X and, as such, it inherits a norm.

If $\gamma : [a, b] \to M$ is a differentiable curve on a continuous Finsler manifold (M, F), we define

$$\text{length}(\gamma) := \int_a^b F(\gamma(t), \dot{\gamma}(t)) \, dt.$$

This definition can be extended in the obvious way to piecewise-differentiable curves. If x and y are two points in M, we define the *distance between x and y* as the infimum of the lengths of all piecewise-differentiable curves that join them. Thus, continuous Finsler manifolds are metric spaces and metric techniques can be used to study them.

Each definition of volume on normed spaces gives a definition of volume on continuous Finsler manifolds: if we are given a volume definition μ and an n-dimensional continuous Finsler manifold (M, F), then the map that assigns to every point m the norm $\mu_{T_m M}$ on $\Lambda^n T_m M$ is a volume density on M. Notice that, in particular, a definition of volume on normed spaces immediately yields a way to measure the volumes of submanifolds of a normed space X because, as remarked above, the tangent space $T_m M$ of such a submanifold can be regarded as a subspace of X and so inherits both a norm and a volume. This will be studied from an extrinsic viewpoint and in much more detail in the next section.

EXERCISE 3.22. Show that if a definition of volume on normed spaces is used to define a volume on Finsler manifolds, it satisfies the following two properties:

(1) If the Finsler manifold is Riemannian, its volume is the standard Riemannian volume;
(2) If $\varphi : M \to N$ is a short map (*i.e.,* does not increase distances) between two Finsler manifolds of the same dimension, then φ does not increase volumes.

Extending our four volume definitions from normed spaces to continuous Finsler manifolds, we may speak of the Busemann, Holmes–Thompson, mass, and mass* definition of volumes on continuous Finsler manifolds. To end the section we relate the Busemann and Holmes–Thompson definitions with well-known geometric constructions.

As was previously remarked, Finsler manifolds are metric spaces and, as such, we can define their volume as their Hausdorff measure: if (M, F) is an n-dimensional Finsler manifold and $r > 0$, we cover M by a family of metric balls of radius at most r, $B(m_1, r_1)$, $B(m_2, r_2)$, ..., and consider the quantity

$\varepsilon_n(r_1^n + r_2^n + \cdots)$. We now take the infimum of this quantity over all possible covering families and take the limit as r tends to zero. The resulting number is the *n-dimensional Hausdorff measure of* (M, F).

THEOREM 3.23 [Busemann 1947]. *The Busemann volume of a continuous Finsler manifold is equal to its Hausdorff measure.*

To explain the second construction, we need to recall some standard facts about the geometry of cotangent bundles.

If $\pi : T^*M \to M$ is the canonical projection, we define the *canonical 1-form* α on T^*M by the formula

$$\alpha(\boldsymbol{v}_p) := p(\pi_*(\boldsymbol{v}_p)).$$

In standard coordinates $(x_1, \ldots, x_n, p_1, \ldots, p_n)$, $\alpha := \sum_{i=1}^n p_i dq_i$. The *canonical symplectic form* on T^*M is defined as $\omega := -d\alpha$ and the *Liouville volume* is defined by $\Omega := \omega^n/n!$.

If (M, F) is a continuous Finsler manifold, each tangent space $T_m M$ carries the norm $F(m, \cdot)$ and, hence, each cotangent space T_m^*M carries the dual norm $F^*(m, \cdot)$. Let us denote the unit ball in T_m^*M by B_m^* and define the *unit co-disc bundle* of M as the set

$$B^*(M) := \bigcup_{m \in M} B_m^* \subset T^*M.$$

PROPOSITION 3.24. *The Holmes–Thompson volume of an n-dimensional, continuous Finsler manifold is equal to the Liouville volume of its unit co-disc bundle divided by the volume of the n-dimensional Euclidean unit ball.*

PROOF. It suffices to verify the result on normed spaces where it easily follows from the definitions. □

THEOREM 3.25 [Duran 1998]. *If M is a Finsler manifold, then the Holmes–Thompson volume of M is less than or equal to its Hausdorff measure with equality if and only if M is Riemannian.*

PROOF. By the Blaschke–Santaló inequality, at each point $m \in M$ we have that $\mu_{T_m M}^{\mathrm{ht}} \leq \mu_{T_m M}^{\mathrm{b}}$ with equality if and only if $T_m M$ is Euclidean. The result now follows immediately from Theorems 3.23 and 3.24 . □

4. k-Volume Densities

The theory of volumes and areas on Euclidean and Riemannian spaces is based on the fact that a Euclidean structure on a vector space induces natural Euclidean structures on its exterior powers: if $\boldsymbol{x}_1, \ldots, \boldsymbol{x}_n$ is an orthonormal basis of a Euclidean space X, then the vectors

$$\boldsymbol{x}_{i_1} \wedge \boldsymbol{x}_{i_2} \wedge \cdots \wedge \boldsymbol{x}_{i_k}, \quad 1 \leq i_1 < i_2 < \cdots < i_k \leq n,$$

form an orthonormal basis of $\Lambda^k X$. If we want to know the area of the parallel-ogram formed by the vectors $\boldsymbol{x}$ and $\boldsymbol{y}$, we need to compute the norm of $\boldsymbol{x} \wedge \boldsymbol{y}$ in $\Lambda^2 X$. In normed and Finsler spaces these simple algebraic constructions, which should be seen as functors from the category of Euclidean spaces onto itself, can-not be reproduced and we need to understand their geometry to see how they may be naturally extended to these spaces.

The first important remark is that in order to compute k-dimensional volumes (k-volumes from now on), we do not need to define a norm on all of $\Lambda^k X$. It suffices to define the magnitude of vectors of the form $\boldsymbol{v}_1 \wedge \boldsymbol{v}_2 \wedge \cdots \wedge \boldsymbol{v}_k$, where $\boldsymbol{v}_1, \ldots, \boldsymbol{v}_k$ are vectors in X. In this paper we shall refer to these k-vectors as *simple* and denote the set of all simple k-vectors in X by $\Lambda_s^k X$. Note that for $k = 1, n-1$ every k-vector is simple, which makes the study of $(n-1)$-volume in n-dimensional normed spaces a richer and more approachable subject than the study of volumes in higher codimension. Indeed, when $k \neq 1, n-1$, $\Lambda_s^k X$ is not a vector subspace of $\Lambda^k X$, but just an algebraic cone.

EXERCISE 4.1. Let $\mathcal{Q} : \Lambda^2 \mathbb{R}^4 \to \Lambda^4 \mathbb{R}^4$ be the quadratic form defined by $\mathcal{Q}(a) = a \wedge a$. Show that $\Lambda_s^2 \mathbb{R}^4$ is the quadric $\mathcal{Q} = 0$ and use this to prove that the intersection of $\Lambda_s^2 \mathbb{R}^4$ with the (Euclidean) unit sphere in $\Lambda^2 \mathbb{R}^4$ is a product of two 2-dimensional spheres.

In general, the intersection of $\Lambda_s^k \mathbb{R}^n$ with the Euclidean unit sphere in $\Lambda^k \mathbb{R}^n$ is the Plücker embedding of the Grassmannian of oriented k-planes in $\mathbb{R}^n$, $G_k^+(\mathbb{R}^n)$. Let us recall that this Grassmannian is a smooth manifold of dimension $k(n-k)$.

Following our first remark, we see that in order to compute k-volumes in a normed space X, we need to define a "norm" on the cone of simple k-vectors of X. The fact that $\Lambda_s^k X$ is not a vector space complicates matters since it is not clear how to write the triangle inequality, and, even if an apparent analogue could be found, it would have to be justified in terms of its geometric significance. Nevertheless, the homogeneity and positivity of a norm are easy to generalize:

DEFINITION 4.2. A k-*density* on a vector space X is a continuous function

$$\phi : \Lambda_s^k X \longrightarrow \mathbb{R}$$

that is homogeneous of degree one (*i.e.*, $\phi(\lambda a) = |\lambda| \phi(a)$). A k-density ϕ is said to be a k-*volume density* if $\phi(a) \geq 0$ with equality if and only if $a = 0$.

4.1. Examples of k-volume densities. In the previous chapter we studied four natural volume definitions on normed spaces. Each one of these definitions yields natural constructions of k-volume densities on the spaces of simple k-vectors of a normed space X.

Given a volume definition μ and a k-vector a in a k-dimensional normed space Y, we can compute $\mu(a)$. To be perfectly rigorous, we should include the normed space as a variable in μ, for example, by writing $\mu_Y(a)$. If a is a simple k-vector in an n-dimensional normed space X, then we may consider it as a k-vector on the k-dimensional normed space "spanned by a",

$$\langle a \rangle := \{ \boldsymbol{x} \in X : a \wedge \boldsymbol{x} = 0 \} \subset X,$$

(provided with the induced norm), and compute $\mu_{\langle a \rangle}(a)$. Thus, once we have chosen a way to define volume in *all* finite-dimensional normed spaces, we have a way to associate to each norm on an n-dimensional vector space X a family of k-volume densities, with $1 \leq k \leq n$.

The Busemann k-volume densities. Let X be a normed space of dimension n with unit ball B, and let k be a positive integer less than n. The Busemann k-volume density on X is defined by the formula

$$\mu^b(a) := \frac{\varepsilon_k}{\mathrm{vol}(B \cap \langle a \rangle; a)}.$$

The Holmes–Thompson k-volume densities. Let X be a normed space of dimension n, and let k be a positive integer less than n. If a is a simple k-vector spanning the k-dimensional subspace $\langle a \rangle$, we consider the inclusion of $\langle a \rangle$ into X and the dual projection $\pi : X^* \to \langle a \rangle^*$. Regarding a as a volume form on $\langle a \rangle^*$ we define

$$\mu^{\mathrm{ht}}(a) := \varepsilon_k^{-1} \int_{\pi(B_X^*)} |a|.$$

The mass k-volume densities. Let $(X, \| \cdot \|)$ be a normed space of dimension n, and let k be a positive integer less than n. The mass k-volume density on X is defined by the formula

$$\mu^m(a) := \inf \left\{ \prod_{i=1}^{k} \| \boldsymbol{x}_i \| : \boldsymbol{x}_1 \wedge \boldsymbol{x}_2 \wedge \cdots \wedge \boldsymbol{x}_k = a \right\}.$$

The mass∗ k-volume densities. According to the characterization of mass∗ given in Exercise 3.9, we may describe the mass∗ k-volume densities as follows:

Let X be a normed space and let $W \subset X$ be a k-dimensional subspace. If a is a simple k-vector on X, we define $\mu^{m*}(a)$ as the supremum of the numbers $|\boldsymbol{\xi}_1 \wedge \boldsymbol{\xi}_2 \wedge \cdots \wedge \boldsymbol{\xi}_k(a)|$, where $\boldsymbol{\xi}_1, \ldots, \boldsymbol{\xi}_k$ ranges over all short bases of $\langle a \rangle^*$.

However, there is a simpler description:

EXERCISE 4.3. Using the Hahn–Banach theorem and the notation above, show that $\mu^{m*}(a)$ is the supremum of the numbers $|\boldsymbol{\xi}_1 \wedge \boldsymbol{\xi}_2 \wedge \cdots \wedge \boldsymbol{\xi}_k(a)|$, where $\boldsymbol{\xi}_1, \ldots, \boldsymbol{\xi}_k \in B_X^*$.

In the study of volumes and areas on Finsler manifolds, we shall also need to work with k-densities and smooth k-densities on manifolds. For this purpose we introduce the bundle of simple tangent k-vectors on a manifold M, $\Lambda_s^k TM$. This is a subbundle of algebraic cones of the vector bundle $\Lambda^k TM$, and if we omit the zero section it is a smooth manifold.

DEFINITION 4.4. A k-density ϕ (*resp.* k-volume density) on a manifold M is a continuous function $\phi : \Lambda_s^k TM \to \mathbb{R}$ such that at each point m, the restriction of ϕ to $\Lambda_s^k T_m M$ is a k-density (*resp.* k-volume density). If the function ϕ is smooth outside the zero section, we shall say that the density is smooth.

Since every tangent space of a Finsler manifold (M, F) is a normed space, we may define the Busemann, Holmes–Thompson, mass, and mass$*$ k-volume densities on (M, F) by assigning to each tangent space its respective k-density. It is easy to show that if F is a smooth Finsler metric, then the Busemann and Holmes–Thompson k-volume densities are smooth. This is probably not the case with mass and mass$*$, but we have no explicit examples to illustrate this.

Just like differential k-forms, k-densities can be pulled back: if $f : N \to M$ is a smooth map and ϕ is a k-density on M, then

$$f^*\phi(\boldsymbol{v}_1 \wedge \boldsymbol{v}_2 \wedge \cdots \wedge \boldsymbol{v}_k) = \phi\big(Df(\boldsymbol{v}_1) \wedge Df(\boldsymbol{v}_2) \wedge \cdots \wedge Df(\boldsymbol{v}_k)\big).$$

Remark that if $f : N \to M$ is an immersion and ϕ is a k-volume density on M, then $f^*\phi$ is a k-volume density on N.

Also like differential forms, k-densities can be integrated over k-dimensional submanifolds: if $N \subset M$ is a k-dimensional submanifold of M and $i : N \to M$ is the inclusion map, then $i^*\phi$ is a volume density on N, and its integral over N was defined in Section 3. This integral is independent of the parameterization and orientation of N. In the same way, we may define the integral of a k-density over a k-chain.

For the rest of the chapter, we associate to a given k-volume density ϕ on a vector space X the functional

$$N \longmapsto \int_N \phi,$$

and investigate the relationship between the behavior of the functional and certain convexity properties of ϕ. The easiest case is when ϕ is an $(n-1)$-volume density in an n-dimensional vector space.

4.2. Convexity of $(n-1)$-volume densities.

This case is special because every $(n-1)$-vector in an n-dimensional vector space, X, is simple and we may impose the condition that an $(n-1)$-volume density be a norm in $\Lambda^{n-1} X$. This is, for example, satisfied by $(n-1)$-volume densities for the Busemann, Holmes–Thompson, and mass$*$ definitions of volume. Nevertheless, it remains to see why such a condition is desirable.

The next result is the first of four characterizations of norms on $\Lambda^{n-1} X$.

THEOREM 4.5. *Let ϕ be an $(n-1)$-volume density on an n-dimensional vector space X. The following conditions on ϕ are equivalent:*

- ϕ *is a norm;*
- *If* $P \subset X$ *is a closed* $(n{-}1)$*-dimensional polyhedron in* X, *then the area of any one of its facets is less than or equal to the sum of the areas of the remaining facets.*

Before proving this theorem, we need to introduce a classical construction that associates to any k-dimensional polyhedral surface on X a set of simple k-vectors. This set will be called the *Gaussian image* of the polyhedron (see also [Burago and Ivanov 2002], where the almost identical notion of *Gaussian measure* is used).

If $P \subset X$ is a polyhedron with facets $F_1, \ldots, F_m$ we associate to each facet F_i the simple k-vector a_i such that $\langle a_i \rangle$ is parallel to F_i and such that $\mathrm{vol}(F_i; a_i) = 1$. The Gaussian image of P is the set $\{a_1, \ldots, a_m\} \subset \Lambda_s^k X$. If ϕ is a k-volume density in X, the k-volume of P (with respect to ϕ) is just $\phi(a_1) + \cdots + \phi(a_m)$.

EXERCISE 4.6. Show that if $\{a_1, \ldots, a_m\} \subset \Lambda_s^k X$ is the Gaussian image of a closed polyhedron in X, then $a_1 + \cdots + a_m = 0$.

In general, the condition that the sum of a set of simple k-vectors be zero, does not imply that it is the Gaussian image of a closed k-dimensional polyhedron in X. However, in codimension one we have the following celebrated theorem of Minkowski.

THEOREM 4.7 (MINKOWSKI). *A set of* $(n{-}1)$*-vectors of an* n*-dimensional vector space* X *is the Gaussian image of a closed, convex polyhedron if and only if the* $(n{-}1)$*-vectors span* $\Lambda^{n-1}X$ *and their sum equals zero.*

To prove Theorem 4.5, we shall need an easy particular case of Minkowski's result:

EXERCISE 4.8. Let X be an n-dimensional vector space and let $a_1, \ldots, a_n$ be a basis of $\Lambda^{n-1}X$. Show that there exists a simplex in X whose Gaussian image is the set $\{a_1, \ldots, a_n, -(a_1 + \cdots + a_n)\}$.

PROOF OF THEOREM 4.5. Assume that ϕ is a norm, and let $P \subset X$ be an $(n{-}1)$-dimensional closed polyhedron with Gaussian image $\{a_0, a_1, \ldots, a_m\}$. Since the sum of the a_i's is zero, we may use the triangle inequality to write

$$\phi(a_0) = \phi(a_1 + \cdots + a_m) \le \phi(a_1) + \cdots + \phi(a_m).$$

In other words, the area of the facet corresponding to a_0 is less than or equal to the sum of the areas of the remaining facets.

To prove the converse, we take any two $(n{-}1)$-vectors a_1 and a_2, which we assume to be linearly independent, and use them as part of a basis $a_1, \ldots, a_n$ of $\Lambda^{n-1}X$. By Exercise 4.8, the set $\{a_1, \ldots, a_n, -(a_1 + \cdots + a_n)\}$ is the Gaussian image of a simplex in X. Then, by assumption,

$$\phi(a_1 + \cdots + a_n) \le \phi(a_1) + \cdots + \phi(a_n).$$

By letting $a_3, \ldots, a_n$ tend to zero in the above inequality we obtain the triangle
inequality $\phi(a_1 + a_2) \le \phi(a_1) + \phi(a_2)$, and, therefore, ϕ is a norm.   □

EXERCISE 4.9. Consider the tetrahedron in the normed space ℓ_∞^3 with vertices
$(0,0,0)$, $(-1,1,1)$, $(1,-1,1)$, $(1,1,-1)$, and show that the mass of the facet
opposite the origin is greater than the sum of the masses of the three other
facets. *Hint.* Use the definition of the mass 2-volume density in terms of minimal
circumscribed parallelograms.

By Theorem 4.5, the previous exercise shows that the mass $(n-1)$-volume density
of a normed space X is not necessarily a norm in $\Lambda^{n-1}X$. As we shall see in
the rest of this chapter, this is a good reason to disqualify mass as a satisfactory
definition of volume on normed spaces.

   Our second characterization of norms in $\Lambda^{n-1}X$ is another variation on the
theme of *flats minimize*.

THEOREM 4.10. *Let ϕ be an $(n-1)$-volume density on an n-dimensional vector
space X. The following conditions on ϕ are equivalent*:

- *ϕ is a norm*;
- *Whenever C and C' are $(n-1)$-chains with real coefficients such that $\partial C = \partial C'$ and the image of C is contained in a hyperplane, then the area of C is less than or equal to the area of C'.*

In order to prove this theorem we need to introduce the concept of *calibration*
formalized by Harvey and Lawson [1982].

DEFINITION 4.11. A closed k-form ω is said to *calibrate* a k-density ϕ if for all
simple k-vectors a in TM we have that $\omega(a) \le \phi(a)$ and equality is attained on
a nonempty subset of $\Lambda_s^k TM$.

The homogeneity of ω and ϕ allows us to consider the set where they coincide as
a subset E of the bundle of oriented k-dimensional subspaces of TM, $G_k^+(TM)$.

PROPOSITION 4.12 [Harvey and Lawson 1982]. *Let ϕ be a k-volume density on
a manifold M, let ω be a closed k-form on M that calibrates ϕ and let $E \subset
G_k^+(TM)$ be the set where ϕ and ω coincide. If $N \subset M$ is a k-dimensional
oriented submanifold all of whose tangent planes belong to E, and N' is another
submanifold of M homologous to N, then*

$$\int_N \phi \le \int_{N'} \phi.$$

PROOF. Using that $\phi = \omega$ on the tangent spaces of N and Stokes' formula, we
have

$$\int_N \phi = \int_N \omega = \int_{N'} \omega \le \int_{N'} \phi. \qquad \square$$

PROOF OF THEOREM 4.10. Assume that ϕ is a norm, let C and C' be as in the statement of the theorem, and let a be an $(n-1)$-vector on X such that $\phi(a) = 1$ and the subspace $\langle a \rangle$ is parallel to the hyperplane containing the image of C. Since the unit sphere in $(\Lambda^{n-1}X, \phi)$ is a convex hypersurface, it has a supporting hyperplane that touches it at a. This hyperplane can be given as the set $\omega = 1$, where ω is a constant-coefficient $(n-1)$-form on X. Since the unit sphere lies in the half-space $\omega \leq 1$, we have $\omega \leq \phi$ and, thus, ω calibrates ϕ.

Using that $\omega = \phi$ on C, that $d\omega = 0$, and that $C + (-C')$ is a closed chain homologous to zero, we have

$$\int_C \phi = \int_C \omega = \int_{C'} \omega \leq \int_{C'} \phi.$$

To prove the converse, note that the second condition in the theorem immediately implies that the $(n-1)$-volume of the facet of a closed polyhedron is less than or equal to the sum of the $(n-1)$-volumes of the remaining facets. By Theorem 4.5, this implies that ϕ is a norm.   □

In Euclidean geometry, the orthogonal projection onto a k-dimensional subspace is area-decreasing. This can be generalized as follows:

THEOREM 4.13. *Let ϕ be an $(n-1)$-volume density on an n-dimensional vector space X. The following conditions on ϕ are equivalent:*

- ϕ *is a norm;*
- *For every $(n-1)$-dimensional subspace $W \subset X$ there is a ϕ-decreasing linear projection $P_W : X \to W$.*

The proof of this theorem rests on a simple lemma in multi-linear algebra.

LEMMA 4.14. *Let X be an n-dimensional vector space and let $W \subset X$ be a k-dimensional subspace. If $w_1, \ldots, w_k$ is a basis of W and $\omega \in \Lambda^k X^*$ is such that $\omega(w_1 \wedge w_2 \wedge \cdots \wedge w_k) = 1$, then the linear map*

$$P\boldsymbol{x} := \sum_{i=1}^{k} (-1)^i \omega(\boldsymbol{x} \wedge \boldsymbol{w}_1 \wedge \cdots \wedge \hat{\boldsymbol{w}}_i \wedge \cdots \wedge \boldsymbol{w}_k) \boldsymbol{w}_i$$

is a projector with range W. Moreover, ω is simple if and only if for any vectors $\boldsymbol{x}_1, \ldots, \boldsymbol{x}_k \in X$,

$$P\boldsymbol{x}_1 \wedge P\boldsymbol{x}_2 \wedge \cdots \wedge P\boldsymbol{x}_k = \omega(\boldsymbol{x}_1 \wedge \boldsymbol{x}_2 \wedge \cdots \wedge \boldsymbol{x}_k)\, \boldsymbol{w}_1 \wedge \boldsymbol{w}_2 \wedge \cdots \wedge \boldsymbol{w}_k.$$

The proof of the lemma is left as an exercise to the reader.

PROOF OF THEOREM 4.13. Assume that ϕ is a norm in $\Lambda^{n-1}X$ and let $W \subset X$ be an $(n-1)$-dimensional subspace. Choose a basis of W, $\boldsymbol{w}_1, \ldots, \boldsymbol{w}_{n-1}$, such that $\phi(\boldsymbol{w}_1 \wedge \boldsymbol{w}_2 \wedge \cdots \wedge \boldsymbol{w}_{n-1}) = 1$ and consider the support hyperplane of the unit sphere of $(\Lambda^{n-1}X, \phi)$ at the point $\boldsymbol{w}_1 \wedge \boldsymbol{w}_2 \wedge \cdots \wedge \boldsymbol{w}_{n-1}$. This hyperplane is given by an equation of the form $\omega = 1$, where ω is an $(n-1)$-form with constant coefficients. In other words, $\omega \in \Lambda^{n-1} X^*$.

We claim that the linear projection

$$P\boldsymbol{x} := \sum_{i=1}^{n-1} (-1)^i \omega(\boldsymbol{x} \wedge \boldsymbol{w}_1 \wedge \cdots \wedge \hat{\boldsymbol{w}}_i \wedge \cdots \wedge \boldsymbol{w}_{n-1}) \boldsymbol{w}_i$$

is ϕ-decreasing. Indeed, since ω is an $(n-1)$-form on an n-dimensional space, it is simple. Using the second part of Lemma 4.14, we have, for any $(n-1)$-vector $a := \boldsymbol{x}_1 \wedge \boldsymbol{x}_2 \wedge \cdots \wedge \boldsymbol{x}_{n-1}$,

$$\phi(P\boldsymbol{x}_1 \wedge P\boldsymbol{x}_2 \wedge \cdots \wedge P\boldsymbol{x}_{n-1}) = |\omega(a)| \phi(\boldsymbol{w}_1 \wedge \boldsymbol{w}_2 \wedge \cdots \wedge \boldsymbol{w}_{n-1}) = |\omega(a)| \leq \phi(a).$$

To prove the converse, we note that the existence of a ϕ-decreasing linear projection onto any given hyperplane implies that the $(n-1)$-volume of the facet of any closed $(n-1)$-dimensional polyhedron is less than or equal to the sum of the areas of the remaining facets. The argument is quite simple: if the closed polyhedron has facets $F_0, F_1, \ldots, F_m$, we use the ϕ-decreasing projection P to project the whole polyhedron onto the hyperplane containing F_0. Note that $P(F_1) \cup \cdots \cup P(F_m)$ contains $P(F_0) = F_0$ and, therefore, the sum of the $(n-1)$-volumes of the $P(F_i)$, $1 \leq i \leq m$, is greater than or equal to the $(n-1)$-volume of F_0. Since P is ϕ-decreasing, this gives us that the sum of the $(n-1)$-volumes of the F_i, $1 \leq i \leq m$, is greater than or equal to the $(n-1)$-volume of F_0.   □

We now state the fourth and last of our characterizations of norms on the space of $(n-1)$-vectors in an n-dimensional normed space.

THEOREM 4.15. *Let ϕ be an $(n-1)$-volume density on an n-dimensional vector space X. The following conditions on ϕ are equivalent*:

- ϕ *is a norm*;
- *If $K \subset K'$ are two nested convex bodies in X, then the area of ∂K is less than or equal to the area of $\partial K'$ with equality if and only if K equals K'.*

The proof of this theorem is a simple consequence of the relation between norms in the space of $(n-1)$-vectors and the theory of mixed volumes that will be developed in Section 6.

At the beginning of this section we mentioned that for any n-dimensional normed space X the Busemann, Holmes–Thompson, and mass∗ $(n-1)$-volume densities on X are norms in $\Lambda^{n-1} X$. For the Busemann definition this is a celebrated theorem of Busemann [1949a]. For the mass∗ definition this result is due to Benson [1962]. We shall follow [Gromov 1983] and give a proof of a much stronger result later in this section. For the Holmes–Thompson definition, the result — under a different formulation — goes back to Minkowski.

THEOREM 4.16 (MINKOWSKI). *The Holmes–Thompson $(n-1)$-volume density of an n-dimensional normed space X is itself a norm in $\Lambda^{n-1} X$.*

In order to prove the convexity of the Holmes–Thompson $(n-1)$-volume density, we shall first give an integral representation for it. This representation depends, in turn, on two classical constructions: the *Gauss map* and the *surface-area measure*. Our approach follows [Fernandes 2002].

Let X be an n-dimensional vector space and let ϕ be an $(n-1)$-volume density on X. If $N \subset X$ is an oriented hypersurface and $n \in N$, we define $\mathcal{G}_\phi(n)$ as the unique $(n-1)$-vector in $\Lambda^{n-1}T_nN \subset \Lambda^{n-1}X$ that is positively oriented and satisfies $\phi(\mathcal{G}_\phi(n)) = 1$. Notice that when N is a strictly convex hypersurface, the Gauss map

$$\mathcal{G}_\phi : N \longrightarrow \Sigma := \{a \in \Lambda^{n-1}X : \phi(a) = 1\}$$

is a diffeomorphism. In this case, we define the *surface-area measure* of N as the $(n-1)$-volume density $dS_N := \mathcal{G}_\phi^{-1*}\phi$ on Σ.

LEMMA 4.17.   *Let $\pi : X \to Y$ be a surjective linear map between an n-dimensional vector space X and an $(n-1)$-dimensional vector space Y, and let ϕ be an $(n-1)$-volume density on X with unit sphere $\Sigma \subset \Lambda^{n-1}X$. If $N \subset X$ is a smooth, strictly convex hypersurface and ω is a volume form on Y, then*

$$\int_{\pi(N)} |\omega| = \frac{1}{2} \int_{a \in \Sigma} |\pi^*\omega(a)| \, dS_N.$$

PROOF. By the definition of the Gauss map, if $n \in N$ and $\boldsymbol{x}_1 \wedge \boldsymbol{x}_2 \wedge \cdots \wedge \boldsymbol{x}_{n-1} \in \Lambda^{n-1}T_nN$,

$$\boldsymbol{x}_1 \wedge \boldsymbol{x}_2 \wedge \cdots \wedge \boldsymbol{x}_{n-1} = \phi(\boldsymbol{x}_1 \wedge \boldsymbol{x}_2 \wedge \cdots \wedge \boldsymbol{x}_{n-1})\mathcal{G}_\phi(n).$$

Therefore, $\pi^*|\omega|(\boldsymbol{x}_1 \wedge \boldsymbol{x}_2 \wedge \cdots \wedge \boldsymbol{x}_{n-1}) = |\pi^*\omega(\mathcal{G}_\phi(n))|\phi(\boldsymbol{x}_1 \wedge \boldsymbol{x}_2 \wedge \cdots \wedge \boldsymbol{x}_{n-1})$. Then

$$\int_{\pi(N)} |\omega| = \frac{1}{2} \int_N \pi^*|\omega| = \frac{1}{2} \int_{n \in N} |\pi^*\omega(\mathcal{G}_\phi(n))|\phi$$

$$= \frac{1}{2} \int_\Sigma \mathcal{G}_\phi^{-1*} |\pi^*\omega(\mathcal{G}_\phi(n))|\phi = \frac{1}{2} \int_{a \in \Sigma} |\pi^*\omega(a)|\mathcal{G}_\phi^{-1*}\phi$$

$$= \frac{1}{2} \int_{a \in \Sigma} |\pi^*\omega(a)| \, dS_N. \qquad \square$$

PROOF OF THEOREM 4.16.   By a standard approximation argument, it suffices to consider the case where the dual unit sphere $\partial B^* \subset X^*$ is smooth and strictly convex. Applying the previous lemma to the surface $N = \partial B^*$ and to an arbitrary $(n-1)$-volume density on X^*, we have

$$\mu^{\mathrm{ht}}(a) = \varepsilon_{n-1}^{-1} \int_{\pi(B^*)} |a| = \varepsilon_{n-1}^{-1} \int_{\xi \in \Sigma} |a(\xi)| \, dS_{\partial B^*}.$$

Since the surface-area measure $dS_{\partial B^*}$ is nonnegative,

$$
\begin{aligned}
\mu^{\text{ht}}(a+b) &= \varepsilon_{n-1}^{-1} \int_{\xi \in \Sigma} |a(\xi) + b(\xi)| \, dS_{\partial B^*} \\
&\leq \varepsilon_{n-1}^{-1} \int_{\xi \in \Sigma} |a(\xi)| \, dS_{\partial B^*} + \varepsilon_{n-1}^{-1} \int_{\xi \in \Sigma} |b(\xi)| \, dS_{\partial B^*} \\
&= \mu^{\text{ht}}(a) + \mu^{\text{ht}}(b). \qquad \square
\end{aligned}
$$

EXERCISE 4.18. Let X be an n-dimensional vector space and let ϕ be an $(n-1)$-volume density on X. Show that if ϕ is a norm, then compact hypersurfaces cannot by minimal.

4.3. Convexity properties of k-volume densities. We now pass to the more delicate subject that Busemann, Ewald, and Shephard studied extensively under the heading of *convexity on Grassmannians*. Most of what follows can be found in their papers, "Convex bodies and convexity on Grassmannian cones" I–XI, but we have tried to make the language and proofs more accessible.

We shall see that there are several notions and degrees of convexity for k-volume densities. These are closely related to the concept of *ellipticity* in geometric measure theory and, historically, to the generalization of the Legendre condition for variational problems.

Weakly convex k-densities. Let X be an n-dimensional vector space and let $\Lambda_s^k X$, $1 \leq k \leq n-1$, be the cone of simple k-vectors on X. If Y is a $(k+1)$-dimensional subspace of X, then the subspace $\Lambda^k Y \subset \Lambda^k X$ lies inside $\Lambda_s^k X$. This motivates a definition:

DEFINITION 4.19. A k-volume density ϕ on an n-dimensional vector space X, $n > k$, is said to be *weakly convex* if for any linear subspace Y of dimension $k+1$, the restriction of ϕ to the linear space $\Lambda^k Y$ is a norm.

From the previous section, we know that the k-volume densities of any normed space for the Busemann, Holmes–Thompson, or mass∗ definitions of volume are weakly convex.

EXERCISE 4.20. Show that a k-volume density in a vector space X is weakly convex if for every $(k+1)$-dimensional simplex in X the area of any one facet is less than or equal to the sum of the areas of the remaining facets.

Extendibly convex k-volume densities

DEFINITION 4.21. A k-volume density ϕ on an n-dimensional vector space X, $n > k$, is said to be *extendibly convex* if it is the restriction of a norm on $\Lambda^k X$ to the cone of simple k-vectors in X.

Equivalently, ϕ is extendibly convex if and only if there is a support hyperplane for the unit sphere

$$
\mathcal{S} := \{ a \in \Lambda_s^k X : \phi(a) = 1 \}
$$

passing through any of its points.

THEOREM 4.22. *If ϕ is an extendibly convex k-volume density on a vector space X, then any k-chain with real coefficients whose image is contained in a k-dimensional flat is ϕ-minimizing.*

The proof — by the method of calibrations — is nearly identical to the proof of Theorem 4.10 and is left as an exercise for the reader. Notice that a corollary to Theorem 4.22 is that if $P \subset X$ is a closed k-dimensional polyhedron, the area of any of its facets is less than or equal to the sum of the areas of the remaining facets.

The problem of determining whether the Busemann k-volume densities are extendibly convex was posed by Busemann in several of his papers as a major problem in convexity. So far, there are no results in this direction.

PROBLEM. *Is the Busemann 2-volume density of a 4-dimensional normed space extendibly convex?*

In the case of the Holmes–Thompson definition, Busemann, Ewald, and Shephard have given explicit examples of norms for which the k-volume densities, $1 < k < n - 1$, are not extendibly convex (see [Busemann et al. 1963]). A simpler example has been given recently by Burago and Ivanov:

THEOREM 4.23 [Burago and Ivanov 2002]. *Consider the norm $\| \cdot \|$ on $\mathbb{R}^4$ whose dual unit ball in $\mathbb{R}^{4*}$ is the convex hull of the curve*

$$\gamma(t) := (\sin t, \cos t, \sin 3t, \cos 3t), \quad 0 \le t \le 2\pi.$$

The Holmes–Thompson 2-volume density for $(\mathbb{R}^4, \| \cdot \|)$ is not extendibly convex.

Despite these examples, in many important cases the Holmes–Thompson k-volume densities are extendibly convex.

THEOREM 4.24. *The Holmes–Thompson k-volume densities of a hypermetric normed space are extendibly convex.*

In order to prove this result, we shall derive a formula for the Holmes–Thompson k-volume densities of a Minkowski space in terms of the Fourier transform of its norm. In a somewhat different guise, this formula was first obtained by W. Weil [1979]. In the present form it was rediscovered by Álvarez and Fernandes in [1999], where it was shown to follow from the Crofton formula for Minkowski spaces.

Let ϕ be a smooth, even, homogeneous function of degree one on an n-dimensional vector space X, let $e_1, \ldots, e_n$ be a basis of X, and let $\boldsymbol{\xi}_1, \ldots, \boldsymbol{\xi}_n$ be the dual basis in X^*. Using the basis $e_1, \ldots, e_n$ and its dual to identify both X and X^* with $\mathbb{R}^n$, we can compute the (distributional) Fourier transform of ϕ,

$$\hat{\phi}(\boldsymbol{\xi}) := \int_{\mathbb{R}^n} e^{i\boldsymbol{\xi} \cdot \boldsymbol{x}} \phi(\boldsymbol{x}) \, dx.$$

The form $\hat{\phi}\, d\xi_1 \wedge \cdots \wedge d\xi_n$ does not depend on the choice of basis in X. Up to a constant factor, we define the form $\check{\phi}$ as the contraction of this n-form with the Euler vector field, $X_E(\boldsymbol{\xi}) = \boldsymbol{\xi}$, in X^*:

$$\check{\phi} := \frac{-1}{4(2\pi)^{n-1}} \hat{\phi}\, d\xi_1 \wedge \cdots \wedge d\xi_n \rfloor X_E.$$

It is known (see [Hörmander 1983, pages 167–168]) that $\hat{\phi}$ is smooth on $X^* \setminus \mathbf{0}$ and homogeneous of degree $-n-1$; therefore $\check{\phi}$ is a smooth differential form on $X^* \setminus \mathbf{0}$ that is homogeneous of degree -1.

Denoting by $\check{\phi}^k$ the product form in the product space $(X^* \setminus \mathbf{0})^k$, we have the following result:

THEOREM 4.25. *Let (X, ϕ) be an n-dimensional Minkowski space. For any simple k-vector a on X, $1 \leq k < n$, we have*

$$\mu^{\mathrm{ht}}(a) = \frac{1}{\varepsilon_k} \int_{(\boldsymbol{\xi}_1, \ldots \boldsymbol{\xi}_k) \in S^{*k}} |\boldsymbol{\xi}_1 \wedge \cdots \wedge \boldsymbol{\xi}_k \cdot a| \check{\phi}^k,$$

where S^* is any closed hypersurface in $X^* \setminus \mathbf{0}$ that is star-shaped with respect to the origin.

Notice that this formula allows us to extend the definition of the Holmes–Thompson k-volume density of any Minkowski space to all of $\Lambda^k X$. It remains to see when this extension is a norm.

PROOF OF THEOREM 4.24. It is enough to prove convexity in the case the hypermetric space (X, ϕ) is also a Minkowski space. This allows us to use the integral representation given above. Since X is hypermetric, Theorem 2.5 tells us that the form $\hat{\phi}\, d\xi_1 \wedge \cdots \wedge d\xi_n$ is a volume form, and, therefore, the restriction of $\check{\phi}^k$ to the manifold S^{*k} defines a nonnegative measure. Then for any two k-vectors a and b we have

$$\mu^{\mathrm{ht}}(a + b) = \int_{S^{*k}} |\boldsymbol{\xi}_1 \wedge \cdots \wedge \boldsymbol{\xi}_k \cdot (a + b)| \check{\phi}^k$$

$$\leq \int_{S^{*k}} |\boldsymbol{\xi}_1 \wedge \cdots \wedge \boldsymbol{\xi}_k \cdot a| \check{\phi}^k + \int_{S^{*k}} |\boldsymbol{\xi}_1 \wedge \cdots \wedge \boldsymbol{\xi}_k \cdot b| \check{\phi}^k$$

$$= \mu^{\mathrm{ht}}(a) + \mu^{\mathrm{ht}}(b). \qquad \square$$

Totally convex k-densities

DEFINITION 4.26. A k-density ϕ on an n-dimensional vector space X, $n > k$, is said to be *totally convex* if through every point of the unit sphere of $\Lambda^k X$ there passes a supporting hyperplane of the form $\xi = 1$ with ξ a simple k-vector in $\Lambda^k X^*$.

Total convexity implies extendible convexity and, in turn, weak convexity. The following result, stated in [Busemann 1961] gives an important characterization of totally convex k-densities in terms of what Gromov [1983] calls the *compressing property*.

THEOREM 4.27. *A k-density ϕ on an n-dimensional vector space X is totally convex if and only if for every k-dimensional linear subspace there exists a ϕ-decreasing linear projection onto that subspace.*

The proof, using Lemma 4.14, is nearly identical to the proof of Theorem 4.13.

Of all the four volume definitions we have studied, mass∗ has by far the strongest convexity property:

THEOREM 4.28. *The mass∗ k-volume densities of an n-dimensional normed space X, $1 \le k \le n-1$, are totally convex.*

PROOF. By Theorem 4.27, it is enough to show that given any k-dimensional subspace W, there exists a linear projection $P : X \to W$ that is mass∗-decreasing.

Choose a basis $\boldsymbol{\xi}_1, \ldots, \boldsymbol{\xi}_k$ of W^* which satisfies two properties:

(1) It is short (*i.e.,* $|\boldsymbol{\xi}_i(\boldsymbol{x})| \le 1$ for all $\boldsymbol{x} \in B \cap W$);
(2) The integral of the volume density $|\boldsymbol{\xi}_1 \wedge \boldsymbol{\xi}_2 \wedge \cdots \wedge \boldsymbol{\xi}_k|$ over $B \cap W$ is maximal among all short bases.

Notice that for any basis $\boldsymbol{w}_1, \ldots, \boldsymbol{w}_k$ of W, we have that

$$\mu^{m*}(\boldsymbol{w}_1 \wedge \boldsymbol{w}_2 \wedge \cdots \wedge \boldsymbol{w}_k) = |\boldsymbol{\xi}_1 \wedge \boldsymbol{\xi}_2 \wedge \cdots \wedge \boldsymbol{\xi}_k \, (\boldsymbol{w}_1 \wedge \boldsymbol{w}_2 \wedge \cdots \wedge \boldsymbol{w}_k)|,$$

and that if $\boldsymbol{w}_1, \ldots, \boldsymbol{w}_k$ is dual to $\boldsymbol{\xi}_1, \ldots, \boldsymbol{\xi}_k$, then $\mu^{m*}(\boldsymbol{w}_1 \wedge \boldsymbol{w}_2 \wedge \cdots \wedge \boldsymbol{w}_k) = 1$.

By the Hahn–Banach theorem, there exist covectors $\hat{\boldsymbol{\xi}}_1, \ldots, \hat{\boldsymbol{\xi}}_k \in X^*$ such that

(1) $|\hat{\boldsymbol{\xi}}_i(\boldsymbol{x})| \le 1$ for all $\boldsymbol{x} \in B$ and for all i, $1 \le i \le k$;
(2) the restriction of $\hat{\boldsymbol{\xi}}_i$ to W equals $\boldsymbol{\xi}_i$ for all i, $1 \le i \le k$.

We may now define the projection $P : X \to W$ by the formula

$$P(\boldsymbol{x}) := \sum_{i=1}^{k} \hat{\boldsymbol{\xi}}_i(\boldsymbol{x})\boldsymbol{w}_i,$$

and show that it is μ^{m*}-decreasing. Indeed, if $a = \boldsymbol{v}_1 \wedge \boldsymbol{v}_2 \wedge \cdots \wedge \boldsymbol{v}_k$ is a simple k-vector in X,

$$P(\boldsymbol{v}_1) \wedge P(\boldsymbol{v}_2) \wedge \cdots \wedge P(\boldsymbol{v}_k) = \hat{\boldsymbol{\xi}}_1 \wedge \hat{\boldsymbol{\xi}}_2 \wedge \cdots \wedge \hat{\boldsymbol{\xi}}_k \, (a)\boldsymbol{w}_1 \wedge \boldsymbol{w}_2 \wedge \cdots \wedge \boldsymbol{w}_k,$$

and therefore

$$\mu^{m*}(P(\boldsymbol{v}_1) \wedge P(\boldsymbol{v}_2) \wedge \cdots \wedge P(\boldsymbol{v}_k)) = |\hat{\boldsymbol{\xi}}_1 \wedge \hat{\boldsymbol{\xi}}_2 \wedge \cdots \wedge \hat{\boldsymbol{\xi}}_k \, (a)|.$$

Since the restriction of $\hat{\boldsymbol{\xi}}_i$, $1 \le i \le k$, to $\langle a \rangle$ form a short basis of $\langle a \rangle^*$, we have

$$\mu^{m*}(P(\boldsymbol{v}_1) \wedge P(\boldsymbol{v}_2) \wedge \cdots \wedge P(\boldsymbol{v}_k)) = |\hat{\boldsymbol{\xi}}_1 \wedge \hat{\boldsymbol{\xi}}_2 \wedge \cdots \wedge \hat{\boldsymbol{\xi}}_k \, (a)| \le \mu^{m*}(a). \quad \square$$

We end the section with an exercise and an open problem:

EXERCISE 4.29. Show that the sum of two totally convex 2-volume densities in $\mathbb{R}^4$ is not necessarily totally convex. On the other had, show that the maximum of two totally convex k-volume densities is a totally convex k-volume density.

PROBLEM. *For what (hypermetric) normed spaces are the Holmes–Thompson k-volume densities totally convex?*

5. Length and Area in Two-Dimensional Normed Spaces

Before going into the rich and beautiful theory of (hypersurface) area on finite-dimensional normed spaces, we shall sharpen our intuition by carefully considering the case of two-dimensional normed spaces. This case is fundamentally simpler because the notion of hypersurface area coincides with that of length and is thus independent of our volume definition. Nevertheless, we shall see that the theory of length and area on two-dimensional normed spaces is far from trivial and provides a platform from which to jump to higher dimensions.

We start with two theorems that involve solely the notion of length:

THEOREM 5.1 [Gołąb 1932]. *The perimeter of the unit circle of a two-dimensional normed space is between six and eight. Moreover, the length is equal to six if and only if the unit ball is an affine regular hexagon and is equal to eight if and only if it is a parallelogram.*

Full proofs can be found in [Schäffer 1967] and [Thompson 1996]. We stress that the length of the unit circle is measured with the definition of length given by the norm:

If $\gamma : [a, b] \to X$ is a continuous curve on the normed space $(X, \| \cdot \|)$, the length of γ is defined as the supremum of the quantities

$$\sum_{i=0}^{n-1} \|\gamma(t_{i+1}) - \gamma(t_i)\|$$

taken over all partitions $a = t_0 < t_1 < \cdots < t_n = b$ of the interval $[a, b]$. Notice that if γ is differentiable, we can also compute its length by the integral

$$\ell(\gamma) = \int_a^b \|\dot{\gamma}(t)\| \, dt.$$

It is convenient to denote the length of a curve γ on the normed space X with unit disc B by $\ell_B(\gamma)$. Note that $\ell_B(\partial B)$ is an affine invariant of the convex body B.

THEOREM 5.2 [Schäffer 1973]. *If B and D are unit balls of two norms in a two-dimensional space X and if B^* and D^* are the dual balls in X^* then*

$$\ell_D(\partial B) = \ell_{B^*}(\partial D^*).$$

In particular,

$$\ell_B(\partial B) = \ell_{B^*}(\partial B^*).$$

A complete proof is available in [Thompson 1996].

For those who like simply stated open problems, we pass on the following question of Schäffer (private communication):

PROBLEM. *Given an arbitrary convex body $B \subset \mathbb{R}^3$ that is symmetric with respect to the origin, does there always exist a plane Π passing through the origin and for which $\ell_{\Pi \cap B}(\partial(\Pi \cap B))$ is less than or equal to 2π?*

Next we discuss the relation between length and area on two-dimensional normed spaces. The first important question that arises is the isoperimetric problem: Of all convex bodies in a two-dimensional normed space X with a given perimeter find those that enclose the largest area.

The solution of this problem passes through the representation of the length as a mixed volume (in this case a mixed area). This permits the use of Brunn–Minkowski theory to solve the isoperimetric problem and to also give further properties of the length functional. The reader is referred to [Schneider 1993] for a complete discussion of the theory, but our needs can be met in just a few paragraphs.

Let X be an n-dimensional vector space and let λ be a Lebesgue measure on X. If K and L are two subsets of X, the *Minkowski sum* of K and L is the set

$$K + L := \{x + y \in X : x \in K, y \in L\}.$$

If L is the unit ball of a norm in X, we may think of $K + L$ as the set of all points in X whose distance from K is less than or equal to one. In other words, the tube of radius one about the set K.

The *mixed volume* $V(K[n-1], L)$ of two closed, bounded convex sets K and L in X is defined as a "directional derivative" of the Lebesgue measure:

$$V(K[n-1], L) = \frac{1}{n} \lim_{t \to +0} \frac{\lambda(K + tL) - \lambda(K)}{t}.$$

In the two-dimensional, case $V(K, L) := V(K[1], L)$ is linear and monotonic in each variable. The key result in the solution of the isoperimetric problem in normed spaces is the *Minkowski mixed volume inequality:*

$$V(K[n-1], L) \geq \lambda(K)^{n-1} \lambda(L).$$

Moreover, if K and L are convex bodies, then equality holds if and only if K is obtained from L by translation and dilation.

Back to the two-dimensional case, if we're given a centered convex body B, we may define the magnitude of a vector x in two different ways:

(1) Take B to be the unit ball of a norm $\| \cdot \|$ on X and set the magnitude of x to be $\|x\|$.

(2) Let $[x] \subset X$ denote the line segment from the origin to x and define the magnitude of x as $V([x], B)$.

EXERCISE 5.3. Show that for any convex body B that is symmetric with respect to the origin, the map $x \mapsto V([x], B)$ is a norm, but that in general its unit disc is *different* from B.

The first step in solving the isoperimetric problem in a normed space X is to find a centrally symmetric convex body I such that

$$\|\boldsymbol{x}\| = V([\boldsymbol{x}], I), \text{ for all } \boldsymbol{x} \in X.$$

Of course, I will also depend on the choice of Lebesgue measure λ used to define the mixed volume. However, given a volume definition the body I will be defined intrinsically in terms of the norm.

The construction of I is extremely simple: Let B be the unit ball of X and let Ω be the volume form on X that satisfies $\Omega(\boldsymbol{x} \wedge \boldsymbol{y}) = \lambda(\boldsymbol{x} \wedge \boldsymbol{y})$ for all positive bases $\boldsymbol{x}, \boldsymbol{y}$ of X (we are forced to take an orientation of X at this point, but the result will not depend on the choice). If

$$i_\Omega : X \longrightarrow X^*$$

is defined by $i_\Omega(\boldsymbol{v})(\boldsymbol{w}) := \Omega(\boldsymbol{v} \wedge \boldsymbol{w})$, the set I is given by $(i_\Omega B)^*$.

Summarizing:

PROPOSITION 5.4. *Let X be a two-dimensional normed space with unit ball B and volume form Ω. If I denotes the body $(i_\Omega B)^*$, then*

$$\|\boldsymbol{x}\| = V([\boldsymbol{x}], I)$$

for all vectors $\boldsymbol{x} \in X$.

The proof will be postponed to the next section where we will treat the n-dimensional version of the proposition.

Notice that if the orientation of X is changed, the form Ω changes sign, but the symmetry of the unit disc B implies that the body I stays the same.

EXERCISE 5.5. Show that if K is a convex body in X, its perimeter equals $2V(K, I)$ and that, in particular, the perimeter of I is twice its area. *Hint:* Try first with bodies whose boundaries are polygons and use the previous proposition.

The representation of length as a mixed volume gives an easy proof of the following monotonicity property of length in two-dimensional normed spaces.

PROPOSITION 5.6. *If $K_1 \subset K_2$ are nested convex bodies in a two-dimensional normed space, then $\ell(\partial K_1) \leq \ell(\partial K_2)$.*

The proof is left as a simple exercise to the reader. The following related exercise is, perhaps, somewhat harder.

EXERCISE 5.7. Show that a Finsler metric on the plane satisfies the monotonicity property in the previous proposition if and only if its geodesics are straight lines.

THEOREM 5.8. *Let X be a two-dimensional normed space with unit disc B and area form $\Omega \in \Lambda^2 X^*$. Of all convex bodies in X with a given perimeter the one that encloses the largest area is, up to translations, a dilate of $I := (i_\Omega B)^*$.*

PROOF. Let $K \subset X$ be a convex body and let

$$\ell_B(\partial K) = 2V(K, I)$$

be its perimeter. By *Minkowski's mixed volume inequality*, we have

$$\frac{\ell_B(\partial K)^2}{4} = V(K, I)^2 \geq \lambda(K)\lambda(I)$$

with equality if an only if K and I are homothetic. Thus, the area enclosed by K is maximal for a given perimeter if and only if K is a dilate of I.   □

DEFINITION 5.9. Let $Y \mapsto (\Lambda^2 Y, \mu_Y)$ be a volume definition for two-dimensional normed spaces. If X is a two-dimensional normed space with unit ball B, the *isoperimetrix* of X corresponding to the volume definition μ is the body $\mathbb{I}_X := (i_{\Omega_X}(B))^*$, where Ω_X is a 2-form on X satisfying $|\Omega_X| = \mu_X$.

We shall denote the isoperimetrices of a two-dimensional normed space X with respect to the Busemann, Holmes–Thompson, mass, and mass∗ volume definitions by $\mathbb{I}_X^b$, $\mathbb{I}_X^{ht}$, $\mathbb{I}_X^m$, and $\mathbb{I}_X^{m*}$.

If $T : X \to X$ is an invertible linear transformation, the isoperimetrix, with respect to any volume definition, of the norm with unit ball $T(B_X)$ is $T(\mathbb{I}_X)$.

EXERCISE 5.10. If X is a two-dimensional normed space with unit ball B and if μ is a particular choice of volume definition, then

$$\ell_B(\partial \mathbb{I}_X) = 2\mu_X(\mathbb{I}_X) \text{ and } \mu_X(\mathbb{I}_X) = \mu_X^*(B^*).$$

Using this exercise, we can give sharp estimates on the area and perimeter of the isoperimetrix of a two-dimensional normed space for Busemann, Holmes–Thompson, and mass∗ volume definitions.

Indeed, it follows trivially from the exercise that $\mu_X^b(\mathbb{I}_X^b) = \mathrm{vp}(B_X)/\pi$ and that $\mu_X^{ht}(\mathbb{I}_X^{ht}) = \pi$. Using the Mahler and Blaschke–Santaló inequalities, we have

$$\frac{8}{\pi} \leq \mu_X^b(\mathbb{I}_X^b) \leq \pi.$$

The fact that $\mu_X^{m*}(\mathbb{I}_X^{m*}) \leq \pi$ with equality if and only if X is Euclidean is equivalent to the inequality $\mu^{m*} \geq \mu^{ht}$ for two-dimensional normed spaces.

EXERCISE 5.11. Find the sharp lower bound for $\mu_X^{m*}(\mathbb{I}_X^{m*})$.

It is interesting to note that the Blaschke–Santaló inequality implies that

$$\mu_X^b(B_X) \geq \mu_X^b(\mathbb{I}_X^b) \quad \text{and} \quad \mu_X^{ht}(B_X) \leq \mu_X^{ht}(\mathbb{I}_X^{ht}),$$

with equality in both cases if and only if B is an ellipse. Of course this implies that for both the Busemann and Holmes–Thompson definitions $B_X = \mathbb{I}_X$ if and only if X is Euclidean. Notice that whether a unit disc is equal to its isoperimetrix depends on the volume definition we are using. However, whether the unit disc is a dilate of its isoperimetrix does not depend on such a choice.

DEFINITION 5.12. Let X be a two-dimensional normed space. If B_X is a dilate of $\mathbb{I}_X$ for one (and, therefore, any) volume definition, the unit circle, ∂B_X, is said to be a *Radon curve*.

For comparison with the higher-dimensional case we summarize the properties of the map $\mathbb{I}$ that sends a unit disc B_X to $\mathbb{I}_X$. This maps sends convex bodies to convex bodies; it is a bijection; it commutes with linear maps in the sense that $T(B_X)$ is sent to $T(\mathbb{I}_X)$ for all invertible linear maps T; it maps polygons to polygons, smooth bodies to strictly convex bodies and strictly convex bodies to smooth bodies; and the only fixed points for the μ^b and μ^{ht} normalizations are ellipses.

A good, very elementary account of the construction of the isoperimetrix from first principles and its relationship to physics and symplectic geometry (the ball is used for measuring position and the isoperimetrix for measuring velocity) is given by Wallen [1995].

Finally, we explore the relationship between the perimeter and area of the unit ball. The motivation is that $\ell(\partial\mathbb{I}) = 2\mu(\mathbb{I}_B)$ and that in the Euclidean case this holds for the ball.

THEOREM 5.13. *If X is a two-dimensional normed space with unit ball B then*

$$2\mu^m(B) \leq \ell(\partial B) \leq 2\mu^{m*}(B)$$

with equality on the left if and only if ∂B is a Radon curve and on the right if and only if ∂B is an equiframed curve.

For the definition of equiframed curves and a proof of the theorem we refer the reader to [Martini et al. 2001] where the history of this result is also discussed.

EXERCISE 5.14. Use this result and properties of $\mathbb{I}_X$ to show that $B_X = \mathbb{I}_X^m$ if and only if ∂B_X is a Radon curve; and that $B_X = \mathbb{I}_X^{m*}$ if and only if ∂B_X is equiframed.

There is a further recent result in this direction.

THEOREM 5.15 [Moustafaev]. *If X is a two-dimensional normed space, then*

$$2\mu_X^{\mathrm{ht}}(B_X) \leq \ell(\partial B_X),$$

with equality if and only if X is Euclidean.

PROOF. By definition of the isoperimetrix and Minkowski's mixed volume inequality, we have

$$\ell(B_X)^2 = 4V(B_X, \mathbb{I}_X^{\mathrm{ht}}) \geq 4\mu_X^{\mathrm{ht}}(B_X)\mu_X^{\mathrm{ht}}(\mathbb{I}_X^{\mathrm{ht}}).$$

Using that $\mu_X^{\mathrm{ht}}(\mathbb{I}_X^{\mathrm{ht}}) = \pi$ and that $\mu_X^{\mathrm{ht}}(B_X)/\pi \leq 1$, we have

$$\ell(B_X)^2 \geq 4\pi\mu_X^{\mathrm{ht}}(B_X) \geq 4\pi\mu_X^{\mathrm{ht}}(B_X)\frac{\mu_X^{\mathrm{ht}}(B_X)}{\pi} = 4\mu_X^{\mathrm{ht}}(B_X)^2. \qquad \square$$

EXERCISE 5.16. If X is a two-dimensional normed space, show that

$$2 \leq \mu_X^m(B_X) \leq \pi,$$
$$3 \leq \mu_X^{m*}(B_X) \leq 4,$$

and (using inequalities from Section 3)

$$8/\pi \leq \mu_X^{\mathrm{ht}}(B_X) \leq \pi.$$

Give the equality cases.

6. Area on Finite-Dimensional Normed Spaces

In Section 4, we saw that the Busemann, Holmes–Thompson, and mass* volume definitions induce k-volume densities that are weakly convex. In the special case where the dimension of the normed space X is $n = k + 1$, then the $(n-1)$-volume densities are norms on the space $\Lambda^{n-1}X$.

It follows from the properties of the volume definitions that, in all three cases, the map that assigns to the normed space X the normed space $\Lambda^{n-1}X$ has the following properties:

(1) If $T : X \to Y$ is a short linear map between normed spaces X and Y, then the induced map $T_* : \Lambda^{n-1}X \to \Lambda^{n-1}Y$ is also short.
(2) The map $X \mapsto \Lambda^{n-1}X$ is continuous with respect to the topology induced by the Banach–Mazur distance.
(3) If X is a Euclidean space, then the $(n-1)$-volume density is the standard Euclidean area on X.
(4) If the dimension of X is two, the map $X \mapsto \Lambda^1 X$ is the identity.

Notice that property (1) states that for the Busemann, Holmes–Thompson, and mass* definitions, the map that takes the normed space X to the normed space $\Lambda^{n-1}X$ is a covariant functor in the category $\mathcal{N}$ of finite-dimensional normed spaces.

DEFINITION 6.1. A *definition of area* on normed spaces assigns to every n-dimensional, $n \geq 2$, normed space X a normed space $(\Lambda^{n-1}X, \sigma_X)$ in such a way that properties (1)–(4) above are satisfied.

For simplicity, we shall speak of the Busemann, Holmes–Thompson, and mass* definitions of area to refer to the definitions of area induced, respectively, by the Busemann, Holmes–Thompson, and mass* volume definitions.

Definitions of area in normed spaces are related to important constructions in convex geometry such as intersection bodies, projection bodies, and Wulff shapes. However, let us start by posing a few natural questions that arise whenever we have a definition of area. The answer to some of these questions, once specialized to the Busemann and Holmes–Thompson definitions, are deep results in the theory of convex bodies. Other questions are long-standing open problems, and yet others seem to be new.

Given a definition of area $X \mapsto (\Lambda^{n-1}X, \sigma_X)$ on normed spaces, we may ask: Is the map $X \mapsto (\Lambda^{n-1}X, \sigma_X)$ injective? What is its range? Does it send crystalline norms to crystalline norms? Does it send Minkowski spaces to Minkowski spaces? In what numeric range is the area of the unit sphere of an n-dimensional normed space?

Other problems arise when we consider the relationship between length, area, and volume, but, for now, let us concentrate on the questions we have just posed.

6.1. Injectivity and range of the area definition Let us start the study of the injectivity and range of the Busemann definition of area by describing the unit ball of the $(n-1)$-volume density in terms of a well-known construction in convex geometry.

Busemann area and intersection bodies. Consider $\mathbb{R}^n$ with its Euclidean structure and its unit sphere S^{n-1}. If $K \subset \mathbb{R}^n$ is a star-shaped body containing the origin, the *intersection body* of K, IK, is defined by the following simple construction: if $\boldsymbol{x} \in \mathbb{R}^n$ is a unit vector, let $A(K \cap \boldsymbol{x}^\perp)$ denote the area of the intersection of K with the hyperplane perpendicular to $\boldsymbol{x}$, and let IK be the star-shaped body enclosed by the surface

$$\{\boldsymbol{x}/A(K \cap \boldsymbol{x}^\perp) \in \mathbb{R}^n : \boldsymbol{x} \in S^{n-1}\}.$$

A celebrated theorem of Busemann, which is equivalent to the weak convexity of the Busemann volume definition, states that if K is a centered convex body, then IK is also a centered convex body.

Let X be an n-dimensional normed space. Choose a basis of X and use it to identify X with $\mathbb{R}^n$. Take the Euclidean structure in $\mathbb{R}^n$ for which the basis is orthonormal and use the resulting Euclidean structure to identify the spaces X^* and $\Lambda^{n-1}X$, as well as to define the unit sphere S^{n-1} in $\mathbb{R}^n$.

EXERCISE 6.2. Show that with all these identifications, the convex body $\{\boldsymbol{x} \in \mathbb{R}^n : \sigma_X^b(\boldsymbol{x}) \leq 1\}$ is ε_{n-1} times the intersection body of B_X; (here σ_X^b is the norm induced on X^* by the norm on $\Lambda^{n-1}X$).

Notice that we can now write the question of whether the Busemann definition of area is injective in the following classical form: Is a centered convex body determined uniquely by the area of its intersections with hyperplanes passing through the origin? The answer is affirmative (see [Lutwak 1988] and [Gardner 1995]), and so we have the following result:

THEOREM 6.3. *The Busemann area definition is injective.*

Determining the range of the Busemann area definition is somewhat trickier. Thanks to the efforts of R. Gardner, G. Zhang, and others in the solution of the first of the Busemann–Petty problems, it is known (see [Gardner 1994], [Gardner et al. 1999], and [Zhang 1999] and the references therein) that in dimensions two, three, and four every convex body symmetric with respect to the origin is the intersection body *of some star-shaped body.* It is not clear at this point whether those bodies that are intersection bodies of centered convex bodies can be characterized effectively. For dimensions greater than four, not every centered convex body is an intersection body ([Gardner et al. 1999]). For further information about intersection bodies see, for example, [Gardner 1995] and [Lutwak 1988].

Examples in [Thompson 1996] show that the Busemann area definition does not take crystalline norms to crystalline norms. We don't know whether it takes Minkowski norms to Minkowski norms.

Let us now pass to the Holmes–Thompson definition.

Holmes–Thompson area and projection bodies. Consider $\mathbb{R}^n$ with its Euclidean structure and its unit sphere S^{n-1}. If $K \subset \mathbb{R}^n$ is a convex body, the *projection body* of K, ΠK, is given by the following simple construction: if $x \in \mathbb{R}^n$ is a unit vector, let $A(K|x^{\perp})$ denote the area of the orthogonal projection of K onto the hyperplane perpendicular to x, and let the *polar* of ΠK be the body enclosed by the surface

$$\{A(K|x^{\perp})x \in \mathbb{R}^n : x \in S^{n-1}\}.$$

As in the case of the Busemann definition of area, identifying a normed space X with $\mathbb{R}^n$ allows us to write the unit ball for the $(n-1)$-volume density in terms of this nonintrinsic construction.

EXERCISE 6.4. Show that by identifying a normed space X with $\mathbb{R}^n$ as in the previous exercise, the convex body $\{x \in \mathbb{R}^n : \sigma_X^{\mathrm{ht}}(x) \leq 1\}$ is $1/\varepsilon_{n-1}$ times the polar of the projection body of B_X^*.

The question of the injectivity of the Holmes–Thompson definition of area can now be formulated in classical terms: Is a centered convex body determined uniquely by the area of its orthogonal projections onto hyperplanes? The answer, in the affirmative, follows from a celebrated result of Alexandrov [1933] (see also [Gardner 1995]). We then have the following result:

THEOREM 6.5. *The Holmes–Thompson area definition is injective.*

It is known, basically from the time of Minkowski, that a centered convex body B is the projection body of another if and only if it is a zonoid (see [Gardner 1995]). By Theorem 2.12, this means that for any n-dimensional normed space X the normed space $(\Lambda^{n-1}X, \sigma_X^{\mathrm{ht}})$ is hypermetric.

Moreover, because of the integral formula for the Holmes–Thompson $(n-1)$-volume density in terms of the surface area measure of the dual sphere given in the proof of Theorem 4.16, the problem of reconstructing the norm from the Holmes–Thompson $(n-1)$-volume density is precisely the famous *Minkowski problem:* Reconstruct a convex body from the knowledge of its Gauss curvature as a function of its unit normals. The next two theorems follow directly from the work of Minkowski, Pogorelov, and Nirenberg (see [Pogorelov 1978] for a detailed presentation).

THEOREM 6.6. *The range of the Holmes–Thompson area definition is the set of hypermetric normed spaces.*

THEOREM 6.7. *Let X be an n-dimensional vector space and let $\sigma : \Lambda^{n-1}X \to [0, \infty)$ be a Minkowski norm. If $(\Lambda^{n-1}X, \sigma)$ is hypermetric, then there exists a unique Minkowski norm $\| \cdot \|$ on X such that σ is the Holmes–Thompson $(n-1)$-volume density of the normed space $(X, \| \cdot \|)$.*

Another important feature of the Holmes–Thompson area is the following (for a proof see [Thompson 1996]):

THEOREM 6.8. *The Holmes–Thompson area definition takes Minkowski spaces to Minkowski spaces and crystalline norms to crystalline norms.*

Mass* area and wedge bodies. Let B be a centered convex body in an n-dimensional vector space X, and let B^{n-1} be the $(n-1)$-fold product of B in the $n(n-1)$-dimensional space X^{n-1}. If Alt $: X^{n-1} \to \Lambda^{n-1}X$ denotes the (nonlinear) map

$$(\boldsymbol{x}_1, \ldots, \boldsymbol{x}_{n-1}) \longmapsto \boldsymbol{x}_1 \wedge \boldsymbol{x}_2 \wedge \cdots \wedge \boldsymbol{x}_{n-1},$$

we define the *wedge body* of B, denoted by $\mathcal{W}B$, as the convex hull of $\text{Alt}(B^{n-1})$ in $\Lambda^{n-1}X$.

We remark that even if $B^{n-1} \subset X^{n-1}$ is a centered convex body, $\text{Alt}(B^{n-1})$ is not necessarily convex.

THEOREM 6.9. *The unit ball in $\Lambda^{n-1}X$ for the mass* $(n-1)$-volume density of a normed space X is the body $(\mathcal{W}B_X^*)^*$.*

PROOF. By Exercise 4.3, we have

$$\sigma_X^{m*}(a) = \sup\{|\boldsymbol{\xi}_1 \wedge \boldsymbol{\xi}_2 \wedge \cdots \wedge \boldsymbol{\xi}_{n-1} \cdot a| : \boldsymbol{\xi}_1, \ldots, \boldsymbol{\xi}_{n-1} \in B_X^*\}.$$

But this is just the supremum of $|\eta \cdot a|$, where $\eta \in \text{Alt}\big((B_X^*)^{n-1}\big)$. Therefore σ_X^{m*} is the dual to the norm in $\Lambda^{n-1}X^*$ whose unit ball is $\mathcal{W}B_X^*$.   □

It is quite easy to do calculations for $\mathcal{W}B^*$ in the case when the centered convex body B is a simple object. The following statements are based on such calculations, the details of which are left as exercises (see also [Thompson 1999]).

PROPOSITION 6.10. *The mass* area definition is not injective.*

SKETCH OF THE PROOF.. All we must do is find two centered convex bodies B and K such that $\mathcal{W}B^* = \mathcal{W}K^*$, but $B \neq K$.

Let B be the cube with vertices at $(\pm 1, \pm 1, \pm 1)$. In this case, B^* is the octahedron with vertices $(\pm 1, 0, 0)$, $(0, \pm 1, 0)$, $(0, 0, \pm 1)$ and $\mathcal{W}B^* = B^*$.

Let K be the cuboctahedron with vertices $(\pm 1, \pm 1, 0)$, $(\pm 1, 0, \pm 1)$, $(0, \pm 1, \pm 1)$. The dual ball K^* is the rhombic dodecahedron with vertices $\pm (1, 0, 0)$, $\pm (0, 1, 0)$, $\pm (0, 0, 1)$ and $(\pm \frac{1}{2}, \pm \frac{1}{2}, \pm \frac{1}{2})$. A simple calculation shows that $\mathcal{W} B^* = \mathcal{W} K^*$.

In fact, if L is any centered convex body that lies between the cube and the cube-octahedron then $\mathcal{W} L^* = \mathcal{W} B^*$. $\square$

While it seems unlikely that the wedge body of the unit ball in a Minkowski space is the ball of a Minkowski space, it is not hard to show that the wedge body of a polytope is a polytope. Then:

PROPOSITION 6.11 [Thompson 1999]. *The mass∗ area definition takes crystalline norms to crystalline norms.*

The question of determining the range for the mass∗ area definition is completely open. Is it possible that any centered convex body is a wedge body?

6.2. Area of the unit sphere. In this section we give the higher-dimensional analogues (as far as we know them) of the theorems of Schäffer and Gołąb discussed in Section 5.

The Holmes–Thompson definition was designed originally to yield a generalization of Schäffer's result and we have the following theorem.

THEOREM 6.12 [Holmes and Thompson 1979]. *If B and K are the unit balls of two norms $\| \cdot \|_B$ and $\| \cdot \|_K$ in the vector space X, the Holmes–Thompson area of ∂K in the normed space $(X, \| \cdot \|_B)$ equals the Holmes–Thompson area of ∂B^* in the normed space $(X^*, \| \cdot \|_K^*)$.*

Notice that in particular, the Holmes–Thompson area of the unit sphere of a normed space equals the Holmes–Thompson area of the unit sphere of its dual. Simple calculations show that neither the Busemann, the mass∗, nor the mass definition have this property. In fact, Daniel Hug (private communication) has shown that Theorem 6.12 characterizes the Holmes–Thompson definition. However, the following question remains open.

PROBLEM [Thompson 1996]. *Is the Holmes–Thompson definition of volume characterized by the fact that the area of the unit sphere of a normed space equals the area of the unit sphere of its dual?*

The first result extending Gołąb's theorem to higher dimension is the following sharp upper bound for the Busemann area of a unit sphere.

THEOREM 6.13 [Busemann and Petty 1956]. *The Busemann area of the unit sphere of an n-dimensional normed space is at most $2n\varepsilon_{n-1}$ with equality if and only if B is a parallelotope.*

For $n \geq 3$ no sharp lower bound for the Busemann area of the unit sphere of an n-dimensional normed space has been proved. It is conjectured that the minimum is $n\varepsilon_n$ attained by the Euclidean ball. However, when $n = 3$ it is also attained by the rhombic dodecahedron.

Since $\mu^b \geq \mu^{\mathrm{ht}}$, an upper bound for the Busemann area is also an upper bound for the Holmes–Thompson area.

COROLLARY 6.14. *The Holmes–Thompson area of the unit sphere of an n-dimensional normed space is less than* $2n\varepsilon_{n-1}$.

While the sharp upper bound for the Holmes–Thompson area of the unit sphere in any dimension greater than two is not known, the sharp lower bound in dimension three is given by the following unpublished result of Álvarez, Ivanov, and Thompson:

THEOREM 6.15. *The Holmes–Thompson area of the unit sphere of a three-dimensional normed space is at least* $36/\pi$. *Moreover, equality holds if the unit ball is a cuboctahedron or a rhombic dodecahedron.*

Since $\mu^b \geq \mu^h t$ and $\mu^{m*} \geq \mu^{\mathrm{ht}}$, we have the following lower bound for the Busemann and mass∗ areas of the unit sphere of a three-dimensional normed space.

COROLLARY 6.16. *The Busemann and mass∗ areas of unit sphere of a three-dimensional normed space is greater than* $36/\pi$.

Although these bounds are not sharp, they are the best bounds known so far.

It is possible to use a variety of inequalities including the Petty projection inequality (in the case of σ^{ht}) and the Busemann intersection inequality (in the case of σ^b) to give nonsharp lower bounds. The reader is referred to [Thompson 1996] for examples of what one can get.

6.3. Mixed volumes and the isoperimetrix. We now pass to questions concerning the relationship between areas and volumes, and, in particular, to the solution of the isoperimetric problem in finite-dimensional normed spaces. The subject is classical and has been studied from different viewpoints by convex geometers, geometric measure theorists, and crystallographers (see, for example, [Busemann 1949b], [Taylor 1978], and [Ambrosio and Kirchheim 2000]). Nevertheless, being interested in a particular intrinsic viewpoint and relations to area on normed and Finsler spaces that are not treated elsewhere, we shall give a short account of the subject.

Let X be an n-dimensional vector space and let λ be a Lebesgue measure on X. If $I \subset X$ is a centered convex body, we can define an $(n-1)$-volume density on X by the following construction: given $n-1$ linearly independent vectors $\boldsymbol{x}_1, \ldots, \boldsymbol{x}_{n-1} \in X$, we denote the parallelotope they define by $[\boldsymbol{x}_1, \ldots, \boldsymbol{x}_{n-1}]$ and set

$$\sigma_I(\boldsymbol{x}_1 \wedge \boldsymbol{x}_2 \wedge \cdots \wedge \boldsymbol{x}_{n-1}) := \frac{1}{n} \lim_{t \to +0} \frac{\lambda([\boldsymbol{x}_1, \ldots, \boldsymbol{x}_{n-1}] + tI) - \lambda([\boldsymbol{x}_1, \ldots, \boldsymbol{x}_{n-1}])}{t}.$$

It is easy to see that σ_I is well defined and that by changing λ for another Lebesgue measure on X we simply multiply σ_I by a constant. Note also that although the measure of $[\boldsymbol{x}_1, \ldots, \boldsymbol{x}_{n-1}]$ is zero, we have included it in the formula to stress its relationship with the n-dimensional mixed volume of two bodies,

$$V(K[n-1], L) := \frac{1}{n} \lim_{t \to +0} \frac{\lambda(K + tL) - \lambda(K)}{t}.$$

With this definition, if K is a convex body in X,

$$\int_{\partial K} \sigma_I = nV(K[n-1], I).$$

EXERCISE 6.17. Show that the $(n-1)$-volume density σ_I constructed above is a norm on $\Lambda^{n-1}(X)$, and that

$$\int_{\partial I} \sigma_I = n\lambda(I).$$

We would also like to reverse this construction: Starting from a norm $\sigma : \Lambda^{n-1}X \to [0, \infty)$ and a Lebesgue measure λ on X construct a convex body $I \subset X$ such that $\sigma = \sigma_I$. The construction is quite simple: Let Ω be a volume form on X such that $|\Omega| = \lambda$ and consider the linear isomorphism

$$i_\Omega : \Lambda^{n-1}X \longrightarrow X^*$$

defined by $i_\Omega(\boldsymbol{x}_1 \wedge \boldsymbol{x}_2 \wedge \cdots \wedge \boldsymbol{x}_{n-1})(\boldsymbol{x}) = \Omega(\boldsymbol{x}_1 \wedge \boldsymbol{x}_2 \wedge \cdots \wedge \boldsymbol{x}_{n-1} \wedge \boldsymbol{x})$. The body I is given by $(i_\Omega \mathcal{B})^*$, where $\mathcal{B} \subset \Lambda^{n-1}X$ is the unit ball of σ.

In terms of mixed volumes, we have the following result:

PROPOSITION 6.18. *Let X be an n-dimensional vector space, let σ be a norm on $\Lambda^{n-1}X$ with unit ball $\mathcal{B}$ and let λ be a Lebesgue measure on X. Using the notation above, if $I := (i_\Omega \mathcal{B})^*$, we have*

$$\int_{\partial K} \sigma = nV(K[n-1], I)$$

for all convex bodies $K \subset X$.

To prove the proposition, let us give a simpler, more visual relationship between σ and $I := (i_\Omega \mathcal{B})^*$ that is of independent interest. Given a nonzero $(n-1)$-vector $a \in \Lambda^{n-1}X$, we shall say that a vector $\boldsymbol{v} \in X$ is *normal to a with respect to I* if $\boldsymbol{v} \in \partial I$, the hyperplane parallel to $\langle a \rangle$ and passing through $\boldsymbol{v}$ supports I, and $\Omega(a \wedge \boldsymbol{v}) > 0$. When I is smooth and strictly convex the normal is unique, but this is of no importance to what follows. Notice, and this is important, that $\boldsymbol{v}$ is constructed in such a way that

$$\Omega(a \wedge \boldsymbol{v}) = \sup\{|\Omega(a \wedge \boldsymbol{x})| : \boldsymbol{x} \in I\}.$$

LEMMA 6.19. *Let X be an n-dimensional vector space, let σ be a norm on $\Lambda^{n-1}X$ with unit ball $\mathcal{B}$ and let $\Omega \in \Lambda^n X^*$ be a volume form on X. If a is a nonzero $(n-1)$-vector on X and $\boldsymbol{v} \in X$ is normal to a with respect to $I := (i_\Omega \mathcal{B})^*$, then*

$$\sigma(a) = \Omega(a \wedge \boldsymbol{v}).$$

PROOF. Let $\|\cdot\|^*$ denote the norm in X^* whose unit ball is $I^* = i_\Omega\mathcal{B}$. Trivially, we have $\sigma(a) = \|i_\Omega(a)\|^*$ for any $a \in \Lambda^{n-1}X$. Therefore,

$$\sigma(a) = \sup\{|\Omega(a \wedge \boldsymbol{x})| : \boldsymbol{x} \in I\} = \Omega(a \wedge \boldsymbol{v}). \qquad \square$$

In other terms, if $\boldsymbol{x}_1, \ldots, \boldsymbol{x}_{n-1}$ are linearly independent vectors in X and $\boldsymbol{v}$ is normal to $\boldsymbol{x}_1 \wedge \boldsymbol{x}_2 \wedge \cdots \wedge \boldsymbol{x}_{n-1}$ with respect to I, then the volume of the n-dimensional parallelotope $[\boldsymbol{x}_1, \ldots, \boldsymbol{x}_{n-1}, \boldsymbol{v}]$ is the area of the $(n-1)$-dimensional parallelotope $[\boldsymbol{x}_1, \ldots, \boldsymbol{x}_{n-1}]$.

PROOF OF PROPOSITION 6.18. Let $\boldsymbol{x}_1, \ldots, \boldsymbol{x}_{n-1}$ be linearly independent vectors in X and let $[\boldsymbol{x}_1, \ldots, \boldsymbol{x}_{n-1}]$ denote the parallelotope spanned by them. Notice that if $\boldsymbol{v}$ is normal to $\boldsymbol{x}_1 \wedge \boldsymbol{x}_2 \wedge \cdots \wedge \boldsymbol{x}_{n-1}$ with respect to I, then for any $t > 0$, the union of the n-dimensional parallelotopes

$$[\boldsymbol{x}_1, \ldots, \boldsymbol{x}_{n-1}, t\boldsymbol{v}] \text{ and } [\boldsymbol{x}_1, \ldots, \boldsymbol{x}_{n-1}, -t\boldsymbol{v}],$$

which we denote by $P(t)$, is contained in the set $[\boldsymbol{x}_1, \ldots, \boldsymbol{x}_{n-1}] + tI$. Moreover, since up to terms of order 2 and higher in t the volumes of $P(t)$ and $[\boldsymbol{x}_1, \ldots, \boldsymbol{x}_{n-1}] + tI$ are the same, we have

$$\frac{1}{n}\lim_{t \to +0} \frac{\lambda([\boldsymbol{x}_1, \ldots, \boldsymbol{x}_{n-1}] + tI)}{t} = \frac{1}{n}\lim_{t \to +0} \frac{\lambda(P(t))}{t} = \sigma(\boldsymbol{x}_1 \wedge \boldsymbol{x}_2 \wedge \cdots \wedge \boldsymbol{x}_{n-1}),$$

and this concludes the proof. $\qquad \square$

We are now ready to solve the isoperimetric problem for convex bodies:

THEOREM 6.20. *Let X be an n-dimensional vector space, let σ be a norm on $\Lambda^{n-1}X$ with unit ball $\mathcal{B}$, and let $\Omega \in \Lambda^n X^*$ be a volume form on X. Of all convex bodies in X with a given surface area the one that encloses the largest volume is, up to translations, a dilate of $I := (i_\Omega\mathcal{B})^*$.*

PROOF. Let $K \subset X$ be a convex body and let

$$\int_{\partial K} \sigma = nV(K[n-1], I)$$

be its surface area. By Minkowski's mixed volume inequality, we have

$$\left(\int_{\partial K} \sigma\right)^n = n^n V(K[n-1], I)^n \geq n^n \lambda(K)^{n-1}\lambda(I)$$

with equality if an only if K and I are homothetic. Thus, the volume enclosed by K is maximal for a given surface area if and only if K is a dilate of I. $\qquad \square$

We shall denote the isoperimetrices of a normed space X with respect to the Busemann, Holmes–Thompson, and mass∗ definitions by $\mathbb{I}_X^b$, $\mathbb{I}_X^{ht}$, and $\mathbb{I}_X^{m*}$, respectively. In the case of the Busemann and Holmes–Thompson definitions, the isoperimetrices can be given, nonintrinsically, in terms of intersection and projection bodies.

EXERCISE 6.21. Using Exercises 6.2 and 6.4, and the construction of the isoperimetrix, show that

$$\mathbb{I}_X^b = \frac{\varepsilon_{n-1}}{\varepsilon_n}\lambda(B_X)(IB_X)^* \text{ and } \mathbb{I}_X^{ht} = \frac{\varepsilon_n}{\varepsilon_{n-1}}\lambda^*(B_X^*)^{-1}\Pi B_X^*,$$

where λ and λ^* are, respectively, the Euclidean volumes on X and X^* given by their identification with $\mathbb{R}^n$.

EXERCISE 6.22. Describe the isoperimetrix $\mathbb{I}_X^{m*}B$ in terms of wedge bodies.

6.4. Geometry of the isoperimetrix. We now turn our attention to problems relating the unit ball of a normed space and its isoperimetrix with respect to some volume definition. Let us start with the deceptively simple problem of estimating the volume of the isoperimetrix.

Identifying the normed space X with $\mathbb{R}^n$ as in Exercise 6.21, we see that the Holmes–Thompson volume of $\mathbb{I}_X^{ht}$ is

$$\mu_X^{ht}(\mathbb{I}_X^{ht}) = \varepsilon_n^{-1}\left(\frac{\varepsilon_n}{\varepsilon_{n-1}}\right)^n \lambda^*(B_X^*)^{-n+1}\lambda(\Pi B_X^*).$$

The statement that this quantity is greater than or equal to ε_n with equality if and only if X is Euclidean is known as *Petty's conjectured projection inequality*, and is one of the major open problems in the theory of affine geometric inequalities.

Sharp lower bounds for $\mu_X^b(\mathbb{I}_X^b)$ and $\mu_X^{m*}(\mathbb{I}_X^{m*})$ are also unknown, although as observed in [Thompson 1996] the inequality $\mu_X^b(\mathbb{I}_X^b) \geq \varepsilon_n$ for $n \geq 3$ would easily yield (exercise!) that the Busemann area of unit sphere of a normed space of dimension n is at least $n\varepsilon_n$.

Another interesting affine invariant involving the isoperimetrix is the symplectic volume of $B_X \times \mathbb{I}_X^*$ in $X \times X^*$. In the two-dimensional case this simply yields the square of the area of the unit disc, but in higher dimension it is a much more interesting invariant:

EXERCISE 6.23. Pick up either [Gardner 1995] or [Thompson 1996] and, using Exercise 6.21, prove that the inequality

$$\text{svol}(B_X \times \mathbb{I}_X^*) \leq \varepsilon_n \mu_X(B_X)$$

is true for the Busemann (resp. Holmes–Thompson) definition by showing that it is equivalent to Busemann's intersection inequality (resp. Petty's projection inequality).

It would be interesting to complete the picture by having a sharp upper bound for $\text{svol}(B_X \times (\mathbb{I}_X^{m*})^*)$ in terms of $\mu^{m*}{}_X(B_X)$.

We finish the paper by considering some questions relating length, area, and volume. In terms of the isoperimetrix they have very simple statements: Given a volume definition, when is the isoperimetrix equal to the unit ball, when is it a multiple of the ball, and when is it inside the ball?

These simple questions are really about the existence of a coarea formula or inequality for the different definitions of volume on normed and Finsler spaces. Many Riemannian and Euclidean results depend, or seem to depend, on the simple fact that volume $=$ base $\times$ height. To what extent is this true in normed and Finsler spaces?

In order to relate the coarea formula and inequality with the geometry of the isoperimetrix, let us first define the height of a parallelotope $[\boldsymbol{x}_1, \ldots, \boldsymbol{x}_{n-1}, \boldsymbol{x}_n]$ in a vector space X with respect to a centered convex body $B \subset X$ by the following construction: let $\boldsymbol{\xi} \in X^*$ be a covector in ∂B^* such that $\boldsymbol{\xi}(\boldsymbol{x}_i) = 0$ for all i between 1 and $n - 1$. The quantity $|\boldsymbol{\xi}(\boldsymbol{x}_n)|$, which is independent of the choice of $\boldsymbol{\xi}$, will be called the *height* of $[\boldsymbol{x}_1, \ldots, \boldsymbol{x}_{n-1}, \boldsymbol{x}_n]$ with respect to B.

By the construction of the isoperimetrix, we know that the volume of the parallelotope $[\boldsymbol{x}_1, \ldots, \boldsymbol{x}_{n-1}, \boldsymbol{x}_n]$ equals the area of its base, $[\boldsymbol{x}_1, \ldots, \boldsymbol{x}_{n-1}]$, times its height with respect to the isoperimetrix. Therefore, if the volume of every parallelotope in a normed space equals the area of its base times its height with respect to the unit ball, the ball equals the isoperimetrix. If the volume is greater than the area of the base times the height with respect to the unit ball, then the isoperimetrix is contained in the ball, and so on.

The first clear sign that the relationship between length, area and volume may not go smoothly on normed and Finsler spaces is the following result of Thompson:

PROPOSITION 6.24 [Thompson 1996]. *The isoperimetric of a normed space X for the Holmes–Thompson definition is contained in the unit ball if and only if the space is Euclidean. In which case, the ball and the isoperimetric are equal.*

In other words, the coarea equality or inequality "volume $\geq$ base $\times$ height" for the Holmes–Thompson definition is true only for Euclidean spaces.

PROOF. If $\mathbb{I}_X^{\mathrm{ht}} \subset B_X$, then $B_X^* \subset (\mathbb{I}_X^{\mathrm{ht}})^*$ and, therefore,

$$\mathrm{svol}(B_X \times (\mathbb{I}_X^{\mathrm{ht}})^*) \geq \mathrm{svol}(B_X \times B_X^*) = \varepsilon_n \mu_X^{\mathrm{ht}}(B_X).$$

By Exercise 6.23, the only way this can happen is if X is Euclidean. $\square$

However, for the mass$*$ definition the coarea inequality is always true:

THEOREM 6.25 [Gromov 1983]. *If X is a finite-dimensional normed space, then $\mathbb{I}_X^{m*} \subset B_X$.*

PROOF. We must show that if $[\boldsymbol{v}_1, \ldots, \boldsymbol{v}_n]$ is a parallelotope,

$$\mu^{m*}(\boldsymbol{v}_1 \wedge \boldsymbol{v}_2 \wedge \cdots \wedge \boldsymbol{v}_n) \geq \sigma^{m*}(\boldsymbol{v}_1 \wedge \boldsymbol{v}_2 \wedge \cdots \wedge \boldsymbol{v}_{n-1})|\boldsymbol{\xi}(\boldsymbol{v}_n)|,$$

where $\boldsymbol{\xi} \in \partial B_X^*$ and $\boldsymbol{\xi}(\boldsymbol{v}_i) = 0$, $1 \leq i \leq n - 1$.

Without loss of generality we may suppose that $\boldsymbol{v}_1, \boldsymbol{v}_2, \ldots \boldsymbol{v}_{n-1}$ is an extremal basis in the subspace $V \subset X$ they span, *i.e.* each vector $\boldsymbol{v}_i$ is a point of contact between $B_X \cap V$ and a minimal circumscribing parallelotope for $B \cap V$. Let $\boldsymbol{u}$

be such that $\|u\| = \boldsymbol{\xi}(u) = 1$ and set $v_n = \alpha u + x$ where $x \in V$. The right hand side of the above inequality is $|\alpha|$.

Let $\boldsymbol{\xi}_1, \boldsymbol{\xi}_2, \ldots, \boldsymbol{\xi}_{n-1}$ be the dual basis to the v_i's in V and extend these to the whole of X by setting $\boldsymbol{\xi}_i(u) = 0$. Then $\boldsymbol{\xi}_1, \boldsymbol{\xi}_2, \ldots, \boldsymbol{\xi}_{n-1}, \boldsymbol{\xi}$ are all of norm 1 and form the dual basis to $v_1, v_2, \ldots, v_{n-1}, u$. Now

$$\mu_X^{m*}(v_1 \wedge v_2 \wedge \cdots \wedge v_n) = |\alpha| \mu_X^{m*}(v_1 \wedge v_2 \wedge \cdots \wedge v_{n-1} \wedge u)$$
$$= |\alpha| (\mu_{X*}^m (\boldsymbol{\xi}_1 \wedge \boldsymbol{\xi}_2 \wedge \cdots \wedge \boldsymbol{\xi}_{n-1} \wedge \boldsymbol{\xi}))^{-1}$$
$$\geq |\alpha| (\|\boldsymbol{\xi}\| \prod \|\boldsymbol{\xi}_i\|)^{-1} = |\alpha|.$$

The inequality comes from the definition of mass. $\qquad\square$

As we have said, the problem of determining for what normed spaces metric balls are solutions to the isoperimetric problem, *i.e.* when is the isoperimetrix a multiple of the unit ball, is completely open for all three definitions of volume in dimensions greater than two.

Acknowledgments

It is a pleasure to acknowledge the hospitality and wonderful working conditions of the Research in Pairs program at the Mathematisches Forschungsinstitut Oberwolfach, where the authors did a substantial part of the research going into this paper. Álvarez also thanks the MAPA Institute at the Université Catholique de Louvain for its hospitality and technical support in the final stages of this work. Thompson is grateful to NSERC for its support through grant #A-4066.

Many of the ideas in this paper concerning the abstract framework of the subject were first developed in the Geometry Seminar at the Université Catholique de Louvain and were greatly influenced by discussions with Emmanuel Fernandes and Gautier Berck. The thesis [Fernandes 2002] was for us a stepping-stone for many of the subjects developed in this work.

References

[Alexandrov 1933] A. D. Aleksandrov, "A theorem on convex polyhedra" (Russian), *Trudy Fiz.-Mat. Inst. Steklov.* **4** (1933), 87.

[Alvarez 2002] J. C. Álvarez Paiva, "Dual mixed volumes and isosystolic inequalities", preprint, 2002.

[Alvarez and Fernandes 1998] J. C. Álvarez Paiva and E. Fernandes, "Crofton formulas in projective Finsler spaces", *Electronic Research Announcements of the Amer. Math. Soc.* **4** (1998), 91–100.

[Alvarez and Fernandes 1999] J. C. Álvarez Paiva and E. Fernandes, "Fourier transforms and the Holmes–Thompson volume of Finsler manifolds", *Int. Math. Res. Notices* **19** (1999), 1032–104.

[Ambrosio and Kirchheim 2000] L. Ambrosio and B. Kirchheim, "Currents in metric spaces", *Acta Math.* **185** (2000), 1–80.

[Benson 1962] R. V. Benson, "The geometry of affine areas", Ph.D. Thesis, University of Southern California, Los Angeles, (1962). (University Microfilms Inc., Ann Arbor, Michigan 62-6037).

[Bolker 1969] E. D. Bolker, "A class of convex bodies", *Trans. Amer. Math. Soc.* **145** (1969), 323–345.

[Bourgain and Lindenstrauss 1988] J. Bourgain and J. Lindenstrauss, "Projection bodies", pp. 250–270 in *Geometric aspects of functional analysis* (Tel Aviv, 1986/87), edited by J. Lindenstrauss and V. D. Milman, Lecture Notes in Math. **1317**, Springer, Berlin, 1988.

[Burago and Ivanov 2002] D. Burago and S. Ivanov, "On asymptotic volume of Finsler tori, minimal surfaces in normed spaces, and symplectic filling volume", *Ann. of Math.* (2) **156**:3 (2002), 891–914.

[Busemann 1947] H. Busemann, "Intrinsic area", *Ann. of Math.* (2) **48** (1947), 234–267.

[Busemann 1949a] H. Busemann, "A theorem on convex bodies of the Brunn–Minkowski type", *Proc. Nat. Acad. Sci. USA* **35** (1949), 27–31.

[Busemann 1949b] H. Busemann, "The isoperimetric problem for Minkowski area", *Amer. J. Math.* **71** (1949), 743–762.

[Busemann 1961] H. Busemann, "Convexity on Grassmann manifolds", *Enseign. Math.* (2) **7** (1961), 139–152.

[Busemann 1969] H. Busemann, "Convex bodies and convexity on Grassmann cones: XI", *Math. Scand.* **24** (1969), 93–101.

[Busemann and Petty 1956] H. Busemann and C. M. Petty, "Problems on convex bodies", *Math. Scand.* **4** (1956), 88–94.

[Busemann et al. 1963] H. Busemann, G. Ewald, and G. C. Shephard, "Convex bodies and convexity on Grassmann cones: I–IV", *Math. Ann.* **151** (1963), 1–41.

[Busemann et al. 1962] H. Busemann, G. Ewald, and G. C. Shephard, "Convex bodies and convexity on Grassmann cones: V", *Arch. Math.* **13** (1962), 512–526.

[Busemann and Sphephard 1965] H. Busemann and G. C. Shephard, "Convex bodies and convexity on Grassmann cones: X", *Ann. Mat. Pura Appl.* **70** (1965), 271–294.

[Cohn 1980] D. L. Cohn, *Measure theory*, Birkhäuser, Basel, 1980.

[Duran 1998] C. E. Durán, "A volume comparison theorem for Finsler manifolds", *Proc. Amer. Math. Soc.* **126** (1998), 3079–3082.

[Ewald 1964a] G. Ewald, "Convex bodies and convexity on Grassmann cones: VII", *Abh. Math. Sem. Univ. Hamburg* **27** (1964), 167–170.

[Ewald 1964b] G. Ewald, "Convex bodies and convexity on Grassmann cones: IX", *Math. Ann.* **157** (1964), 219–230.

[Fernandes 2002] E. Fernandes, "Double fibrations: a modern approach to integral geometry and Crofton formulas in projective Finsler spaces", Ph.D. Thesis, Université Catholique de Louvain, 2002.

[Gardner 1995] R. J. Gardner, *Geometric Tomography*, Encyclopedia of Mathematics and its Applications **58**, Cambridge University Press, Cambridge, 1995.

[Gardner 1994] R. J. Gardner, "A positive answer to the Busemann–Petty problem in three dimensions", *Ann. of Math.* (2) **140**:2 (1994), 435–447.

[Gardner et al. 1999] R. J. Gardner, A. Koldobsky, and T. Schlumprecht, "An analytic solution to the Busemann–Petty problem on sections of convex bodies", *Ann. of Math.* **149**:2 (1999), 691–703.

[Gołąb 1932] S. Gołąb, "Quelques problèmes métriques de la géometrie de Minkowski", *Trav. l'Acad. Mines Cracovie* **6** (1932), 1–79.

[Goodey and Weil 1993] P. Goodey and W. Weil, "Zonoids and generalisations", pp. 1297–1326 in *Handbook of convex geometry*, Vol. B, edited by P. M. Gruber et al., North-Holland, Amsterdam, 1993.

[Gromov 1967] M. Gromov, "On a geometric hypothesis of Banach" (Russian), *Izv. Akad. Nauk SSSR Ser. Mat.* **31** (1967), 1105–1114.

[Gromov 1983] M. Gromov, "Filling Riemannian manifolds", *J. Diff. Geom.* **18** (1983), 1–147.

[Harvey and Lawson 1982] R. Harvey and H. B. Lawson, "Calibrated geometries", *Acta Math.* **148** (1982), 47–157.

[Holmes and Thompson 1979] R. D. Holmes and A. C. Thompson, "N-dimensional area and content in Minkowski spaces", *Pacific J. Math.* **85** (1979), 77–110.

[Hörmander 1994] L. Hörmander, *Notions of convexity*, Progress in Mathematics **127**, Birkhäuser, Boston, 1994.

[Hörmander 1983] L. Hörmander, *The Analysis of Linear Partial Differential Operators* I, Grundlehren der mathematischen Wissenschaften **256**, Springer, Berlin, 1983.

[Ivanov 2002] S. Ivanov, "On two-dimensional minimal fillings", *St. Petersburg Math. J.* **13** (2002), 17–25.

[John 1948] F. John, "Extremum problems with inequalities as subsidiary conditions", pp. 187–204 in *Studies and essays presented to R. Courant on his 60th birthday*, edited by K. O. Friedrichs, O. Neuegebauer, and J. J. Stoker, Interscience, New York, 1948.

[Lutwak 1988] E. Lutwak, "Intersection bodies and dual mixed volumes", *Adv. Math.* **71** (1988), 232–261.

[Mahler 1939] K. Mahler, "Ein Minimalproblem für konvexe Polygone", *Mathematica* (Zutphen) **B7** (1939), 118–127.

[Martini et al. 2001] H. Martini, K. J. Swanepoel, and Weiß, G., "The geometry of Minkowski spaces: a survey, part I", *Exposition. Math.* **19** (2001), 97–142.

[McKinney 1974] J. R. McKinney, "On maximal simplices inscribed to a central convex set", *Mathematika* **21** (1974), 38–44.

[Moustafaev] Z. Moustafaev, "The ratio of the length of the unit circle to the area of the unit disc in Minkowski planes", to appear in *Proc. Amer. Math. Soc.*

[Nirenberg 1953] L. Nirenberg, "The Weyl and Minkowski problems in differential geometry in the large", *Comm. Pure Appl. Math.* **6** (1953), 337–394.

[Pogorelov 1978] A. V. Pogorelov, *The Minkowski multidimensional problem*, Scripta Series in Mathematics, V. H. Winston, Washington (DC) and Halsted Press, New York, 1978.

[Reisner 1985] S. Reisner, "Random polytopes and the volume-product of symmetric convex bodies", *Math. Scand.* **57** (1985), 386–392.

[Reisner 1986] S. Reisner, "Zonoids with minimal volume product", *Math. Z.* **192** (1986), 339–346.

[Schäffer 1967] J. J. Schäffer, "Inner diameter, perimeter, and girth of spheres", *Math. Ann.* **173** (1967), 59–79.

[Schäffer 1973] J. J. Schäffer, "The self-circumference of polar convex bodies", *Arch. Math.* **24** (1973), 87–90.

[Schneider 1993] R. Schneider, *Convex bodies: the Brunn–Minkowski Theory*, Encyclopedia of Math. and Its Appl. **44**, Cambridge University Press, New York, 1993.

[Schneider 2001] R. Schneider, "On the Busemann area in Minkowski spaces", *Beitr. Algebra Geom.* **42** (2001), 263–273.

[Schneider 2001] R. Schneider, "Crofton formulas in hypermetric projective Finsler spaces", Festschrift: Erich Lamprecht. *Arch. Math.* (Basel) 77:1 (2001), 85–97.

[Schneider 2002] R. Schneider, "On integral geometry in projective Finsler spaces", *Izv. Nats. Akad. Nauk Armenii Mat.* **37** (2002) 34–51.

[Schneider and Weil 1983] R. Schneider and W. Weil, "Zonoids and related topics", pp. 296–317 in *Convexity and its applications*, edited by P. M. Gruber and J. M. Wills, Birkhäuser, Basel, 1983.

[Schneider and Wieacker 1997] R. Schneider and J. A. Wieacker, "Integral geometry in Minkowski spaces", *Adv.in Math.* **129** (1997), 222–260.

[Shephard 1964a] G. C. Shephard, "Convex bodies and convexity on Grassmann cones: VI", *J. London Math. Soc.* **39** (1964), 307–319.

[Shephard 1964b] G. C. Shephard, "Convex bodies and convexity on Grassmann cones: VIII", *J. London Math. Soc.* **39** (1964), 417–423.

[Taylor 1978] J. Taylor, "Crystalline variational problems", *Bull. Amer. Math. Soc.* **84**:4 (1978), 568–588.

[Thompson 1996] A. C. Thompson, *Minkowski Geometry*, Encyclopedia of Mathematics and its Applications **63**, Cambridge University Press, New York, 1996.

[Thompson 1999] A. C. Thompson, "On Benson's definition of area in Minkowski space", *Canad. Math. Bull.* **42**:2 (1999) 237–247.

[Wallen 1995] L. J. Wallen, "Kepler, the taxicab metric and beyond: an isoperimetric primer", *College J. Math.* **26** (1995), 178–190.

[Weil 1979] W. Weil, "Centrally symmetric convex bodies and distributions II", *Israel J. Math.* **32** (1979), 173–182.

[Zhang 1999] G. Zhang, "A positive solution to the Busemann–Petty problem in $\mathbf{R}^4$". *Ann. of Math.* (2) **149**:2 (1999), 535–543.

J. C. ÁLVAREZ PAIVA
DEPARTMENT OF MATHEMATICS
POLYTECHNIC UNIVERSITY
SIX METROTECH CENTER
BROOKLYN, NEW YORK, 11201
UNITED STATES
jalvarez duke.poly.edu

A. C. THOMPSON
DEPARTMENT OF MATHEMATICS AND STATISTICS
DALHOUSIE UNIVERSITY
HALIFAX, NOVA SCOTIA
CANADA B3H 3J5
tony@mathstat.dal.ca

Riemann–Finsler Geometry
MSRI Publications
Volume **50**, 2004

Anisotropic and Crystalline
Mean Curvature Flow

GIOVANNI BELLETTINI

CONTENTS

1. Introduction

Motion by mean curvature of an embedded smooth hypersurface without boundary has been the subject of several recent papers, because of its geometric interest and of its application to different areas, see for instance the pioneering book [Brakke 1978], or the papers [Allen and Cahn 1979], [Huisken 1984], [Osher and Sethian 1988], [Evans and Spruck 1991], [Almgren et al. 1993]. A smooth boundary ∂E of an open set $E = E(0) \subset \mathbb{R}^n$ flows by mean curvature if there exists a time-dependent family $(\partial E(t))_{t \in [0,T]}$ of smooth boundaries satisfying the following property: the normal velocity of any point $x \in \partial E(t)$ is equal to the sum of the principal curvatures of $\partial E(t)$ at x. One can show that, at each time t, such an evolution process reduces the area of $\partial E(t)$ as fast as possible. Mean curvature flow has therefore a variational character, since it can be interpreted as the gradient flow associated with the area functional $\partial E \to \mathcal{H}^{n-1}(\partial E)$, where $\mathcal{H}^{n-1}$ indicates the $(n-1)$-dimensional Hausdorff measure in $\mathbb{R}^n$.

In several physical processes (for instance in certain models of dendritic growth and crystal growth, see [Cahn et al. 1992], or in statistical physics (see for example [Spohn 1993]) it turns out, however, that the evolution of the surface is not simply by mean curvature, but is an anisotropic evolution. From the energy point of view, this means that the functional of which we are taking the gradient flow is not the area of ∂E anymore, but is a weighted area, which can be derived by looking at $\mathbb{R}^n$ as a normed space. Let $\phi : \mathbb{R}^n \to [0,+\infty[$ be a norm on $\mathbb{R}^n$. One of the most common area measures of ∂E in the normed space $(\mathbb{R}^n, \phi)$ is the so-called Minkowski content $\mathcal{M}_{d_\phi}^{n-1}(\partial E)$ of ∂E induced by ϕ. Denoting by $d_\phi(x,y) := \phi(y-x)$ the distance on $\mathbb{R}^n$ induced by ϕ and by $\mathcal{H}_{d_\phi}^n$ the n-dimensional Hausdorff measure with respect to d_ϕ, $\mathcal{M}_{d_\phi}^{n-1}$ is defined as

$$\mathcal{M}_{d_\phi}^{n-1}(\partial E) := \lim_{\rho \to 0^+} \frac{1}{2\rho} \mathcal{H}_{d_\phi}^n \left(\{z \in \mathbb{R}^n : \inf_{x \in \partial E} d_\phi(z,x) < \rho\} \right). \qquad (1\text{--}1)$$

Since it is possible to prove that $\mathcal{H}_{d_\phi}^n$ coincides with the Lebesgue measure $|\cdot|$ multiplied by the factor

$$c_{n,\phi} := \frac{\omega_n}{|\{\xi \in \mathbb{R}^n : \phi(\xi) \le 1\}|},$$

ω_n being a normalizing constant, we have

$$\mathcal{M}_{d_\phi}^{n-1}(\partial E) = c_{n,\phi} \lim_{\rho \to 0^+} \frac{1}{2\rho} \left| \{z \in \mathbb{R}^n : \inf_{x \in \partial E} d_\phi(z,x) < \rho\} \right|. \qquad (1\text{--}2)$$

Therefore, $\mathcal{M}_\phi^{n-1}(\partial E)$ measures (for small $\rho > 0$) the ratio between the volume of a ρ-tubular neighborhood of ∂E and ρ. Definition (1–1) can be made more explicit, since it turns out that

$$\mathcal{M}_{d_\phi}^{n-1}(\partial E) = c_{n,\phi} \int_{\partial E} \phi^o(\nu^E) \, d\mathcal{H}^{n-1}. \qquad (1\text{--}3)$$

Here ν^E is the Euclidean unit normal to ∂E pointing outside of E, and the function $\phi^o : \mathbb{R}^n \to [0,+\infty[$ is the dual norm of ϕ. Physically, $\phi^o(\nu)$ plays the rôle of a surface tension of a flat surface whose normal is ν, and can be considered as the anisotropy. The functional (1–3) is the above mentioned weighted area, whose gradient flow gives raise to the so-called anisotropic motion by mean curvature. In the *regular* case, that is, when ϕ^2 is smooth and strictly convex, the relevant quantity is the so-called Cahn–Hoffman vector field n_ϕ^E on ∂E, which is the image of $\nu_\phi^E := \nu^E / \phi^o(\nu^E)$ through the map $\frac{1}{2} \nabla((\phi^o)^2)$, and whose divergence is the anisotropic mean curvature of ∂E (denoted by κ_ϕ^E). The anisotropic mean curvature is derived (again by a variational principle) in the computation of the first variation of (1–3). Therefore, for any time t, anisotropic mean curvature flow is defined in such a way to decrease $\mathcal{M}_{d_\phi}^{n-1}(\partial E(t))$ as fast as possible.

Besides the regular case, other anisotropies can be considered; we are partic- ularly interested in the *crystalline case*, when the function ϕ is piecewise linear (or equivalently when the unit ball $B_\phi := \{\phi \leq 1\}$ is a polytope). From the mathematical point of view, this field of research was initiated by the work of J. Taylor [Taylor 1978], [Taylor 1986], [Taylor 1991], [Cahn et al. 1992], [Taylor 1992], [Taylor 1993], [Almgren and Taylor 1995]. See also the papers [Hoffman and Cahn 1972], [Cahn and Hoffman 1974], [Cahn et al. 1993]. Recently, several authors contributed to the subject: see for instance

- [Girao and Kohn 1994], [Girao 1995], [Rybka 1997], [Ishii and Soner 1999] and [Giga and Giga 2000] for general properties of the crystalline flow in two dimensions and for the convergence of a crystalline algorithm;
- [Stancu 1996] for self-similar solutions of the crystalline flow in two dimen- sions;
- [Fukui and Giga 1996], [Giga and Giga 1997], [Giga and Giga 1998b], [Giga and Giga 1999] for the crystalline evolution of graphs in two dimensions;
- [Giga and Gurtin 1996] for a comparison theorem for crystalline evolutions in two dimensions;
- [Roosen and Taylor 1994] for the crystalline evolution in a diffusion field, and [Giga and Giga 1998a] for the crystalline flow with a driving force in two dimensions;
- [Ambrosio et al. 2002] for some regularity properties of solutions to crystalline variational problems in two dimensions;
- [Yunger 1998] and [Paolini and Pasquarelli 2000] for some properties of the crystalline flow in three dimensions.

In two dimensions (that is, for crystalline curvature evolution of curves) the situation is essentially understood, since the notion of crystalline curvature is clear, as well as the corresponding geometric evolution law. For instance, for polygonal initial curves (whose geometry is compatible with the geometry of ∂B_ϕ) a comparison principle is available and the flow admits local existence and uniqueness. It turns out that each edge of the curve translates in normal direction during the flow, and the evolution can be described by a system of ordinary differential equations.

However, in three space dimensions the situation is not so clear. As in the two- dimensional case, it is necessary to redefine what is a smooth boundary ∂E, in order to assign to our interface some notion of ϕ-mean curvature. To this purpose we recall a (rather strong) notion of smoothness, the Lipschitz ϕ-regularity. Even with this notion at our disposal, the definition of the crystalline mean curvature is quite involved. In addition, once the crystalline mean curvature κ_ϕ^E is defined on ∂E, one realizes that, in general, it is not constant on two-dimensional facets F of ∂E. This fact is a source of difficulties, since (being κ_ϕ^E identified with the normal velocity of ∂E at the initial time) facets can split in several pieces, or can even bend (forming curved regions) during the subsequent evolutionary process. These phenomena (which probably should not be considered as singularities of the flow) partially explain why when $n = 3$, a short time existence theorem for crystalline mean curvature flow is still missing (even for convex initial data E). Concerning this kind of behavior, we refer also to the work [Yunger 1998].

Before illustrating the plan of the paper, we observe that other choices of area measures are possible in $(\mathbb{R}^n, \phi)$, which are as natural as the ϕ-Minkowski content. For example, one could consider the $(n-1)$-dimensional Hausdorff measure $\mathcal{H}_{d_\phi}^{n-1}(\partial E)$ of ∂E with respect to d_ϕ. Also for this notion of area, an integral representation theorem is available, which shows in particular that $\mathcal{M}_{d_\phi}^{n-1}(\partial E)$ and $\mathcal{H}_{d_\phi}^{n-1}(\partial E)$ may differ. Therefore, taking the first variation of $\mathcal{H}_{d_\phi}^{n-1}$ would give a notion of mean curvature (see [Shen 1998]) which is different from κ_ϕ^E. This, in turn, implies that the gradient flow of the d_ϕ-Hausdorff measure functional is a different geometric evolution process. Our viewpoint will be to work with the ϕ-Minkowski content of ∂E. We remark that all the theory that we develop can be similarly constructed for $\mathcal{H}_{d_\phi}^{n-1}$: indeed, what is really relevant is that $\mathcal{H}_{d_\phi}^{n-1}(\partial E)$ can also be represented as an integral on ∂E, by means of an integrand (weighting ν^E) which is *convex*. Finally, we recall that the mean curvature obtained from the first variation of the volume form for the Minkowski content on regular hypersurfaces in a Finsler manifold is considered in [Shen 2001].

The content of the paper is the following. In Section 2 we give some notation. In particular, in Subsection 2.1 we introduce the norm ϕ, its unit ball B_ϕ and the induced distance d_ϕ. In Subsection 2.2 we recall the main properties of the dual ϕ^o of ϕ and of the duality maps T_ϕ and T_{ϕ^o}. In particular, we discuss the geometric properties of such maps, also in the crystalline case. In Subsection 3.1 we discuss the integral representation of $\mathcal{H}_{d_\phi}^{n-1}$ (Theorem 3.3). In Subsection 3.2 we discuss the integral representation of $\mathcal{M}_{d_\phi}^{n-1}$ (Theorem 3.7). The relations between $\mathcal{H}_{d_\phi}^{n-1}$ and $\mathcal{M}_{d_\phi}^{n-1}$ are considered in Subsection 3.3. Section 3 is based on the results proved in [Bellettini et al. 1996]. In Section 4 we recall the first variation of area in the Euclidean case, and the main definitions and properties of Euclidean mean curvature flow. We rely heavily on the notion of oriented distance function from ∂E. In Subsection 4.2 we focus attention on the regular case and on the first variation of the weighted area. The definition of ϕ-mean curvature is given in (4–9). Also here the oriented ϕ-distance function plays a crucial rôle. In Subsection 4.3 we state some generalizations of the previous results when the norm is space-dependent. Subsections 4.2 and 4.3 are based on the results proved in [Bellettini and Paolini 1996]. The crystalline case is deepened in Section 5. Lipschitz ϕ-regularity is introduced in Definition 5.2, and illustrated with examples. The geometry of a facet $F \subset \partial E$ is studied in Section

6, which is preliminary to the definition of crystalline mean curvature on F (Definition 7.2). We will restrict for simplicity to polyhedral Lipschitz ϕ-regular sets. After some examples, in Subsection 7.1 we illustrate some properties of those facets having constant crystalline mean curvature. Sections 5, 6 and 7 are based on the results originally proved in [Bellettini et al. 1999], [Bellettini et al. 2001a], [Bellettini et al. 2001b], [Bellettini et al. 2001c]. We conclude the paper with Section 8, where we briefly summarize the main ideas and motivations behind our approach.

2. Notation

Given two vectors $v, w \in \mathbb{R}^n$, $n \geq 2$, we denote by $\langle v, w \rangle$ the scalar product between v and w. We also set $|v| := \sqrt{\langle v, v \rangle}$ and $S^{n-1} := \{v \in \mathbb{R}^n : |v| = 1\}$. Given an integer $k \in [0, n]$, we denote by $\mathcal{H}^k$ the k-dimensional Hausdorff measure in $\mathbb{R}^n$ (see [Federer 1969] and [Ambrosio et al. 2000]). If $B \subset \mathbb{R}^n$ is a Borel set, we let $|B|$ to be the Lebesgue measure of B (which equals $\mathcal{H}^n(B)$) and $\operatorname{dist}(x, B)$ to be the distance of the point $x \in \mathbb{R}^n$ from B, defined as $\inf\{|y - x| : y \in B\}$. Even if B is a smooth hypersurface, it is not difficult to realize that the distance function $\operatorname{dist}(\,\cdot\,, B)$ is not differentiable on B. We let $\omega_m := \pi^{m/2} / \int_0^{+\infty} s^{m/2} e^{-s}\, ds$, which turns out to be the Lebesgue measure of $\{x \in \mathbb{R}^m : |x| \leq 1\}$, for an integer $m \in [0, n]$.

We say that the set $M \subset \mathbb{R}^n$ is an $(n-1)$-dimensional Lipschitz manifold if M can be written, locally, as the graph of a Lipschitz function (with respect to a suitable orthogonal system of coordinates) defined on an open subset of $\mathbb{R}^{n-1}$. We say that the open set $E \subset \mathbb{R}^n$ is Lipschitz (or that its topological boundary ∂E is Lipschitz) if ∂E can be written, locally, as the graph of a Lipschitz function of $(n-1)$ variables (with respect to a suitable orthogonal system of coordinates) and E is locally the subgraph. We recall (see [Federer 1969]) that, if E has Lipschitz boundary, then at $\mathcal{H}^{n-1}$-almost every $x \in \partial E$, the unit (Euclidean) normal vector to ∂E pointing toward $\mathbb{R}^n \setminus E$ is well defined and, in the sequel, will be denoted by $\nu^E(x)$.

If f is a smooth function defined on an open subset of $\mathbb{R}^n$, we denote by $\nabla f = (\partial f / \partial x_1, \ldots, \partial f / \partial x_n)$ the gradient of f and by Δf the Laplacian of f.

2.1. The norm ϕ and the distance d_ϕ. In what follows we indicate by $\phi : \mathbb{R}^n \to [0, +\infty[$ a *convex* function which satisfies the properties

$$\phi(\xi) \geq \lambda |\xi|, \quad \xi \in \mathbb{R}^n, \tag{2-1}$$

for a suitable constant $\lambda \in \,]0, +\infty[$, and

$$\phi(a\xi) = |a|\,\phi(\xi), \quad \xi \in \mathbb{R}^n, \ a \in \mathbb{R}. \tag{2-2}$$

Notice that our assumptions ensure that there exists a constant $\Lambda \geq \lambda$ such that $\phi(\xi) \leq \Lambda |\xi|$ for any $\xi \in \mathbb{R}^n$. The function ϕ is a norm on $\mathbb{R}^n$, called a Minkowski norm (or Minkowski metric). The vector space $\mathbb{R}^n$ endowed with ϕ is a normed space called Minkowski space and is probably the simplest example of a Finsler manifold: in this case, the manifold is $\mathbb{R}^n$, and ϕ is a norm (independent of the position) on its tangent space which obviously is a copy of $\mathbb{R}^n$. We set

$$B_\phi := \{\xi \in \mathbb{R}^n : \phi(\xi) \leq 1\}$$

(the unit ball of ϕ), a bounded convex set containing the origin in its interior and centrally symmetric (a so-called symmetric convex body). B_ϕ is usually called the indicatrix (sometimes also Wulff shape). The function ϕ can be identified with B_ϕ, since given a symmetric convex body K, the function $\xi \to \inf\{\alpha > 0 : \xi \in \alpha K\}$ is a convex function satisfying (2–1), (2–2) and having K as unit ball. Clearly ϕ is uniquely determined by its values on the unit sphere S^{n-1}.

REMARK 2.1. If we weaken assumption (2–2) into $\phi(a\xi) = a\phi(\xi)$ for any $\xi \in \mathbb{R}^n$ and any $a \geq 0$, we have that B_ϕ is not centrally symmetric anymore. Some of the results in the next sections can be generalized to nonsymmetric functions ϕ; in the sequel, for simplicity we will restrict to the symmetric case.

It is always useful to keep in mind the Euclidean case, which corresponds to the choice $\phi(\xi) = |\xi|$. The Riemannian case corresponds to a norm ϕ whose B_ϕ is an ellipsoid: $\phi(\xi) := \sqrt{\langle A\xi, \xi \rangle}$ for a real positive definite symmetric matrix A.

DEFINITION 2.2. We say that ϕ is regular if B_ϕ has boundary of class $\mathcal{C}^\infty$ and each principal curvature of ∂B_ϕ is strictly positive at each point of ∂B_ϕ.

EXAMPLE 2.1. Let $p \in [1, +\infty[$, $p \neq 2$ and set $\phi(\xi) := (\sum_{i=1}^n |\xi_i|^p)^{1/p}$. Then ϕ is not regular: indeed, if $p > 2$ then B_ϕ is of class $\mathcal{C}^2$ but the requirement on principal curvatures in Definition 2.2 is not fulfilled. On the other hand, if $p < 2$ then ∂B_ϕ is not of class $\mathcal{C}^2$.

DEFINITION 2.3. We say that ϕ is crystalline if B_ϕ is a polytope.

EXAMPLE 2.2. The norms $\phi(\xi) := \sum_{i=1}^n |\xi_n|$ and $\phi(\xi) := \max\{|\xi_1|, \ldots, |\xi_n|\}$ are crystalline.

It is well known that, given the norm ϕ, we can measure distances in $\mathbb{R}^n$ by "integrating" ϕ as follows: the ϕ-distance $d_\phi(x, y)$ between two points $x, y \in \mathbb{R}^n$ is given by

$$d_\phi(x, y) = \inf\left\{ \int_0^1 \phi(\dot{\gamma})\, dt : \gamma \in AC([0,1]; \mathbb{R}^n), \gamma(0) = x, \gamma(1) = y \right\}$$

$$= \phi(y - x), \tag{2-3}$$

where $AC([0,1]; \mathbb{R}^n)$ is the class of all absolutely continuous curves from $[0,1]$ to $\mathbb{R}^n$. The last equality in (2–3) is, for instance, a consequence of Jensen's inequality. The function d_ϕ is nonnegative, symmetric, vanishes only if $x = y$, and satisfies the triangular property. To be consistent with the beginning of this section, when $\phi(\xi) = |\xi|$ (Euclidean case) we omit the subscript ϕ in the notation of the distance function.

REMARK 2.4. We recall the following interesting fact. Given a distance $d :$ $\mathbb{R}^n \times \mathbb{R}^n \to [0, +\infty[$ on $\mathbb{R}^n$, by differentiation we can construct a new function $\psi_d : \mathbb{R}^n \times \mathbb{R}^n \to [0, +\infty]$ which, under suitable assumptions, turns out to be a Finsler metric (which, in this case, depends on the position x):

$$\psi_d(x, \xi) := \limsup_{t \to 0} \frac{d(x, x + t\xi)}{t}, \quad x \in \mathbb{R}^n, \ \xi \in \mathbb{R}^n.$$

Some of the properties of the functions ψ_d, d_{ψ_d} have been investigated for instance in the papers [De Giorgi 1989], [De Giorgi 1990], [Venturini 1992], [Bellettini et al. 1996], [Amar et al. 1998] (see also [De Cecco and Palmieri 1993], [De Cecco and Palmieri 1995] for related results on Lipschitz manifolds).

2.2. The dual norm ϕ^o. The duality maps. Given the norm ϕ acting on vectors, we define the dual norm $\phi^o : \mathbb{R}^n \to [0, +\infty[$ of ϕ (acting on covectors) as

$$\phi^o(\xi^o) := \sup\{\langle \xi, \xi^o \rangle : \xi \in B_\phi\}, \quad \xi^o \in \mathbb{R}^n. \tag{2-4}$$

It is not difficult to verify that ϕ^o is a norm on (the dual of) $\mathbb{R}^n$, and that $(\phi^o)^o = \phi$. It is possible to prove that if ϕ is regular then also ϕ^o is regular, and that if ϕ is crystalline then ϕ^o is crystalline. The function ϕ^o is strictly related to the Legendre–Fenchel transform ϕ^* of ϕ, defined as $\phi^*(\xi^o) := \sup\{\langle \xi, \xi^o \rangle - \phi(\xi) : \xi \in \mathbb{R}^n\}$. Indeed $\phi^*(\xi^o) = +\infty$ if $\xi \notin B_{\phi^o}$, and $\phi^*(\xi^o) = 0$ if $\xi \in B_{\phi^o}$.

EXAMPLE 2.3. Figure 1 describes how to construct B_{ϕ^o} starting from ϕ. Assume we have been given a smooth symmetric convex body $\{\phi \leq 1\}$ as (the ellipse) in Figure 1. Let $\nu \in S^{n-1}$ be a unit vector. By definition and using the homogeneity, computing $\phi^o(\nu)$ is equivalent to solve the maximum problem $\max\{\langle \nu, z \rangle : z \in \partial B_\phi\}$.

The vector $\xi = \xi(\nu)$ in Figure 1 is the solution, hence $\phi^o(\nu) = \langle \nu, \xi \rangle$. Observe that the strict convexity of B_ϕ ensures uniqueness of the solution of the maximum problem (2–4); it is clear that if ∂B_ϕ contains some flat region, problem (2–4) has in general more than one solution.

REMARK 2.5. Notice that

$$\phi^o(\nu) = \frac{1}{2} \mathcal{H}^1(\mathrm{pr}_\nu(B_\phi)), \tag{2-5}$$

where $\mathrm{pr}_\nu(B_\phi)$ denotes the orthogonal projection of B_ϕ onto the line $\mathbb{R}\nu$; see Figure 1.

The map described in Figure 1 associating to the vector $\xi \in \partial B_\phi$ the vector $\nu/\phi^o(\nu) \in \partial B_{\phi^o}$ (extended in a one-homogeneous way on the whole of $\mathbb{R}^n$) is called the duality map, and can also be defined in the crystalline case.

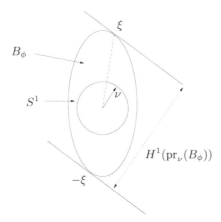

Figure 1. ν is a unit vector. The vector $\xi \in \partial B_\phi$ (dotted line) is such that $\phi^o(\nu) = \langle \nu, \xi \rangle$. Observe that the line tangent to ∂B_ϕ at ξ is orthogonal to ν. The distance between this line and its parallel line tangent to ∂B_ϕ at $-\xi$ is the length of the orthogonal projection of B_ϕ onto the one-dimensional subspace $\mathbb{R}\nu$, and equals $2\phi^o(\nu)$.

DEFINITION 2.6. By T_ϕ and T_{ϕ^o} we denote the (possibly multivalued) duality maps defined as

$$T_\phi(\xi) := \{\xi^o \in \mathbb{R}^n : \langle \xi^o, \xi \rangle = \phi(\xi)^2 = (\phi^o(\xi^o))^2\}, \quad \xi \in \mathbb{R}^n,$$
$$T_{\phi^o}(\xi^o) := \{\xi \in \mathbb{R}^n : \langle \xi, \xi^o \rangle = (\phi^o(\xi^o))^2 = \phi(\xi)^2\}, \quad \xi^o \in \mathbb{R}^n. \tag{2-6}$$

Possibly adopting the conventions on multivalued mappings (see for instance [Brezis 1973]) one can check that $T_\phi(a\xi) = |a| T_\phi(\xi)$ for any $\xi \in \mathbb{R}^n$ and any $a \in \mathbb{R}$, and similarly for T_{ϕ^o}. Moreover T_ϕ takes ∂B_ϕ onto ∂B_{ϕ^o}, T_{ϕ^o} takes ∂B_{ϕ^o} onto ∂B_ϕ, and $T_\phi T_{\phi^o} = T_{\phi^o} T_\phi = \mathrm{Id}$.

REMARK 2.7. Let ϕ be regular. Then T_ϕ and T_{ϕ^o} are single valued. Moreover

$$T_\phi = \frac{1}{2}\nabla((\phi)^2) = \phi\nabla\phi, \quad T_{\phi^o} = \frac{1}{2}\nabla((\phi^o)^2) = \phi^o\nabla\phi^o. \tag{2-7}$$

EXAMPLE 2.4. When $\phi(\xi) = |\xi|$ (Euclidean case), then $T_\phi = \mathrm{Id}$. When $\phi(\xi) = \sqrt{\langle A\xi, \xi \rangle}$ is Riemannian, then $T_\phi(\xi) = A\xi$.

Figure 2 illustrates how to construct T_ϕ in a regular case. First of all, since T_ϕ is one-homogeneous, it is enough to evaluate T_ϕ on ∂B_ϕ.

The point ξ belongs to ∂B_ϕ; since $T_\phi(\xi) = \nabla\phi(\xi)$ and ∂B_ϕ is a level set of ϕ, it is clear that $T_\phi(\xi)$ is orthogonal to ξ. In addition, $\phi^o(\nabla\phi(\xi)) = 1 = \langle \xi, \nabla\phi(\xi) \rangle$, which implies $|T_\phi(\xi)| = \langle \xi, \nabla\phi(\xi)/|\nabla\phi(\xi)| \rangle^{-1}$.

In what follows, it is important to keep in mind that the duality maps in (2–6) are still well defined under our assumptions on a Minkowski norm, in particular in the crystalline case. If ϕ is not regular, the equalities in (2–7) become

$$T_\phi = \tfrac{1}{2}\nabla^-((\phi)^2) = \phi\nabla^-\phi, \quad T_{\phi^o} = \tfrac{1}{2}\nabla^-((\phi^o)^2) = \phi^o\nabla^-\phi^o, \tag{2-8}$$

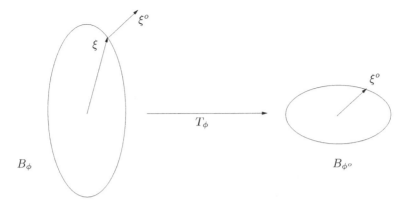

Figure 2. The point ξ belongs to ∂B_ϕ. The point $T_\phi(\xi) := \xi^o \in \partial B_{\phi^o}$ is the unit normal $\nu^{B_\phi}(\xi)$ to ∂B_ϕ at ξ, multiplied by the factor $\langle \xi, \nu^{B_\phi}(\xi) \rangle^{-1}$.

where ∇^- denotes the usual subdifferential in convex analysis (see for instance [Rockafellar 1972]); the main feature of the maps T_ϕ and T_{ϕ^o} is that they are no longer one-to-one.

Geometrically, if $\xi \in \partial B_\phi$, then $T_\phi(\xi)$ is the intersection of the closed outward normal cone to ∂B_ϕ with ∂B_{ϕ^o}. In Figure 3 we show an example of B_ϕ and of its dual body B_{ϕ^o}: B_ϕ is the Cartesian product of a planar regular hexagon with the interval $[-1, 1]$. If $\xi \in \partial B_\phi$ is a point in the relative interior of a facet, then the normal cone $T_\phi(\xi)$ to ∂B_ϕ at ξ is a singleton (a vertex in ∂B_{ϕ^o}); if $\xi \in \partial B_\phi$ is a point in the relative interior of an edge, then $T_\phi(\xi)$ is a one-dimensional closed segment (a closed edge in ∂B_{ϕ^o}); if $\xi \in \partial B_\phi$ is a vertex, then $T_\phi(\xi)$ is a closed triangle (a closed facet in ∂B_{ϕ^o}).

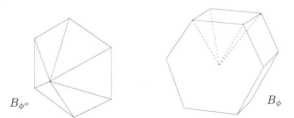

Figure 3. Dual polytopes. Duality maps take vertices into closed facets, points in the relative interior of a facet into vertices, and points in the relative interior of an edge into closed edges. The B_{ϕ^o} depicted here is supposed to look like an umbrella viewed from above.

3. Area Measures in $(\mathbb{R}^n, \phi)$

3.1. The $(n{-}1)$-dimensional Hausdorff measure. We recall the definition of Hausdorff measure in the metric space $(\mathbb{R}^n, d_\phi)$; see [Federer 1969].

DEFINITION 3.1. *If $A \subseteq \mathbb{R}^n$ is a Borel subset of $\mathbb{R}^n$ and $m \in \{n-1, n\}$ we set*

$$\mathcal{H}_{d_\phi}^m(A) := \frac{\omega_m}{2^m} \lim_{\delta \to 0^+} \inf \left\{ \sum_{i=1}^{+\infty} (\mathrm{diam}_{d_\phi}(S_i))^m : A \subseteq \bigcup_{i=1}^{+\infty} S_i, \ \mathrm{diam}_{d_\phi}(S_i) < \delta \right\},$$
(3–1)

where $\mathrm{diam}_{d_\phi}(S_i) := \sup\{d_\phi(s, \sigma) : (s, \sigma) \in S_i \times S_i\}$ is the diameter of the set S_i with respect to d_ϕ.

It is not difficult to prove that the limit in (3–1) exists. Consistent with the notation in Section 2, when ϕ is the Euclidean norm we omit the subscript d_ϕ in the notation of the Hausdorff measure. The following integral representation result provides an explicit formula for computing $\mathcal{H}_{d_\phi}^n$ and $\mathcal{H}_{d_\phi}^{n-1}$.

THEOREM 3.2. *Let ϕ be a norm on $\mathbb{R}^n$. Then*

$$\mathcal{H}_{d_\phi}^n(A) = \int_A \frac{\omega_n}{|B_\phi|} \, dx = \frac{\omega_n}{|B_\phi|} |A|,$$

for any Borel set $A \subseteq \mathbb{R}^n$.

Given $\nu \in S^{n-1}$, denote by

$$S_\nu(B_\phi) := \{\xi \in B_\phi : \langle \nu, \xi \rangle = 0\}$$

the section of B_ϕ with the hyperplane orthogonal to ν passing through the origin; moreover, set

$$I\phi(\nu) := \frac{1}{\mathcal{H}^{n-1}(S_\nu(B_\phi))}.$$
(3–2)

THEOREM 3.3. *Let ϕ be a norm on $\mathbb{R}^n$. Let M be a $(n-1)$-dimensional Lipschitz manifold in $\mathbb{R}^n$. Then*

$$\mathcal{H}_{d_\phi}^{n-1}(A \cap M) = \omega_{n-1} \int_{A \cap M} I\phi(\nu^M) \, d\mathcal{H}^{n-1},$$
(3–3)

where $\nu^M(x)$ is a unit vector normal to M at x and $A \subseteq \mathbb{R}^n$ is a Borel set.

The measure $\mathcal{H}_{d_\phi}^{n-1}$ is also called Busemann surface measure, see the books [Thompson 1996], [Schneider 1993] for detailed information on this topic and for complete references.

We will still denote by $I\phi$ the one-homogeneous extension of the function in (3–2) on the whole of $\mathbb{R}^n$. We then have that the unit ball of $I\phi$ can be written as

$$\{\xi \in \mathbb{R}^n : I\phi(\xi) \leq 1\} = \{\xi \in \mathbb{R}^n : |\xi| \leq \mathcal{H}^{n-1}(S_{\xi/|\xi|}(B_\phi))\},$$

that is, $\{I\phi \leq 1\}$ is the so-called intersection body $I(B_\phi)$ of B_ϕ. An interesting result of Busemann ensures that $I(B_\phi)$ is convex (see [Busemann 1947], [Busemann 1949], [Thompson 1996], [Schneider 1993] and references therein). This, in turn, is essentially equivalent to the following semicontinuity property of $\mathcal{H}_{d_\phi}^{n-1}$:

if $\{E_k\}_k$ is a sequence of finite perimeter sets whose characteristic functions converge in $L^1(\mathbb{R}^n)$ to the characteristic function of a finite perimeter set E, then $\mathcal{H}_{d_\phi}^{n-1}(\partial E) \leq \liminf_{k \to +\infty} \mathcal{H}_{d_\phi}^{n-1}(\partial E_k)$.

3.2. The $(n-1)$-dimensional Minkowski content.

DEFINITION 3.4. Let M be a $(n-1)$-dimensional Lipschitz manifold. We define the $(n-1)$-dimensional Minkowski content $\mathcal{M}_{d_\phi}^{n-1}(M)$ of M with respect to d_ϕ as

$$\mathcal{M}_{d_\phi}^{n-1}(M) := \lim_{\rho \to 0^+} \frac{\mathcal{H}_{d_\phi}^n\left(\{z \in \mathbb{R}^n : \mathrm{dist}_\phi(z, M) < \rho\}\right)}{2\rho}. \tag{3-4}$$

Under our regularity assumption on ∂E it is possible to prove that the limit in (3–4) exists (see [Federer 1969], [Ambrosio et al. 2000]).

In the Euclidean case $\phi(\xi) = |\xi|$, the $(n-1)$-dimensional Minkowski content coincides with the $(n-1)$-dimensional Hausdorff measure.

Observe that $\mathcal{H}_{d_\phi}^n$ and $\mathcal{M}_\phi^{n-1}$ are invariant under isometries between the normed ambient spaces, while $c_{n,\phi}$ is not invariant.

REMARK 3.5. The Minkowski content $\mathcal{M}_{d_\phi}^{n-1}$ provides a notion of surface measure which is constructed by means of the orthogonal projections of B_ϕ onto the (one-dimensional) normal spaces to the manifold M; on the other hand the Hausdorff measure $\mathcal{H}_{d_\phi}^{n-1}$ is constructed by means of the intersections of B_ϕ with the $((n-1)$-dimensional) tangent spaces to M. Other notions of surface measure, different in general from these two notions, can be considered, such as the Holmes–Thompson measure, see [Thompson 1996], or the definitions introduced in [De Giorgi 1995] (see [Ambrosio and Kirchheim 2000]).

REMARK 3.6. It can be proved that, for a Lipschitz set E, $\mathcal{M}_{d_\phi}^{n-1}(\partial E)$ coincides with the perimeter of the set E with respect to ϕ, whose definition is given in a distributional way.

The following representation result provides an explicit integral formula for computing the Minkowski content of a sufficiently smooth set.

THEOREM 3.7. Let M be a $(n-1)$-dimensional Lipschitz manifold. Then

$$\mathcal{M}_{d_\phi}^{n-1}(M) = c_{n,\phi} \int_M \phi^o(\nu^M) \, d\mathcal{H}^{n-1}. \tag{3-5}$$

The validity of Theorem 3.7 can be explained as follows: the measure $\mathcal{H}_{d_\phi}^n$ is $c_{n,\phi}$ times the Lebesgue measure. Moreover the Lebesgue measure of the ρ-tubular neighborhood (in the distance d_ϕ) in (3–4) is approximately $\mathcal{H}^{n-1}(M)$ multiplied by the 1-dimensional length of the orthogonal projection of $\{\phi \leq \rho\}$ in the direction ν, see Figure 4. This length equals $2\rho\phi^o(\nu)$, hence (3–5) follows.

Since ϕ^o is convex, the following semicontinuity property of $\mathcal{M}_{d_\phi}^{n-1}$ holds: if $\{E_k\}_k$ is a sequence of finite perimeter sets whose characteristic functions converge in $L^1(\mathbb{R}^n)$ to the characteristic function of a finite perimeter set E, then $\mathcal{M}_{d_\phi}^{n-1}(\partial E) \leq \liminf_{k \to +\infty} \mathcal{M}_{d_\phi}^{n-1}(\partial E_k)$.

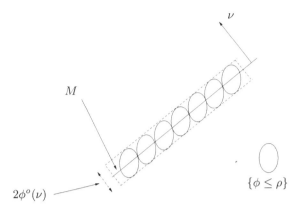

Figure 4. We assume for simplicity that M is flat, and that $\nu \in S^{n-1}$ is orthogonal to M. The ρ-tubular neighborhood (in the distance d_ϕ) is obtained by centering rescaled sets $\{\phi \le \rho\}$ of $\{\phi \le 1\}$ at points of M.

3.3. From the Hausdorff measure to the Minkowski content. As already remarked in the Introduction, in what follows we will work with $\mathcal{M}_{d_\phi}^{n-1}$ and this will affect the value of the ϕ-mean curvature. However, one could consider $\mathcal{H}_{d_\phi}^{n-1}$ as well, and change all subsequent definitions by replacing ϕ^o with $I\phi$. We refer to [Shen 1998] for a definition of mean curvature obtained by considering the intrinsic Hausdorff measure.

The following result follows from Theorems 3.3 and 3.7.

PROPOSITION 3.8. *Let ϕ be a norm on $\mathbb{R}^n$. Then*

$$\mathcal{H}_{d_\phi}^{n-1} = \mathcal{M}_{d_\psi}^{n-1}, \quad \psi := \left(\frac{\omega_{n-1} |B_{(I\phi)^o}|}{\omega_n} \right)^{1/(n-1)} (I\phi)^o. \qquad (3\text{--}6)$$

4. First Variation of Area and Mean Curvature Flow. Regular Case

In this section we define the anisotropic ϕ-mean curvature. We first recall some facts concerning the Euclidean case.

4.1. Preliminaries on the Euclidean case. Let $E \subset \mathbb{R}^n$ be an open set with smooth compact boundary. It is known (see for instance [Giusti 1984], [Ambrosio 1999]) that, under these assumptions, there exists a tubular neighborhood U of ∂E such that the oriented distance function d^E from ∂E negative inside E, defined as

$$d^E(z) := \operatorname{dist}(z, E) - \operatorname{dist}(z, \mathbb{R}^n \setminus E), \quad z \in \mathbb{R}^n,$$

is smooth on U, and $|\nabla d^E(z)| = 1$ for any $z \in U$ (eikonal equation). Hence, given any $x \in \partial E$, $\nabla d^E(x)$ is the outer unit normal $\nu^E(x)$ to ∂E at x. In addition $\Delta d^E(x)$ is the sum of the principal curvatures (the mean curvature) of ∂E at x. Therefore $-\Delta d^E(x)\nabla d^E(x)$ is the mean curvature vector of ∂E at x.

In order to compute the first variation of area, we need to introduce a class of admissible variations. We define a family Ψ_λ of compactly supported diffeomorphisms of the ambient space $\mathbb{R}^n$ as follows. Denote by $\Psi : \mathbb{R}^{n+1} \to \mathbb{R}^n$ a smooth vector field. Given any $\lambda \in \mathbb{R}$, define $\Psi_\lambda : \mathbb{R}^n \to \mathbb{R}^n$ as $\Psi_\lambda(x) := \Psi(x, \lambda)$. Assume that $\Psi_0 = \mathrm{Id}$ and that $\Psi_\lambda - \mathrm{Id}$ has compact support for any $\lambda \in \mathbb{R} \setminus \{0\}$. The following theorem is the classical result on the first variation of area, see for instance [Giusti 1984].

THEOREM 4.1. *For any $\lambda \in \mathbb{R}$ define $E_\lambda := \Psi_\lambda(E)$. Then*

$$\frac{d}{d\lambda} \mathcal{H}^{n-1}(\partial E_\lambda)_{|\lambda=0} = \int_{\partial E} \Delta d^E \langle X, \nabla d^E \rangle \, d\mathcal{H}^{n-1}, \qquad (4\text{--}1)$$

where $X := (\partial \Psi_\lambda / \partial \lambda)_{|\lambda=0}$.

Observe that only the normal component of X enters in formula (4–1).

We are now in a position to define the smooth mean curvature flow starting from a given ∂E.

DEFINITION 4.2. Let $E \subset \mathbb{R}^n$ be an open set with smooth compact boundary. Let $T > 0$ and, for any $t \in [0, T]$, let $E(t)$ be a set with compact boundary. We say that $(E(t))_{t \in [0,T]}$ is a smooth mean curvature flow in $[0, T]$ starting from $E = E(0)$ if the following conditions hold:

(i) there exists an open set $A \subset \mathbb{R}^n$ containing $\partial E(t)$ for any $t \in [0, T]$ such that, if we set

$$\bar{d}(z, t) := \mathrm{dist}(z, E(t)) - \mathrm{dist}(z, \mathbb{R}^n \setminus E(t)), \quad z \in \mathbb{R}^n, \ t \in [0, T],$$

we have $\bar{d} \in \mathcal{C}^\infty(A \times [0, T])$;

(ii) $$\frac{\partial}{\partial t} \bar{d}(x, t) = \Delta \bar{d}(x, t), \quad x \in \partial E(t), \ t \in [0, T]. \qquad (4\text{--}2)$$

Condition (i) implies that each $\partial E(t)$ is a smooth boundary, smoothly evolving in time. The vector $-(\partial \bar{d}/\partial t)\nabla \bar{d}(x, t)$ is the projection of the velocity of the point x on the normal space to $\partial E(t)$ at x (see for instance [Ambrosio 1999]).

EXAMPLE 4.1. The main explicit example of mean curvature flow is the one of the sphere $E = \{|z| < R_0\}$, which shrinks self-similarly. Indeed in this case $\bar{d}(z, t) = |z| - R(t)$, and the equation (4–2) becomes $\dot{R} = -(n-1)/R$. Its solution represents the evolving sphere of radius $R(t) = \sqrt{R_0^2 - 2(n-1)t}$ for $t \in [0, \frac{1}{2}R_0^2/(n-1)]$, which disappears for times larger than $\frac{1}{2}R_0^2/(n-1)$.

It is customary to say that the evolution law (4–2) is the gradient flow of the area functional $\mathcal{H}^{n-1}(\partial E)$; the idea is that, at each time, the set $E(t)$ evolves in such a way to make $\mathcal{H}^{n-1}(\partial E(t))$ as small as possible. This assertion has been made rigorous in the paper [Almgren et al. 1993], where it is shown that the correct (nonsymmetric) distance between sets to use in order to obtain the mean curvature flow is

$$(E, F) \subset \mathbb{R}^n \times \mathbb{R}^n \to \int_{(E \setminus F) \cup (F \setminus E)} \text{dist}(x, \partial F) \, dx. \qquad (4\text{--}3)$$

Mean curvature flow has been the subject of several recent papers, of which we list some. We refer to:

- [Gage and Hamilton 1986] and [Evans and Spruck 1992a] for a local in time existence and uniqueness theorem of a smooth solution;
- [Ecker and Huisken 1989] and [Ecker and Huisken 1991] for the evolution of graphs;
- [Huisken 1984] for the evolution of convex sets;
- [Grayson 1987] and [Angenent 1991a] for local and global properties of the flow of curves;
- [Barles et al. 1993], [Evans and Spruck 1992b], [Evans and Spruck 1995], [White 1995], [White 2000], [White 2003] for some qualitative properties of the flow.
- Concerning global in time solutions, defined after the onset of singularities, we refer to:
- [Brakke 1978], where a geometric measure theory approach is introduced;
- [Evans and Spruck 1991] and [Chen et al. 1991] for the level set method and viscosity solutions;
- [Ilmanen 1992], [Ilmanen 1993a], [De Giorgi 1993], [De Giorgi 1994] and [Bellettini and Paolini 1995] for the barriers method (see also [Soner 1993]);
- [Almgren et al. 1993] for a variational approach based on time discretization;
- [Ilmanen 1993b] for the approximation of mean curvature flow by means of a sequence of reaction-diffusion equations;
- [Ilmanen 1994] for the elliptic regularization method;
- [Angenent 1991a], [Soner and Souganidis 1993], [Altschuler et al. 1995] and [Bellettini and Paolini 1994] for the analysis of some kind of singularity of the flow;
- [Fierro and Paolini 1996], [Paolini and Verdi 1992], [Angenent et al. 1995] for numerical simulations of certain singularities.

4.2. The anisotropic regular case. We assume in this subsection that ϕ is a regular norm. Let E be a set with smooth compact boundary. Also in this case it is possible to prove that there exists a tubular neighborhood U of ∂E such that the oriented ϕ-distance function d_ϕ^E from ∂E negative inside E, defined as

$$d_\phi^E(z) := \text{dist}_\phi(z, E) - \text{dist}_\phi(z, \mathbb{R}^n \setminus E), \quad z \in \mathbb{R}^n,$$

is smooth on U, and $\phi^o(\nabla d_\phi^E(z)) = 1$ for any $z \in U$ (anisotropic eikonal equation). Therefore

$$T_{\phi^o}(\nabla d_\phi^E(z)) = \phi^o(\nabla d_\phi^E(z)) \nabla \phi^o(\nabla d_\phi^E(z)) = \nabla \phi^o(\nabla d_\phi^E(z)), \quad z \in U. \qquad (4\text{--}4)$$

In particular, given any $x \in \partial E$, we have

$$\nabla d_\phi^E(x) = \frac{\nu^E(x)}{\phi^o(\nu^E(x))} =: \nu_\phi^E(x). \qquad (4\text{--}5)$$

The following result is a generalization of Theorem 4.1, and shows also the rôle of the duality map T_{ϕ^o}. Let Ψ_λ and E_λ be as in Subsection 4.1.

THEOREM 4.3. *For any* $\lambda \in \mathbb{R}$ *define* $E_\lambda := \Psi_\lambda(E)$. *Then*

$$\frac{d}{d\lambda} \mathcal{M}_{d_\phi}^{n-1}(\partial E_\lambda)_{|\lambda=0} = c_{n,\phi} \int_{\partial E} \mathrm{div}\, \mathrm{n}_\phi^E \, \langle X, \nabla d_\phi^E \rangle \, \phi^o(\nu^E) d\mathcal{H}^{n-1}, \qquad (4\text{--}6)$$

where $X := \frac{\partial}{\partial \lambda} \Psi_\lambda{}_{|\lambda=0}$, *and*

$$\mathrm{n}_\phi^E(z) := T_{\phi^o}(\nabla d_\phi^E(z)), \quad z \in U. \qquad (4\text{--}7)$$

The vector field n_ϕ is sometimes called the Cahn–Hoffman vector field, and satisfies

$$\phi(\mathrm{n}_\phi^E) = 1 = \langle \nabla d_\phi^E, \mathrm{n}_\phi^E \rangle \quad \text{on } U. \qquad (4\text{--}8)$$

In the Euclidean case $\phi(\xi) = |\xi|$ we have $\mathrm{n}_\phi^E = \nu_\phi^E = \nu^E$ on ∂E.

REMARK 4.4. The left hand side of (4–6) depends on the values of ϕ^o only on S^{n-1} (recall (3–5)). Hence also the right hand side of (4–6), written in terms of the one-homogeneous extension of ϕ^o on the whole of $\mathbb{R}^n$, must depend only on the values of ϕ^o on S^{n-1}.

Observe that

$$\frac{d}{d\lambda} \mathcal{M}_{d_\phi}^{n-1}(\partial E_\lambda)_{|\lambda=0} = \int_{\partial E} \mathrm{div}\, \mathrm{n}_\phi^E \, \langle X, \nabla d_\phi^E \rangle \, d\mathcal{P}_\phi^{n-1},$$

where $d\mathcal{P}_\phi^{n-1}$ is the measure on ∂E having $c_{n,\phi}\phi^o(\nu^E)$ as density with respect to $\mathcal{H}^{n-1}$.

We are in a position to give the following definition.

DEFINITION 4.5. Let E be an open set with smooth compact boundary. We define the ϕ-mean curvature κ_ϕ^E of ∂E as

$$\kappa_\phi^E(x) := \mathrm{div}\, \mathrm{n}_\phi^E(x), \quad x \in \partial E. \qquad (4\text{--}9)$$

It is possible to prove that $\kappa_\phi^E(x)$ is also the tangential divergence of n_ϕ^E evaluated at $x \in \partial E$. Indeed, define $f(z) := \langle \nu^E(x), \mathrm{n}_\phi^E(z) \rangle$ for any $z \in U$. Thanks to (4–8), f has a maximum at x (with value $\phi^o(\nu^E(x))$). Therefore ∇f vanishes at x, that is, $\nu^E(x)\nabla \mathrm{n}_\phi^E(x) = 0$. This implies that the tangential divergence of n_ϕ^E at x equals $\mathrm{div}\, \mathrm{n}_\phi^E(x)$.

EXAMPLE 4.2. The ϕ-mean curvature of ∂B_ϕ is constantly equal to $n-1$. Indeed $\nabla d_\phi^{B_\phi}(z) = \nabla \phi(z)$ and $T_{\phi^o}(\nabla d_\phi^{B_\phi}(z)) = z/\phi(z)$ for any z in $\mathbb{R}^n \setminus \{0\}$. Then a computation gives $\mathrm{div}(z/\phi(z)) = n - 1$ on ∂B_ϕ.

EXAMPLE 4.3. Let $n = 2$, and write $\phi^o(\xi^o) = |\xi^o|\phi^o(\xi^o/|\xi^o|) =: \rho\psi(\theta)$, where $\xi_o = (\xi_1^o, \xi_2^o) = (\rho\cos\theta, \rho\sin\theta)$. Then $\kappa_\phi^E = \kappa^E(\psi + \psi'')$, where κ^E is the Euclidean curvature of ∂E.

EXAMPLE 4.4. Observe that from (4–4) we derive $\kappa_\phi^E = \text{tr}(\nabla^2\phi^o(\nabla d_\phi^E)\nabla^2 d_\phi^E)$ on ∂E, where ∇^2 denotes the Hessian matrix and tr is the trace operator.

REMARK 4.6. In the paper [Bellettini and Fragalà 2002] the second variation of $\mathcal{M}_{d_\phi}^{n-1}$ is computed. What replaces the squared length of the second fundamental form of ∂E is the term $\text{tr}(\nabla n_\phi^E \nabla n_\phi^E)$. In the same paper, a sort of Laplace–Beltrami operator is introduced, see also [Bao et al. 2000] (for references therein), [Shen 2001], [Mugnai 2003].

We are now in a position to define what is a smooth anisotropic mean curvature flow, for a regular anisotropy ϕ.

DEFINITION 4.7. Let $E \subset \mathbb{R}^n$ be an open set with smooth compact boundary. Let $T > 0$ and, for any $t \in [0, T]$, let $E(t)$ be a set with compact boundary. We say that $(E(t))_{t \in [0,T]}$ is a smooth ϕ-mean curvature flow in $[0, T]$ starting from $E = E(0)$ if the following conditions hold:

(i) there exists an open set $A \subset \mathbb{R}^n$ containing $\partial E(t)$ for any $t \in [0, T]$ such that, if we set

$$\bar{d}_\phi(z, t) := \text{dist}_\phi(z, E(t)) - \text{dist}_\phi(z, \mathbb{R}^n \setminus E(t)), \quad z \in \mathbb{R}^n, \ t \in [0, T],$$

we have $d_\phi \in \mathcal{C}^\infty(A \times [0, T])$;

(ii) $$\frac{\partial}{\partial t}d_\phi(x, t) = \text{divn}_\phi^{E(t)}(x), \quad x \in \partial E(t), \ t \in [0, T]. \qquad (4\text{–}10)$$

EXAMPLE 4.5. We show that $\{\xi \in \mathbb{R}^n : \phi(\xi) < R_0\}$ shrinks self-similarly under the flow (4–10). We have in this case $\bar{d}_\phi(z, t) = \phi(z) - R(t)$, and (4–10) (thanks to Example 4.2) becomes $\dot{R} = -(n-1)/R$. Its solution represents the evolving set $\{\xi \in \mathbb{R}^n : \phi(\xi) < R(t)\}$, where $R(t) = \sqrt{R_0^2 - 2(n-1)t}$ for $t \in [0, \frac{1}{2}R_0^2/(n-1)]$, which disappears for times larger than $\frac{1}{2}R_0^2/(n-1)$.

The evolution law (4–10) is the gradient flow of $\mathcal{M}_{d_\phi}^{n-1}(\partial E)$. This can be seen, for instance, by using the (nonsymmetric) distance between sets in (4–3), where $\text{dist}(x, \partial F)$ is replaced by $\text{dist}_\phi(x, \partial F)$.

For what concerns anisotropic mean curvature flows, we refer to the following (largely incomplete) list of papers: [Hoffman and Cahn 1972], [Angenent 1991b], [Spohn 1993], [Cahn et al. 1993], [Gage 1994], [Giga and Goto 1992] (see also [Giga et al. 1991] for weak solutions to a large class of anisotropic equations).

4.3. On space dependent norms. In this subsection we list some generalization of the previous results. Assume that $\phi = \phi(x, \xi) : \mathbb{R}^n \times \mathbb{R}^n \to [0, +\infty[$ depends on the position x, ϕ^2 is smooth, $\phi^2(x, \cdot)$ is strictly convex (in the sense of Definition 2.2) and ϕ satisfies (2–1), (2–2) for any $x \in \mathbb{R}^n$, with λ independent of

x. Define $\phi^o(x, \xi^o) := \sup\{\langle \xi, \xi^o \rangle : \xi \in B_\phi(x)\}$, $B_\phi(x) := \{\xi \in \mathbb{R}^n : \phi(x, \xi) \leq 1\}$, and d_ϕ as in the first equality of (2–3), with $\phi(\gamma, \dot{\gamma})$ in place of $\phi(\dot{\gamma})$. T_ϕ and T_{ϕ^o} are defined as in (2–8) taking x fixed. Then (3–3) holds with

$$I\phi(x, \nu^M(x)) = \frac{1}{\mathcal{H}^{n-1}(S_\nu(B_\phi(x))}$$

in place of $I\phi(\nu^M(x))$. The n-dimensional d_ϕ-Hausdorff measure $\mathcal{H}^n_{d_\phi}$ in $\mathbb{R}^n$ has the integral representation

$$\mathcal{H}^n_{d_\phi}(A) = \omega_n \int_A \frac{1}{|B_\phi(x)|}\, dx,$$

for any Borel set $A \subseteq \mathbb{R}^n$. Then (3–5) becomes

$$\mathcal{M}^{n-1}_{d_\phi}(M) = \omega_n \int_M \phi^o(x, \nu^M(x)) \frac{1}{|B_\phi(x)|}\, d\mathcal{H}^{n-1}(x).$$

In addition, the function ψ in (3–6) becomes $\psi(x, \xi) = f(x)(I\phi)^o(x, \xi)$, where f depends only on x and has the expression

$$f(x) = \left(\frac{\omega_{n-1}|B_{(I\phi)^o}(x)|}{\omega_n} \right)^{1/(n-1)}.$$

Concerning the first variation of area and ϕ-mean curvature, (4–6) of Theorem 4.3 reads as

$$\frac{d}{d\lambda}\mathcal{M}^{n-1}_{d_\phi}(\partial E_\lambda)_{|\lambda=0} = \omega_n \int_{\partial E} \operatorname{div}_\phi \mathrm{n}^E_\phi \langle X, \nabla d^E_\phi \rangle \frac{1}{|B_\phi(x)|} \phi^o(x, \nu^E)d\mathcal{H}^{n-1},$$

where the operator div_ϕ acts on a smooth vector field v as follows:

$$\operatorname{div}_\phi v := \operatorname{div} v + \left\langle v,\ \nabla\left(\log \frac{1}{|B_\phi(x)|} \right) \right\rangle,$$

and

$$\mathrm{n}^E_\phi(z) := \nabla\phi^o(z, \nabla d^E_\phi(z)),$$

$\nabla\phi^o$ being the gradient of ϕ^o with respect to the ξ variable.

5. The Crystalline Case: Lipschitz ϕ-Regular Sets

In this section we assume that ϕ is a crystalline norm. The main difficulties when trying to generalize the notion of ϕ-curvature given in (4.5) to the crystalline case are due to the loss of regularity, both of ∂E and of the norm ϕ^o. Observe that the explicit computation of κ^E_ϕ in (4.5) requires the computation of the hessian of ϕ^o which, in the crystalline case, is just a (nonnegative) measure. Recall also that, in the crystalline case, the duality maps are not single valued anymore. This will force us to consider inclusions in place of equalities, and suitable selection principles will be required. Finally, we have to keep in mind that, whichever definition of smoothness we choose, the set ∂B_ϕ must be smooth.

Unlike the regular case, in the crystalline case we have to redefine what is a smooth boundary. The idea is to define smoothness of ∂E by requiring the existence of at least one Lipschitz selection of a normal (in a suitable sense) vector field.

Before giving formal definitions, we recall some notation. At points $x \in \partial E$ where $\nu^E(x)$ exists we set $\nu^E_\phi(x) := \nu^E(x)/\phi^o(\nu^E(x))$. We indicate by $\mathrm{Lip}(\partial E; \mathbb{R}^n)$ the class of all Lipschitz vector fields defined on the Lipschitz boundary ∂E.

DEFINITION 5.1. If $E \subset \mathbb{R}^n$ is Lipschitz we define

$$\mathrm{Nor}_\phi(\partial E; \mathbb{R}^n) := \{N \in L^\infty(\partial E; \mathbb{R}^n) : N(x) \in T^o(\nu^E_\phi(x)) \text{ for } \mathcal{H}^{n-1}\text{a.e. } x \in \partial E\},$$

$$\mathrm{Lip}_{\nu,\phi}(\partial E; \mathbb{R}^n) := \mathrm{Lip}(\partial E; \mathbb{R}^n) \cap \mathrm{Nor}_\phi(\partial E; \mathbb{R}^n).$$

As we shall see, in general smooth sets E in the usual sense do not admit even one element in the class $\mathrm{Lip}_{\nu,\phi}(\partial E; \mathbb{R}^n)$. The best of smoothness that we can hope is described by the following definition [Bellettini and Novaga 1998]:

DEFINITION 5.2. Let $E \subset \mathbb{R}^n$ be an open set with compact boundary. We say that E is Lipschitz ϕ-regular if ∂E is Lipschitz continuous and there exists a vector field $\eta \in \mathrm{Lip}_{\nu,\phi}(\partial E; \mathbb{R}^n)$. With a small abuse of notation, the pair (E, η) will also denote a Lipschitz ϕ-regular set.

Observe that, unlike ν^E_ϕ, the vector field η is defined everywhere on ∂E.

To better understand the meaning of Definition 5.2, we consider examples.

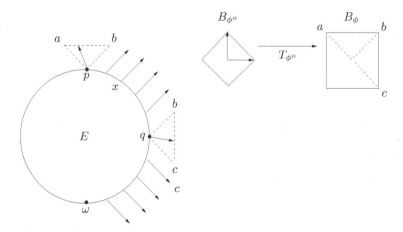

Figure 5. The open ball E is *not* Lipschitz ϕ-regular when we endow $\mathbb{R}^2$ with the norm $\phi(\xi) = \max\{|\xi_1|, |\xi_2|\}$.

EXAMPLE 5.1. Let $n = 2$ and $\phi(\xi) := \max\{|\xi_1|, |\xi_2|\}$, in such a way that B_ϕ is the square of side 2, see Figure 5. Let $E := \{z \in \mathbb{R}^2 : |z| < 1\}$ be the open unit disk. Then E is *not* Lipschitz ϕ-regular. Indeed, to see that E is Lipschitz ϕ-regular, we have to compute $T^o(\nu_\phi^E(x))$, for $x \in \partial E$, and to show that we can produce a vector field η on ∂E which is a Lipschitz selection for the multivalued map $x \in \partial E \to T_{\phi^o}(\nu_\phi^E(x))$. Observe now that $T^o(\nu_\phi^E(p))$ is the upper horizontal segment $[a, b]$ of ∂B_ϕ; we depict therefore a corresponding dotted triangle at p. Similarly, $T_{\phi^o}(\nu_\phi^E(q))$ is the right vertical segment $[b, c]$ of ∂B_ϕ, and again we depict the corresponding dotted triangle at q. On the other hand, any point x on ∂E lying in the (relatively) open arc A between p and q is such that $T_{\phi^o}(\nu_\phi^E(x)) = b$. We deduce that $\eta \equiv b$ on A, and $\eta \equiv c$ on the open arc on ∂E between q and ω. Hence, *any* vector we choose inside the dotted triangles (for instance, the triangle at q) will produce a *discontinuity* in the vector field η (at q). We can conclude that the circle, considered in $(\mathbb{R}^2, \phi)$, is not Lipschitz ϕ-regular, and that it takes the rôle of the square in the usual Euclidean plane.

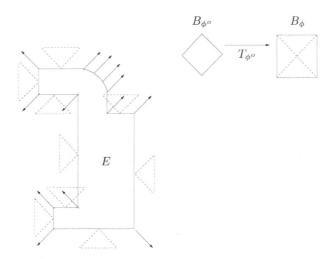

Figure 6. Example of a Lipschitz ϕ-regular set E when $\phi(\xi) = \max\{|\xi_1|, |\xi_2|\}$. The values of η are uniquely determined at the vertices and on the curved arc of ∂E. Any Lipschitz extension of these values on the interior of the edges, which lies in the dotted triangles, produces a Lipschitz vector field satisfying the required properties (that is, making E Lipschitz ϕ-regular). Examples are depicted in Figures 10 and 13.

EXAMPLE 5.2. Let $\phi(\xi) := \max\{|\xi_1|, |\xi_2|\}$. In Figure 6 we show an example of a Lipschitz ϕ-regular set E.

At the vertices of ∂E the vector ν_ϕ^E is not defined. Let v be a vertex of ∂E, and let F_1 and F_2 be the two arcs of ∂E having v as a vertex (arcs can also be flat, i.e., segments). For any x in the relative interior of F_i, the closed convex set $T_{\phi^o}(\nu_\phi^E(x))$ is either a segment or a singleton; in both cases is independent of x and depends only on F_i. Denote it by K_i. The crucial property that makes E Lipschitz ϕ-regular is that the intersection $\cap_{i=1}^2 K_i$ is a singleton, see Figure 6. This produces a unique vector at each vertex of ∂E; at this point, it is easy to realize that we can construct infinitely many vector fields $\eta \in \mathrm{Lip}_{\nu,\phi}(\partial E; \mathbb{R}^2)$ lying inside the dotted triangles with the assigned values at the vertices (see for instance Figures 10 and 13).

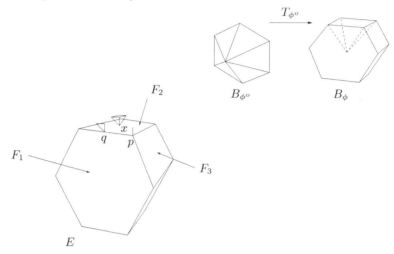

Figure 7. An example of Lipschitz ϕ-regular set E for the norm ϕ whose unit ball is the product of an hexagon with $[-1, 1]$.

EXAMPLE 5.3. In Figure 7 it is shown an example of Lipschitz ϕ-regular set in $\mathbb{R}^3$ when the unit ball of ϕ is prism with hexagonal basis. Observe that the vector ν_ϕ^E is not defined on the vertices and on the edges of ∂E. Let $p \in \partial E$ be a vertex of ∂E and let F_1, F_2, F_3 be the three (relatively) closed facets of ∂E having p as a common vertex. For any x in the relative interior of F_i, the convex set $T_{\phi^\circ}(\nu_\phi^E(x))$ is a closed facet of ∂B_ϕ which is independent of x and depends only on F_i. Denote it by K_i. The intersection $\cap_{i=1}^3 K_i$ is a singleton (and is the corresponding vertex of ∂B_ϕ). In Figure 7 we have depicted such an intersection as a vector at the point p. On the other hand, if F_1 and F_2 have in common the segment S, and q is a point in the relative interior of S, then $\cap_{i=1}^2 K_i$ is the corresponding edge of ∂B_ϕ. We have depicted this set as a triangle. Finally, if x is a point in the relative interior of a facet (for instance, the top facet F), then $\nu_\phi^E(x)$ coincides with the top vertex of the ∂B_{ϕ°, and therefore $T_{\phi^\circ}(\nu_\phi(F))$ is the top facet of ∂B_ϕ, and we have depicted this set on the interior of F as a pyramid. Showing that E is Lipschitz ϕ-regular means to exhibit a Lipschitz vector field $\eta : \partial E \to \mathbb{R}^3$ which on the vertices of ∂E is fixed (to be the corresponding vertices of ∂B_ϕ), on the relative interior of an edge of ∂E is constrained to lie in the corresponding segment of ∂B_ϕ, and in the relative interior of a facet of ∂E is constrained to lie in the corresponding facet of ∂B_ϕ. It is at this point easy to realize that such a choice can be made (in infinitely many different ways) for the set E in Figure 7.

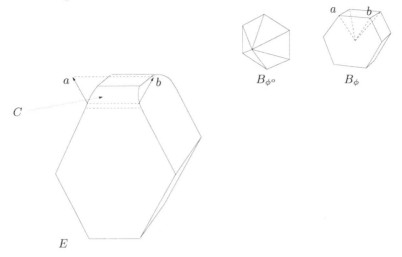

Figure 8. A non polyhedral Lipschitz ϕ-regular set E.

EXAMPLE 5.4. In Figure 8 we show an example of Lipschitz ϕ-regular set E in $\mathbb{R}^3$ which is not polyhedral. The curved region C is *ruled*; if S is any horizontal segment in C (see the dotted lines), any vector field $\eta \in \mathrm{Lip}_{\nu,\phi}(\partial E; \mathbb{R}^3)$ must lie, on S, in the corresponding segment $[a, b]$ of ∂B_ϕ.

In two dimensions the structure of Lipschitz ϕ-regular sets E having a finite number of arcs (arcs can be also segments) can be described as follows. The arcs are located in a precise order consistent with ϕ-regularity, and are divided into two classes. In the first class there are edges which are parallel to some facet of B_ϕ and have the same exterior Euclidean normal vector (and we say that the edge *corresponds to* a facet of B_ϕ). The second class consists of the arcs (some of which can be flat) not belonging to the first class, where there is only one possible choice of the vector field η consistent with the ϕ-regularity. The arcs of the second class have therefore zero ϕ-curvature (see [Taylor 1993], [Giga and Gurtin 1996]).

6. On Facets of Polyhedral Lipschitz ϕ-Regular Sets

We have seen in Example 5.2 that, even in two dimensions, if (E, η) is a Lipschitz ϕ-regular set, there are in general infinitely many vector fields $\bar\eta \in \mathrm{Lip}_{\nu,\phi}(\partial E; \mathbb{R}^2)$ such that $(E, \bar\eta)$ is Lipschitz ϕ-regular. Therefore the divergence of each of these $\bar\eta$ could be considered as a ϕ-mean curvature of ∂E, and in this way the Lipschitz boundary ∂E would have infinitely many different ϕ-mean curvatures. This approach could be pursued to some extent; however, we shall see that, among all vector fields satisfying the required constraints, there are some which are distinguished, have the *same* divergence, and such a uniquely defined divergence is what we can call the ϕ-mean curvature of ∂E, at least from the evolutionary point of view. Indeed, this notion of ϕ-mean curvature should be identified with the initial velocity of the interface under crystalline mean curvature flow.

To simplify notation, in this section we shall consider, when $n \geq 3$, only polyhedral Lipschitz ϕ-regular sets E with a finite number of facets, that will be understood as (relatively) closed connected $(n-1)$ dimensional flats with Lipschitz (polyhedral) boundary. The symbol F will always denote such a facet. This is surely a restriction, since in general facets can produce curved regions during the flow.

6.1. Some notation. If $F \subset \partial E$ is a facet, we denote by ∂F and $\mathrm{int}(F)$ the relative boundary and relative interior of F. An edge of ∂E with vertices p, q will be denoted by $[p, q]$, and its relative interior by $]p, q[$. We denote by Π_F the affine hyperplane spanned by the facet F.

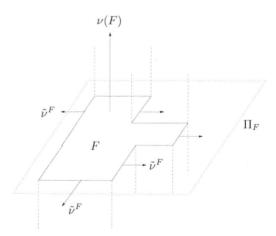

Figure 9. A facet F. The dotted lines delimit the solid set E having F as a facet.

We define $\nu(F)$ to be the unit normal to $\mathrm{int}(F)$ which points outside of E and we set $\nu_\phi(F) := \nu(F)/\phi^o(\nu_\phi(F))$. We indicate by $\tilde{\nu}^F$ the ($\mathcal{H}^{n-2}$-almost everywhere defined) unit normal to ∂F pointing outside of F; see Figure 9. Only facets F such that $T_{\phi^o}(\nu_\phi(F))$ is a facet of B_ϕ (that is., facets of ∂E corresponding to some facet of ∂B_ϕ) will be considered.

DEFINITION 6.1. Let (E, η) be a Lipschitz ϕ-regular set. We define the trace function $c_F \in L^\infty(\partial F)$ as

$$c_F := \langle \eta, \tilde{\nu}^F \rangle. \tag{6-1}$$

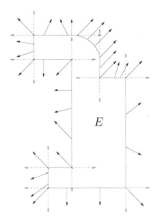

Figure 10. The normal trace of η on the boundary of each one-dimensional facet of ∂E is independent of η itself (among all vector fields making E Lipschitz ϕ-regular).

EXAMPLE 6.1. In Figure 10 we depict a vector field η which makes ∂E Lipschitz ϕ-regular (B_ϕ is the square as in Figure 6). Since the values of η are uniquely determined at the vertices of ∂E, the constants c_F do not depend on the particular choice of η. The dotted vectors at the vertices indicate the unit normals (in the line containing the facet F) pointing outward F (that is, $\tilde{\nu}^F$).

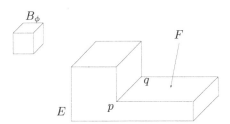

Figure 11. A Lipschitz ϕ-regular set when B_ϕ is the cube.

EXAMPLE 6.2. Consider the Lipschitz ϕ-regular set (E, η) of Figure 11 (B_ϕ is the unit cube). In Figure 12 the bold vectors at the vertices of ∂E are the unique possible values for η. The vector field $\tilde{\nu}^F$ points outside F, and on $]p, q[$ points inside E. The pyramids with vertex on the relative interior of the two facets having $[p, q]$ in common represent the corresponding facets of ∂B_ϕ (for instance, $T_{\phi^o}(\nu_\phi(F))$ for the facet F), that is, the range of admissibility of η. It follows that c_F is negative on $]p, q[$, while c_F is positive on the remaining relatively open edges of ∂F.

Given a Lipschitz ϕ-regular set (E, η), in general it is possible to prove that c_F does not depend on the choice of η in $\mathrm{Lip}_{\nu,\phi}(\partial E; \mathbb{R}^n)$. More precisely, for

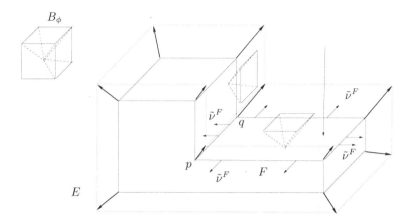

Figure 12. On the relative interior of $[p,q]$ the function c_F is negative (and constant).

$\mathcal{H}^{n-1}$-almost every $x \in \partial F$, we have

$$c_F(x) = \begin{cases} \max\{\langle \xi, \tilde{\nu}^F(x) \rangle : \xi \in T_{\phi^o}(\nu_\phi(F))\} & \text{if } \tilde{\nu}^F(x) \text{ points outside } E, \\ \min\{\langle \xi, \tilde{\nu}^F(x) \rangle : \xi \in T_{\phi^o}(\nu_\phi(F))\} & \text{if } \tilde{\nu}^F(x) \text{ points inside } E. \end{cases} \quad (6\text{--}2)$$

7. ϕ-Mean Curvature on a Facet

In this section we want to define (pointwise almost everywhere) the ϕ-mean curvature on a facet F of a polyhedral Lipschitz ϕ-regular set (E, η). We need some preliminaries. We let

$$\text{Nor}_\phi(F; \Pi_F) := \big\{ N \in L^\infty(\text{int}(F); \Pi_F) : \\ N(x) \in T_{\phi^o}(\nu_\phi(F)) \text{ for } \mathcal{H}^{n-1} \text{ a.e. } x \in \text{int}(F) \big\}.$$

It is possible to prove (see [Giga et al. 1998]) that any $N \in \text{Nor}_\phi(F; \Pi_F)$ with $\text{div} N \in L^2(\text{int}(F))$ admits a normal trace $\langle N, \tilde{\nu}^F \rangle$ on ∂F, for which the Gauss–Green Theorem holds on F (see [Anzellotti 1983]). We set

$$H(F; \Pi_F) := \\ \big\{ N \in \text{Nor}_\phi(F; \Pi_F) : \text{div} N \in L^2(\text{int}(F)), \ \langle N, \tilde{\nu}^F \rangle = c_F \ \mathcal{H}^{n-2} \text{ a.e. on } \partial F \big\}.$$

REMARK 7.1. Thanks to (6–2), the class $H(F; \Pi_F)$ does not depend on the choice of the vector field η making E Lipschitz ϕ-regular.

We define the functional $\mathcal{F}(\cdot, F) : H(F; \Pi_F) \to [0, +\infty[$ as

$$\mathcal{F}(N, F) := c_{n,\phi} \int_{\text{int}(F)} (\text{div} N)^2 \, \phi^o(\nu^E) d\mathcal{H}^{n-1}. \quad (7\text{--}1)$$

The right hand side of (7–1) equals $c_{n,\phi} \phi^o(\nu(F)) \int_F (\text{div} N)^2 \, d\mathcal{H}^{n-1}$, since F is flat.

Our definition of ϕ-mean curvature is based on the following result: the minimum problem

$$\inf\{\mathcal{F}(N,F) : N \in H(F;\Pi_F)\} \tag{7-2}$$

admits a solution, and any two minimizers have the *same* divergence.

Denote by $N_{\min}^F$ a solution of problem (7–2); since $\mathrm{div} N_{\min}^F$ is independent of the choice of $N_{\min}^F$ among all minimizers of (7–2), we can give the definition of crystalline mean curvature.

DEFINITION 7.2. We define the ϕ-mean curvature κ_ϕ^F on the relative interior of F as

$$\kappa_\phi^F(x) := \mathrm{div} N_{\min}^F(x), \quad \mathcal{H}^{n-1} \text{ a.e. } x \in \mathrm{int}(F).$$

Observe that κ_ϕ^F is only a function in $L^2(\mathrm{int}(F))$. We then set $\kappa_\phi^E := \kappa_\phi^F$ on each facet F of ∂E: it turns out that the orthogonal projection of minimizing vector fields on the orthogonal to ∂F is continuous on ∂F.

REMARK 7.3. The minimum problem (7–2), which is at the basis of Definition 7.2, arises when looking at the best way to decrease the $\mathcal{M}_{d_\phi}^{n-1}(\partial E)$ through deformations of the ambient space, precisely in the computation of the first variation

$$\liminf_{\lambda \to 0^+} \frac{\mathcal{M}_{d_\phi}^{n-1}(\partial E_\lambda) - \mathcal{M}_{d_\phi}^{n-1}(\partial E)}{\lambda}$$

of $\mathcal{M}_{d_\phi}^{n-1}$ at ∂E. Here, using the notation of Theorem 4.3, we have $E_\lambda = \Psi_\lambda(E)$ and $\Psi_\lambda(x) = x + \lambda X(x)$, where X is a suitable Lipschitz vector field.

The ϕ-mean curvature of ∂B_ϕ is constantly equal to $n-1$. Indeed, the vector field $x/\phi(x)$ has constant divergence on ∂B_ϕ, hence it solves the Euler–Lagrange inequality derived from (7–2). We now use the (strict) convexity in the divergence to show that $x/\phi(x)$ is actually a minimizer of $\mathcal{F}(\cdot,F)$ on any facet $F \subset \partial B_\phi$.

The following example concerning crystalline curvature of curves is enlightening.

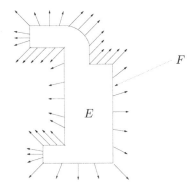

Figure 13. The vector field $\overline{N}_{\min}^E : \partial E \to \mathbb{R}^2$ is, on each facet F, the linear combination of the values of η at the vertices.

EXAMPLE 7.1. Let $n = 2$. We compute explicitly the ϕ-curvature of a two-dimensional Lipschitz ϕ-regular set (E, η). Given a facet $F \subset \partial E$ (in this case F equals a segment $[z, w]$), the minimum problem (7–2) reads as

$$\inf\left\{ \int_{]z,w[} (N'(s))^2 d\mathcal{H}^1(s) : N \in L^2(]z, w[; \Pi_{[z,w]}), \ N' \in L^2(]z, w[), \right.$$
$$\left. N(x) \in T_{\phi^\circ}(\nu_\phi^{[z,w]}(x)) \text{ for a.e. } x \in]z, w[, \ N(z) = c_z, N(w) = c_w \right\},$$

where c_z and c_w are the orthogonal projections of $\eta(z)$ and $\eta(w)$ on the line $\Pi_{[z,w]}$, with the correct sign.

We now observe that the above minimum problem has a *unique* solution $\overline{N}_{\min}^F$, which is simply the linear function connecting c_z at z with c_w at w. Hence, when $n = 2$, not only the divergence of a minimizer is unique, but also the minimizer itself. If we now repeat this procedure for any facet, and on each facet we add to $\overline{N}_{\min}^F$ the proper (constant) normal component to F, we end up with the vector field $\overline{N}_{\min}^E : \partial E \to \mathbb{R}^2$ whose divergence is the ϕ-curvature of ∂E. An example of this vector field is depicted in Figure 13. Curved regions in ∂E have zero ϕ-curvature. On the other hand, if F is a facet of $\partial E \subset \mathbb{R}^2$ and $B_F \subset \partial B_\phi$ is the corresponding facet in ∂B_ϕ, κ_ϕ^F is *constant* on F and

$$\kappa_\phi^F = \delta_F \frac{|B_F|}{|F|} \quad \text{on int}(F), \tag{7–3}$$

where $\delta_F \in \{0, \pm 1\}$ is a convexity factor: $\delta_F = 1, -1$ or 0 depending on whether E is locally convex at F, locally concave at F, or neither.

In two dimensions (7–3) is used to define the curvature flow of a Lipschitz ϕ-regular set (see the references quoted in the Introduction). If ∂E has a finite number of arcs, crystalline curvature flow can be described with a system of ordinary differential equations, since each arc (with nonzero ϕ-curvature) moves in normal direction in the evolution process: it cannot split or curve since, as dictated by (7–3), its normal velocity is constant. On the other hand, arcs or segments with zero ϕ-curvature stay still, and are progressively eaten by the other evolving arcs.

When the space dimension n is larger than or equal to 3, the computation of the ϕ-mean curvature on a facet is not, in general, an easy problem. As already mentioned in the Introduction, a short time existence theorem of a crystalline mean curvature flow is still missing. Concerning the comparison principle, only an indirect proof is available, in a certain class of crystalline evolutions, see [Bellettini et al. 2000].

DEFINITION 7.4. We say that E is convex at F if E, locally around F, lies on one side of Π_F.

The following results show that the ϕ-curvature enjoys some additional regularity properties.

THEOREM 7.5. $\kappa_\phi^F \in L^\infty(\mathrm{int}(F))$. *Moreover,* κ_ϕ^F *has bounded variation on* $\mathrm{int}(F)$. *Finally, if* F *is convex and* E *is convex at* F, *then* κ_ϕ^F *is convex on* $\mathrm{int}(F)$.

Since the jump set of a function with bounded variation is well defined (see for instance [Ambrosio et al. 2000]), this theorem makes it possible to speak of the jump set of κ_ϕ^F on $\mathrm{int}(F)$, which should describe, at time zero, where the facet splits under crystalline mean curvature flow. For small times in the evolution problem, F is expected to translate parallel to itself if κ_ϕ^F is constant on $\mathrm{int}(F)$ or to bend if κ_ϕ^F is continuous but not constant on $\mathrm{int}(F)$. Facets with constant ϕ-mean curvature have been isolated and studied in [Bellettini et al. 1999], [Bellettini et al. 2001c], where the following notation was introduced.

DEFINITION 7.6. *We say that* F *is* ϕ-*calibrable if* κ_ϕ^F *is constant on* $\mathrm{int}(F)$.

More explicitly, F is ϕ-calibrable provided there exists a vector field $N : \mathrm{int}(F) \to \Pi_F$ which solves the following problem:

$$
\begin{cases}
N \in L^\infty(\mathrm{int}(F); \Pi_F), \\
N(x) \in T_{\phi^\circ}(\nu_\phi(F)) \text{ for } \mathcal{H}^{n-1} \text{ a.e. } x \in \mathrm{int}(F), \\
\langle N, \tilde{\nu}^F \rangle = c_F \quad \mathcal{H}^{n-2} \text{ a.e. on } \partial F, \\
\mathrm{div}N = \dfrac{1}{|F|} \displaystyle\int_{\partial F} c_F \, d\mathcal{H}^{n-1}.
\end{cases} \tag{7-4}
$$

Observe that the constant on the right hand side of the differential equation in (7–4) is determined by using the Gauss–Green theorem on F. The complete characterization of ϕ-calibrable facets F is not yet available. We conclude the paper by pointing out some known results in this direction.

7.1. Characterization of ϕ-calibrable facets in special cases.
Assume that $n = 3$. Let $B_F \subset \partial B_\phi$ be the facet corresponding to F. If necessary, we identify B_F with its orthogonal projection on the plane parallel to Π_F passing through the origin of $\mathbb{R}^3$. We also assume that B_F contains the origin in its interior and that it is symmetric (this latter assumption can be weakened). Therefore B_F can be considered as the unit ball of a norm (in $\mathbb{R}^2$), which we denote by $\tilde{\phi}$. We assume that F is Lipschitz $\tilde{\phi}$-regular. Denote by $\kappa_{\tilde{\phi}}^F$ the $\tilde{\phi}$-curvature of ∂F and by $\tilde{\phi}^\circ$ the dual of $\tilde{\phi}$. The following result holds.

THEOREM 7.7. *Let* $n = 3$. *Assume that* F *is convex and that* E *is convex at* F. *Then* F *is* ϕ-*calibrable if and only if*

$$
\sup_{\partial F} \kappa_{\tilde{\phi}}^F \le \frac{1}{|F|} \int_{\partial F} \tilde{\phi}^\circ(\tilde{\nu}^F) \, d\mathcal{H}^1. \tag{7-5}
$$

The sup in (7–5) is an essential supremum, since $\kappa_{\tilde{\phi}}^F$ is a function in $L^\infty(\partial F)$. Hence, under the assumptions of Theorem 7.7, problem (7–4) is solvable if and only if the $\tilde{\phi}$-curvature of ∂F is bounded above by the constant on the right hand side of (7–5); this means, roughly speaking, that the edges of ∂F cannot be too "short".

Finally, let us mention that examples of facets which are not ϕ-calibrable are given in [Bellettini et al. 1999], and that the problem of calibrability when B_ϕ is a portion of a cylinder (hence not in a crystalline setting) has been recently considered, under rather mild assumptions, in the papers [Bellettini et al. 2002] and [Bellettini et al. 2003].

8. Concluding Remarks

Fix a norm ϕ on $\mathbb{R}^n$ and denote by d_ϕ the distance induced by ϕ. As a starting point of our approach let us consider the $(n-1)$-dimensional measure $\mathcal{M}_{d_\phi}^{n-1}$, defined as in (1–1) on compact and sufficiently smooth boundaries ∂E of solid sets E. Such a notion is called the ϕ-Minkowski content of the manifold ∂E, and is a geometric invariant under isometries of the ambient space (because the n-dimensional Hausdorff measure $\mathcal{H}_{d_\phi}^n$ with respect to d_ϕ is invariant and the tubular neighborhoods are computed with respect to the distance d_ϕ). On the other hand $c_{n,\phi}$ is not invariant (recall that ω_n is a normalizing constant).

We recall that, in the generic (finite dimensional) normed space $(\mathbb{R}^n, \phi)$, there are other meaningful notions of surface measure, such as for instance the $(n-1)$-dimensional Hausdorff measure $\mathcal{H}_{d_\phi}^{n-1}$ with respect to d_ϕ, the Holmes–Thompson measure and the measure considered in [De Giorgi 1995]. Even in $n = 2$ dimensions, there are examples of norms ϕ on $\mathbb{R}^2$ for which $\mathcal{M}_{d_\phi}^1$ and $\mathcal{H}_{d_\phi}^1$ are different. Roughly speaking, this can be explained as follows. $\mathcal{M}_{d_\phi}^{n-1}(\partial E)$ is constructed by taking the projections of $B_\phi := \{\phi \leq 1\}$ onto the (one-dimensional) normal spaces to ∂E. On the other hand, $\mathcal{H}_{d_\phi}^{n-1}(\partial E)$ is constructed by taking the intersections of B_ϕ with the $((n-1)$-dimensional) tangent spaces to ∂E. We notice that, in any case, $\mathcal{H}_{d_\phi}^{n-1}$ can be seen as the Minkowski content with respect to another norm ψ.

Beside its geometric interest, our choice of working with $\mathcal{M}_{d_\phi}^{n-1}$ is motivated also by the physics of phase transitions, where it happens that some relevant phenomena are concentrated in a very thin tubular neighborhood of the interface (sometimes called diffuse interface), and lead in the limit to the Minkowski content.

The next step in our approach consists in the definition of the ϕ-mean curvature κ_ϕ^E of ∂E. This concept, which as usual depends also on the immersion of the manifold, is obtained by computing the first variation of $\mathcal{M}_{d_\phi}^{n-1}$ on ∂E, and is identified with the normal velocity of the initial datum $\partial E = \partial E(0)$ under the anisotropic mean curvature flow.

The computation of the first variation of $\mathcal{M}_{d_\phi}^{n-1}$ is rather direct when ∂B_ϕ is smooth and all its principal curvatures are strictly positive (regular case), but becomes more involved in the crystalline case (that is, when B_ϕ a polytope). In this latter situation, assuming that ∂E is polyhedral and has a geometry locally resembling the geometry of ∂B_ϕ, we compute κ_ϕ^E on its facets, by applying a suitable minimization principle. We then discuss the problem of characterizing ϕ-calibrable facets of ∂E in $n = 3$ dimensions, that is, those facets $F \subset \partial E$ for which κ_ϕ^E is constant on the relative interior int(F) of F.

We conclude by recalling that (a scalar multiple of) B_ϕ is a solution of the so-called isoperimetric problem, that is, the problem of minimizing $\mathcal{M}_{d_\phi}^{n-1}(\partial E)$ among all finite perimeter sets E with $\mathcal{H}_{d_\phi}^n(E)$ fixed. This is in agreement with the fact that ∂B_ϕ has constant ϕ-mean curvature, precisely equal to $n - 1$.

References

[Allen and Cahn 1979] S. M. Allen and J. W. Cahn, "A macroscopic theory for antiphase boundary motion and its application to antiphase domain coarsening", *Acta Metall. Mater.* **27** (1979), 1085–1095.

[Almgren and Taylor 1995] F. J. Almgren and J. E. Taylor, "Flat flow is motion by crystalline curvature for curves with crystalline energies", *J. Differential Geom.* **42** (1995), 1–22.

[Almgren et al. 1993] F. J. Almgren, J. E. Taylor and L. Wang, "Curvature-driven flows: a variational approach", SIAM J. Control Optim. **31** (1993), 387–437.

[Altschuler et al. 1995] S. Altschuler, S. Angenent and Y. Giga, "Mean curvature flow through singularities for surfaces of rotation", *J. Geom. Anal.* **5** (1995), 293–358.

[Amar et al. 1998] M. Amar, G. Bellettini and S. Venturini, "Integral representation of functionals defined on curves of $W^{1,p}$", *Proc. Roy. Soc. Edinb.* **128A** (1998), 193–217.

[Ambrosio 1999] L. Ambrosio, "Lecture notes on geometric evolution problems, distance function and viscosity solutions", pp. 5–94 in *Calculus of variations and partial differential equations: topics on geometrical evolution problems and degree theory*, edited by G. Buttazzo et al., Springer, Berlin, 1999.

[Ambrosio et al. 2000] L. Ambrosio, N. Fusco and D. Pallara, *Functions of bounded variation and free discontinuity problems*, Clarendon Press, Oxford, 2000.

[Ambrosio and Kirchheim 2000] L. Ambrosio and B. Kirchheim, "Currents in metric spaces", *Acta Math.* **185** (2000), 1–80.

[Ambrosio et al. 2002] L. Ambrosio, M. Novaga and E. Paolini, "Some regularity results for minimal crystals", *ESAIM Control Optim. Calc. Var.* **8** (2002), 69–103.

[Angenent 1991a] S. Angenent, "On the formation of singularities in the curve shortening flow", *J. Differential Geom.* **33** (1991) 601–633.

[Angenent 1991b] S. Angenent, "Parabolic equations for curves and surfaces Part II. Intersections, blow-up and generalized solutions", *Ann. of Math.* **133** (1991), 171–215.

[Angenent et al. 1995] S. Angenent, D. L. Chopp and T. Ilmanen, "A computed example of nonuniqueness of mean curvature flow in $\mathbb{R}^3$", *Comm. Partial Differential Equations* **20** (1995), 1937–1958.

[Anzellotti 1983] G. Anzellotti, "Pairings between measures and bounded functions and compensated compactness", *Ann. Mat. Pura Appl.* **135** (1983), 293–318.

[Bao et al. 2000] D. Bao, S.-S. Chern, and Z. Shen, *An introduction to Riemann–Finsler geometry*, Graduate Texts in Mathematics **200**, Springer, New York, 2000.

[Barles et al. 1993] G. Barles, H.-M. Soner and P. E. Souganidis, "Front propagation and phase field theory", *SIAM J. Control Optim.* **31** (1993), 439–469.

[Bellettini and Fragalà 2002] G. Bellettini and I. Fragala, "Elliptic approximations of prescribed mean curvature surfaces in Finsler geometry", *Asymptot. Anal.* **22** (2002), 87–111.

[Bellettini and Novaga 1998] G. Bellettini and M. Novaga, "Approximation and comparison for non-smooth anisotropic motion by mean curvature in $\mathbb{R}^N$", *Math. Mod. Meth. Appl. Sc.* **10** (2000), 1–10.

[Bellettini and Paolini 1994] G. Bellettini and M. Paolini, "Two examples of fattening for the curvature flow with a driving force", *Atti Accad. Naz. Lincei Cl. Sci. Fis. Mat. Natur. Rend. Lincei* (9) Mat. **5** (1994), 229–236.

[Bellettini and Paolini 1995] G. Bellettini and M. Paolini, "Some results on minimal barriers in the sense of De Giorgi applied to driven motion by mean curvature", *Rend. Atti Acc. Naz. Sci. XL Mem. Mat.* **XIX** (1995), 43–67. Errata in **XXVI** (2002), 161–165.

[Bellettini and Paolini 1996] G. Bellettini and M. Paolini, "Anisotropic motion by mean curvature in the context of Finsler geometry", *Hokkaido Math. J.* **25** (1996), 537–566.

[Bellettini et al. 1996] G. Bellettini and M. Paolini and S. Venturini, "Some results on surface measures in Calculus of Variations", *Ann. Mat. Pura Appl.* (4) **IV** (1996), 329–359.

[Bellettini et al. 1999] G. Bellettini, M. Novaga and M. Paolini, "Facet-breaking for three-dimensional crystals evolving by mean curvature", *Interfaces Free Bound.* **1** (1999), 39–55.

[Bellettini et al. 2000] G. Bellettini, R. Goglione and M. Novaga, "Approximation to driven motion by crystalline curvature in two dimensions", *Adv. Math. Sci. Appl.* **10** (2000), 467–493.

[Bellettini et al. 2001a] G. Bellettini, M. Novaga and M. Paolini, "On a crystalline variational problem, part I: first variation and global L^∞-regularity", *Arch. Rational Mech. Anal.* **157** (2001), 165–191.

[Bellettini et al. 2001b] G. Bellettini, M. Novaga and M. Paolini, "On a crystalline variational problem, part II: BV-regularity and structure of minimizers on facets", *Arch. Rational Mech. Anal.* **157** (2001), 193–217.

[Bellettini et al. 2001c] G. Bellettini, M. Novaga and M. Paolini, "Characterization of facets–breaking for nonsmooth mean curvature flow in 3D in the convex case", *Interfaces Free Bound.* **3** (2001), 415–446.

[Bellettini et al. 2002] G. Bellettini, V. Caselles and M. Novaga, "The total variation flow in $\mathbb{R}^n$", *J. Diff. Equations* **184** (2002), 475–525.

[Bellettini et al. 2003] G. Bellettini, V. Caselles and M. Novaga, "Explicit solutions of the eigenvalue problem $-\operatorname{div}(Du/|Du|) = u$", preprint, Univ. Pisa, 2003.

[Brakke 1978] K. A. Brakke, *The motion of a surface by its mean curvature*, Princeton Univ. Press, Princeton, 1978.

[Brezis 1973] H. Brezis, *Opérateurs maximaux monotones et semi-groupes de contractions dans les espaces de Hilbert*, North-Holland, Amsterdam, 1973.

[Busemann 1947] H. Busemann, "Intrinsic area", *Ann. Math.* **48** (1947), 234–267.

[Busemann 1949] H. Busemann, "A theorem on convex bodies of the Brunn–Minkowski type", *Proc. Nat. Acad. Sci. USA* **35** (1949), 27–31.

[Cahn and Hoffman 1974] J. W. Cahn and D. W. Hoffman, "A vector thermodynamics for anisotropic interfaces, 2: curved and faceted surfaces", *Acta Metall. Mater.* **22** (1974), 1205–1214.

[Cahn et al. 1992] J. W. Cahn, C. A. Handwerker and J. E. Taylor, "Geometric models of crystal growth", *Acta Metall. Mater.* **40** (1992), 1443–1474.

[Cahn et al. 1993] J. W. Cahn and C. A. Handwerker, "Equilibrium geometries of anisotropic surfaces and interfaces", *Mat. Sci. Eng.* **A162** (1993), 83–95.

[Chen et al. 1991] Y.-G. Chen, Y. Giga and S. Goto, "Uniqueness and existence of viscosity solutions of generalized mean curvature flow equations", *J. Differential Geom.* **3** (1991), 749–786.

[De Cecco and Palmieri 1993] G. De Cecco and G. Palmieri, "Distanza intrinseca su una varietà finsleriana di Lipschitz", *Rend. Atti Acc. Naz. Sci. XL Mem. Mat.* (5) **17** (1993), 129–151.

[De Cecco and Palmieri 1995] G. De Cecco and G. Palmieri, "Lip manifolds: from metric to Finslerian structure", *Math. Z.* **218** (1995), 223–237.

[De Giorgi 1989] E. De Giorgi, "Su alcuni problemi comuni all'Analisi e alla Geometria", *Note di Matematica*, *IX Suppl.*, (1989), 59–71.

[De Giorgi 1990] E. De Giorgi, "Conversazioni di matematica", *Quaderni Dip. Mat. Univ. Lecce* **2** (1990).

[De Giorgi 1993] E. De Giorgi, "Congetture riguardanti barriere, superfici minime, movimenti secondo la curvatura media", manuscript, Lecce, November 4, 1993.

[De Giorgi 1994] E. De Giorgi, "Barriers, boundaries, motion of manifolds", Conference held at Department of Mathematics of Pavia, March 18, 1994.

[De Giorgi 1995] E. De Giorgi, Problema di Plateau generale e funzionali geodetici, *Atti Sem. Mat. Fis. Univ. Modena* **43** (1995), 285–292.

[Ecker and Huisken 1989] K. Ecker and G. Huisken, "Mean curvature flow of entire graphs", *Ann. of Math.* (2) **130** (1989), 453–471.

[Ecker and Huisken 1991] K. Ecker and G. Huisken, "Interior estimates for hypersurfaces moving by mean curvature", *Invent. Math.* **103** (1991), 547–569.

[Evans and Spruck 1991] L. C. Evans and J. Spruck, "Motion of level sets by mean curvature I", *J. Differential Geom.* **33** (1991), 635–681.

[Evans and Spruck 1992a] L. C. Evans and J. Spruck, "Motion of level sets by mean curvature II", *Trans. Amer. Math. Soc.* **330** (1992), 321–332.

[Evans and Spruck 1992b] L. C. Evans and J. Spruck, "Motion of level sets by mean curvature III", *J. Geom. Anal.* **2** (1992), 121–160.

[Evans and Spruck 1995] L. C. Evans and J. Spruck, "Motion of level sets by mean curvature IV", *J. Geom. Anal.* **5** (1995), 77–114.

[Federer 1969] H. Federer, *Geometric measure theory*, Springer, Berlin, 1969.

[Fierro and Paolini 1996] F. Fierro and M. Paolini, "Numerical evidence of fattening for the mean curvature flow", *Math. Models Methods Appl. Sci.* **6** (1996), 793–813.

[Fukui and Giga 1996] T. Fukui and Y. Giga, "Motion of a graph by nonsmooth weighted curvature", pp. 47–56 in *World Congress of Nonlinear Analysis '92*, edited by V. Lakshmikantham, de Gruyter, Berlin, 1996.

[Gage 1994] M. E. Gage, "Evolving plane curves by curvature in relative geometries, II", *Duke Math. J.* **75** (1994), 79–98.

[Gage and Hamilton 1986] M. E. Gage and R. S. Hamilton, "The heat equation shrinking convex plane curves", *J. Differential Geom.* **23** (1986), 69–96.

[Giga and Giga 1997] M.-H. Giga and Y. Giga, "Remarks on convergence of evolving graphs by nonlocal curvature", pp. 99–116 in *Progress in partial differential equations* (Pont-á-Mousson, 1997), vol. 1, edited by H. Amann et al., Pitman research notes in mathematics **383**, Longman, Harlow (Essex), 1998.

[Giga and Giga 1998a] M.-H. Giga and Y. Giga, "A subdifferential interpretation of crystalline motion under nonuniform driving force", pp. 276–287 in *Dynamical systems and differential equations* (Springfield, MO, 1996), vol. 1, edited by Wenxiong Chen and Shouchuan Hu, *Discrete Contin. Dynam. Systems*, added volume I (1998).

[Giga and Giga 1998b] M.-H. Giga and Y. Giga, "Evolving graphs by singular weighted curvature", *Arch. Rational Mech. Anal.* **141** (1998), 117–198.

[Giga and Giga 1999] M.-H. Giga and Y. Giga, "Stability of evolving graphs by nonlocal weighted curvature", *Comm. Partial Differential Equations* **24** (1999), 109–184.

[Giga and Giga 2000] M.-H. Giga and Y. Giga, "Crystalline and level set flow-convergence of a crystalline algorithm for a general anisotropic curvature flow in the plane", pp. 64–79 in *Free boundary problems: theory and applications* (Chiba, 1999), vol. I, edited by N. Kenmochi, GAKUTO Intern. Ser. Math. Appl. **13**, Gakkōtosho, Tokyo, 2000.

[Giga and Goto 1992] Y. Giga and s. Goto, "Geometric evolutions of phase-boundaries", pp. 51–66 in *On the evolution of phase-boundaries*, edited by M. Gurtin and Mc Fadden, IMA Vols. Math. Appl. **43**, Springer, Berlin, 1992.

[Giga and Gurtin 1996] Y. Giga and M. E. Gurtin, "A comparison theorem for crystalline evolutions in the plane", *Quart. Appl. Math.* **54** (1996), 727–737.

[Giga et al. 1991] Y. Giga, S. Goto, H. Ishii and M.-H. Sato, "Comparison principle and convexity preserving properties for singular degenerate parabolic equations on unbounded domains", *Indiana Univ. Math. J.* **40** (1991), 443–470.

[Giga et al. 1998] Y. Giga, M. E. Gurtin and J. Matias, "On the dynamics of crystalline motion", *Japan J. Indust. Appl. Math.* **15** (1998), 7–50.

[Girao 1995] P. M. Girao, "Convergence of a crystalline algorithm for the motion of a simple closed convex curve by weighted curvature", *SIAM J. Numer. Anal.* **32** (1995), 886–899.

[Girao and Kohn 1994] P. M. Girao and R. V. Kohn, "Convergence of a crystalline algorithm for the heat equation in one dimension and for the motion of a graph", *Numer. Math.* **67** (1994), 41–70.

[Giusti 1984] E. Giusti, *Minimal surfaces and functions of bounded variation*, Birkhäuser, Boston, 1984.

[Grayson 1987] M. A. Grayson, "The heat equation shrinks embedded plane curves to round points", *J. Differential Geom.* **26** (1987), 285–314.

[Hoffman and Cahn 1972] D. W. Hoffman and J. W. Cahn, "A vector thermodynamics for anisotropic surfaces, I: fundamentals and application to plane surface junctions", *Surf. Sci.* **31** (1972), 368–388.

[Huisken 1984] G. Huisken, "Flow by mean curvature of convex surfaces into spheres", *J. Differential Geom.* **20** (1984), 237–266.

[Ilmanen 1992] T. Ilmanen, "Generalized flow of sets by mean curvature on a manifold", *Indiana Univ. Math. J.* **41** (1992), 671–705.

[Ilmanen 1993a] T. Ilmanen, "The level set flow on a manifold", pp. 193–204 in *Differential geometry: partial differential equations on manifolds* (Los Angeles, 1990), edited by Robert Greene and S.-T. Yau, Proc. Sympos. Pure Math. **54**, Part 1, Amer. Math. Soc., Providence, RI, 1993.

[Ilmanen 1993b] T. Ilmanen, "Convergence of the Allen–Cahn equation to Brakke's motion by mean curvature", *J. Differential Geom.* **38** (1993), 417–461.

[Ilmanen 1994] T. Ilmanen, "Elliptic regularization and partial regularity for motion by mean Curvature", Mem. Amer. Math. Soc. **108** (1994).

[Ishii and Soner 1999] K. Ishii and H. M. Soner, "Regularity and convergence of crystalline motion", *SIAM J. Math. Anal.* **30** (1999), 19–37.

[Mugnai 2003] L. Mugnai, "Relaxation and variational approximation for curvature-depending functionals in 2D", Ph.D. Thesis, Univ. of Pisa, 2003.

[Osher and Sethian 1988] S. Osher and J. A. Sethian, "Fronts propagating with curvature dependent speed: algorithms based on Hamilton–Jacobi formulations", *J. Comput. Phys.* **79** (1988), 12–49.

[Paolini and Pasquarelli 2000] M. Paolini and F. Pasquarelli, "Numerical simulations of crystalline curvature flow in 3D by interface diffusion", pp. 376–389 in: *Free boundary problems: theory and applications* (Chiba, 1999), vol. II, edited by N. Kenmochi, GAKUTO Intern. Ser. Math. Sci. Appl. **14** Gakkōtosho, Tokyo, 2000.

[Paolini and Verdi 1992] M. Paolini and C. Verdi, "Asymptotic and numerical analyses of the mean curvature flow with a space-dependent relaxation parameter", *Asymptot. Anal.* **5** (1992), 553–574.

[Rockafellar 1972] R. T. Rockafellar, *Convex analysis*, Princeton Univ. Press, Princeton, 1972.

[Roosen and Taylor 1994] A. R. Roosen and J. Taylor, "Modeling crystal growth in a diffusion field using fully faceted interfaces", *J. Comput. Phys.* **114** (1994), 444-451.

[Rybka 1997] P. Rybka, "A crystalline motion: uniqueness and geometric properties", *SIAM J. Appl. Math.* **57** (1997), 53–72.

[Schneider 1993] R. Schneider, *Convex Bodies: the Brunn–Minkowski theory*, Encyclopedia of Math. and its Appl. **4**, Cambridge Univ. Press, New York, 1993.

[Shen 1998] Z. Shen, "On Finsler geometry of submanifolds", *Math. Ann.* **311** (1998), 549–576.

[Shen 2001] Z. Shen, *Lectures on Finsler geometry*, World Scientific, Singapore, 2001.

[Soner 1993] H.-M. Soner, "Motion of a set by the curvature of its boundary", *J. Differential Equations* **101** (1993), 313–372.

[Soner and Souganidis 1993] H.-M. Soner and P. E. Souganidis, "Singularities and uniqueness of cylindrically symmetric surfaces moving by mean curvature", *Comm. Partial Differential Equations* **18** (1993), 859–894.

[Spohn 1993] H. Spohn, "Interface motion in models with stochastic dynamics", *J. Stat. Phys.* **71** (1993), 1081–1132.

[Stancu 1996] A. Stancu, "Uniqueness of self-similar solutions for a crystalline flow", *Indiana Univ. Math. J.* **45** (1996), 1157–1174.

[Taylor 1978] J. E. Taylor, "Crystalline variational problems", *Bull. Amer. Math. Soc. (N. S.)* **84** (1978), 568–588.

[Taylor 1986] J. E. Taylor, "Complete catalog of minimizing embedded crystalline cones", pp. 379–403 in *Geometric measure theory and the calculus of variations* (Arcata, CA, 1984), edited by William K. Allard and Frederick J. Almgren, Jr., Proc. Symposia Math. **44**, Amer. Math. Soc., Providence, 1986.

[Taylor 1991] J. E. Taylor, "Constructions and conjectures in crystalline nondifferential geometry", pp. 321–336 in *Differential geometry: a symposium in honour of Manfredo do Carmo*, edited by Blaine Lawson and Keti Tenenblat, Pitman monographs and surveys in pure and applied mathematics **52**, Longman, Harlow (Essex), 1991.

[Taylor 1992] J. E. Taylor, "Mean curvature and weighted mean curvature", *Acta Metall. Mater.* **40** (1992), 1475–1485.

[Taylor 1993] J. E. Taylor, "Motion of curves by crystalline curvature, including triple junctions and boundary points", pp. 417–438 in *Differential Geometry: partial differential equations on manifolds* (Los Angeles, 1990), edited by Robert Greene and S.-T. Yau, Proc. Sympos. Pure Math. **54**, Part 1, Amer. Math. Soc., Providence, RI, 1993.

[Thompson 1996] A. C. Thompson, *Minkowski geometry*, Encyclopedia of Math. and its Appl. **63**, Cambridge Univ. Press, New York, 1996.

[Venturini 1992] S. Venturini, "Derivation of distance functions on $\mathbb{R}^n$", preprint, 1992.

[White 1995] B. White, "The topology of hypersurfaces moving by mean curvature", *Comm. Anal. Geom.* **3** (1995), 317–333.

[White 2000] B. White, "The size of the singular set in mean curvature flow of mean-convex sets", *J. Amer. Math. Soc.* **13** (2000), 665–695.

[White 2003] B. White, "The nature of singularities in mean curvature flow", *J. Amer. Math. Soc.* **16** (2003), 123–138.

[Yunger 1998] J. Yunger, "Facet stepping and motion by crystalline curvature", Ph.D. thesis, Rutgers University, 1998.

GIOVANNI BELLETTINI
DIPARTIMENTO DI MATEMATICA
UNIVERSITÀ DI ROMA "TOR VERGATA"
VIA DELLA RICERCA SCIENTIFICA
00133 ROMA
ITALY
    belletti@mat.uniroma2.it

Riemann–Finsler Geometry
MSRI Publications
Volume **50**, 2004

Finsler Geometry on Complex Vector Bundles

TADASHI AIKOU

CONTENTS

1. Introduction

A Finsler metric of a manifold or vector bundle is defined as a smooth assignment for each base point a norm on each fibre space, and thus the class of Finsler metrics contains Riemannian metrics as a special sub-class. For this reason, Finsler geometry is usually treated as a generalization of Riemannian geometry. In fact, there are many contributions to Finsler geometry which contain Riemannian geometry as a special case (see e.g., [Bao et al. 2000], [Matsumoto 1986], and references therein).

On the other hand, we can treat Finsler geometry as a special case of Riemannian geometry in the sense that Finsler geometry may be developed as differential geometry of fibred manifolds (e.g., [Aikou 2002]). In fact, if a Finsler metric in the usual sense is given on a vector bundle, then it induces a Riemannian inner product on the vertical subbundle of the total space, and thus, Finsler geometry is translated to the geometry of this Riemannian vector bundle.

It is natural to question why we need Finsler geometry at all. To answer this question, we shall describe a few applications of complex Finsler geometry to some subjects which are impossible to study via Hermitian geometry.

The notion of complex Finsler metric is old and goes back at least to Carathéodory who introduced the so-called *Carathéodory metric*. The geometry of complex Finsler manifold, via tensor analysis, was started by [Rizza 1963], and afterwards, the connection theory on complex Finsler manifolds has been developed by [Rund 1972], [Icijyō 1994], [Fukui 1989], and [Cao and Wong 2003], etc..

Recently, from the viewpoint of the geometric theory of several complex variables, complex Finsler metric has become an interesting subject. In particular, an intrinsic metric on a complex manifold, namely the *Kobayashi metric*, is a holomorphic invariant metric on a complex manifold. The Kobayashi metric is, by its definition, a pseudo Finsler metric. However, by the fundamental work of [Lempert 1981], the Kobayashi metric on a smoothly bounded strictly convex domain in $\mathbb{C}^n$ is a smooth pseudoconvex Finsler metric.

The interest in complex Finsler geometry also arises from the study of holomorphic vector bundles. The characterization of ample (or negative) vector bundles due to Kobayashi [Kobayashi 1975] shows the importance of Finsler geometry. In fact, he has proved that E is ample if and only if its dual E^* admits a "negatively curved" pseudoconvex Finsler metric (Theorem 3.2). The meaning of the term "negatively curved" is defined by using the curvature tensor of the Finsler connection on a Finsler bundle (E, F).

Another example of interest in complex Finsler geometry arises from the geometry of *geometrically ruled surfaces* $\mathcal{X}$. A geometrically ruled surface $\mathcal{X}$ is, by definition (see [Yang 1991]), an algebraic surface with a holomorphic projection $\phi : \mathcal{X} \to M$, M a compact Riemann surface, such that each fibre is isomorphic to the complex projective line $\mathbb{P}^1$. Every geometrically ruled surface is isomorphic to $\mathbb{P}(E)$ for some holomorphic vector bundle $\pi : E \to M$ of rank$(E) = 2$. Then, every geometrically ruled surface $\mathcal{X} = \mathbb{P}(E)$ is also a compact Kähler manifold by Lemma 6.37 in [Shiffmann and Sommese 1985], and any Kähler metric $g_{\mathcal{X}}$ on $\mathcal{X}$ induces a Finsler metric F, which is *not* a Hermitian metric in general, on the bundle E. Thus the geometry of $(\mathcal{X}, g_{\mathcal{X}})$ is translated to the geometry of the Finsler bundle (E, F).

In general, an algebraic curve (or polarized manifold) $\varphi : \mathcal{X} \to \mathbb{P}^N$ has a Kähler metric $\omega_{\mathcal{X}} = \varphi^* \omega_{FS}$ induced from the Fubini–Study metric ω_{FS} on $\mathbb{P}^N$. An interesting subject in complex geometry is to investigate how metrics of this kind are related to constant curvature metrics, and moreover, it is interesting to investigate how constant scalar curvature metrics should be related to algebro-geometric stability. LeBrun [LeBrun 1995] has investigated minimal ruled surfaces $\mathcal{X} = \mathbb{P}(E)$ over a compact Riemann surface M of genus $g(M) \geq 2$ with constant scalar curvature. He showed that, roughly speaking, $\mathcal{X}$ admits such a Kähler metric $g_{\mathcal{X}}$ if and only if the bundle E is semi-stable in the sense of Mumford–Takemoto. By the statement above, an arbitrary Kähler metric on a minimal ruled surface $\mathcal{X}$ determines a Finsler metric F on E by the identity $\omega_{\mathcal{X}} = \sqrt{-1} \, \partial \bar{\partial} \log F$ for the Kähler form $\omega_{\mathcal{X}}$. The geometry of such a minimal ruled surface can also be investigated by the study of the Finsler bundle (E, F) (see [Aikou 2003b]).

In this article, we shall report on the geometry of complex vector bundles with Finsler metrics, i.e., *Finsler bundles*. Let F be a Finsler metric on a holomorphic vector bundle $\pi : E \to M$ over a complex manifold M. The geometry of a Finsler bundle (E, F) is the study of the vertical bundle $V_E = \ker \pi_*$ with a Hermitian metric G_{V_E} induced from the given Finsler metric.

The main tool of the investigation in Finsler geometry is the *Finsler connection*. The connection is a unique one on the Hermitian bundle (V_E, g_{V_E}), satisfying some geometric condition (see definition below). Although it is natural to investigate (V_E, g_{V_E}) by using the Hermitian connection of (V_E, g_{V_E}), it is convenient to use the Finsler connection for investigating some special Finsler metrics. For example, the flatness of the Hermitian connection of (V_E, g_{V_E}) implies that the Finsler metric F is reduced to a flat Hermitian metric. However, if the Finsler connection is flat, then the metric F belongs to an important class, the so-called *locally Minkowski metrics* (we simply call these special metrics *flat Finsler metrics*). If the Finsler connection is induced from a connection on E, then the metric F belongs to another important class, the so-called *Berwald metrics* (sometimes a Berwald metric is said to be *modeled on a Minkowski space*). In this sense, the big difference between Hermitian geometry and Finsler geometry is the connection used for the investigation of the bundle (V_E, g_{V_E}).

2. Ampleness

2.1. Ample line bundles. Let L be a holomorphic line bundle over a compact complex manifold M. We denote by $\mathcal{O}(L)$ the sheaf of germs of holomorphic sections of L. Since M is compact, $\dim_{\mathbb{C}} H^0(M, \mathcal{O}(L))$ is finite. Let $\{f^0, \ldots, f^N\}$ be a set of linearly independent sections of L, from the complex vector space of global sections. The vector space spanned by these sections is called a *linear system* on M. If the vector space consists of all global sections of L, it is called a *complete linear system* on M. Using these sections $\{f^0, \ldots, f^N\}$, a rational map $\varphi_{|L|} : M \to \mathbb{P}^N$ is defined by

$$\varphi_{|L|}(z) = [f^0(z) : \cdots : f^N(z)]. \tag{2.1}$$

This rational map is defined on the open set in M which is complementary to the common zero-set of the sections f^i $(0 \leq i \leq N)$. It is verified that the rational map $\tilde{\varphi}_{|L|}$ obtained from another basis $\{\tilde{f}^0, \ldots, \tilde{f}^N\}$ is transformed by an automorphism of $\mathbb{P}^N$.

DEFINITION 2.1. A line bundle L over M is said to be *very ample* if the rational map $\varphi_{|L|} : M \to \mathbb{P}^N$ determined by its complete linear system $|L|$ is an embedding. L is said to be *ample* if there exists some integer $m > 0$ such that $L^{\otimes m}$ is very ample.

Let L be a very ample line bundle over a compact complex manifold M, and $\{f^0, \ldots, f^N\}$ a basis of $H^0(M, \mathcal{O}(L))$ which defines an embedding $\varphi_{|L|} : M \to \mathbb{P}^N$. Embedding M into $\mathbb{P}^N$, we may consider the line bundle as the hyperplane bundle over $M \subset \mathbb{P}^N$. We define an open covering $\{U_{(j)}\}$ of M by $U_{(j)} = \{z \in M : f^j(z) \neq 0\}$. With respect to this covering, the local trivialization φ_j over $U_{(j)} \times \mathbb{C}$ is given by $\varphi_j(f^i) = (z^\alpha_{(j)}, f^i_{(j)})$. The transition cocycle $\{l_{jk} : U_{(j)} \cap U_{(k)} \to \mathbb{C}^*\}$ is given by

$$l_{jk}(z) = \frac{f^i_{(k)}}{f^i_{(j)}}. \tag{2.2}$$

Let $\{h_{jk}\}$ be the transition cocycle of the hyperplane bundle $\mathbb{H}$ with respect to the standard covering $\{U_{(j)}\}$ of $\mathbb{P}^N$. Then, $\{h_{jk}\}$ is given by $h_{jk} = \xi^k/\xi^j$ in terms of the homogeneous coordinate system $[\xi^0 : \cdots : \xi^N]$ of $\mathbb{P}^N$. Since (2.2) implies

$$h_{jk} \circ \varphi_{|L|} = \frac{f^i_{(k)}}{f^i_{(j)}} = l_{jk},$$

we obtain $L = \varphi^*_{|L|} \mathbb{H}$.

LEMMA 2.1. *Let L be a very ample line bundle over a complex manifold M. Then L is isomorphic to the pullback bundle $\varphi^*_{|L|} \mathbb{H}$ of the hyperplane bundle $\mathbb{H}$ over the target space $\mathbb{P}^N$ of $\varphi_{|L|}$.*

EXAMPLE 2.1. (1) The hyperplane bundle $\mathbb{H}$ over $\mathbb{P}^N$ is very ample.

(2) Let E be a holomorphic line bundle and L an ample line bundle over a compact complex manifold M. For some sufficiently large integer k, the line bundle $E \oplus L^{\otimes k}$ is very ample (see [Griffith and Harris 1978, p. 192]).

As we can see from the above, it is an algebro-geometric issue to determine whether a holomorphic line bundle is ample or not. However, the Kodaira embedding theorem provides a differential geometric way to check ampleness; see Theorem 2.1 and Proposition 2.1 below. The key idea is to relate ampleness to the notion of positivity, defined as follows.

DEFINITION 2.2. A holomorphic line bundle L is said to be *positive* if its Chern class $c_1(L) \in H^2(M, \mathbb{R})$ is represented by a positive real $(1,1)$-form. A holomorphic line bundle L is said to be *negative* if its dual L^* is positive. Since $c_1(L^*) = -c_1(L)$, the holomorphic line bundle L is negative if $c_1(L)$ is represented by a negative real $(1,1)$-form.

A Hermitian metric g on L is given by the family $\{g_{(j)}\}$ of local positive functions $g_{(j)} : U_{(j)} \to \mathbb{R}$, satisfying $g_{(k)} = |l_{jk}|^2 g_{(j)}$ on $U_{(j)} \cap U_{(k)}$ for the transition cocycle $\{l_{jk}\}$ of L. Since l_{jk} are holomorphic, we have $\bar{\partial}\partial \log g_{(j)} = \bar{\partial}\partial \log g_{(k)}$, and thus $\{\bar{\partial}\partial \log g_{(j)}\}$ defines a global $(1,1)$-form on M, which will be denoted by $\bar{\partial}\partial \log g$, and is just the curvature form of (L, g). The Chern form $c_1(L, g)$ defined by $c_1(L, g) = \sqrt{-1}\, \bar{\partial}\partial \log g$ is a representative of the Chern class $c_1(L)$.

By the definition above, a holomorphic line bundle L is positive if and only if L admits a Hermitian metric g whose Chern form $c_1(L, g)$ is positive-definite. A compact complex manifold M is called a *Hodge manifold* if there exists a positive line bundle L over M. If M is a Hodge manifold, then there exists a Hermitian line bundle (L, g) whose Chern form $c_1(L, g)$ is positive-definite, and thus $c_1(L, g)$ defines a Kähler metric on M. Consequently, every Hodge manifold is Kähler.

The hyperplane bundle $\mathbb{H}$ over $\mathbb{P}^N$ is positive. In fact, if we define a function $g_{(j)}$ on $V_{(j)} = \{[\xi^0 : \cdots : \xi^N] \in \mathbb{P}^N : \xi^j \neq 0\}$ by

$$g_{(j)} = \frac{|\xi^j|^2}{\sum |\xi^k|^2}, \tag{2.3}$$

the family $\{g_{(j)}\}_{j=0,\dots,N}$ satisfies $g_{(k)} = |h_{jk}|^2 g_{(j)}$ on $V_{(j)} \cap V_{(k)}$, and thus it determines a Hermitian metric $g_{\mathbb{H}}$ on $\mathbb{H}$. Then we have

$$c_1(\mathbb{H}) = \left[\frac{\sqrt{-1}}{2\pi} \bar{\partial}\partial \log g_{\mathbb{H}} \right] > 0. \tag{2.4}$$

The closed real $(1,1)$-form representing $c_1(\mathbb{H})$ induces a Kähler metric on $\mathbb{P}^N$, which is called the *Fubini–Study metric* g_{FS} with the Kähler form

$$\omega_{FS} = \sqrt{-1}\partial\bar{\partial}\log \|\xi\|^2,$$

where we put $\|\xi\|^2 = \sum |\xi^i|^2$.

The following well-known theorem shows that every Hodge manifold M is algebraic, i.e., M is holomorphically embedded in a projective space $\mathbb{P}^N$.

THEOREM 2.1 (KODAIRA'S EMBEDDING THEOREM). *Let L be a holomorphic line bundle over a compact complex manifold M. If L is positive, then it is ample, i.e., there exists some integer $n_0 > 0$ such that for all $m \geq n_0$ the map $\varphi_{|L^{\otimes m}|} : M \to \mathbb{P}^N$ is a holomorphic embedding.*

Conversely, we suppose that L is ample. Then, by definition, there exists a basis $\{f^0, \dots, f^N\}$ of $H^0(M, \mathcal{O}(L^{\otimes m}))$ such that $\varphi_{|L^{\otimes m}|} : M \to \mathbb{P}^N$ defined by (2.1) is an embedding. By Lemma 2.1, the line bundle $L^{\otimes m}$ is identified with $\varphi_{|L^{\otimes m}|}^* \mathbb{H}$. Thus $L^{\otimes m}$ admits a Hermitian metric $g = \varphi_{|L^{\otimes m}|}^* g_{\mathbb{H}}$, and $c_1(L^{\otimes m})$ is given by

$$c_1(L^{\otimes m}) = mc_1(L) = \left[\frac{\sqrt{-1}}{2\pi} \bar{\partial}\partial \log g \right].$$

Since $\mathbb{H}$ is positive, the $(1,1)$-form $\sqrt{-1}\bar{\partial}\partial \log g$ is positive, and thus

$$c_1(L) = \frac{1}{m} \left[\frac{\sqrt{-1}}{2\pi} \bar{\partial}\partial \log g \right]$$

is positive. Consequently:

PROPOSITION 2.1. *A holomorphic line bundle L over a compact complex manifold M is ample if and only if L is positive.*

Let M be a compact Riemann surface. The integer $g(M)$ defined by

$$g(M) := \dim_{\mathbb{C}} H^1(M, \mathcal{O}_M) = \dim_{\mathbb{C}} H^0(M, \mathcal{O}(K_M))$$

is called the *genus* of M, where $K_M = T_M^*$ is the canonical line bundle over M, and $\mathcal{O}_M$ is the sheaf of germs of holomorphic functions on M. The *degree* of a line bundle L is defined by

$$\deg L = \int_M c_1(L) \in \mathbb{Z}.$$

Applying the *Riemann–Roch theorem*

$$\dim_{\mathbb{C}} H^0(M, \mathcal{O}(L)) - \dim_{\mathbb{C}} H^1(M, \mathcal{O}(L)) = \deg L + 1 - g(M)$$

to the case of $L = K_M$, we have

$$\dim_{\mathbb{C}} H^1(M, \mathcal{O}(K_M)) = \dim_{\mathbb{C}} H^0(M, \Omega^1(K_M^*)) = \dim_{\mathbb{C}} H^0(M, \mathcal{O}_M) = 1,$$

since M is compact. Consequently we have $\deg K_M = 2g(M) - 2$, and the Euler characteristic $\chi(M)$ is given by

$$\chi(M) = \int_M c_1(T_M) = -\deg K_M = 2 - 2g(M).$$

By the uniformisation theorem (e.g., Theorem 4.41 in [Jost 1997]), any compact Riemann surface M is determined completely by its genus $g(M)$. If $g(M) = 0$, then M is isomorphic to the Riemannian sphere $S^2 \cong \mathbb{P}^1$ and its holomorphic tangent bundle T_M is ample. In the case of $g(M) = 1$, then M is isomorphic to a torus $T = \mathbb{C}/\Lambda$, where Λ is a module over $\mathbb{Z}$ of rank two, and T_M is trivial. In the last case of $g(M) \geq 2$, it is well-known that M is *hyperbolic*, i.e., M admits a Kähler metric of negative constant curvature, and T_M is negative since $c_1(T_M) < 0$.

In the case of $\dim_{\mathbb{C}} M \geq 2$, Hartshone's conjecture ("If the tangent bundle T_M is ample, then M is bi-holomorphic to the projective space $\mathbb{P}^n$") was solved affirmatively by an algebro-geometric method ([Mori 1979]). Then, it is natural to investigate complex manifolds with negative tangent bundles. We next discuss the *negativity* and *ampleness* of holomorphic vector bundles.

2.2. Ample vector bundles. Let $\pi : E \to M$ be a holomorphic vector bundle of $\mathrm{rank}(E) = r + 1 \; (\geq 2)$ over a compact complex manifold M, and $\phi : \mathbb{P}(E) = E^\times / \mathbb{C}^\times \to M$ the *projective bundle* associated with E. Here and in the sequel, we put $E^\times = E - \{0\}$ and $\mathbb{C}^\times = \mathbb{C} - \{0\}$. We also denote by $\mathbb{L}(E)$ the *tautological line bundle* over $\mathbb{P}(E)$, i.e.,

$$\mathbb{L}(E) = \{(V, v) \in \mathbb{P}(E) \times E \mid v \in V\}.$$

The dual line bundle $\mathbb{H}(E) = \mathbb{L}(E)^*$ is called the *hyperplane bundle* over $\mathbb{P}(E)$.

Since $\mathbb{L}(E)$ is obtained from E by blowing up the zero section of E to $\mathbb{P}(E)$, the manifold $\mathbb{L}(E)^\times$ is biholomorphic to $E^\times$. This biholomorphism is given by the holomorphic map

$$\tau : E^\times \ni v \to ([v], v) \in \mathbb{P}(E) \times E^\times. \tag{2.5}$$

Then, for a arbitrary Hermitian metric $g_{\mathbb{L}(E)}$ on $\mathbb{L}(E)$, we define the norm $\|v\|_E$ of $v \in E^\times$ by

$$\|v\|_E = \sqrt{g_{\mathbb{L}(E)}(\tau(v))}. \tag{2.6}$$

Extending this definition to the whole of E continuously, we obtain a function $F : E \to \mathbb{R}$ by

$$F(v) = \|v\|_E^2 \tag{2.7}$$

for every $v \in E$. This function satisfies the following conditions.

(F.1) $F(v) \geq 0$, and $F(v) = 0$ if and only if $v = 0$,
(F.2) $F(\lambda v) = |\lambda|^2 F(v)$ for any $\lambda \in \mathbb{C}^\times = \mathbb{C} \backslash \{0\}$,
(F.3) $F(v)$ is smooth outside of the zero-section.

DEFINITION 2.3. Let $\pi : E \to M$ be a holomorphic vector bundle over a complex manifold M. A real valued function $F : E \to \mathbb{R}$ satisfying the conditions $(F1) \sim (F3)$ is called a *Finsler metric* on E, and the pair (E, F) is called a *Finsler bundle*. If a Finsler metric F satisfies, in addition,

(F.4) the real $(1,1)$-form $\sqrt{-1}\,\partial\bar{\partial}F$ is positive-definite on each fibre E_z,

then F is said to be *pseudoconvex*. (Note: it's $\sqrt{-1}\,\partial\bar{\partial}F$, not $\sqrt{-1}\,\partial\bar{\partial}\log F$.)

This discussion shows that any Hermitian metric on $\mathbb{L}(E)$ defines a Finsler metric on E. Conversely, an arbitrary Finsler metric F on E determines a Hermitian metric $g_{\mathbb{L}(E)}$ on $\mathbb{L}(E)$, i.e., we obtain

PROPOSITION 2.2 [Kobayashi 1975]. *There exists a one-to-one correspondence between the set of Hermitian metrics on* $\mathbb{L}(E)$ *and the set of Finsler metrics on* E.

DEFINITION 2.4 [Kobayashi 1975]. A holomorphic vector bundle $\pi : E \to M$ over a compact complex manifold M is said to be *negative* if its tautological line bundle $\mathbb{L}(E) \to \mathbb{P}(E)$ is negative, and E is said to be *ample* if its dual E^* is negative.

The Chern class $c_1(\mathbb{L}(E))$ is represented by the closed real $(1,1)$-form

$$c_1(\mathbb{L}(E), F) = \frac{\sqrt{-1}}{2\pi} \bar{\partial}\partial \log F$$

for a Finsler metric F on E. Thus, E is negative if and only if E admits a Finsler metric F satisfying $c_1(\mathbb{L}(E), F) < 0$. Consequently:

PROPOSITION 2.3. *Let E be a holomorphic vector bundle over a compact complex manifold M. Then E is negative if and only if E admits a Finsler metric F satisfying $\sqrt{-1}\,\bar{\partial}\partial \log F < 0$.*

Given any negative holomorphic vector bundle E over M, we shall construct a pseudoconvex Finsler metric F on E, with $c_1(\mathbb{L}(E), F) < 0$ (see [Aikou 1999] and [Wong 1984]). By definition the line bundle $\mathbb{L}(E)$ is negative, and so $\mathbb{L}(E)^*$ is ample. Hence there exists a sufficiently large positive $m \in \mathbb{Z}$ such that $L := \mathbb{L}(E)^{*\otimes m}$ is very ample. By the definition of very ampleness, we can take $f^0, \ldots, f^N \in H^0(\mathbb{P}(E), L)$ such that

$$\varphi_{|L|} : \mathbb{P}(E) \ni [v] \rightarrow [f^0([v]) : \cdots : f^N([v])] \in \mathbb{P}^N$$

defines a holomorphic embedding $\varphi_{|L|} : \mathbb{P}(E) \rightarrow \mathbb{P}^N$. Then, by Lemma 2.1, we have $L \cong \varphi_{|L|}^* \mathbb{H}$ for the hyperplane bundle $\mathbb{H} \rightarrow \mathbb{P}^N$. Since $c_1(\mathbb{H})$ is given by (2.4), we have $c_1(L) = c_1(\varphi_{|L|}^* \mathbb{H}) = \varphi_{|L|}^* c_1(\mathbb{H}) > 0$, and the induced metric g_L is given by $g_L = \varphi_{|L|}^* g_{\mathbb{H}}$ for the metric $g_{\mathbb{H}}$ on $\mathbb{H}$ defined by (2.3). Since $L = (\mathbb{L}(E)^*)^{\otimes m}$, we have $g_L = g_{\mathbb{L}(E)}^{-m}$, and thus the induced metric $g_{\mathbb{L}(E)}$ on $\mathbb{L}(E)$ is given by

$$g_{\mathbb{L}(E)} = \left[\varphi_{|L|}^* g_{\mathbb{H}}\right]^{-1/m} = \left[\frac{1}{\varphi_{|L|}^* g_{\mathbb{H}}}\right]^{1/m}.$$

Because of (2.3), the metric g_L is locally given by

$$g_{(j)} = \frac{|f^j|^2}{\sum |f^k|^2},$$

and the Finsler metric F on E corresponding to $g_{\mathbb{L}(E)}$ is given by

$$F(v) = \left[(\varphi_{|L|}^* g_{\mathbb{H}})(\tau(v))\right]^{-1/m} = \left[\frac{\sum |f^k([v])|^2}{|f^j([v])|^2}\right]^{1/m} |v^j|^2 \qquad (2.8)$$

for $v = (v^1, \ldots, v^{r+1}) \in E_z$. The Finsler metric F obtained as above satisfies the condition $c_1(\mathbb{L}(E), F) < 0$. The pseudoconvexity of F will be shown by more local computations (see Theorem 3.2).

REMARK 2.1. Every pseudoconvex Finsler metric on a holomorphic vector bundle E is obtained from a pseudo-Kähler metric on $\mathbb{P}(E)$ (Propositions 4.1 and 4.2).

In a later section, we shall show a theorem of Kobayashi's (Theorem 3.2) which characterizes negative vector bundles in terms of the curvature of Finsler metrics. For this purpose, in the next section, we shall discuss the theory of Finsler connections on (E, F).

3. Finsler Connections

Let $\pi : E \to M$ be a holomorphic vector bundle of $\mathrm{rank}(E) = r + 1$ over a complex manifold M. We denote by T_M the holomorphic tangent bundle of M. We also denote by T_E the holomorphic tangent bundle of the total space E. Then we have an exact sequence of holomorphic vector bundles

$$0 \longrightarrow V_E \stackrel{i}{\longrightarrow} T_E \stackrel{\pi_*}{\longrightarrow} \pi^* T_M \longrightarrow 0, \tag{3.1}$$

where $V_E = \ker \pi_*$ is the *vertical subbundle* of T_E. A connection of the bundle $\pi : E \to M$ is a smooth splitting of this sequence.

DEFINITION 3.1. A *connection* of a fibre bundle $\pi : E \to M$ is a smooth V_E-valued $(1,0)$-form $\theta_E \in \Omega^{1,0}(V_E)$ satisfying

$$\theta_E(Z) = Z \tag{3.2}$$

for all $Z \in V_E$. A connection θ_E defines a smooth splitting

$$T_E = V_E \oplus H_E \tag{3.3}$$

of the sequence (3.1), where $H_E \subset T_E$ is a the *horizontal subbundle* defined by $H_E = \ker \theta_E$.

The complex general linear group $GL(r + 1, \mathbb{C})$ acts on E in a natural way. A connection θ_E is called a *linear connection* if the horizontal subspace at each point is $GL(r+1, \mathbb{C})$-invariant. A Hermitian metric on E defines a unique linear connection θ_E.

On the other hand, the multiplier group $\mathbb{C}^\times = \mathbb{C} \backslash \{0\} \cong \{c \cdot I \mid c \in \mathbb{C}^\times\} \subset GL(r + 1, \mathbb{C})$ also acts on the total space E by multiplication $L_\lambda : E \ni v \to \lambda v \in E$ on the fibres for all $v \in E$ and $\lambda \in \mathbb{C}^\times$. In this paper, we assume that a connection θ_E is $\mathbb{C}^\times$-invariant. We denote by $\mathcal{E} \in \mathcal{O}_E(V_E)$ the *tautological section* of V_E generated by the action of $\mathbb{C}^\times$, i.e., $\mathcal{E}$ is defined by

$$\mathcal{E}(v) = (v, v)$$

for all $v \in E$. The invariance of θ_E under the action of $\mathbb{C}^\times$ is equivalent to

$$\mathcal{L}_\mathcal{E} H_E \subset H_E, \tag{3.4}$$

where $\mathcal{L}_\mathcal{E}$ denotes the Lie derivative by $\mathcal{E}$. In this sense, a $\mathbb{C}^\times$-invariant connection θ_E is called a *non-linear connection*.

EXAMPLE 3.1. Let E be a holomorphic vector bundle with a Hermitian metric h. Let $\bar{\nabla}$ be the Hermitian connection of h, and $\nabla = \pi^* \bar{\nabla}$ the induced connection on V_E. If we define $\theta \in \Omega^{1,0}(V_E)$ by

$$\theta_E = \bar{\nabla} \mathcal{E}, \tag{3.5}$$

then θ_E defines a linear connection on the bundle $\pi : E \to M$.

If a pseudoconvex Finsler metric F is given on E, then it defines a canonical non-linear connection θ_E (see below).

If a connection θ_E is given on E, we put

$$X^V = \theta_E(X), \ \ X^H = X - X^V,$$

and

$$(d^V f)(X) = df\left(X^V\right), \ \ (d^H f)(X) = df\left(X^H\right)$$

for every $X \in T_E$ and $f \in C^\infty(E)$. These operators can be decomposed as $d^V = \partial^V + \bar{\partial}^V$ and $d^H = \partial^H + \bar{\partial}^H$. Furthermore the partial derivatives are also decomposed as $\partial = \partial^H + \partial^V$ and $\bar{\partial} = \bar{\partial}^H + \bar{\partial}^V$.

3.1. Finsler connection. We define a partial connection $\nabla^H : V_E \to \Omega^1(V_E)$ on V_E by

$$\nabla^H_Y Z = \theta_E\left([Y^H, Z]\right) \tag{3.6}$$

for all $Y \in T_E$ and $Z \in V_E$, where $[\cdot, \cdot]$ denotes the Lie bracket on T_E. This operator ∇^H is linear in X and satisfies the Leibnitz rule

$$\nabla^H(fZ) = (d^H f) \otimes Z + f \nabla^H Z$$

for all $f \in C^\infty(E)$.

On the other hand, since E is a fibre bundle over M, the projection map $\pi : E \to M$ can be used to pullback the said fibre bundle, generating a $\pi^* E$ which sits over E. Note that $V_E \cong \pi^* E$. Thus V_E admits a canonical relatively flat connection $\nabla^V : V_E \to \Omega^1(V_E)$ defined by $\nabla^V_X(\pi^* s) = 0$ for every local holomorphic section s of E, i.e.,

$$\nabla^V = d^V. \tag{3.7}$$

Then a connection $\nabla : V_E \to \Omega^1(V_E)$ is defined by

$$\nabla Z = \nabla^H Z + d^V Z \tag{3.8}$$

for every $Y \in T_E$ and $Z \in V_E$. We note here that the connection ∇ is determined uniquely from the connection θ_E on the bundle $\pi : E \to M$.

PROPOSITION 3.1. *Let* $\nabla : V_E \to \Omega^1(V_E)$ *be the connection on* V_E *defined by* (3.8), *from a connection* θ_E *on the bundle* $\pi : E \to M$. *Then* ∇ *satisfies*

$$\nabla \mathcal{E} = \theta_E. \tag{3.9}$$

PROOF. Since θ_E is invariant by the action of $\mathbb{C}^\times$, we have $\nabla^H \mathcal{E} = 0$. In fact,

$$\nabla^H_X \mathcal{E} = \theta_E\left([X^H, \mathcal{E}]\right) = -\theta_E\left(\mathcal{L}_\mathcal{E} X^H\right) = 0.$$

Furthermore, since $\left(d^V \mathcal{E}\right)(X) = X^V$, we obtain

$$\nabla_X \mathcal{E} = \nabla^H_X \mathcal{E} + \left(d^V \mathcal{E}\right)(X) = X^V = \theta_E(X)$$

for every $X \in T_E$. Hence we have (3.9). $\qquad\square$

For the rest of this paper, we shall use the following local coordinate system on M and E. Let U be an open set in M with local coordinates $(z^1, \ldots, z^n)$, and $s = (s_0, \ldots, s_r)$ a holomorphic local frame field on U. By setting $v = \sum \xi^i s_i(z)$ on for each $v \in \pi^{-1}(U)$, we take $(z, \xi) = (z^1, \ldots, z^n, \xi^0, \ldots, \xi^r)$ as a local coordinate system on $\pi^{-1}(U) \subset E$. We use the notation

$$\partial_\alpha = \frac{\partial}{\partial z^\alpha} \quad \text{and} \quad \partial_j = \frac{\partial}{\partial \xi^j}.$$

We also denote by $\partial_{\bar\alpha}$ and $\partial_{\bar j}$ their conjugates.

We suppose that a pseudoconvex Finsler metric F is given on E. Then, by definition in the previous section, the form $\omega_E = \sqrt{-1}\,\partial\bar\partial F$ is a closed real $(1,1)$-form on the total space E such that the restriction ω_z on each fibre $E_z = \pi^{-1}(z)$ defines a Kähler metric G_z on E_z, and ω_E defines a Hermitian metric G_{V_E} on the bundle V_E. Thus we shall investigate the geometry of the Hermitian bundle $\{V_E, G_{V_E}\}$.

A Finsler metric F on E is pseudoconvex if and only if the Hermitian matrix $(G_{i\bar j})$ defined by

$$G_{i\bar j}(z, \xi) = \partial_i \partial_{\bar j} F \tag{3.10}$$

is positive-definite. Each fibre E_z may be considered as a Kähler manifold with Kähler form $\omega_z = \sqrt{-1}\sum G_{i\bar j} d\xi^i \wedge d\bar\xi^j$. The family $\{E_z, \omega_z\}_{z \in M}$ is considered as a family of Kähler manifolds and the bundle is considered as the associated fibred manifold. The Hermitian metric G_{V_E} on V_E is defined by

$$G_{V_E}(s_i, s_j) = G_{i\bar j}, \tag{3.11}$$

where we consider $s = (s_0, \ldots, s_r)$ as a local holomorphic frame field for $V_E \cong \pi^* E$. We denote by $\| \cdot \|_E$ the norm defined by the Hermitian metric G_{V_E}. Then, because of the homogeneity $(F2)$, we have

$$\|\mathcal{E}\|_E^2 = G_{V_E}(\mathcal{E}, \mathcal{E}) = F(z, \xi) \tag{3.12}$$

and

$$\mathcal{L}_\mathcal{E} G_{V_E} = G_{V_E} \tag{3.13}$$

for the tautological section $\mathcal{E}(z, \xi) = \sum \xi^i s_i(z)$.

Let $\theta = (\theta^1, \ldots, \theta^r)$ be the dual frame field for the dual bundle V_E^*, i.e., $\theta^i(s_j) = \delta_j^i$. A connection θ_E for the bundle $\pi : E \to M$ is written as

$$\theta_E = s \otimes \theta = \sum s_i \otimes \theta^i.$$

PROPOSITION 3.2. *Let F be a pseudoconvex Finsler metric on a holomorphic vector bundle E. Then there exists a unique connection θ_E such that (3.3) is the orthogonal splitting with respect to ω_E.*

PROOF. We can easily show that the required $(1,0)$-form θ^i is defined by $\theta^i = d\xi^i + \sum N_\alpha^i \, dz^\alpha$, where the local functions $N_\alpha^i(z, \xi)$ are given by

$$N_\alpha^i = \sum G^{i\bar{m}} \partial_\alpha \partial_{\bar{m}} F$$

for the inverse matrix $\left(G^{i\bar{m}}\right)$ of $(G_{i\bar{m}})$. By the homogeneity $(F2)$, these functions satisfy $N_\alpha^i(z, \lambda\xi) = \lambda N_\alpha^i(z, \xi)$ for every $\lambda \in \mathbb{C}^\times$, which implies that θ_E is $\mathbb{C}^\times$-invariant.   □

For the rest of this paper, we shall adopt the connection θ_E obtained in Proposition 3.2 on a pseudoconvex Finsler bundle (E, F).

PROPOSITION 3.3 [Aikou 1998]. *The connection θ_E satisfies*

$$\partial^H \circ \partial^H \equiv 0. \tag{3.14}$$

Such a connection θ_E determines a unique connection ∇ on V_E.

DEFINITION 3.2. The connection $\nabla : V_E \to \Omega^1(V_E)$ on (V_E, G_{V_E}) defined by (3.8) is called the *Finsler connection* of (E, F).

The connection ∇ defined by (3.8) is canonical in the following sense.

PROPOSITION 3.4 [Aikou 1998]. *Let ∇ be the Finsler connection on a pseudoconvex Finsler bundle (E, F). Then $\nabla = \nabla^H + d^V$ satisfies the following metrical condition.*

$$d^H G_{V_E}(Y, Z) = G_{V_E}(\nabla^H Y, Z) + G_{V_E}(Y, \nabla^H Z) \tag{3.15}$$

for all $Y, Z \in V_E$.

The *connection form* $\omega = \left(\omega_j^i\right)$ of ∇ with respect to a local holomorphic frame field $s = (s_0, \ldots, s_r)$ is defined by $\nabla s_j = \sum s_i \otimes \omega_j^i$. By the identity (3.15), ω is given by

$$\omega = G^{-1} \partial^H G. \tag{3.16}$$

3.2. Curvature. Let ∇ be the Finsler connection on (E, F). We also denote by $\nabla : \Omega^k(V_E) \to \Omega^{k+1}(V_E)$ the covariant exterior derivative defined by ∇.

DEFINITION 3.3. The section $R = \nabla \circ \nabla \in \Omega^2\left(\mathrm{End}(V_E)\right)$ is called the *curvature* of ∇.

With respect to the local frame field $s = (s_0, \ldots, s_r)$, we put

$$R(s_j) = \sum s_i \otimes \Omega_j^i.$$

In matrix notation, the *curvature form* $\Omega = \left(\Omega_j^i\right)$ of ∇ is given by

$$\Omega = d\omega + \omega \wedge \omega. \tag{3.17}$$

The curvature form Ω is decomposed as $\Omega = d^H \omega + \omega \wedge \omega + d^V \omega$, which can be simplified to

$$\Omega = \bar{\partial}^H \omega + d^V \omega. \tag{3.18}$$

This is made possible by

PROPOSITION 3.5. *The horizontal $(2,0)$-part of Ω vanishes, i.e.,*

$$\partial^H \omega + \omega \wedge \omega \equiv 0. \tag{3.19}$$

PROOF. Since $\omega = G^{-1}\partial^H G$, (3.14) implies

$$\begin{aligned} \partial^H \omega + \omega \wedge \omega &= \partial^H \left(G^{-1}\partial^H G\right) + \omega \wedge \omega \\ &= -G^{-1}\partial^H G \wedge G^{-1}\partial^H G + G^{-1}\partial^H \circ \partial^H G + \omega \wedge \omega \\ &= G^{-1}\partial^H \circ \partial^H G = 0. \end{aligned}$$ $\square$

We give the definition of flat Finsler metrics.

DEFINITION 3.4. A Finsler metric F is said to be *flat* if there exists a holomorphic local frame field $s = (s_0, \ldots, s_r)$ around every point of M such that $F = F(\xi)$, i.e., F is independent of the base point $z \in M$.

THEOREM 3.1 [Aikou 1999]. *A Finsler metric F is flat if and only if the curvature R vanishing identically.*

Let R^H be the curvature of the partial connection ∇^H, i.e., $R^H = \nabla^H \circ \nabla^H$. From (3.18), the curvature form Ω^H of R^H is given by $\Omega^H = \bar{\partial}^H \omega$. If we put

$$\bar{\partial}^H \omega_j^i = \sum R_{j\alpha\bar{\beta}}^i \, dz^\alpha \wedge d\bar{z}^\beta,$$

the curvature R^H is given by

$$R^H (s_j) = \sum s_i \otimes \left(R_{j\alpha\bar{\beta}}^i \, dz^\alpha \wedge d\bar{z}^\beta\right).$$

For the curvature form Ω^H of ∇^H, we define a horizontal $(1,1)$-form Ψ by

$$\Psi(X,Y) = \frac{G_{V_E}\left(R_{XY}^H(\mathcal{E}), \mathcal{E}\right)}{\|\mathcal{E}\|_E^2} = \frac{1}{F}\sum R_{i\bar{j}\alpha\bar{\beta}}(z,\xi)\xi^i \bar{\xi}^j X^\alpha \bar{Y}^\beta$$

for any horizontal vector fields X, Y at $(z, \xi) \in E$, where we put $R_{i\bar{j}\alpha\bar{\beta}} = \sum G_{m\bar{j}} R_{i\alpha\bar{\beta}}^m$. We set

$$\Psi_{\alpha\bar{\beta}}(z,\xi) = \frac{1}{F}\sum R_{i\bar{j}\alpha\bar{\beta}}\xi^i \bar{\xi}^j. \tag{3.20}$$

In [Kobayashi 1975], this $(1,1)$-form $\Psi = \sum \Psi_{\alpha\bar{\beta}}\, dz^\alpha \wedge d\bar{z}^\beta$ is also called the *curvature* of F.

DEFINITION 3.5. If the curvature form Ψ satisfies the negativity condition, i.e., $\Psi(Y,Z) < 0$ for all $Y, Z \in H_E$, then we say that (E, F) is *negatively curved.*

Direct computation gives:

PROPOSITION 3.6 [Aikou 1998]. *For a pseudoconvex Finsler metric F on a holomorphic vector bundle E, the real $(1,1)$ form $\sqrt{-1}\,\partial\bar{\partial}\log F = -c_1\left(\mathbb{L}(E), F\right)$ is given by*

$$\sqrt{-1}\,\partial\bar{\partial}\log F = \sqrt{-1}\begin{pmatrix} -\Psi_{\alpha\bar{\beta}} & \mathrm{O} \\ \mathrm{O} & \partial_i\partial_{\bar{j}}\log F \end{pmatrix}$$

with respect to the orthogonal decomposition (3.3), *i.e.*,

$$\sqrt{-1}\,\partial\bar{\partial}\log F = \sqrt{-1}\left(-\sum\Psi_{\alpha\bar{\beta}}\,dz^{\alpha}\wedge d\bar{z}^{\beta} + \sum\partial_i\partial_{\bar{j}}\left(\log F\right)\theta^i\wedge\bar{\theta}^j\right). \quad (3.21)$$

Analyzing the negativity of the form $c_1\left(\mathbb{L}(E), F\right)$, we have the following theorem of Kobayashi.

THEOREM 3.2 [Kobayashi 1975]. *A holomorphic vector bundle* $\pi : E \to M$ *over a compact complex manifold* M *is negative if and only if* E *admits a negatively curved pseudoconvex Finsler metric.*

PROOF. By Proposition 2.3, E is negative if and only if there exists a Finsler metric F such that $\sqrt{-1}\,\partial\bar{\partial}\log F < 0$. Since ∂ and $\bar{\partial}$ anti-commute, this characterization is equivalent to $\sqrt{-1}\,\partial\bar{\partial}\log F > 0$. Thus, E is negative if and only if the right hand side of (3.21) is positive.

Denote by F_z the restriction of F to each fibre $E_z = \pi^{-1}(z)$. Then, we have

$$\sqrt{-1}\,\partial\bar{\partial}F_z = \sqrt{-1}\,F_z\left(\partial\bar{\partial}\log F_z + \partial\log F_z \wedge \bar{\partial}\log F_z\right).$$

If (3.21) has positive right hand side, then $\sqrt{-1}\,\Psi$ must be negative, and $\sqrt{-1}\,\partial\bar{\partial}\log F_z$ must be positive. The latter, in conjunction with the displayed formula, implies the positivity of $\sqrt{-1}\,\partial\bar{\partial}F_z$. Thus F is pseudoconvex and negatively curved.

Conversely, suppose F is pseudoconvex and negatively curved. That is, we have $\sqrt{-1}\,\partial\bar{\partial}F_z > 0$ and $\sqrt{-1}\,\Psi < 0$. Now, the pseudoconvexity of F implies that the second term on the right hand side of (3.21) is positive-definite (see section 4.1 for details). Thus the entire right hand side of (3.21) is positive. □

REMARK 3.1. In this section, the horizontal $(1,1)$-part R^H of R plays an important role. In Finsler geometry, there are other important tensors. The V_E-valued 2-form T_E defined by

$$T_E = \nabla\theta_E \quad (3.22)$$

is called the *torsion form* of ∇, which is expressed by

$$T_E = s \otimes (d\theta + \omega \wedge \theta) = \sum s_i \otimes \left(d\theta^i + \sum\omega^i_j \wedge \theta^j\right).$$

Because of (3.9), the torsion form T_E is also given by

$$T_E = R(\mathcal{E}).$$

The torsion form T_E vanishes if and only if the horizontal subbundle H_E defined by θ_E is holomorphic and integrable (see [Aikou 2003b]).

On the other hand, the mixed part R^{HV} of R defined by

$$R^{HV}(s_j) = \sum s_i \otimes d^V\omega^i_j$$

is also an important curvature form. The vanishing of R^{HV} shows that (E, F) is modeled on a complex Minkowski space, i.e., there exists a Hermitian metric h_F

on E such that the Finsler connection ∇ on (E, F) is obtained by $\nabla = \pi^* \bar{\nabla}$ for the Hermitian metric $\bar{\nabla}$ of (E, h_F) (see [Aikou 1995]). Hence a Finsler metric F is flat if and only if (E, F) is modeled on a complex Minkowski space and the associated Hermitian metric h_F is flat.

3.3. Holomorphic sectional curvature. We now study the holomorphic tangent bundle T_M of a complex manifold with a pseudoconvex Finsler metric $F : T_M \to \mathbb{R}$. The pair (M, F) is called a *complex Finsler manifold*. This is the special case of $E = T_M$ in Definition 2.3, and we have the natural identifications $V_E \cong H_E \cong \pi^* T_M$.

Let $\Delta(r) = \{\zeta \in \mathbb{C} : |\zeta| < r\}$ be the disk of radius r in $\mathbb{C}$ with the *Poincaré metric*

$$g_r = \frac{4r^2}{(r^2 - |\zeta|^2)^2} \, d\zeta \otimes d\bar{\zeta}.$$

For every point $(z, \xi) \in T_M$, there exists a holomorphic map $\varphi : \Delta(r) \to M$ satisfying $\varphi(0) = z$ and

$$\varphi_*(0) := \varphi_* \left(\left(\frac{\partial}{\partial \zeta} \right)_{\zeta=0} \right) = \xi. \tag{3.23}$$

Then, the pullback $\varphi^* F$ defines a Hermitian metric on $\Delta(r)$ by

$$\varphi^* F = E(\zeta) \, d\zeta \otimes d\bar{\zeta},$$

where we put $E(\zeta) = F(\varphi(\zeta), \varphi_*(\zeta))$. The Gauss curvature $K_{\varphi^* F}(z, \xi)$ is defined by

$$K_{\varphi^* F}(z, \xi) = - \left(\frac{1}{E} \frac{\partial^2 \log E}{\partial \zeta \, \partial \bar{\zeta}} \right)_{\zeta=0}.$$

DEFINITION 3.6 [Royden 1986]. The *holomorphic sectional curvature* K_F of (M, F) at $(z, \xi) \in T_M$ is defined by

$$K_F(z, \xi) = \sup_{\varphi} \{ K_{\varphi^* F}(z, \xi) : \varphi(0) = z, \varphi_*(0) = \xi \},$$

where φ ranges over all holomorphic maps from a small disk into M satisfying $\varphi(0) = z$ and (3.23).

Then K_F has a computable expression in terms of the curvature tensor of the Finsler connection ∇.

PROPOSITION 3.7 [Aikou 1991]. *The holomorphic sectional curvature K_F of a complex Finsler manifold (M, F) is given by*

$$K_F(z, \xi) = \frac{\Psi(\mathcal{E}, \mathcal{E})}{\|\mathcal{E}\|_E^2} = \frac{1}{F^2} \sum R_{i\bar{j}k\bar{l}}(z, \xi) \xi^i \bar{\xi}^j \xi^k \bar{\xi}^l, \tag{3.24}$$

where $R_{i\bar{j}k\bar{l}} = \sum G_{m\bar{j}} R^m_{ik\bar{l}}$ is the curvature tensor of the Finsler connection ∇ on (T_M, F).

Then we have a Schwarz-type lemma:

PROPOSITION 3.8 [Aikou 1991]. *Let F be a pseudoconvex Finsler metric on the holomorphic tangent bundle of a complex manifold M. Suppose that its holomorphic sectional curvature $K_F(z, \xi)$ at every point $(z, \xi) \in T_M$ is bounded above by a negative constant $-k$. Then, for every holomorphic map $\varphi : \Delta(r) \to M$ satisfying $\varphi(0) = z$ and (3.23), we have*

$$g_r \geq k\varphi^*F. \tag{3.25}$$

The *Kobayashi metric* F_M on a complex manifold M is a positive semidefinite pseudo metric defined by

$$F_M(z, \xi) = \inf_\varphi \left\{ \tfrac{1}{r} : \varphi(0) = z, \ \varphi_*(0) = \xi \right\}. \tag{3.26}$$

In general, F_M is not smooth. F_M is only upper semi-continuous, i.e., for every $X \in T_M$ and every $\varepsilon > 0$ there exists a neighborhood U of X such that $F_M(Y) < F_M(X) + \varepsilon$ for all $Y \in U$ (see [Kobayashi 1998], [Lang 1987]). Even though F_M is not a Finsler metric in our sense, the *decreasing principle* shows the importance of the Kobayashi metric, i.e., for every holomorphic map $\varphi : N \to M$, we have the inequality

$$F_N(X) \geq F_M(\varphi_*(X)). \tag{3.27}$$

This principle shows that F_M is holomorphically invariant, i.e., if $\varphi : N \to M$ is biholomorphic, then we have $F_N = \varphi^* F_M$. In this sense, F_M is an intrinsic metric on complex manifolds.

A typical example of Kobayashi metrics is the one on a domain in $\mathbb{C}^n$. It is well-known that, if M is a strongly convex domain with smooth boundary in $\mathbb{C}^n$, then F_M is a pseudoconvex Finsler metric in our sense (see [Lempert 1981]).

A complex manifold M is said to be *Kobayashi hyperbolic* if its Kobayashi metric F_M is a metric in the usual sense. If M admits a pseudoconvex Finsler metric F whose holomorphic sectional curvature K_F is bounded above by a negative constant $-k$, then (3.25) implies the inequality

$$F_M^2 \geq kF, \tag{3.28}$$

and thus M is Kobayashi hyperbolic.

THEOREM 3.3 [Kobayashi 1975]. *Let M be a compact complex manifold. If its holomorphic tangent bundle T_M is negative, then M is Kobayashi hyperbolic.*

PROOF. We suppose that T_M is negative. Then, Theorem 3.2 implies that there exists a pseudoconvex Finsler metric F on T_M with negative-definite Ψ. By the definition (3.20), the negativity of Ψ and (3.24) imply

$$K_F(z, \xi) = \frac{\Psi(\mathcal{E}, \mathcal{E})}{\|\mathcal{E}\|_E^2} < 0.$$

Since M is compact, $\mathbb{P}(E)$ is also compact. Moreover, since K_F is a function on $\mathbb{P}(E)$, the negativity of K_F shows that K_F is bounded by a negative constant $-k$. Hence we obtain (3.28), and M is Kobayashi hyperbolic. $\qquad\square$

REMARK 3.2. Recently Cao and Wong [Cao and Wong 2003] have introduced the notion of "mixed holomorphic bisectional curvature" for Finsler bundles (E, F), which equals the usual holomorphic bisectional curvature in the case of $E = T_M$. They also succeeded in showing that a holomorphic vector bundle E of rank$(E) \geq 2$ over a compact complex manifold M is ample if and only if E admits a Finsler metric with positive mixed holomorphic bisectional curvature.

4. Ruled Manifolds

4.1. Projective bundle. Let $\phi : \mathbb{P}(E) \to M$ be the projective bundle associated with a holomorphic vector bundle E over M.

DEFINITION 4.1. A locally $\partial\bar{\partial}$-exact real $(1,1)$-form $\omega_{\mathbb{P}(E)}$ on the total space $\mathbb{P}(E)$ is called a *pseudo-Kähler metric* on $\mathbb{P}(E)$ if its restriction to each fibre defines a Kähler metric on $\mathbb{P}(E_z) \cong \mathbb{P}^r$.

If a pseudo-Kähler metric $\omega_{\mathbb{P}(E)}$ is given on $\mathbb{P}(E)$, then its restriction to each fibre $\phi^{-1}(z) = \mathbb{P}(E_z)$ is a Kähler form on $\mathbb{P}(E_z)$. We shall show that a pseudoconvex Finsler metric on E defines a pseudo-Kähler metric on $\mathbb{P}(E)$. For this purpose, we use the so-called Euler sequence (e.g., [Zheng 2000]).

We denote by $\rho : E^{\times} \to \mathbb{P}(E)$ the natural projection. We also denote by $V_{\mathbb{P}(E)} := \ker \phi_*$ the vertical subbundle of $T_{\mathbb{P}(E)}$. Let $s = (s_0, \ldots, s_r)$ be a holomorphic local frame field of E on an open set $U \subset M$, which is naturally considered as a holomorphic local frame field of V_E on $\pi^{-1}(U)$. Then, the vertical subbundle $V_{\mathbb{P}(E)} \subset T_{\mathbb{P}(E)}$ is locally spanned by $\{\rho_* s_0, \ldots, \rho_* s_r\}$ with the relation

$$\rho_* \mathcal{E} = 0. \tag{4.1}$$

Then the Euler sequence

$$0 \longrightarrow \mathbb{L}(E) \xrightarrow{i} \phi^* E \longrightarrow \mathbb{L}(E) \otimes V_{\mathbb{P}(E)} \longrightarrow 0$$

implies the following exact sequence of vector bundles:

$$0 \longrightarrow 1_{\mathbb{P}(E)} \xrightarrow{i} \mathbb{H}(E) \otimes \phi^* E \xrightarrow{\mathcal{P}} V_{\mathbb{P}(E)} \longrightarrow 0, \tag{4.2}$$

where $\mathbb{H}(E) = \mathbb{L}(E)^*$ is the hyperplane bundle over $\mathbb{P}(E)$ and the bundle morphism $\mathcal{P} : \mathbb{H}(E) \otimes \phi^* E \to V_{\mathbb{P}(E)}$ is defined as follows. Since any section Z of $\mathbb{H}(E) \otimes \phi^* E$ can be naturally identified with a section $Z = \sum Z^j s_j$ of V_E satisfying the homogeneity $Z^j(\lambda v) = \lambda Z^j(v)$, the definition $\rho_* Z(v) = (\rho_* Z)([v])$ makes sense. Then $\mathcal{P}$ is defined by

$$\mathcal{P}(Z) = \rho_* \left(\sum Z^j s_j \right). \tag{4.3}$$

Moreover, since $\rho_* \mathcal{E} = 0$, $1_{\mathbb{P}(E)}(= \ker \mathcal{P})$ is the trivial line bundle spanned by $\mathcal{E}$. Then, since $\ker \mathcal{P} = 1_{\mathbb{P}(E)}$ is spanned by $\mathcal{E}$, the morphism $\mathcal{P}$ is expressed as

$$\mathcal{P}(Z) = Z - \frac{G_{V_E}(Z, \mathcal{E})}{\|\mathcal{E}\|_E^2} \mathcal{E}$$

for a Hermitian metric G_{V_E} on V_E. Since $\mathcal{P}$ is surjective, for any sections $\tilde{Z}$ and $\tilde{W}$ of $V_{\mathbb{P}(E)}$, there exist sections Z and W of V_E satisfying $\mathcal{P}(Z) = \tilde{Z}$ and $\mathcal{P}(W) = \tilde{W}$. Then, a Hermitian metric $G_{V_{\mathbb{P}(E)}}$ on $V_{\mathbb{P}(E)}$ is defined by

$$G_{V_{\mathbb{P}(E)}}(\tilde{Z}, \tilde{W}) = \frac{\|\mathcal{E}\|_E^2 \, G_{V_E}(Z, W) - G_{V_E}(Z, \mathcal{E}) G_{V_E}(\mathcal{E}, W)}{\|\mathcal{E}\|_E^4} \tag{4.4}$$

for the Hermitian metric G_{V_E} on V_E defined by (3.11), which induces the orthogonal decomposition

$$\mathbb{H}(E) \otimes \phi^* E = 1_{\mathbb{P}(E)} \oplus V_{\mathbb{P}(E)}.$$

Because of (4.1) and (4.4), the components of the metric $G_{V_{\mathbb{P}(E)}}$ with respect to the local frame $\{\rho_*(s_j)\}$ is given by

$$G_{V_{\mathbb{P}(E)}}(\rho_* s_i, \rho_* s_j) = \partial_i \partial_{\bar{j}} (\log F). \tag{4.5}$$

Consequently we have

PROPOSITION 4.1. *If F is a pseudoconvex Finsler metric on a holomorphic vector bundle E, then the real $(1,1)$-form $\sqrt{-1}\,\partial\bar{\partial}\log F$ defines a pseudo-Kähler metric on $\mathbb{P}(E)$.*

Conversely:

PROPOSITION 4.2. *If $\omega_{\mathbb{P}(E)}$ is a pseudo-Kähler metric on $\mathbb{P}(E)$, then $\omega_{\mathbb{P}(E)}$ defines a pseudoconvex Finsler metric F on E such that $\rho^* \omega_{\mathbb{P}(E)} = \sqrt{-1}\,\partial\bar{\partial}\log F$. Such a pseudoconvex Finsler metric F is unique up to a local positive function σ_U on $U \subset M$.*

PROOF. On each open set $U_{(j)} = \{[v] = (z, [\xi]) \in \phi^{-1}(U) : \xi^j \neq 0\}$ of $\mathbb{P}(E)$, we express the pseudo-Kähler metric $\omega_{\mathbb{P}(E)}$ by

$$\omega_{\mathbb{P}(E)}|_{U_{(j)}} = \sqrt{-1}\,\partial\bar{\partial} g_{(j)},$$

where $\{g_{(j)}\}$ is a family of local smooth functions $g_{(j)} : U_{(j)} \to \mathbb{R}$. Since the restriction of this form to each fibre $\mathbb{P}_z \subset U_{(j)}$ is a Kähler form ω_z on $\mathbb{P}_z$, we may put

$$\omega_z = \sqrt{-1}\,\partial\bar{\partial} g_{z,(j)},$$

where the local functions $g_{z,(j)} = g_{(j)}|_{\mathbb{P}_z}$ depend on $z \in U$ smoothly. Then we define a function $F_z : E_z^\times \to \mathbb{R}$ by

$$F_z(\xi) = |\xi^j|^2 \exp g_{z,(j)}.$$

Since F_z also depends on $z \in U$ smoothly, we extend this function to a smooth function $F : E^\times \to \mathbb{R}$ by $F(z, \xi) = F_z(\xi)$. It is easily verified that F defines a pseudoconvex Finsler metric on E.

We note that another Kähler potential $\{\tilde{g}_{(j)}\}$ for $\omega_{\mathbb{P}(E)}$ which induces the Kähler metric ω_z on each $\mathbb{P}_z$ is given by

$$\tilde{g}_{(j)}(z,[\xi]) = \sigma_U(z) + g_{(j)}(z,[\xi]) \tag{4.6}$$

for some functions $\sigma_U(z)$ defined on U. Hence the Finsler metric $\tilde{F}$ determined from the potential $\{\tilde{g}_j\}$ is connected to the function F by the relation $\tilde{F} = e^{\sigma_U(z)}F$ on each U. $\square$

Similar to Definition 3.4, we say a pseudo-Kähler metric $\omega_{\mathbb{P}(E)}$ on $\mathbb{P}(E)$ is *flat* if there exists an open cover $\{U,s\}$ of E so that we can choose Kähler potentials $g_{(j)}$ for $\omega_{\mathbb{P}(E)}$ which are independent of the base point $z \in M$. Now we define the projective-flatness of Finsler metrics.

DEFINITION 4.2. A Finsler metric on E is said to be *projectively flat* if it is obtained from a flat pseudo-Kähler metric on $\mathbb{P}(E)$.

In a previous paper, we proved this result:

THEOREM 4.1 [Aikou 2003a]. *A pseudoconvex Finsler metric is projectively flat if and only if the trace-free part of the curvature form Ω vanishes identically, i.e.,*

$$\Omega = A(z) \otimes \mathrm{Id} \tag{4.7}$$

for some $(1,1)$-form A on M.

REMARK 4.1. A Finsler metric F is projectively flat if and only if there exists a local function $\sigma_U(z)$ on U such that F is of the form

$$F(z,\xi)|_U = \exp \sigma_U(z) \cdot |\xi^j|^2 \exp g_{(j)}([\xi]) \tag{4.8}$$

on each U. In other words, a Finsler metric F is projectively flat if and only if there exists a local function $\sigma_U(z)$ on U such that the local metric $e^{-\sigma_U(z)}F$ is a flat Finsler metric on U. In the previous paper [Aikou 1997], such a Finsler metric F is said to be *conformally flat*.

We suppose that a pseudoconvex Finsler metric F is projectively flat. Then, since the curvature form Ω is given by (3.18), we have $R^{HV} \equiv 0$, and thus (E,F) is modeled on a complex Minkowski space. We can easily show that the associated Hermitian metric h_F is also projectively flat. Hence:

THEOREM 4.2. *A holomorphic vector bundle E of $\mathrm{rank}(E) = r+1$ admits a projectively flat Finsler metric if and only if $\mathbb{P}(E)$ is flat, i.e.,*

$$\mathbb{P}(E) = \tilde{M} \times_\rho \mathbb{P}^r, \tag{4.9}$$

where $\tilde{M}$ is the universal cover of M, and $\rho : \pi_1(M) \to PU(r)$ is a representation of the fundamental group $\pi_1(M)$ in the projective unitary group $PU(r)$.

4.2. Ruled manifolds. An algebraic surface $\mathcal{X}$ is said to be *ruled* if it is birational to $M \times \mathbb{P}^1$, where M is a compact Riemann surface. An algebraic surface $\mathcal{X}$ is said to be *geometrically ruled* if there exists a holomorphic projection $\phi : \mathcal{X} \to M$ such that every fibre $\phi^{-1}(z) = \mathcal{X}_z$ is holomorphically isomorphic to $\mathbb{P}^1$. As is well known, a geometrically ruled surface is ruled (see [Beauville 1983]), and every geometrically ruled surface $\mathcal{X}$ is holomorphically isomorphic to $\mathbb{P}(E)$ for some holomorphic vector bundle $\pi : E \to M$ of rank$(E) = 2$ (see [Yang 1991], for example).

An algebraic manifold $\mathcal{X}$ is said to be a *ruled manifold* if $\mathcal{X}$ is a holomorphic $\mathbb{P}^r$-bundle with structure group $PGL(r + 1, \mathbb{C}) = GL(r + 1, \mathbb{C})/\mathbb{C}^\times$. Any holomorphic $\mathbb{P}^r$-bundle over M is classified by $H^1(M, PGL(r + 1, \mathcal{O}_M))$, and any rank $r + 1$ holomorphic vector bundle over M is classified by the elements of $H^1(M, GL(r + 1, \mathcal{O}_M))$. The exact sequence

$$0 \to \mathcal{O}_M^* \to GL(r + 1, \mathcal{O}_M) \to PGL(r + 1, \mathcal{O}_M) \to 0$$

implies the sequence of cohomology groups:

$$\cdots \to H^1(M, GL(r+1, \mathcal{O}_M)) \to H^1(M, PGL(r+1, \mathcal{O}_M)) \to H^2(M, \mathcal{O}_M^*) \to \cdots$$

Since $H^2(M, \mathcal{O}_M^*) = 0$, the following is obtained.

PROPOSITION 4.3. *Every ruled manifold $\mathcal{X}$ over a compact Riemann surface M is holomorphically isomorphic to $\mathbb{P}(E)$ for some holomorphic vector bundle $\pi : E \to M$ of* rank$(E) = r + 1$. *Such a bundle E is uniquely determined up to tensor product with a holomorphic line bundle.*

If E is a holomorphic vector bundle over a compact Kähler manifold M, then $\mathbb{P}(E)$ is also a compact Kähler manifold. In fact, we can construct a Kähler metric $\omega_{\mathbb{P}(E)}$ on $\mathbb{P}(E)$ of the form $\omega_{\mathbb{P}(E)} = \phi^* \omega_M + \varepsilon \eta$. Here, ω_M is a Kähler metric on M, ε is a small positive constant, and η is a closed $(1,1)$-form on $\mathbb{P}(E)$ such that η is positive-definite on the fibres of ϕ (see Lemma (6.37) in [Shiffmann and Sommese 1985]). Thus every ruled manifold $\mathcal{X}$ over a compact Riemann surface M is a compact Kähler manifold, and $\phi : \mathcal{X} = \mathbb{P}(E) \to M$ is a holomorphic submersion from a compact Kähler manifold $\mathcal{X}$ to M. Then we have

THEOREM 4.3. *Let $\mathcal{X}$ be a ruled manifold over a compact Riemann surface M with a Kähler metric $\omega_{\mathcal{X}}$. Then there exists a negative vector bundle $\pi : E \to M$ such that $\mathcal{X} = \mathbb{P}(E)$, and a negatively curved pseudoconvex Finsler metric F on E satisfying $\rho_* \omega_{\mathcal{X}} = \sqrt{1} \partial \bar{\partial} \log F$.*

PROOF. Let $\omega_{\mathcal{X}}$ be a Kähler metric on $\mathcal{X}$. Propositions 4.3 and 4.2 imply that there exists a holomorphic vector bundle E satisfying $\mathcal{X} = \mathbb{P}(E)$ with a pseudoconvex Finsler metric F. Then, since

$$\sqrt{-1}\, \partial \bar{\partial} \log F = \omega_{\mathcal{X}} > 0,$$

F is negatively curved, and hence Theorem 2.3 implies that E is negative.   □

LeBrun [LeBrun 1995] has investigated minimal ruled surfaces (i.e., geometrically ruled surface) over a compact Riemann surface of genus $g(M) \geq 2$ with constant negative scalar curvature. Roughly speaking, he proved that such a minimal ruled surface $\mathcal{X}$ is obtained by a semi-stable vector bundle over M so that $\mathcal{X} = \mathbb{P}(E)$. Since the semi-stability of vector bundles over a compact Riemann surface is equivalent to the existence of a projectively flat Hermitian metric on E, such a surface is written in the form (4.9).

On the other hand, by Theorem 4.3, the geometry of a minimal ruled surface $\mathcal{X}$ is naturally translated to the geometry of a negative vector bundle E with a negatively curved pseudoconvex Finsler metric F. From this viewpoint, we have also investigated minimal ruled surfaces, and we have concluded that each minimal ruled surface $\phi : \mathcal{X} \to M$ over a compact Riemann surface of genus $g(M) \geq 2$ with constant negative scalar curvature is a Kähler submersion with isometric fibres (see [Aikou 2003b]).

References

[Abate–Patrizio 1994] M. Abate and G. Patrizio, *Finsler metrics: a global approach*, Lecture Notes in Math. **1591**, Springer, Berlin, 1994.

[Aikou 1991] T. Aikou, "On complex Finsler manifolds", *Rep. Fac. Sci. Kagoshima Univ.* **24** (1991), 9–25.

[Aikou 1995] T. Aikou, "Complex manifolds modeled on a complex Minkowski space", *J. Math. Kyoto Univ.* **35** (1995), 83–111.

[Aikou 1996] T. Aikou, "Some remarks on locally conformal complex Berwald spaces", pp. 109–120 in *Finsler geometry* (Seattle, 1995), edited by David Bao et al., Contemporary Math. **196**, Amer. Math. Soc., Providence, 1996.

[Aikou 1997] T. Aikou, "Einstein–Finsler vector bundles", *Publ. Math. Debrecen* **51** (1997), 363–384.

[Aikou 1998] T. Aikou, "A partial connection on complex Finsler bundles and its applications", *Illinois J. Math.* **42** (1998), 481–492.

[Aikou 1999] T. Aikou, "Conformal flatness of complex Finsler structures", *Publ. Math. Debrecen* **54** (1999), 165–179.

[Aikou 2000] T. Aikou, "Some remarks on the conformal equivalence of complex Finsler structures", pp. 35–52 in *Finslerian geometry: a meeting of minds*, edited by P. Antonelli, Kluwer, Amsterdam, 2000.

[Aikou 2002] T. Aikou, "Applications of Bott connection to Finsler geometry", pp. 1–13 in *Steps in differential geometry*, University of Debrecen Institute of Mathematics and Informatics, Debrecen, 2001.

[Aikou 2003a] T. Aikou, "Projective flatness of complex Finsler metrics", *Publ. Math. Debrecen* **63** (2003), 343–362.

[Aikou 2003b] T. Aikou, "A note on projectively flat complex Finsler metrics", preprint, 2003.

[Bao et al. 1996] D. Bao, S.-S. Chern and Z. Shen (editors), *Finsler geometry*, Contemporary Math. **196**, Amer. Math. Soc., Providence, 1996.

[Bao et al. 2000] D. Bao, S.-S. Chern and Z. Shen, *An introduction to Riemann–Finsler geometry*, Graduate Texts in Math. **200**, Springer, New York, 2000.

[Beauville 1983] A. Beauville, *Complex algebraic surfaces*, Cambridge Univ. Press, Cambridge, 1983.

[Cao and Wong 2003] J.-K. Cao and P.-M. Wong, "Finsler geometry of projectivized vector bundles", to appear in *J. Math. Kyoto Univ.*

[Chern et al. 1999] S.-S. Chern, W. H. Chen and K. S. Lam, *Lectures on differential geometry*, Series on Univ. Math. **1**, World Scientific, Singapore, 1999.

[Dineen 1989] S. Dineen, *The Schwarz Lemma*, Clarendon Press, Oxford, 1989.

[Duchamp and Kalka 1988] T. Duchamp and M. Kalka, "Invariants of complex foliations and the Monge–Ampère equation", *Michigan Math. J.* **35** (1988), 91–115.

[Faran 1990] J. J. Faran, "Hermitian Finsler metrics and the Kobayashi metric", *J. Differential Geometry* **31** (1990), 601–625.

[Fukui 1989] M. Fukui, "Complex Finsler manifolds", *J. Math. Kyoto Univ.* **29** (1989), 609–624.

[Griffith and Harris 1978] G. Griffith and J. Harris, *Principles of algebraic geometry*, Wiley, New York, 1978.

[Icijyō 1988] Y. Icijyō, "Finsler metrics on almost complex manifolds", *Riv. Mat. Univ. Parma* **14** (1988), 1–28.

[Icijyō 1994] Y. Icijyō, "Kaehlerian Finsler manifolds", *J. Math. Tokushima Univ.* **28** (1994), 19–27.

[Icijyō 1998] Y. Icijyō, "Kaehlerian Finsler manifolds of Chern type", *Research Bull. Tokushima Bunri Univ.* **57** (1999), 9–16.

[Jost 1997] J. Jost, *Compact Riemann surfaces*, Universitext, Springer, Berlin, 1997.

[Kobayashi 1975] S. Kobayashi, "Negative vector bundles and complex Finsler structures", *Nagoya Math. J.* **57** (1975), 153–166.

[Kobayashi 1987] S. Kobayashi, *Differential geometry of complex vector bundles*, Iwanami, Tokyo, and Princeton Univ. Press, Priceton, 1987.

[Kobayashi 1996] S. Kobayashi, "Complex Finsler vector bundles", pp. 145–153 in *Finsler geometry* (Seattle, 1995), edited by David Bao et al., Contemporary Math. **196**, Amer. Math. Soc., Providence, 1996.

[Kobayashi 1998] S. Kobayashi, *Hyperbolic complex spaces*, Springer, Berlin, 1998.

[Kobayashi and Wu 1970] S. Kobayashi and H.-H. Wu, "On holomorphic sections on certain Hermitian vector bundles", *Math. Ann.* **189** (1970), 1–4.

[Kobayashi and Nomizu 1963] *Foundations of differential geometry*, vol. II, Wiley, New York, 1963.

[Lang 1987] S. Lang, *Introduction to complex hyperbolic spaces*, Springer, 1987.

[LeBrun 1995] C. LeBrun, "Polarized 4-manifolds, extremal Kähler metrics, and Seiberg–Witten theory", *Math. Res. Lett.* **2** (1995), 653–662.

[Lempert 1981] L. Lempert, "La métrique de Kobayashi et la représentation des domaines sur la boule", *Bull. Soc. Math. France* **109** (1981), 427–474.

[Matsumoto 1986] M. Matsumoto, *Foundations of Finsler geometry and special Finsler spaces*, Kaiseisha Press, Otsu, Japan, 1986.

[Mori 1979] S. Mori, "Projective manifolds with ample tangent bundles", *Ann. of Math.* (2) **110** (1979), 593–606.

[Pang 1992] M.-Y. Pang, "Finsler metrics with properties of the Kobayashi metric of convex domains", *Publicacions Matemàtiques* (Barcelona) **36** (1992), 131–155.

[Patrizio and Wong 1983] G. Patrizio and P.-M. Wong, "Stability of the Monge–Ampère Foliation", *Math. Ann.* **263** (1983), 13-29.

[Rizza 1963] G. B. Rizza, "Strutture di Finsler di tipo quasi Hermitiano", *Riv. Mat. Univ. Parma* **4** (1963), 83–106.

[Royden 1986] H. L. Royden, "Complex Finsler metrics", pp. 261–271 in *Complex differential geometry and nonlinear differential equations*, edited by Yum-Tong Siu, Contemporary Math. **49**, Amer. Math. Soc., Providence, 1986.

[Rund 1959] H. Rund, *The differential geometry of Finsler spaces*, Springer, Berlin, 1959.

[Rund 1972] H. Rund, "The curvature theory of direction-dependent connections on complex manifolds", *Tensor (N.S.)* **24** (1972), 189–205.

[Shiffmann and Sommese 1985] B. Shiffmann and A. J. Sommese, *Vanishing theorems on complex manifolds*, Birkhäuser, Boston, 1985.

[Yang 1991] K. Yang, *Complex algebraic geometry: an introduction to curves and surfaces*, Dekker, New York, 1991.

[Watt 1996] C. Watt, "Complex sprays, Finsler metrics and horizontal complex curves", Ph.D. Thesis, Trinity College Dublin, 1996.

[Wong 1977] B. Wong, "On the holomorphic curvature of some intrinsic metrics", *Proc. Amer. Math. Soc.* **65** (1977), 57–61.

[Wong 1984] B. Wong, "A class of compact complex manifolds with negative tangent bundles", *Math. Z.* **185** (1984), 217–223.

[Wu 1973] B. Wu, "A remark on holomorphic sectional curvature", *Indiana Univ. Math. J.* **22** (1973), 1103–1108.

[Zheng 2000] F. Zheng, *Complex differential geometry*, AMS/IP studies in advanced mathematics **18**, Amer. Math. Soc., Providence, 2000.

TADASHI AIKOU
DEPARTMENT OF MATHEMATICS AND COMPUTER SCIENCE
FACULTY OF SCIENCE
KAGOSHIMA UNIVERSITY
KAGOSHIMA 890-0065
JAPAN
aikou@sci.kagoshima-u.ac.jp

Riemann–Finsler Geometry
MSRI Publications
Volume **50**, 2004

Finsler Geometry of Holomorphic Jet Bundles

KAREN CHANDLER AND PIT-MANN WONG

CONTENTS

Introduction

A complex manifold X is *Brody hyperbolic* if every holomorphic map $f :$ $\mathbb{C} \to X$ is constant. For compact complex manifolds this is equivalent to the condition that the *Kobayashi pseudometric* κ_1 (see (1.12)) is a positive definite Finsler metric. One may verify the hyperbolicity of a manifold by exhibiting a Finsler metric with negative *holomorphic sectional curvature*. The construction of such a metric motivates the use of parametrized jet bundles, as defined by Green–Griffiths. (The theory of these bundles goes back to [Ehresmann 1952].) We examine the algebraic-geometric properties (ample, big, nef, spanned and the dimension of the base locus) of these bundles that are relevant toward the metric's existence. To do this, we start by determining (and computing) basic invariants of jet bundles. Then we apply Nevanlinna theory, via the construction of an appropriate singular Finsler metric of logarithmic type, to obtain precise extensions of the classical Schwarz Lemma on differential forms toward jets. Particularly, this allows direct control over the analysis of the jets $j^k f$ of a holomorphic map $f : \mathbb{C} \to X$; namely, the image of $j^k f$ must be contained in the base locus of the jet differentials. For an algebraically nondegenerate holomorphic map we show by means of reparametrization that the algebraic closure of $j^k f$ is quite large while, under appropriate conditions, the base locus is relatively small. This contradiction shows that the map f must be algebraically degenerate. We apply this method to verify that a generic smooth hypersurface of $\mathbb{P}^3$, of degree $d \geq 5$, is hyperbolic (see Corollary 7.21). Using this we show also the existence of a smooth curve C of degree $d = 5$ in $\mathbb{P}^2$ such that $\mathbb{P}^2 \setminus C$ is *Kobayashi hyperbolic*.

In the classical theory of curves (Riemann surfaces) the most important invariant is the *genus*. The genus g of a curve is the number of independent global regular 1-forms: $g = h^0(\mathcal{K}_X) = \dim H^0(\mathcal{K}_X)$, where $\mathcal{K}_X = T^*X$ is the canonical bundle (which in the case of curves is also the cotangent bundle). A curve is hyperbolic if and only if $g \geq 2$. One way to see this is to take a basis $\omega_1, \ldots, \omega_g$ of regular 1-forms and define a metric ρ on the tangent bundle by setting

$$\rho(v) = \left(\sum_{i=1}^{g} |\omega_i(v)|^2 \right)^{1/2}, \quad v \in TX. \qquad (*)$$

For $g = 1$ the metric is flat, that is, the Gaussian or *holomorphic sectional curvature* (hsc) is zero. Hence X is an elliptic curve. For $g \geq 2$ the curvature of this metric is strictly negative which, by the classical Poincaré–Schwarz Lemma, implies that X is hyperbolic. Algebraic geometers take the dual approach by interpreting ρ as defining a metric along the fibers of the dual $T^*X = \mathcal{K}_X$ and, for $g \geq 2$, the Chern form $c_1(\mathcal{K}_X, \rho)$ is positive, that is, the canonical bundle is ample. Indeed the following four conditions are equivalent:

(i) $g \geq 2$;
(ii) X is hyperbolic;
(iii) T^*X is *ample*; and
(iv) There exists a negatively curved metric.

For a complex compact manifold of higher dimension the number of independent 1-forms $g = h^0(T^*X)$ is known as the *irregularity* of the manifold. If $g \geq 1$, we may define ρ as in $(*)$. More generally, for each m, we may choose a basis $\omega_1, \ldots, \omega_{g_m}$ of $H^0(\bigodot^m T^*X)$, where $\bigodot^m T^*X$ is the m-fold *symmetric* product, and define, if $g_m \geq 1$,

$$\rho(v) = \sum_{i=1}^{g_m} |\omega_i(v)|^{1/m}. \qquad (**)$$

In dimension 2 or higher, ρ cannot, in general, be positive definite and it is only a *Finsler* rather than a *hermitian* metric. However the holomorphic sectional curvature may be defined for a Finsler metric and the condition that the curvature is negative implies that X is hyperbolic. It is known [Cao and Wong 2003] that the ampleness of T^*X is equivalent to the existence of a Finsler metric with negative *holomorphic bisectional curvature* (hbsc):

THEOREM [Aikou 1995; 1998; Cao and Wong 2003]. T^*X *is ample* $\Longleftrightarrow$ *Finsler metric has negative hbsc* $\Longrightarrow$ *Finsler metric has negative hsc* $\Longrightarrow$ X *is hyperbolic.*

In our view the fundamental problems in hyperbolic geometry are the following.

PROBLEM 1.   *Find an algebraic geometric characterization of the concept of negative hsc.*

PROBLEM 2. *Find an algebraic geometric and a differential geometric characterization of hyperbolicity.*

It is known that there are hyperbolic hypersurfaces in $\mathbb{P}^n$ (for each n). On the other hand, there are no global regular 1-forms on hypersurfaces in $\mathbb{P}^n$ for $n \geq 3$; indeed $h^0(\bigodot^m T^*X) = 0$ for all m. These hyperbolic hypersurfaces are discovered using, in one form or another, the Second Main Theorem of Nevanlinna Theory, which involves higher-order information; see for example [Wong 1989; Stoll and Wong 1994; Fujimoto 2001].

This leads us to the concept of the (parametrized) jet bundles [Ehresmann 1952], formalized (for complex manifolds) and studied by Green and Griffiths [1980]. Observe that a complex tangent v at a point x of a manifold may be represented by the first order derivative $f'(0)$ of a local holomorphic map f : $\Delta_r \to X, f(0) = x$ for some disc Δ_r of radius r in the complex plane $\mathbb{C}$ (more precisely, v is the equivalence class of such maps, as different maps may have the same derivative at the origin). A k-*jet* is defined as the equivalence class of the first k-th order derivatives of local holomorphic maps and the k-jet bundle, denoted $J^k X$, is just the collection of all (equivalence classes of) k-jets. Note that $J^1 X = TX$. For $k \geq 2$ these bundles are $\mathbb{C}^*$ bundles but not vector bundles. The nonlinear structure is reflected in reparametrization. Namely, given a k-jet $j^k f(0) = (f(0), f'(0), \ldots, f^{(k)}(0))$ we obtain another k-jet by composing f with another local holomorphic self map ϕ in $\mathbb{C}$ that preserves the origin, then taking $j^k(f \circ \phi)(0)$. In particular, if ϕ is given by multiplication by a complex number λ we see that $j^k(f \circ \phi)(0) = (f(0), \lambda f'(0), \lambda^2 f''(0), \ldots, \lambda^k f^{(k)}(0))$. Equivalence under this action is denoted by $\lambda \cdot j^f(0)$ and this is the $\mathbb{C}^*$-action on $J^k X$; in general *there is no vector bundle structure on $J^k X$*. We write:

$$\lambda \cdot (v_1, \ldots, v_k) = (\lambda v_1, \lambda^2 v_2, \ldots, \lambda^k v_k), \ (v_1, \ldots, v_k) \in J^k X \qquad (***)$$

and assign the weight i to the variable v_i. A 1-form ω may be regarded as a holomorphic function on the tangent bundle $\omega : TX \to \mathbb{C}$ satisfying the condition $\omega(\lambda \cdot v) = \lambda \omega(v)$, that is, linearity along the fibers. More generally, an element $\omega \in H^0(\bigodot^m T^*X)$ is a holomorphic function on the tangent bundle $\omega : TX \to \mathbb{C}$ that is a homogeneous polynomial of degree m along the fibers. Analogously we define a k-*jet differential ω of weight m* to be a holomorphic function on the k-jet bundle $\omega : J^k X \to \mathbb{C}$ which is a *weighted* homogeneous polynomial of degree m along the fibers. The sheaf of k-jet differentials of weight m will be denoted by

$\mathcal{J}_k^m X$. Taking a basis $\omega_1, \ldots, \omega_N$ of $H^0(\mathcal{J}_k^m X)$ we define

$$\rho(v_1, \ldots, v_k) = \sum_{i=1}^{N} |\omega_i(v_1, \ldots, v_k)|^{1/m},$$

and from $(***)$ we see (since each ω_i is a weighted homogeneous polynomial of degree m) that ρ is a Finsler pseudometric, that is,

$$\rho(\lambda \cdot (v_1, \ldots, v_k)) = \sum_{i=1}^{N} \left(|\lambda^m \omega_i(v_1, \ldots, v_k)| \right)^{1/m}$$

$$= |\lambda| \sum_{i=1}^{N} \left(|\omega_i(v_1, \ldots, v_k)| \right)^{1/m} = |\lambda| \, \rho(v_1, \ldots, v_k).$$

The positive definiteness is a separate issue that one must deal with in higher dimension. The algebraic geometric concept that is equivalent to the positive definiteness of ρ is that the sheaf $\mathcal{J}_k^m X$ is *spanned* (meaning that global sections span the fiber at each point). Other relevant concepts here are whether such a sheaf is *ample*, *nef* (numerically effective), or *big*. These concepts are intimately related to the Chern numbers of the sheaf $\mathcal{J}_k^m X$ and the dimensions of the cohomology groups $h^i(\mathcal{J}_k^m X)$, $0 \le i \le n = \dim X$. The starting point here is the computation of the Euler characteristic in the case of a manifold of *general type* by the Riemann–Roch Formula. An asymptotic expansion of $\chi(\mathcal{J}_k^m X)$ was given in [Green and Griffiths 1980] with a sketch of the proof. Often in articles making reference to this result readers questioned the validity of the statement. A detailed proof, in the case of general type surfaces (complex dimension 2), of this formula was given in [Stoll and Wong 2002] using a different approach to that given in by Green and Griffiths. Indeed explicit formulas, not merely asymptotic expansions, were given for $\mathcal{J}_k^m X$, $k = 2$ and 3. The method of computation also shows that $\mathcal{J}_k^m X$ is not *semistable* (see Section 3, Remark 3.5) in the sense of Mumford–Takemoto despite the fact (see [Maruyama 1981; Tsuji 1987; 1988]) that all tensor products $\otimes T^* X$ and symmetric products $\bigodot^m T^* X$ are semistable if X is of general type. In this article we also introduce an analogue of semistability in the sense of Gieseker–Maruyama (see [Okonek et al. 1980]) and show that $\mathcal{J}_k^m X$ is not semistable (see Section 7) in this sense either.

We have (see Section 5 for the reason in choosing the weight $k!$ below)

$T^* X$ is ample $\implies \mathcal{J}_k^{k!} X$ is ample for all k

$\qquad\qquad \implies \mathcal{J}_k^{k!} X$ is ample for some k

$\qquad\qquad \iff$ there exists Finsler metric on $J^k X$ with negative hbsc

$\qquad\qquad \implies$ there exists Finsler metric on $J^k X$ with negative hsc

$\qquad\qquad \implies X$ is hyperbolic.

The condition that $\mathcal{J}_k^{k!}X$ is ample is much stronger than hyperbolicity of X. A weaker condition is that $\mathcal{J}_k^{k!}X$ is *big*; this says that

$$h^0(\mathcal{J}_k^{k!m}X) = \dim(X, \mathcal{J}_k^{k!m}X) = O(m^{n(k+1)-1}),$$

where $n = \dim X$. From the differential geometric point of view, this means that there is a pseudo-Finsler metric on J^kX that is generically positive definite and has generically negative hbsc (as defined wherever the metric is positive definite). The condition that $\mathcal{J}_k^{k!}X$ is big implies that, for any ample divisor D on X, there exists $m_0 = m_0(D)$ such that $\mathcal{J}_k^{k!m_0}X \otimes [-D]$ (the sheaf of k-jet differentials of weight $k!m_0$ vanishing along D) is big. This, however, is not quite enough to guarantee hyperbolicity; the problem is that the base locus of $\mathcal{J}_k^{k!}X \otimes [-D]$ may be "too big". As a natural correction, we verify, using the Schwarz Lemma for jet differentials (see Theorem 6.1, Corollaries 6.2 and 6.3), that the assumption $\mathcal{J}_k^{k!m_0}X \otimes [-D]$ is big and spanned (that is, the base locus is empty) does imply hyperbolicity.

However, the condition that a sheaf is big and spanned may be difficult to verify (unless, perhaps, it is already ample and we know of no hypersurfaces in $\mathbb{P}^3$ satisfying this condition). To alleviate this, we refine the form of Schwarz's Lemma (see Theorem 6.4 and Corollary 6.5) to establish the result that every holomorphic map $f : \mathbb{C} \to \cdot X$ is algebraically degenerate if the dimension of the base locus of $\mathcal{J}_k^{k!m_0}X \otimes [-D]$ in the projectivized jet bundle $\mathbb{P}(J^kX)$ is no more than $n + k - 1$. From this we show in Section 7, using the explicit computation of the invariants of the jet bundles in the first 3 sections (see Theorem 3.9 and Corollary 3.10), that the dimension of the base locus of a generic hypersurface of degree ≥ 5 in $\mathbb{P}^3$ is, indeed, at most $n + k - 1 = k + 1$ ($n = 2$ in this case) and consequently, hyperbolic. *The key ingredient is the extension of the inductive cutting procedure of the base locus, of* [Lu and Yau 1990] *and* [Lu 1991] *(see also* [Dethloff et al. 1995a; 1995b]) *in the case of* 1-*jets, to* k-*jets.* There is a delicate point in the cutting procedure, namely that *intersections of irreducible varieties may not be irreducible*. We show, again using the Schwarz Lemma, that *under the algebraically nondegenerate assumption on* f, *there is no loss of generality in assuming that the intersection is irreducible* (see the proof of Theorems 7.18 and 7.20).

The crucial analytic tools here are the Schwarz Lemma for jet differentials and its refined form. These are established using Nevanlinna Theory. We remark that jet differentials are used routinely in Nevanlinna Theory without a priori knowledge of whether regular jet differentials exist at all. The main idea of the proof of the Schwarz Lemma is to use jet differentials with logarithmic poles; to determine conditions under which the sheaf of such jet differentials is *spanned* and provides a *singular Finsler metric that is positive definite in the extended sense*. The classical Nevanlinna Theory is seen to work well with nonhermitian Finsler metrics with logarithmic poles on account of the fundamental principle (the Lemma of logarithmic derivatives of Nevanlinna) that logarithmic poles are relatively harmless (see the proof of Theorem 6.1 in Section 6 for details).

The article is organized as follows. We describe the parametrized jet bundles of Green–Griffiths, which differ from the usual jet bundles; for example, they are $\mathbb{C}^*$ bundles but in general not vector bundles. The definitions are recalled in Section 1. For the usual jet bundles there is the question of interpolation: Find all varieties with prescribed jets, say, at a finite number of points. This problem, for 1-jets, is equivalent to the Waring problem concerning when a general homogeneous polynomial is the sum of powers of linear forms. The Waring problem is related to the *explicit* construction (not merely existence) of hyperbolic hypersurfaces in $\mathbb{P}^n$ for any n. Limitation of space does not allow us to discuss this problem in this article. Solutions of the interpolation problem for a collection of points can be found in [Alexander and Hirschowitz 1992a; 1992b; 1995; Chandler 1995; 1998a; 2002]. The analogous problem concerning the Green–Griffiths jet bundles is still open.

In Section 2 we give a fairly detailed account of the jet bundles of curves. We calculate the Chern number $c_1(\mathcal{J}_k^m X)$ and the invariants $h^0(\mathcal{J}_k^m X)$ and $h^1(\mathcal{J}_k^m X)$. We show, by examples, how to construct jet differentials explicitly, in terms of the defining polynomial, in the case of curves of degree $d \geq 4$ in $\mathbb{P}^2$. Jet bundles may also be defined for varieties defined over fairly general fields (even in positive characteristic). The explicit construction of sections of powers of the canonical bundle, $\mathcal{K}_X^m$, was useful in the solution of the "strong uniqueness polynomial problem" (see Section 2 and the articles [An et al. 2004] in the complex case and [An et al. 2003a; 2003b] in the case of positive characteristic).

The formulas for invariants of the jet differentials for surfaces (the Chern numbers, the index, the Euler characteristic, the dimensions of cohomology groups) are given in Sections 3, 4 and 7. The calculations are similar to those over curves, though combinatorially much more complicated. We provide computations in special cases; the details are given in [Stoll and Wong 2002]. For example the explicit computation in Section 7 (see Theorem 7.7) shows that, for a smooth hypersurface in $\mathbb{P}^3$ the Euler characteristic $\chi(\mathcal{J}_2^m X)$ is big if and only if the degree is ≥ 16.

In Section 6 we prove a Schwarz Lemma for jet differentials. This is the generalization of the classical result for differential forms on curves: if

$$\omega \in H^0(X, \mathcal{K}_X \otimes [-D]),$$

that is, if ω is a regular 1-form vanishing on an effective ample divisor D in the curve X, then $f^*(\omega) \equiv 0$ for any holomorphic map $f : \mathbb{C} \to X$. This says that f' vanishes identically, that is, f is a constant. The proof given in Section 6 of the Schwarz Lemma for jets $j^k f$ has been in circulation since 1994 but was never formally published; it was used, for example, in the thesis of Jung [1995] and by Cherry–Ru in the context of p-adic jet differentials.

For surfaces of general type the fact that $\mathcal{J}_k^{k!}X$ is big for $k \gg 0$ is equivalent to the "hyperplane" line bundle $\mathcal{L}_k^{k!}$ being big on $\mathbb{P}(J^k X)$. (Note that $\mathcal{L}_k^m$ is locally free only if m is divisible by $k!$; see Section 5 for more details.) Schwarz's Lemma then implies that the lifting $[j^k f] : \mathbb{C} \to \mathbb{P}(J^k X)$ of a holomorphic map $f : \mathbb{C} \to X$ must be contained in some divisor $Y \subset \mathbb{P}(J^k X)$. The idea is to show that $\mathcal{L}_k^{k!}|_Y$ is again big so that the image of $[j^k f]$ is contained in a divisor Z of Y. Then we show that $\mathcal{L}_k^{k!}|_Z$ is big and continue until we reach the critical dimension $n + k - 1 = k + 1$. For surfaces of general type the sheaf of 1-jet differentials $\mathcal{L}_1^1$ is big if $c_1^2 > c_2$. In order for the restriction of $\mathcal{L}_1^1$ to subvarieties to be big, the condition that the index $c_1^2 - 2c_2$ is positive is required. The proof is based on the intersection theory of the projectivized tangent bundle $\mathbb{P}(TX)$ and the fact that the cotangent bundle T^*X of a surface of general type is semistable (in the sense of Mumford–Takemoto) relative to the canonical class. As remarked earlier the k-jet differentials are not semistable for any $k \geq 2$. However, by our explicit computation, for *minimal* surfaces of general type the index of $\mathcal{J}_k^{k!}X$ is positive for $k \gg 0$. Indeed, we may write the index as $\iota(\mathcal{J}_k^{k!}X) = c(\alpha_k c_1^2 - \beta_k c_2)$ where c, α_k and β_k are positive and we show that $\lim_{k \to \infty} \alpha_k / \beta_k = \infty$ (see Corollary 3.10). This is crucial in showing that $\mathcal{L}_k^{k!}|_Z$ is big in the cutting procedure. For example, for a smooth hypersurface of degree 5 in $\mathbb{P}^3$, $\iota(\mathcal{J}_k^{k!}X)$ is positive and the ratio α_k / β_k must be greater than 11 in order to establish the degeneracy of a map from $\mathbb{C}$ to X. Using the explicit expressions for α_k and β_k we show, with the aid of computer, that this occurs precisely for $k \geq 199$ (see the table at the end of Section 3 and Example 7.6 in Section 7). However, in order for the index of the restriction of the sheaf to subvarieties (in the cutting procedure) to be positive (verifying the hyperbolicity of X), k must be even larger. Using our formulas in the proof of Theorem 7.20, Professor B. Hu, using the computer, checked that $k \geq 2283$ is sufficient.

NOTE. *Experts who are familiar with parametrized jet bundles and are interested mainly in the proof of the Kobayashi conjecture may skip the first five sections (with the exception of Theorem 3.9 and Corollary 3.10) and proceed directly to Sections 6 and 7.*

1. Holomorphic Jet Bundles

SUMMARY. *Two notions of jet bundles, the full and the parametrized bundles, are introduced. The parametrized jet bundle is only a $\mathbb{C}^*$-bundle, not a vector bundle in general. For the resolution of the Kobayashi conjecture, as dictated by analysis, it is necessary to work with the parametrized jet bundle. (See Section 6 on the Schwarz Lemma.) Some basic facts are recalled here, all of which may be found in [Green and Griffiths 1980; Stoll and Wong 2002].*

There are, in the literature, two different concepts of jet bundles of a complex manifold. The first is used by analysts (PDE), algebraic geometers [Chandler

1995; 1998a; 2002] and also by number theorists (see Faltings's work on rational points of an ample subvariety of an abelian variety and integral points of the complement of an ample divisor of an abelian variety [Faltings 1991]); it was used implicitly in [Ru and Wong 1991] (see also [Wong 1993b]) for the proof that there are only finitely many integral points in the complement of $2n+1$ hyperplanes in general position in $\mathbb{P}^n$. The second is the jet bundles introduced by Green and Griffiths [1980] (see also [Stoll and Wong 2002]). The first notion shall henceforth be referred to as the *full jet bundle* and these bundles are holomorphic vector bundles (locally free). The second notion of jet bundle shall be referred to as the *parametrized jet bundle*. These bundles are coherent sheaves that are holomorphic $\mathbb{C}^*$-bundles which, in general, are not locally free.

For a complex manifold X the (locally free) sheaf of germs of holomorphic tangent vector fields (differential operators of order 1) of X shall be denoted by T^1X or simply TX. An element of T^1X acts on the sheaf of germs of holomorphic functions by differentiation:

$$(D, f) \in T^1X \times \mathcal{O}_X \mapsto Df \in \mathcal{O}_X$$

and the action is linear over $\mathbb{C}$; in symbols, $D \in \mathcal{H}om_{\mathbb{C}}(\mathcal{O}_X, \mathcal{O}_X)$. This concept may be extended as follows:

DEFINITION 1.1. Let X be a complex manifold of dimension n. The *sheaf of germs of holomorphic k-jets* (differential operators of order k), denoted T^kX, is the subsheaf of the sheaf of germs of homomorphisms $\mathcal{H}om_{\mathbb{C}}(\mathcal{O}_X, \mathcal{O}_X)$ consisting of elements (differential operators) of the form

$$\sum_{j=1}^{k} \sum_{i_j \in \mathbb{N}} D_{i_1} \circ \cdots \circ D_{i_j},$$

where $D_{i_j} \in T^1X$. In terms of holomorphic coordinates $z_1, \ldots, z_n$ an element of T^kX is expressed as

$$\sum_{j=1}^{k} \sum_{1 \le i_1, \ldots, i_j \le n} a_{i_1, \ldots, i_j} \frac{\partial^j}{\partial z_{i_1} \ldots \partial z_{i_j}},$$

where the coefficients $a_{i_1, \ldots, i_j}$ are symmetric in the indices $i_1, \ldots, i_j$. The bundle T^kX is locally free. One may see this by observing that $T^{k-1}X$ injects into T^kX and there is an exact sequence of sheaves:

$$0 \to T^{k-1}X \to T^kX \to T^kX/T^{k-1}X \to 0, \tag{1.1}$$

where $T^kX/T^{k-1}X \cong \bigodot^k T^1X$ is the sheaf of germs of k-fold symmetric products of T^1X. These exact sequences imply, by induction, that T^kX is locally free as each sheaf $\bigodot^k T^1X$, a symmetric product of the tangent sheaf, is locally free. A proof of (1.1) can be found in [Stoll and Wong 2002].

The parametrized k-jet bundles for complex manifolds are introduced by Green–Griffiths. (These are special cases of the general theory of jets due to Ehresmann [1952] for differentiable manifolds.) These bundles are defined as follows. Let $\mathcal{H}_x, x \in X$, be the sheaf of germs of holomorphic curves: $\{f : \Delta_r \to X, \ f(0) = x\}$ where Δ_r is the disc of radius r in $\mathbb{C}$. For $k \in \mathbb{N}$, define an equivalence relation $\sim_k$ by designating two elements $f, g \in \mathcal{H}_x$ as k-equivalent if $f_j^{(p)}(0) = g_j^{(p)}(0)$ for all $1 \le p \le k$, where $f_i = z_i \circ f$ and $z_1, \dots, z_n$ are local holomorphic coordinates near x. The sheaf of *parametrized k-jets* is defined by

$$J^k X = \bigcup_{x \in X} \mathcal{H}_x / \sim_k . \tag{1.2}$$

Elements of $J^k X$ will be denoted by $j^k f(0) = (f(0), f'(0), \dots, f^{(k)}(0))$. The fact that $J^k X, k \ge 2$, is in general not locally free may be seen from the nonlinearity of change of coordinates:

$$(w_j \circ f)' = \sum_{i=1}^n \frac{\partial w_j}{\partial z_i}(f)(z_i \circ f)',$$

$$(w_j \circ f)'' = \sum_{i=1}^n \frac{\partial w_j}{\partial z_i}(f)(z_i \circ f)'' + \sum_{i,k=1}^n \frac{\partial^2 w_j}{\partial z_i \partial z_k}(f)(z_i \circ f)'(z_k \circ f)'$$

and for each k,

$$(w_j \circ f)^{(k)} = \sum_{i=1}^n \frac{\partial w_j}{\partial z_i}(f)(z_i \circ f)^{(k)} + P\left(\frac{\partial^l w_j}{\partial z_{i_1} \dots \partial z_{i_l}}(f), (w_j \circ f)^{(l)}\right),$$

where P is an integer-coefficient polynomial in $\partial^l w_j / \partial z_{i_1} \dots \partial z_{i_l}$ and $(w_j \circ f)^{(l)}$ for $j = 1, \dots, n$ and $l = 1, \dots, k$. There is, however, a natural $\mathbb{C}^*$-action on $J^k X$ defined via parameterization. Namely, for $\lambda \in \mathbb{C}^*$ and $f \in \mathcal{H}_x$ a map $f_\lambda \in \mathcal{H}_x$ is defined by $f_\lambda(t) = f(\lambda t)$. Then $j^k f_\lambda(0) = (f_\lambda(0), f'_\lambda(0), \dots, f_\lambda^{(k)}(0)) = (f(0), \lambda f'(0), \dots, \lambda^k f^{(k)}(0))$. So the C^*-action is given by

$$\lambda \cdot j^k f(0) = (f(0), \lambda f'(0), \dots, \lambda^k f^{(k)}(0)). \tag{1.3}$$

DEFINITION 1.2. The *parametrized k-jet bundle* is defined to be $J^k X$ together with the $\mathbb{C}^*$-action defined by (1.3) and shall simply be denoted by $J^k X$.

It is clear that, for a complex manifold of (complex) dimension n, $J^k X$ is a holomorphic $\mathbb{C}^*$-bundle of rank $r = kn$ and $T^k X$ is a holomorphic vector bundle of rank $r = \sum_{i=1}^k C_i^{n+i-1}$ where C_i^j are the usual binomial coefficients. Although $J^1 X = T^1 X = TX$ these bundles differ for $k \ge 2$. The nonlinearity of the change of coordinates formulas above shows that there is in general no natural way of injecting $J^{k-1} X$ into $J^k X$ as opposed to the case of $T^k X$ (see (1.1)). There is however a natural projection map (the *forgetting map*) $p_{kl} : J^k X \to J^l X$ for any $l \le k$ defined simply by

$$p_{kl}(j^k f(0)) = j^l f(0), \tag{1.4}$$

which then respects the $\mathbb{C}^*$-action defined by (1.3) and so is a $\mathbb{C}^*$-bundle morphism. If $\Phi : X \to Y$ is a holomorphic map between the complex manifolds X and Y then the usual differential $\Phi_* : T^1X \to T^1Y$ is defined. More generally, the k-th order differential $\Phi_{k*} : T^kX \to T^kY$ is given by

$$\Phi_{k*} = (D_1 \circ \cdots \circ D_k)(g) \overset{\text{def}}{=} D_1 \circ \cdots \circ D_k(g \circ \Phi) \qquad (1.5)$$

for any $g \in \mathcal{O}_Y$. The k-th order induced map for the parametrized jet bundle, denoted $J^k\Phi : J^kX \to J^kY$, can also be defined:

$$J^k\Phi(j^kf(0)) \overset{\text{def}}{=} (\Phi \circ f)^{(k)}(0) \qquad (1.6)$$

for any $j^kf(0) \in J^kX$. For the parametrized jet bundle J^kX there is another notion closely related to the differential: the natural lifting of a holomorphic curve. Namely, given any holomorphic map $f : \Delta_r \to X (0 < r \le \infty)$, the lifting $j^kf : \Delta_{r/2} \to J^kX$ is defined by

$$j^kf(\zeta) = j^kg(0), \quad \zeta \in \Delta_{r/2} \qquad (1.7)$$

where $g(\xi) = f(\zeta + \xi)$ is holomorphic for $\xi \in \Delta_{r/2}$.

DEFINITION 1.3. The dual of the full jet bundles T^kX shall be called the *sheaf of germs of k-jet forms* and shall be denoted by T_k^*X. For $m \in \mathbb{N}$ the m-fold symmetric product shall be denoted by $\odot^m T_k^*X$ and its global sections shall be called *k-jet forms of weight m*.

In this article we shall focus on the dual of the parametrized jet bundles defined as follows.

DEFINITION 1.4. The dual of J^kX (i.e., the sheaf associated to the presheaf consisting of holomorphic maps $\omega : j^kX|_U \to \mathbb{C}$ such that $\omega(\lambda \cdot j^kf) = \lambda^m \omega(j^kf)$ for all $\lambda \in \mathbb{C}^*$ and positive integer m) shall be referred to as the *sheaf of germs of k-jet differentials of weight m* and shall be denoted by $\mathcal{J}_k^mX$.

It follows from the definition of the $\mathbb{C}^*$-action on J^kX that a k-jet differential ω of weight m is of the form:

$$\omega(j^kf) = \sum_{|I_1|+2|I_2|+\cdots+k|I_k|=m} a_{I_1,\ldots,I_k}(z)(f')^{I_1} \ldots (f^{(k)})^{I_k}, \qquad (1.8)$$

where $a_{I_1,\ldots I_k}$ are holomorphic functions, $I_j = (i_{1j}, \ldots, i_{nj}), n = \dim X$ are the multi-indices with each i_{lj} being a nonnegative integer and $|I_j| = i_{1j} + \cdots + i_{nj}$. In terms of local coordinates $(z_1, \ldots, z_n)$,

$$(f')^{I_1} \ldots (f^{(k)})^{I_k} = (f_1')^{i_{11}} \ldots (f_n')^{i_{n1}} \ldots (f_1^{(k)})^{i_{1k}} \ldots (f_n^{(k)})^{i_{nk}}.$$

Further, the coefficients $a_{I_1,\ldots I_k}(z)$ are symmetric with respect to the indices in each I_j. More precisely,

$$a_{(i_{\sigma_1(1)1},\ldots,i_{\sigma_1(n)1}),\ldots,(i_{\sigma_k(1)k},\ldots,i_{\sigma_k(n)k})} = a_{(i_{11},\ldots,i_{n1}),\ldots,(i_{1k},\ldots,i_{nk})},$$

where each σ_j, $j = 1, \ldots, n$, is a permutation on n elements. For example, $(f_1')^2(f_2'')^2 + f_1''' f_2' f_2'' + f_1'''' f_2'' + f_1''''' f_2'$ is a 5-jet differential of weight 6.

There are several important naturally defined operators on jet differentials; the first is a *derivation* $\delta : \mathcal{J}_k^m X \to \mathcal{J}_{k+1}^{m+1} X$ defined by

$$\delta\omega(j^{k+1}f) \overset{\text{def}}{=} (\omega(j^k f))'. \tag{1.9}$$

Note that in contrast to the exterior differentiation of differential forms $\delta \circ \delta \neq 0$ on jet differentials. In particular, given a holomorphic function ϕ defined on some open neighborhood U in X, the k-th iteration $\delta^{(k)}$ of δ,

$$\delta^{(k)} \phi(j^k f) = (\phi \circ f)^{(k)}, \tag{1.10}$$

is a k-jet differential of weight k.

Another difference between jet differentials and exterior differential forms is that a lower order jet differential can be naturally associated to a jet differential of higher order. The natural projection $p_{kl} : J^k X \to J^l X$ defined by $p_{kl}(j^k f) = j^l f$, for $k \geq l$, induces an injection $p_{kl}^* : \mathcal{J}_l^m X \to \mathcal{J}_k^m X$ defined by "forgetting" those derivatives higher than l:

$$p_{kl}^* \omega(j^k f) \overset{\text{def}}{=} \omega(p_{kl}(j^k f)) = \omega(j^l f). \tag{1.11}$$

We shall simply write $\omega(j^k f) = \omega(j^l f)$ if no confusion arises.

The wedge (exterior) product of differential forms is replaced by taking symmetric product; the symmetric product of a k-jet differential of weight m and a k'-jet differential of weight m' is a $\max\{k, k'\}$-jet differential of weight $m + m'$.

EXAMPLE 1.5. A 1-jet differential is a differential 1-form $\omega = \sum_{i=1}^n a_i(z)dz_i$. Let $f = (f_1, \ldots, f_n) : \Delta_r \to X$ be a holomorphic map. Then

$$\omega(j^1 f) = \sum_{i=1}^n a_i(f)dz_i(f') = \sum_{i=1}^n a_i(f)f_i'$$

and $\delta\omega$ is a 2-jet differential of weight 2, given by

$$\delta\omega(j^2 f) = (\omega(j^1 f))' = \left(\sum_{i=1}^n a_i(f)f_i' \right)' = \sum_{i,j=1}^n \frac{\partial a_i}{\partial z_j}(f)f_i'f_j' + \sum_{i=1}^n a_i(f)f_i''.$$

Analogously, $\delta^2\omega$ is a 3-jet differential of weight 3, given by

$$\delta^2\omega(j^3 f) = \sum_{i,j=1}^n \frac{\partial^2 a_i}{\partial z_j \partial z_k}(f)f_i'f_j'f_k' + 3\sum_{ij=1}^n \frac{\partial a_i}{\partial z_j}(f)f_i''f_j' + \sum_{i=1}^n a_i(f)f_i'''.$$

The concept of jet bundles extends also to singular spaces. Let us remark on how this may be defined. One may locally embed an open set U of X as a subvariety in a smooth variety $U \subset Y$. At a point $x \in U$ the stalk jet $(J^k Y)_x$ is then defined, as Y is smooth. The stalk $(J^k X)_x$ is defined as the subset

$$\{ j^k f(0) \in (J^k Y)_x \mid f : \Delta_r \to Y \text{ is holomorphic}, \ f(0) = x \text{ and } f(\Delta_r) \subset U \}.$$

From the differential geometric point of view, properties of the full jet bundle $T_k^* X$, as a vector bundle, are reflected by the curvatures of hermitian metrics along its fibers. The parametrized jet bundles, however, are only $\mathbb{C}^*$-bundles hence can only be equipped with *Finsler* metrics. A *Finsler pseudometric* (or a k-jet pseudometric) on X is a map $\rho = \rho_k : J^k X \to \mathbb{R}_{\geq 0}$ satisfying the condition

$$\rho(\lambda \cdot \boldsymbol{j}_k) = |\lambda| \rho(\boldsymbol{j}_k)$$

for all $\lambda \in \mathbb{C}$ and $\boldsymbol{j}_k \in J^k X$. It is said to be a *Finsler metric* if it is positive outside of the zero section. A $(k-1)$-jet (pseudo)-metric ($k \geq 2$) ρ_{k-1} can be considered as a k-jet (pseudo)-metric by the *forgetting map*:

$$\rho_{k-1}(\boldsymbol{j}_k) := \rho_{k-1}(\boldsymbol{j}_{k-1}).$$

where $\boldsymbol{j}_k = j^k f(0)$ and $\boldsymbol{j}_{k-1} = j^{k-1} f(0)$. Define, for $\boldsymbol{j}_k \in J^k M, k \geq 1$,

$$\kappa_k(\boldsymbol{j}_k) = \inf \{1/r\}, \qquad (1.12)$$

where the infimum is taken over all r such that

$$\mathcal{H}_r^k(\zeta) = \{f : \Delta_r \to X \mid f \text{ is holomorphic and } j^k f(0) = \boldsymbol{j}_k\}$$

is nonempty. For $k = 1$ this is the usual Kobayashi–Royden pseudometric on $J^1 X = TX$. Henceforth we shall refer to κ_k as the *k-th infinitesimal Kobayashi–Royden pseudometric*. We shall also say that X is *k-jet hyperbolic* if κ_k is indeed a Finsler metric; that is, $\kappa_k(\boldsymbol{j}_k) > 0$ for each nonzero k-jet $\boldsymbol{j}_k$. Thus 1-jet hyperbolicity is the same as Kobayashi hyperbolicity. Since a holomorphic map $f : \Delta_r \to X$ such that $j^k f(0) = (z, \zeta_1, \ldots, \zeta_k)$ also satisfies $j^{k-1} f(0) = (z, \zeta_1, \ldots, \zeta_{k-1})$, we obtain:

$$\kappa_k(z, \zeta_1, \ldots, \zeta_k) \geq \kappa_{k-1}(z, \zeta_1, \ldots, \zeta_{k-1}). \qquad (1.13)$$

From this we see that $(k-1)$-jet hyperbolicity implies k-jet hyperbolicity.

REMARK 1.6. The notion introduced above is not to be confused with the k-dimensional ($1 \leq k \leq n = \dim X$) Kobayashi pseudometric in the literature (see [Lang 1987], for example); n-dimensional Kobayashi hyperbolicity is more commonly known as measure hyperbolicity.

In general the k-th Kobayashi–Royden metric does not have a good regularity property. It is well-known that κ_1 is upper-semicontinuous (see [Royden 1971] or [Kobayashi 1970]); a similar proof shows that the same is true for κ_k for any k. It is also known that κ_1, in general, is not continuous; however it is continuous if X is complete hyperbolic (that is, the distance function associated to the metric κ_1 is complete). In particular, κ_1 is continuous on a compact hyperbolic manifold X. On the other hand, using a partition of unity one may construct k-jet metrics that are continuous everywhere and smooth outside of the zero section. Consider first the space

$$\mathbb{C}^n_k = \underbrace{\mathbb{C}^n \times \ldots \times \mathbb{C}^n}_{k}$$

with $\mathbb{C}^*$-action $\lambda \cdot (z_1, z_2, \ldots, z_k) \mapsto (\lambda z_1, \lambda^2 z_2, \ldots, \lambda^k z_k)$. Define, for $Z = (z_1, \ldots, z_k) \in \mathbb{C}^n_k$:

$$r_k(Z) = (|z_1|^{2k!} + |z_2|^{2k!/2} + \cdots + |z_k|^{2k!/k})^{1/2k!} \qquad (1.14)$$

where $|z_i|$ is the usual Euclidean norm on $\mathbb{C}^n$. Observe that $r_k(\lambda \cdot Z) = |\lambda| \, r_k(Z)$ and that r_k is continuous on $\mathbb{C}^n_k$, smooth outside of the origin. Indeed $r_k^{2k!}$ is smooth on all of $\mathbb{C}^n_k$. Alternatively we can take

$$r_k(Z) = |z_1| + |z_2|^{1/2} + \cdots + |z_k|^{1/k}, \qquad (1.15)$$

which also satisfies $r_k(\lambda \cdot Z) = |\lambda| \, r_k(Z)$ and is continuous on $\mathbb{C}^n_k$, smooth outside of the set $[z_1 \cdot z_k = 0]$. On a local trivialization $J^k X|_U \cong U \times \mathbb{C}^n_k$ we define simply $\rho_k(z, Z) = r_k(Z)$ on $J^k X|_U$ so that a global k-jet metric is defined via a partition of unity subordinate to a locally finite trivialization cover. This general construction is of limited use as it does not take into account the geometry of the manifold.

In the case of a compact manifold a more useful construction can be carried out by taking a basis $\omega_1, \ldots, \omega_N$ of global holomorphic k-jet differentials (provided that these exist), and defining

$$\rho_k(j^k f) = \left(\sum_{i=1}^{N} |\omega_i(j^k f)|^2 \right)^{1/2}. \qquad (1.16)$$

Then, since a jet differential is a linear functional on the k-jet bundle (that is, $\omega(\lambda \cdot j^k f) = \lambda \omega(j^k f)$), we see readily that $\rho_k(\lambda \cdot j^k f) = |\lambda| \rho_k(j^k f)$. It is clear from the definition that ρ_k is continuous on $J^k X$, real analytic on $J^k X \setminus \{\text{zero section}\}$; indeed, ρ_k^2 is real analytic on $J^k X$. For $k = 1$ use a basis of global holomorphic 1-forms. The number $N = h^0(T^* X)$ is the irregularity of X (for a Riemann surface this is just the genus of X). Thus the invariants $h^0(J^m_k X)$ play an important role in the determination of hyperbolicity.

The jet bundles may be defined, in an analogous way, over fairly general fields. We conclude this section by introducing a very interesting problem:

INTERPOLATION PROBLEM. *Find all subvarieties in $\mathbb{P}^n = \mathbb{P}^n(\mathbb{K})$ (where $\mathbb{K}$ is an infinite field) of a given degree d with prescribed jet spaces at a finite number of points. More precisely, given subspaces $V_1 \subset T^k_{x_1} \mathbb{P}^n, \ldots, V_N \subset T^k_{x_N} \mathbb{P}^n$ (or $V_1 \subset J^k_{x_1} \mathbb{P}^n, \ldots, V_N \subset J^k_{x_N} \mathbb{P}^n$), find all varieties X of degree d such that $V_1 = T^k_{x_1} X, \ldots, V_N \subset T^k_{x_N} X$.*

At this time little is known about the problem for the bundle $J^k X$ however much is known in the case of $T^k X$. For example, the following is known (see [Chandler

1998a; 2002] or else [Alexander 1988; Alexander and Hirschowitz 1992a; 1992b; 1995]).

THEOREM 1.7. *Let Ψ be a general collection of d points in $\mathbb{P}^n$. The codimension in $H^0(\mathcal{O}_{\mathbb{P}^n}(3))$ of the space of sections singular on Ψ is $\min\{(n+1)d, (n+3)!/3!\,n!\}$ unless $n = 4$ and $d = 7$. More generally, the codimension in $H^0(\mathcal{O}_{\mathbb{P}^n}(m))$ of the space of sections singular on Ψ is equal to $\min\{(n+1)d, (n+m)!/m!\,n!\}$ unless $(n, m, d) = (2, 4, 5), (3, 4, 9), (4, 3, 7)$ or $(4, 4, 14)$.*

The problem is related also to the *Waring problem* for linear forms: when can a general degree m form in $n+1$ variables be expressed as a sum of m-th powers of linear forms? Let $PS(n, m, d)$ be the space of homogeneous polynomials in $n+1$ variables expressible as $L_1^m + \cdots + L_d^m$, where $L_1, \ldots, L_d$ are linear forms. Then:

THEOREM 1.8. *With the notation above, we have*

$$\dim PS(n, m, d) = \min\{(n+1)d, (n+m)!/m!\,n!\}$$

unless $(n, m, d) = (2, 4, 5), (3, 4, 9), (4, 3, 7)$ and $(4, 4, 14)$.

For details, see the articles by Chandler and by Alexander and Hirschowitz in the references, as well as [Iarrobino and Kanev 1999].

2. Chern Classes and Cohomology Groups
The Case of Curves

SUMMARY. *The theory of parametrized jet bundles is complicated by their not being vector bundles. This section discusses the case of curves to acquaint readers with the theory in the simplest situation. The theory is based on the fundamental result of Green and Griffiths on the filtration of the parametrized jet bundles (see Theorem 2.3 and Corollary 2.4). The explicit computations of this section have numerous applications (see for example [An et al. 2003a; 2004; 2003b]).*

In this section we compute the Chern numbers and the invariants $h^i(\mathcal{J}_k^m X)$, $i = 0, 1$, of the jet bundles for curves. In the case of curves in $\mathbb{P}^2$ we are interested in finding an explicit expression of a basis for $h^0(\mathcal{J}_k^m X)$. The procedure introduced here for the construction works as well for singular curves and in varieties defined over general differential fields. For applications in this direction to the *strong uniqueness polynomial problem* and the *unique range set problem*; see [An et al. 2004] in the complex case and [An et al. 2003a; 2003b] in the case of fields of positive characteristic.

For the full jet bundles the computation of Chern classes and cohomology groups is straightforward. Dualizing the defining sequence (1.1) we get an exact sequence

$$0 \to \bigodot^k T_1^* X \to T_k^* X \to T_{k-1}^* X \to 0. \tag{2.1}$$

For example, for $k = 3$ the exact sequences

$$0 \to \bigodot^3 T_1^* X \to T_3^* X \to T_2^* X \to 0,$$
$$0 \to \bigodot^2 T_1^* X \to T_2^* X \to T_1^* X \to 0$$

and Whitney's Formula yields

$$c_1(T_3^* X) = c_1(T_2^* X) + c_1(\bigodot^3 T_1^* X) = c_1(T_1^* X) + c_1(\bigodot^2 T_1^* X) + c_1(\bigodot^3 T_1^* X).$$

In general, we have, by induction:

THEOREM 2.1. *The first Chern number of the bundle $T_k^* X$ is given by*

$$c_1(T_k^* X) = \sum_{j=1}^{k} c_1(\bigodot^j T_1^* X).$$

In particular, if X is a Riemann surface,

$$c_1(T_k^* X) = \sum_{j=1}^{k} j c_1(T_1^* X) = \frac{k(k+1)}{2} c_1(\mathcal{K}_X) = k(k+1)(g-1)$$

where $\mathcal{K}_X = T_1^* X$ is the canonical bundle of X and g is the genus.

For a line bundle $\mathcal{L}$ and nonnegative integer i the i-fold tensor product is denoted by $\mathcal{L}^i$ and $\mathcal{L}^{-i}$ is the dual of $\mathcal{L}^i$. (Recall that tensor product and symmetric product on line bundles are equivalent.)

THEOREM 2.2. *Let X be a smooth curve of genus $g \geq 2$. Then $h^0(T_k^* X) = k^2(g-1)+1$ and $h^1(T_k^* X) = 1$.*

PROOF. By Riemann–Roch for curves,

$$h^0(\mathcal{K}_X^i) - h^1(\mathcal{K}_X^i) = \chi(\mathcal{K}_X^i) = \chi(\mathcal{O}_X) + c_1(\mathcal{K}_X^i)$$
$$= h^0(\mathcal{O}_X) - h^1(\mathcal{O}_X) + 2(g-1)i = 1 - g + 2(g-1)i$$

for any nonnegative integer i. Thus $h^0(\mathcal{K}_X^i) = h^1(\mathcal{K}_X^i) + (2i-1)(g-1) = h^0(\mathcal{K}_X^{1-i}) + (2i-1)(g-1)$. As $h^0(\mathcal{K}_X^{1-i}) = 1$ for $i = 1$ and $h^0(\mathcal{K}_X^{1-i}) = 0$ for $i \geq 2$ we get

$$h^0(\mathcal{K}_X^i) = \begin{cases} 0, & i < 0, \\ 1, & i = 0, \\ g, & i = 1, \\ (2i-1)(g-1), & i \geq 2. \end{cases} \qquad (2.2a)$$

By duality, $h^1(\mathcal{K}_X^i) = h^0(\mathcal{K}_X^{1-i})$; hence

$$h^1(\mathcal{K}_X^i) = \begin{cases} 0, & i \geq 2, \\ 1, & i = 1, \\ g, & i = 0, \\ (1-2i)(g-1), & i < 0. \end{cases} \qquad (2.2b)$$

From the short exact sequence (2.1) we get the exact sequence

$$0 \to H^0(\mathcal{K}_X^k) \to H^0(T_k^* X) \to H^0(T_{k-1}^* X) \to H^1(\mathcal{K}_X^k) \to$$
$$\to H^1(T_k^* X) \to H^1(T_{k-1}^* X) \to 0.$$

From (2.2a,b) we deduce that, for $k \geq 2$, $H^1(T_k^* X) = H^1(T_{k-1}^* X)$ and that

$$h^0(T_k^* X) = h^0(T_{k-1}^* X) + h^0(\mathcal{K}_X^k).$$

These imply that

$$h^1(T_k^* X) = h^1(T^* X) = h^1(\mathcal{K}_X) = h^0(\mathcal{O}_X) = 1$$

for all $k \geq 1$ and that

$$h^0(T_k^* X) = \sum_{i=1}^{k} h^0(\mathcal{K}_X^i) = g + \sum_{i=2}^{k} (2i - 1)(g - 1) = k^2(g - 1) + 1. \qquad \square$$

The computation of Chern classes and cohomology groups for the parametrized jet bundles is somewhat more complicated. This depends on the fundamental filtration for these bundles due to Green–Griffiths. Let

$$0 \to \mathcal{S}' \to \mathcal{S} \to \mathcal{S}'' \to 0$$

be an exact sequence of sheaves. Then for any m there is a filtration

$$0 = \mathcal{F}^0 \subset \mathcal{F}^1 \subset \cdots \subset \mathcal{F}^m \subset \mathcal{F}^{m+1} = \bigodot^m \mathcal{S}$$

of the symmetric product $\bigodot^m \mathcal{S}$, such that $\mathcal{F}^i/\mathcal{F}^{i-1} \cong \bigodot^i \mathcal{S}' \otimes \bigodot^{m-i} \mathcal{S}''$. Analogously, for the exterior product $\bigwedge^m \mathcal{S}$ we have a filtration

$$0 = \mathcal{F}^0 \subset \mathcal{F}^1 \subset \cdots \subset \mathcal{F}^m \subset \mathcal{F}^{m+1} = \bigwedge^m \mathcal{S}$$

such that $\mathcal{F}^i/\mathcal{F}^{i-1} \cong \bigwedge^i \mathcal{S}' \otimes \bigwedge^{m-i} \mathcal{S}''$. These filtrations connect the cohomology groups of higher symmetric (resp. exterior) products to the cohomology groups of lower symmetric (resp. exterior) products. The analogue of these is the following theorem of Green and Griffiths (the proof can be found in [Stoll and Wong 2002]):

THEOREM 2.3. *There exists a filtration of $\mathcal{J}_k^m X$:*

$$\mathcal{J}_{k-1}^m X = \mathcal{F}_k^0 \subset \mathcal{F}_k^1 \subset \cdots \subset \mathcal{F}_k^{[m/k]} = \mathcal{J}_k^m X$$

(where $[m/k]$ is the greatest integer smaller than or equal to m/k) such that

$$\mathcal{F}_k^i/\mathcal{F}_k^{i-1} \cong \mathcal{J}_{k-1}^{m-ki} X \otimes (\bigodot^i T^* X).$$

As an immediate consequence [Green and Griffiths 1980], we have:

COROLLARY 2.4. *Let X be a smooth projective variety. Then $\mathcal{J}_k^m X$ admits a composition series whose factors consist precisely of all bundles of the form: $(\bigodot^{i_1} T^*X) \otimes \cdots \otimes (\bigodot^{i_k} T^*X)$ where i_j ranges over all nonnegative integers satisfying $i_1 + 2i_2 + \cdots + ki_k = m$. The first Chern number of $\mathcal{J}_k^m X$ is given by*

$$c_1(\mathcal{J}_k^m X) = \sum_{\substack{i_1 + 2i_2 + \cdots + ki_k = m \\ i_j \in \mathbb{Z}_{\geq 0}}} c_1\big((\bigodot^{i_1} T^*X) \otimes \cdots \otimes (\bigodot^{i_k} T^*X)\big).$$

In particular, if X is a curve then

$$c_1(\mathcal{J}_k^m X) = \sum_{\substack{i_1 + 2i_2 + \cdots + ki_k = m \\ i_j \in \mathbb{Z}_{\geq 0}}} (i_1 + i_2 + \cdots + i_k) c_1(T^*X).$$

EXAMPLE 2.5 [Stoll and Wong 2002]. It is clear that for $m < k$ the filtration degenerates and we have $\mathcal{J}_k^m X = \mathcal{J}_{k-1}^m X = \ldots = \mathcal{J}_m^m X$. In particular, $\mathcal{J}_2^1 X = \mathcal{J}_1^1 X = T^*X$. For $m = k = 2$, the filtration is given by

$$\bigodot^2 T^*X = \mathcal{J}_1^2 X = \mathcal{S}_2^0 \subset \mathcal{S}_2^1 = \mathcal{J}_2^2 X, \quad \mathcal{S}_2^1/\mathcal{S}_2^0 \cong T^*X,$$

so we have the exact sequence

$$0 \to \bigodot^2 T^*X \to \mathcal{J}_2^2 X \to T^*X \to 0.$$

Thus the first Chern numbers are related by the formula

$$c_1(\mathcal{J}_2^2 X) = c_1(\bigodot^2 T^*X) + c_1(T^*X).$$

Analogously, $\mathcal{J}_3^1 X = \mathcal{J}_2^1 X = \mathcal{J}_1^1 X = T^*X$ and $\mathcal{J}_3^2 X = \mathcal{J}_2^2 X$. The filtration of $\mathcal{J}_3^3 X$ is as follows:

$$\mathcal{J}_3^3 X = \mathcal{S}_3^1 \supset \mathcal{S}_3^0 = \mathcal{J}_2^3 X, \quad \mathcal{J}_3^3 X/\mathcal{J}_2^3 X = \mathcal{S}_3^1/\mathcal{S}_3^0 \cong T^*X.$$

Hence we have an exact sequence

$$0 \to \mathcal{J}_2^3 X \to \mathcal{J}_3^3 X \to T^*X \to 0.$$

Now the filtration of $\mathcal{J}_2^3 X$ is

$$\mathcal{J}_2^3 X = \mathcal{S}_2^1 \supset \mathcal{S}_2^0 = \mathcal{J}_1^3 X, \quad \mathcal{J}_2^3 X/\mathcal{J}_1^3 X \cong T^*X \otimes T^*X$$

and, since $\mathcal{J}_1^3 X = \bigodot^3 T^*X$, we have an exact sequence

$$0 \to \bigodot^3 T^*X \to \mathcal{J}_2^3 X \to T^*X \otimes T^*X \to 0.$$

From these two exact sequences we get

$$c_1(\mathcal{J}_3^3 X) = c_1(T^*X) + c_1(T^*X \otimes T^*X) + c_1(\bigodot^3 T^*X).$$

From basic representation theory (or linear algebra in this special case) we have $T^*X \otimes T^*X = \bigodot^2 T^*X \oplus \bigwedge^2 T^*X$ hence

$$c_1(\mathcal{J}_3^3 X) = c_1(T^*X) + c_1(\bigodot^2 T^*X) + c_1(\bigodot^3 T^*X) + c_1(\bigwedge^2 T^*X).$$

In representation theory $\bigwedge^2 T^* X$ is the *Weyl module* $W_{1,1}^* X$ associated to the partition $\{1, 1\}$ (see [Fulton and Harris 1991]). Thus we have

$$c_1(\mathcal{J}_3^3 X) = \sum_{j=1}^3 c_1(\bigodot^j T^* X) + c_1(W_{1,1}^* X).$$

In the special case of a Riemann surface $\bigwedge^2 T^* X$ is the zero-sheaf. Thus for a curve we have

$$c_1(\mathcal{J}_3^3 X) = (1 + 2 + 3)c_1(T^* X) = 6c_1(T^* X).$$

For $m = k = 4$, we have the filtrations $\mathcal{J}_4^4 X = S_4^1 \supset S_4^0 = \mathcal{J}_3^4 X$, $\mathcal{J}_4^4 X / \mathcal{J}_3^4 X = S_4^1 / S_4^0 \cong T^* X$, $\mathcal{J}_3^4 X = S_3^1 \supset S_3^0 = \mathcal{J}_2^4 X$, $\mathcal{J}_3^4 X / \mathcal{J}_2^4 X = S_3^1 / S_3^0 \cong T^* X \otimes T^* X$, and $\mathcal{J}_2^4 X = S_2^2 \supset S_2^1 \supset S_2^0 = \mathcal{J}_1^4 X$, with

$$\mathcal{J}_2^4 X / S_2^1 = \bigodot^2 T^* X, \quad S_2^1 / S_2^0 \cong T^* X \otimes (\bigodot^2 T^* X).$$

Thus the Chern number is given by

$$c_1(\mathcal{J}_4^4 X) = c_1(T^* X) + c_1(T^* X \otimes T^* X) + c_1(\bigodot^2 T^* X)$$
$$+ c_1(T^* X \otimes (\bigodot^2 T^* X)) + c_1(\bigodot^4 T^* X).$$

From elementary representation theory we obtain

$$T^* X \otimes (\bigodot^k T^* X) = W_{k,1}^* X \oplus (\bigodot^{k+1} T^* X)$$

where $W_{k,1}^*$ is the Weyl module associated to the partition $\{k, 1\}$ so that

$$c_1(\mathcal{J}_4^4 X) = c_1(\bigodot^2 T^* X) + \sum_{i=1}^4 c_1(\bigodot^i T^* X) + \sum_{i=1}^2 c_1(W_{j,1}^* X).$$

In particular, if X is a curve,

$$c_1(\mathcal{J}_4^4 X) = (1 + 2 + 2 + 3 + 4)c_1(T^* X) = 12c_1(T^* X).$$

The procedure can be carried out further; for instance,

$$c_1(\mathcal{J}_5^5 X) = \sum_{j=2}^3 c_1(\bigodot^j T^* X)$$
$$+ \sum_{j=1}^5 c_1(\bigodot^j T^* X) + \sum_{j=1}^2 c_1(W_{j,1}^* X) + \sum_{j=1}^3 c_1(W_{j,1}^* X),$$
$$c_1(\mathcal{J}_6^6 X) = c_1(T^* X) + 3c_1(T^* X \otimes T^* X) + 2c_1(T^* X \otimes (\bigodot^3 T^* X))$$
$$+ c_1(\bigodot^2 T^* X) + c_1(\bigodot^3 T^* X) + c_1((\bigodot^2 T^* X) \otimes (\bigodot^2 T^* X))$$
$$+ c_1(T^* X \otimes (\bigodot^4 T^* X)) + c_1(\bigodot^6 T^* X).$$

So if X is a curve we have

$$c_1(\mathcal{J}_5^5 X) = (1 + 2 + 3 + 2 + 3 + 4 + 5)c_1(T^* X) = 20c_1(T^* X),$$

$$c_1(\mathcal{J}_6^6 X) = (1 + 6 + 8 + 2 + 3 + 4 + 5 + 6)c_1(T^* X) = 35c_1(T^* X).$$

The calculation of the sum

$$\sum_{i_1+2i_2+\cdots+ki_k=m} i_1 + \cdots + i_k \tag{2.3}$$

can be carried out using standard combinatorial results which we now describe.

DEFINITION 2.6. (i) A *maximal set* of mutually conjugate elements of S_m (the symmetric group on m elements) is said to be a *class of S_m*.
(ii) A *partition* of a natural number m is a set of positive integers $i_1, \ldots, i_q$ such that $m = i_1 + \cdots + i_q$.

The following asymptotic result concerning the number of partitions of a positive integer m is well-known in representation theory and in combinatorics [Hardy and Wright 1970]:

THEOREM 2.7. *The number of partitions of m, the number of classes of S_m and the number of (inequivalent) irreducible representations of S_m are equal. This common number $p(m)$ is asymptotically approximated by the formula of Hardy–Ramanujan:*

$$p(m) \sim \frac{e^{\pi\sqrt{2m/3}}}{4m\sqrt{3}}.$$

The first few partition numbers are $p(1) = 1$, $p(2) = 2$, $p(3) = 3$, $p(4) = 5$, $p(5) = 7$, $p(6) = 11$, $p(7) = 15$, $p(8) = 22$, $p(9) = 30$, $p(10) = 42$, $p(11) = 56$, $p(12) = 77$, $p(13) = 101$. Consider first the case of partitioning a number by partitions of a *fixed* length k. Denote by $p_k(m)$ the number of *positive integral solutions* of the equation

$$x_1 + \cdots + x_k = m$$

with the condition that $1 \le x_k \le x_{k-1} \le \ldots \le x_1$. This number is equal to the number of *integer solutions* of the equation

$$y_1 + \cdots + y_k = m - k$$

with the condition that the solutions be *nonnegative* and $0 \le y_k \le y_{k-1} \le \ldots \le y_1$. If exactly i of the integers $\{y_1, \ldots, y_k\}$ are positive then these are the solutions of $x_1 + \cdots + x_i = m - k$ and so there are $p_i(m-k)$ of such solutions. Consequently we have (see [Stoll and Wong 2002] for more details):

LEMMA 2.8. *With the notation above we have: $p(m) = \sum_{k=1}^{m} p_k(m)$, where*

$$p_k(m) = \sum_{i=0}^{k} p_i(m-k),$$

for $1 \le k \le m$ and with the convention that $p_0(0) = 1, p_0(m) = 0$ if $m > 0$ and $p_k(m) = 0$ if $k > m$. Moreover, the number $p_k(m)$ satisfies the following recursive relation:

$$p_k(m) = p_{k-1}(m-1) + p_k(m-k).$$

EXAMPLE 2.9. We shall compute $p(6)$ and $p(7)$ using the preceding lemma. We have $p_1(m) = p_m(m) = 1$ and $p_2(m) = m/2$ or $(m-1)/2$ according to m being even or odd; thus $p_1(6) = 1, p_2(6) = 3, p_6(6) = 1$. Analogously, we have:

$$p_3(m) = p_2(m-1) + p_3(m-3),$$
$$p_4(m) = p_3(m-1) + p_4(m-4),$$
$$p_5(m) = p_4(m-1) + p_5(m-5),$$

so that, for example:

$$p_3(6) = p_2(5) + p_3(3) = 2 + 1 = 3,$$
$$p_4(6) = p_3(5) + p_4(2) = p_2(4) = 2,$$
$$p_5(6) = p_4(5) = p_3(4) = p_2(3) = 1.$$

Since $p(m) = \sum_{k=1}^{m} p_k(m)$ we have

$$p(6) = \sum_{k=1}^{6} p_k(6) = 1 + 3 + 3 + 2 + 1 + 1 = 11.$$

For $m = 7$ we have $p_1(7) = 1$, $p_2(7) = 3$, $p_7(7) = 1$, $p_3(7) = p_2(6) + p_3(4) = p_2(6) + p_2(3) = 4$, $p_4(7) = p_3(6) = 3$, $p_5(7) = p_4(6) = 2$, $p_6(7) = p_5(6) = 1$; hence

$$p(7) = \sum_{k=1}^{7} p_k(7) = 1 + 3 + 4 + 3 + 2 + 1 + 1 = 15.$$

For $k \leq m$ denote by $L_k(m)$ the sum of the lengths of all partitions λ of m whose length l_λ is at most k:

$$L_k(m) = \sum_{\lambda, l_\lambda \leq k} l_\lambda.$$

The next lemma follows from the definitions [Wong 1999; Stoll and Wong 2002]:

LEMMA 2.10. *With notation as above we have*

$$L_k(m) = \sum_{\lambda, l_\lambda \leq k} l_\lambda = \sum_{j=1}^{k} j p_j(m) = \sum_{\lambda, l_\lambda \leq k} \sum_{j=1}^{k} i_j,$$

where the sum on the right is taken over all partitions $\lambda = (\lambda_1, \ldots, \lambda_{\rho_\lambda})$ *of* m, $1 \leq \lambda_{l_\lambda} \leq \ldots \leq \lambda_2 \leq \lambda_1$, $l_\lambda \leq k$ *and* i_j *is the number of* j's *in* $\{\lambda_1, \ldots, \lambda_{l_\lambda}\}$.

For $k = m, L(m) = L_m(m)$ is *the total length of all possible partitions of* m. For example if $m = 6$ then $L(6) = 1 + 6 + 9 + 8 + 5 + 6 = 35$ and for $m = 7, L(7) = 1 + 6 + 12 + 12 + 10 + 6 + 7 = 54$. Indeed we have:

THEOREM 2.11. *If* X *is a nonsingular projective curve, the Chern number of* $\mathcal{J}_m^m X$ *is*

$$c_1(\mathcal{J}_m^m X) = L_m(m) c_1(\mathcal{K}_X) = \sum_{j=1}^{m} j p_j(m) c_1(\mathcal{K}_X),$$

where K_X is the canonical bundle of X.

There is a formula for the asymptotic behavior of $p_k(m)$:

THEOREM 2.12. *For k fixed and $m \to \infty$ the number $p_k(m)$ is asymptotically*

$$p_k(m) \sim \frac{m^{k-1}}{(k-1)!\, k!}.$$

We give below the explicit calculation of the above in the first few cases. For $m = k = 3$, we have $p(3) = 3$ and the possible indices are

	λ	l_λ	d_λ	i_1	i_2	i_3	$\sum_{j=1}^{k} i_j$
1	$(1,1,1)$	3	1	3	0	0	3
2	$(2,1)$	2	2	1	1	0	2
3	(3)	1	1	0	0	1	1

The cases cases correspond to the possible partitions of 3: $1+1+1 = 3$, $2+1 = 3$ and $3 = 3$, of respective lengths 3, 2, and 1. The Chern number $c_1(J_3^3 X)$ of a curve X is obtained by summing the last column: $c_1(J_3^3 X) = (1+2+3) \times c_1(T^*X) = 6c_1(T^*X)$.

For $m = k = 4$ the number of partitions is $p(4) = 5$ and we have

	λ	l_λ	d_λ	i_1	i_2	i_3	i_4	$\sum_{j=1}^{k} i_j$
1	$(1,1,1,1)$	4	1	4	0	0	0	4
2	$(2,1,1)$	3	3	2	1	0	0	3
3	$(3,1)$	2	3	1	0	1	0	2
4	$(2,2)$	2	2	0	2	0	0	2
5	(4)	1	1	0	0	0	1	1

and $c_1(J_4^4 X) = 12c_1(T^*X)$.

For $m = k = 5$, $p(5) = 7$,

	λ	ρ_λ	d_λ	i_1	i_2	i_3	i_4	i_5	$\sum_{j=1}^{k} i_j$
1	$(1,1,1,1,1)$	5	1	5	0	0	0	0	5
2	$(2,1,1,1)$	4	4	3	1	0	0	0	4
3	$(3,1,1)$	3	6	2	0	1	0	0	3
4	$(2,2,1)$	3	5	1	2	0	0	0	3
5	$(4,1)$	2	4	1	0	0	1	0	2
6	$(3,2)$	2	15	0	1	1	0	0	2
7	(5)	1	1	0	0	0	0	1	1

and $c_1(J_5^5 X) = 20c_1(T^*X)$.

For $m = k = 6$, $p(6) = 11$,

	λ	l_λ	d_λ	i_1	i_2	i_3	i_4	i_5	i_6	$\sum_{j=1}^k i_j$
1	$(1,1,1,1,1,1)$	6	1	6	0	0	0	0	0	6
2	$(2,1,1,1,1)$	5	5	4	1	0	0	0	0	5
3	$(3,1,1,1)$	4	10	3	0	1	0	0	0	4
4	$(2,2,1,1)$	4	9	2	2	0	0	0	0	4
5	$(4,1,1)$	3	10	2	0	0	1	0	0	3
6	$(3,2,1)$	3	36	1	1	1	0	0	0	3
7	$(2,2,2)$	3	5	0	3	0	0	0	0	3
8	$(5,1)$	2	30	1	0	0	0	1	0	2
9	$(4,2)$	2	9	0	1	0	1	0	0	2
10	$(3,3)$	2	5	0	0	2	0	0	0	2
11	(6)	1	1	0	0	0	0	0	1	1

and $c_1(\mathcal{J}_6^6 X) = 35 c_1(T^* X)$.

The next few values of $L_k(k)$ are $L_7(7) = 54$, $L_8(8) = 86$, $L_9(9) = 128$, $L_{10}(10) = 192$, $L_{11}(11) = 275$, $L_{12}(12) = 399$, $L_{13}(13) = 556$, $L_{14}(14) = 780$, $L_{15}(15) = 1068$, $L_{16}(16) = 1463$.

Next we deal with the problem of computing the invariants: $h^i(\mathcal{J}_k^m X) = \dim H^i(\mathcal{J}_k^m X)$ for a curve X of genus $g \geq 2$. We have

$$h^0(\mathcal{J}_1^1 X) = h^0(\mathcal{K}_X) = g,$$
$$h^1(\mathcal{J}_1^1 X) = h^0(\mathcal{O}_X) = 1.$$

For curves the filtration of Green–Griffiths takes the form

$$\mathcal{J}_k^m X = \mathcal{S}_k^{[m/k]} \supset \cdots \supset \mathcal{S}_k^0 = \mathcal{J}_{k-1}^m X, \quad \mathcal{S}_k^i / \mathcal{S}_k^{i-1} = \mathcal{K}_X^i \otimes \mathcal{J}_{k-1}^{m-ki}(X).$$

Hence, for $k = 2$, $\mathcal{S}_2^i / \mathcal{S}_2^{i+1} = \mathcal{K}_X^i \otimes \mathcal{J}_1^{m-2i} = \mathcal{K}_X^i \otimes \mathcal{K}_X^{m-2i} = \mathcal{K}_X^{m-i}$. It is clear from the filtration that $\mathcal{J}_1^1 X \cong \mathcal{J}_2^1 X$ (the isomorphism is given by the forgetting map (1.11)). For $\mathcal{J}_2^2 X$ the filtration yields the short exact sequence

$$0 \to \mathcal{K}_X^2 = \odot^2 T^* X \to \mathcal{J}_2^2 X \to T^* X = \mathcal{K}_X \to 0,$$

from which we get the exact sequence

$$0 \to H^0(\mathcal{K}_X^2) \to H^0(\mathcal{J}_2^2 X) \to H^0(\mathcal{K}_X) \to H^1(\mathcal{K}_X^2) \to H^1(\mathcal{J}_2^2 X) \to H^1(\mathcal{K}_X) \to 0.$$

By (2.2b) we have $h^1(\mathcal{K}_X) = 1$ and $h^1(\mathcal{K}_X^2) = 0$ if $d \geq 2$. Hence, as $\mathcal{K}_X = \mathcal{O}_X(3-d)$,

$$h^0(\mathcal{J}_2^2 X) = h^0(\mathcal{K}_X) + h^0(\mathcal{K}_X^2) = 4g - 3,$$
$$h^1(\mathcal{J}_2^2 X) = h^1(T^* X) = h^0(\mathcal{O}_X) = 1.$$

For $\mathcal{J}_2^3 X$ we obtain the short exact sequence from the filtration,

$$0 \to \mathcal{K}_X^3 \to \mathcal{J}_2^3 X \to \mathcal{K}_X^2 \to 0,$$

and the exact cohomology sequence

$$0 \to H^0(\mathcal{K}_X^3) \to H^0(\mathcal{J}_2^3 X) \to H^0(\mathcal{K}_X^2) \to H^1(\mathcal{K}_X^3) \to H^1(\mathcal{J}_2^3 X) \to H^1(\mathcal{K}_X^2) \to 0.$$

Since $h^1(\mathcal{K}_X^2) = h^1(\mathcal{K}_X^3) = 0$ for $g \geq 2$ we find

$$h^1(\mathcal{J}_2^3 X) = 0,$$
$$h^0(\mathcal{J}_2^3 X) = h^0(\mathcal{K}_X^2) + h^0(\mathcal{K}_X^3) = 8(g-1).$$

For $\mathcal{J}_2^4 X$ the filtration is given by

$$\mathcal{J}_2^4 X = \mathcal{S}_2^2 \supset \mathcal{S}_2^1 \supset \mathcal{S}_2^0 = \mathcal{J}_1^4 X = \mathcal{K}_X^4$$

with $\mathcal{S}_2^2 / \mathcal{S}_2^1 = \mathcal{K}_X^2, \mathcal{S}_2^1 / \mathcal{S}_2^0 = \mathcal{K}_X^3$. From the filtration we have two short exact sequences,

$$0 \to \mathcal{S}_2^1 \to \mathcal{J}_2^4 X \to \mathcal{K}_X^2 \to 0 \quad \text{and} \quad 0 \to \mathcal{K}_X^4 \to \mathcal{S}_2^1 \to \mathcal{K}_X^3 \to 0.$$

For $g \geq 2$ we get from the second exact sequence and the fact that $h^1(\mathcal{K}_X^3) = h^1(\mathcal{K}_X^4) = 0$ that $h^1(\mathcal{S}_2^1) = 0$. This and the first exact sequence imply that

$$h^1(\mathcal{J}_2^4 X) = h^1(\mathcal{K}_X^2) = 0 \text{ and}$$
$$h^0(\mathcal{J}_2^4 X) = h^0(\mathcal{K}_X^2) + h^0(\mathcal{K}_X^3) + h^0(\mathcal{K}_X^4) = 15(g-1).$$

We get, inductively:

THEOREM 2.13. *For a smooth curve with genus $g \geq 2$ the following equalities hold:*

(i) $\mathcal{J}_2^1 X = \mathcal{K}_X$; *hence* $h^1(\mathcal{J}_2^1 X) = 1, h^0(\mathcal{J}_2^1 X) = $ *genus of* X;
(ii) $h^1(\mathcal{J}_2^2 X) = 1, \quad h^0(\mathcal{J}_2^2 X) = h^0(\mathcal{K}_X) + h^0(\mathcal{K}_X^2) = 4g - 3$;
(iii) $h^1(\mathcal{J}_2^m X) = 0$, *and for* $m \geq 3$,

$$h^0(\mathcal{J}_2^m X) = \sum_{j=0}^{[m/2]} h^0(\mathcal{K}_X^{m-j}) = (2m - [\tfrac{m}{2}] - 1)([\tfrac{m}{2}] + 1)(g - 1);$$

(iv) *for* $i \geq 1$,

$$h^0(\mathcal{J}_2^1 X \otimes \mathcal{K}_X^i) = h^0(\mathcal{K}_X^{i+1}) = (2i+1)(g-1),$$
$$h^0(\mathcal{J}_2^2 X \otimes \mathcal{K}_X^i) = h^0(\mathcal{K}_X^{i+1}) + h^0(\mathcal{K}_X^{i+2}) = 4(i+1)(g-1),$$

and for $m \geq 3$,

$$h^0(\mathcal{J}_2^m X \otimes \mathcal{K}_X^i) = \sum_{j=0}^{[m/2]} h^0(\mathcal{K}_X^{m+i-j}) = (2m + 2i - [\tfrac{m}{2}] - 1)([\tfrac{m}{2}] + 1)(g - 1).$$

PROOF. Parts (i), (ii) and (iii) are clear. For part (iv), tensoring the exact sequence $0 \to \mathcal{K}_X^2 \to \mathcal{J}_2^2 X \to \mathcal{K}_X \to 0$ by $\mathcal{K}_X^i$ yields the exact sequence $0 \to \mathcal{K}_X^{i+2} \to \mathcal{J}_2^2 X \otimes \mathcal{K}_X^i \to \mathcal{K}_X^{i+1} \to 0$. From the associated long exact cohomology sequence one sees that $h^0(\mathcal{J}_2^2 X \otimes \mathcal{K}_X^i) = h^0(\mathcal{K}_X^{i+1}) + h^0(\mathcal{K}_X^{i+2})$ as claimed. From the exact sequences

$$0 \to \mathcal{S}_{m,2}^{[m/2]-1} \to \mathcal{J}_2^m X \to \mathcal{K}^{[m/2]} \otimes \mathcal{J}_1^{m-2[m/2]} X = \mathcal{K}_X^{m-[m/2]} \to 0,$$

$$0 \to \mathcal{S}_{m,2}^{[m/2]-2} \to \mathcal{S}_{m,2}^{[m/2]-1} \to \mathcal{K}_X^{[m/2]-1} \otimes \mathcal{J}_1^{m-2([m/2]-1)} X = \mathcal{K}_X^{m-[m/2]+1} \to 0,$$

$$\cdots\cdots\cdots\cdots\cdots\cdots\cdots\cdots\cdots\cdots\cdots\cdots\cdots$$

$$0 \to \mathcal{J}_1^m X = \mathcal{K}_X^m \to \mathcal{S}_{m,2}^1 \to \mathcal{K}_X \otimes \mathcal{J}_1^{m-2} X = \mathcal{K}_X^{m-1} \to 0,$$

we obtain, by tensoring with $\mathcal{K}_X^i$, $i \geq 0$, the exact sequences

$$0 \to \mathcal{S}_{m,2}^{[m/2]-1} \otimes \mathcal{K}_X^i \to \mathcal{J}_2^m X \otimes \mathcal{K}_X^i \to \mathcal{K}_X^{m+i-[m/2]} \to 0,$$

$$0 \to \mathcal{S}_{m,2}^{[m/2]-2} \otimes \mathcal{K}_X^i \to \mathcal{S}_{m,2}^{[m/2]-1} \otimes \mathcal{K}_X^i \to \mathcal{K}_X^{m+i-[m/2]+1} \to 0,$$

$$\cdots\cdots\cdots\cdots\cdots\cdots\cdots\cdots\cdots\cdots\cdots\cdots\cdots$$

$$0 \to \mathcal{K}_X^{m+i} \to \mathcal{S}_{m,2}^1 \otimes \mathcal{K}_X^i \to \mathcal{K}_X^{m+i-1} \to 0,$$

from which we deduce that $h^1(\mathcal{J}_2^m X \otimes \mathcal{K}_X^i) = h^1(\mathcal{S}_{m,2}^{[m/2]-j} \otimes \mathcal{K}_X^i) = 0$ for $0 \leq j \leq [m/2]$ and that

$$h^0(\mathcal{J}_2^m X \otimes \mathcal{K}_X^i) = \sum_{j=0}^{[m/2]} h^0(\mathcal{K}_X^{m+i-j}),$$

as claimed.   □

The coefficient of $(g-1)$ in part (iv) of the preceding lemma may be expressed as

$$\alpha(m,i,2) = \begin{cases} \frac{1}{4}(3m^2 + 4m(i+1) + 8i - 4) = \frac{1}{4}(m+2)(3m+4i-2), & m \text{ even}, \\ \frac{1}{4}(3m^2 + 2m(2i+1) + 4i - 1) = \frac{1}{4}(m+1)(3m+4i-1)4, & m \text{ odd}. \end{cases}$$

The coefficient of $(g-1)$ in part (iii) is $\alpha(m,2) = \alpha(m,0,2)$.

COROLLARY 2.14. *For a smooth curve of genus $g \geq 2$ we have, for $m \geq 3$,*

$$h^0(\mathcal{J}_2^m X) = \begin{cases} \frac{1}{4}(3m^2 + 4m - 4)(g-1), & m \text{ even}, \\ \frac{1}{4}(3m^2 + 2m - 1)(g-1), & m \text{ odd}, \end{cases}$$

and

$$c_1(\mathcal{J}_2^m X) = \begin{cases} \frac{1}{4}(3m^2 + 6m)(g-1), & m \text{ even}, \\ \frac{1}{4}(3m^2 + 4m + 1)(g-1), & m \text{ odd}. \end{cases}$$

PROOF. The first formula is given by part (iii) of Theorem 2.13. The second formula is a consequence of the *Riemann–Roch for curves*:

$$h^0(\mathcal{J}_2^m X) - h^1(\mathcal{J}_2^m X) = c_1(\mathcal{J}_2^m X) - (\mathrm{rk}\,\mathcal{J}_2^m X)(g-1),$$

using the fact that $\mathrm{rk}\,\mathcal{J}_2^m X = [m/2] + 1$ and that $h^1(\mathcal{J}_2^m X) = 0$ if $m \geq 3$.   □

We now deal with the case of general k. We shall be content with asymptotic formulas as the general formulas become complicated by the fact that the general formula for sums of powers is only given recursively. However the highest order term is quite simple:

$$\sum_{i=1}^{m} i^d = \frac{m^{d+1}}{d+1} + O(m^d). \tag{2.4}$$

The filtration theorem of Green–Griffiths implies that

$$\operatorname{rk} \mathcal{J}_k^m X = \sum_{I \in \mathcal{I}_{k,m}} \operatorname{rk} \mathcal{S}_I.$$

For a curve, $\mathcal{S}_I = \bigodot^{i_1} T^*X \otimes \cdots \otimes \bigodot^{i_{k-1}} T^*X \otimes \bigodot^{i_k} T^*X = \mathcal{K}_X^{|I|} = \mathcal{K}_X^{i_1 + \cdots + i_k}$. Hence

$$\operatorname{rk} \mathcal{J}_k^m X = \#\mathcal{I}_{k,m}, \quad \mathcal{I}_{k,m} = \Big\{ I = (i_1, \ldots, i_k) \mid \sum_{j=1}^{k} j i_j = m \Big\}.$$

Alternatively, since $\mathcal{S}_I = \mathcal{S}_{I'} \otimes \bigodot^{i_k} T^*X$, where $I' = (i_1, \ldots, i_{k-1}) \in \mathcal{I}_{k-1, m-k i_k}$, we have

$$\operatorname{rk} \mathcal{J}_k^m X = \sum_{i_k=0}^{[m/k]} \operatorname{rk} \left(\mathcal{J}_{k-1}^{m-k i_k} X \otimes \mathcal{K}_X^{i_k} \right) = \sum_{i_k=0}^{[m/k]} \operatorname{rk} \mathcal{J}_{k-1}^{m-k i_k} X;$$

equivalently,

$$\#\mathcal{I}_{k,m} = \sum_{i_k=0}^{[m/k]} \#\mathcal{I}_{k-1, m-k i_k}.$$

THEOREM 2.15. *Let* $\mathcal{I}_{k,m} = \Big\{ I = (i_1, \ldots, i_k) \mid \sum_{j=1}^{k} j i_j = m \Big\}$. *Then, for a curve* X,

$$\operatorname{rk} \mathcal{J}_k^m X = \#\mathcal{I}_{k,m} = \frac{m^{k-1}}{k!(k-1)!} + O(m^{k-2}).$$

PROOF. It is clear that $\operatorname{rk} \mathcal{J}_1^m X = 1$ and we have seen that $\operatorname{rk} \mathcal{J}_2^m X = [m/2]+1$, thus writing $\operatorname{rk} \mathcal{J}_k^m X = a_k m^{k-1} + O(m^{k-2})$ we get, via (2.4),

$$a_k m^{k-1} + O(m^{k-2}) = a_{k-1} \sum_{i_k=0}^{[m/k]} (m - k i_k)^{k-2} + O(m^{k-2})$$

$$= a_{k-1} \sum_{i_k=0}^{[m/k]} (m - k i_k)^{k-2} + O(m^{k-2})$$

$$= a_{k-1} \sum_{j=0}^{k-2} (-1)^j \frac{(k-2)!}{j!(k-2-j)!} m^{k-2-j} k^j \sum_{i_k=0}^{[m/k]} i_k^j + O(m^{k-2})$$

$$= a_{k-1} \sum_{j=0}^{k-2} (-1)^j \frac{(k-2)!}{j!(k-2-j)!} m^{k-2-j} k^j \frac{m^{j+1}}{(j+1)k^{j+1}} + O(m^{k-2})$$

$$= \frac{a_{k-1}}{k} \sum_{j=0}^{k-2} \frac{(-1)^j}{j+1} \frac{(k-2)!}{j!(k-2-j)!} m^{k-1} + O(m^{k-2}).$$

The following formula is easily verified by double induction:

LEMMA 2.16. *For any positive integers $1 \le l \le k$, we have*

$$\sum_{j=0}^{k} \frac{(-1)^j}{j+l} \binom{k}{j} = \frac{(l-1)!\, k!}{(k+l)!}.$$

Using this lemma we obtain a recursive formula for $k \ge 2$:

$$a_k = \frac{a_{k-1}}{k(k-1)}, \qquad a_1 = 1.$$

The first few values of a_k are $a_1 = 1$, $a_2 = 1/2$, $a_3 = 1/2^2 3$, $a_4 = 1/(2^4 3^2)$, $a_5 = 1/(2^6 3^2 5)$, $a_6 = 1/(2^7 3^3 5^2)$. The recursive formula also yields the general formula for a_k:

$$a_k = \frac{1}{\prod_{l=2}^{k}(l-1)l} = \frac{1}{(k-1)!\, k!}. \qquad \square$$

The filtration also yields a formula for a curve of genus g:

$$c_1(\mathcal{J}_k^m X) = \sum_{I \in \mathcal{I}_{k,m}} c_1(\mathcal{S}_I) = \sum_{I \in \mathcal{I}_{k,m}} |I| c_1(\mathcal{K}_X) = 2 \sum_{I \in \mathcal{I}_{k,m}} |I|(g-1),$$

where $|I| = i_1 + \cdots + i_k$. On the other hand we have

$$c_1(\mathcal{J}_k^m X) = \sum_{i_k=0}^{[m/k]} \sum_{I' \in \mathcal{I}_{k-1,m-ki_k}} (c_1(\mathcal{S}_{I'}) + i_k c_1(\mathcal{K}_X))$$

$$= 2 \sum_{i_k=0}^{[m/k]} \sum_{I' \in \mathcal{I}_{k-1,m-ki_k}} (|I'| + i_k)(g-1).$$

It is clear that $c_1(\mathcal{J}_1^m X) = 2m(g-1)$ and we have seen that

$$c_1(\mathcal{J}_2^m X) = \begin{cases} \frac{1}{4}(3m^2 + 6m)(g-1), & m \text{ even}, \\ \frac{1}{4}(3m^2 + 4m + 1)(g-1), & m \text{ odd}. \end{cases}$$

THEOREM 2.17. *For a curve of genus $g \ge 2$ we have, for each $k \ge 2$,*

$$c_1(\mathcal{J}_k^m X) = \left(\frac{2(g-1)}{(k!)^2} \sum_{i=1}^{k} \frac{1}{i} \right) m^k + O(m^{k-1}).$$

PROOF. It is clear that asymptotically $c_1(\mathcal{J}_k^m X) = O(m^k)$. Hhence, writing

$$c_1(\mathcal{J}_k^m X) = 2b_k m^k (g-1) + O(m^{k-1}),$$

we get via Theorem 2.15 that, for $k \geq 3$,

$$2b_k m^k (g-1) + O(m^{k-1})$$

$$= 2 \sum_{i_k=0}^{m/k} \sum_{I' \in \mathcal{I}_{k-1,m-ki_k}} (b_{k-1}(m-ki_k)^{k-1} + i_k)(g-1) + O(m^{k-1})$$

$$= 2(g-1) \sum_{i_k=0}^{m/k} \left(a_{k-1} i_k (m-ki_k)^{k-2} + b_{k-1}(m-ki_k)^{k-1} \right) + O(m^{k-1}).$$

By Lemma 2.16 we have

$$\sum_{i_k=0}^{m/k} i_k (m-ki_k)^{k-2} = \frac{1}{k^2} \sum_{j=0}^{k-2} \frac{(-1)^j}{j+2} \frac{(k-2)!}{j!(k-2-j)!} m^k = \frac{1}{k^3(k-1)}$$

and

$$\sum_{i_k=0}^{m/k} (m-ki_k)^{k-1} = \frac{1}{k} \sum_{j=0}^{k-1} \frac{(-1)^j}{j+1} \frac{(k-1)!}{j!(k-1-j)!} m^k = \frac{1}{k} \frac{(k-1)!}{k!} = \frac{1}{k^2}.$$

From these we obtain

$$2b_k m^k (g-1) + O(m^{k-1}) = 2(g-1)\left(\frac{a_{k-1}}{k^3(k-1)} + \frac{b_{k-1}}{k^2} \right) + O(m^{k-1})$$

and hence the recursive relation:

$$b_k = \frac{a_{k-1}}{k^3(k-1)} + \frac{b_{k-1}}{k^2} = \frac{1}{k^2 k!(k-1)!} + \frac{b_{k-1}}{k^2} = \frac{1}{k^2}\left(b_{k-1} + \frac{1}{k!(k-1)!} \right)$$

with $a_1 = 1$ and $b_1 = 1$. An explicit formula is obtained by repeatedly using the recursion. More precisely, we first apply the recursive formula to b_{k-1}:

$$b_{k-1} = \frac{1}{(k-1)^2}\left(b_{k-2} + \frac{1}{(k-1)!(k-2)!} \right)$$

and substitution yields

$$b_k = \frac{1}{k^2}\left(\frac{1}{(k-1)^2}\left(b_{k-2} + \frac{1}{(k-1)!(k-2)!} \right) + \frac{1}{k!(k-1)!} \right).$$

The procedure above may be repeated until we reach $b_1 = 1$. Induction shows that

$$b_k = \frac{1}{(k!)^2} \sum_{i=1}^{k} \frac{1}{i}. \qquad \square$$

For general $m > k \geq 2$ we get from the filtrations the following $[m/k]$ exact sequences:

$$0 \to \mathcal{S}_{m,k}^{[m/k]-1} \to \mathcal{J}_k^m X \to \mathcal{K}_X^{[m/k]} \otimes \mathcal{J}_{k-1}^{m-k[m/k]} X \to 0$$

$$0 \to \mathcal{S}_{m,k}^{[m/k]-2} \to \mathcal{S}_{m,k}^{[m/k]-1} \to \mathcal{K}_X^{[m/k]-1} \otimes \mathcal{J}_{k-1}^{m-k([m/k]-1)} X \to 0 \qquad (2.5)$$

$$\dotsb\dotsb\dotsb\dotsb\dotsb\dotsb\dotsb\dotsb\dotsb\dotsb\dotsb\dotsb$$

$$0 \to \mathcal{J}_{k-1}^m X \to \mathcal{S}_{m,k}^1 \to \mathcal{K}_X \otimes \mathcal{J}_{k-1}^{m-k} X \to 0.$$

Observe that $m - k[m/k] = 0$ or 1 depending on whether m is divisible by k. By induction, we see that $h^1(\mathcal{K}_X^i \otimes \mathcal{J}_{k-1}^{m-ki} X) = 0$ implies that $h^1(\mathcal{S}_{m,k}^i) = 0$, for $0 \leq i \leq [m/k]$. Hence $h^1(\mathcal{J}_k^m X) = 0$ for any $k \geq 2$, and as a result we also have

$$h^0(\mathcal{J}_k^m X) = h^0(\mathcal{J}_{k-1}^m X) + \sum_{i=1}^{[m/k]} h^0(\mathcal{K}_X^i \otimes \mathcal{J}_{k-1}^{m-ki} X) \quad \text{if } m > k. \qquad (2.6)$$

COROLLARY 2.18. *Let X be a curve of genus $g \geq 2$. Then, $h^1(\mathcal{J}_k^m X) = 0$ if $m \geq k$, and for all $k \geq 2$ we have*

$$h^0(\mathcal{J}_k^m X) = \left(\frac{2(g-1)}{(k!)^2} \sum_{i=1}^k \frac{1}{i} \right) m^k + O(m^{k-1}).$$

PROOF. By Riemann–Roch for curves, we have

$$h^0(\mathcal{J}_k^m X) - h^1(\mathcal{J}_k^m X) = c_1(\mathcal{J}_k^m X) - (\text{rk } \mathcal{J}_k^m X)(g-1).$$

As observed, $h^1(\mathcal{J}_k^m X)$ vanishes. By Theorem 2.15, $(\text{rk } \mathcal{J}_k^m X)(g-1)$ is of lower order, so $h^0(\mathcal{J}_k^m X) = c_1(\mathcal{J}_k^m X)$ and the result follows from Theorem 2.17. $\qquad\square$

By induction we get from (2.6)

$$h^0(\mathcal{J}_k^m X)$$

$$= h^0(\mathcal{J}_2^m X) + \sum_{i=1}^{[m/3]} h^0(\mathcal{K}_X^i \otimes \mathcal{J}_2^{m-3i} X) + \cdots + \sum_{i=1}^{[m/k]} h^0(\mathcal{K}_X^i \otimes \mathcal{J}_{k-1}^{m-ki} X)$$

$$= \sum_{i=0}^{[m/2]} h^0(\mathcal{K}_X^{m-i}) + \sum_{i=1}^{[m/3]} h^0(\mathcal{K}_X^i \otimes \mathcal{J}_2^{m-3i} X) + \cdots + \sum_{i=1}^{[m/k]} h^0(\mathcal{K}_X^i \otimes \mathcal{J}_{k-1}^{m-ki} X).$$

Tensoring (2.5) with $\mathcal{K}_X^i$ yields exact sequences

$$0 \to \mathcal{S}_{m,k}^{[m/k]-1} \otimes \mathcal{K}_X^i \to \mathcal{J}_k^m X \otimes \mathcal{K}_X^i \to \mathcal{K}_X^{[m/k]+i} \otimes \mathcal{J}_{k-1}^{m-k[m/k]} X \to 0,$$

$$0 \to \mathcal{S}_{m,k}^{[m/k]-2} \otimes \mathcal{K}_X^i \to \mathcal{S}_{m,k}^{[m/k]-1} \otimes \mathcal{K}_X^i \to \mathcal{K}_X^{[m/k]+i-1} \otimes \mathcal{J}_{k-1}^{m-k([m/k]-1)} X \to 0,$$

$$\dotsb\dotsb\dotsb\dotsb\dotsb\dotsb\dotsb\dotsb\dotsb\dotsb\dotsb\dotsb$$

$$0 \to \mathcal{J}_{k-1}^m X \otimes \mathcal{K}_X^i \to \mathcal{S}_{m,k}^1 \to \mathcal{K}_X^{i+1} \otimes \mathcal{J}_{k-1}^{m-k} X \to 0.$$

These imply that

$$h^0(\mathcal{J}_k^m X \otimes \mathcal{K}_X^i) = \sum_{j=0}^{[m/k]} h^0(\mathcal{K}_X^{i+j} \otimes \mathcal{J}_{k-1}^{m-kj} X).$$

Thus, for $k = 3$ we get by Theorem 2.13:

$$h^0(\mathcal{J}_3^m X) = \sum_{i=0}^{[m/2]} h^0(\mathcal{K}_X^{m-i}) + \sum_{i=1}^{[m/3]} h^0(\mathcal{K}_X^i \otimes \mathcal{J}_2^{m-3i} X)$$

$$= \sum_{i=0}^{[m/2]} h^0(\mathcal{K}_X^{m-i}) + \sum_{i=1}^{[m/3]} \sum_{j=0}^{[(m-3i)/2]} h^0(\mathcal{K}_X^{m-2i-j}).$$

With this it is possible to write down the explicit formulas. In the case of $\mathcal{J}_2^m X$ there are only two cases depending on the parity of m. For 3-jets there are the following cases: (1a) $m = 3q$, q even; (1b) $m = 3q$, q odd; (2a) $m = 3q+1$, q even; (2b) $m = 3q+1$, q odd; (3a) $m = 3q+1$, q even; and (3b) $m = 3q+2$, q odd. For simplicity we shall only do this for case (1a). First we observe that the rank of $\mathcal{J}_3^m X$ is given by the number

$$\operatorname{rk} \mathcal{J}_3^m X = \left[\frac{m}{2}\right] + 1 + \sum_{i=1}^{[m/3]} \left(\left[\frac{m-3i}{2}\right] + 1\right) = O(m^2).$$

If m is divisible by 3! then

$$\operatorname{rk} \mathcal{J}_3^m X = \frac{m}{2} + 1 + \frac{m}{3} + \sum_{l=1}^{m/6} \frac{m-3(2l-1)}{2} + \sum_{l=1}^{m/6} \frac{m-3(2l)}{2}$$

$$= \tfrac{1}{12}(m+3)(m+4). \tag{2.7}$$

For $k = 3$ we get, by Theorem 2.13,

$$h^0(\mathcal{J}_3^m X)$$

$$= \sum_{i=0}^{[m/2]} h^0(\mathcal{K}_X^{m-i}) + \sum_{i=1}^{[m/3]} h^0(\mathcal{K}_X^i \otimes \mathcal{J}_2^{m-3i} X)$$

$$= \sum_{i=0}^{[m/2]} h^0(\mathcal{K}_X^{m-i}) + \sum_{i=1}^{[m/3]} \sum_{j=0}^{[(m-3i)/2]} h^0(\mathcal{K}_X^{m-2i-j})$$

$$= (g-1) \times$$

$$\left(\left(2m-1-\left[\frac{m}{2}\right]\right)\left(\left[\frac{m}{2}\right]+1\right) + \sum_{i=1}^{[m/3]} \left(2m-4i-1-\left[\frac{m-3i}{2}\right]\right)\left(\left[\frac{m-3i}{2}\right]+1\right)\right).$$

If m is divisible by 3!, both m and $[m/3]$ are even. Then, denoting the second sum above by S,

$$S = \sum_{l=1}^{m/6}\left(2m - 4(2l-1) - 1 - \frac{m-1-3(2l-1)}{2}\right)\left(\frac{m-1-3(2l-1)}{2} + 1\right)$$

$$+ \sum_{l=1}^{m/6}\left(2m - 8l - 1 - \frac{m-6l}{2}\right)\left(\frac{m-6l}{2} + 1\right)$$

$$= \frac{1}{2}\sum_{l=1}^{m/6}(3m^2 + 10m + 6 + 60l^2 - (28m+36)l) = \frac{1}{2^2 3^3}m(11m^2 - 18m - 18).$$

Thus for m divisible by 3! we have

$$h^0(\mathcal{J}_3^m X) = \left(\frac{1}{2^2 3^3}m(11m^2 - 18m - 18) + \frac{1}{2^2}(m+2)(3m-2)\right)(g-1),$$

and, by Riemann–Roch and (2.7),

$$c_1(\mathcal{J}_3^m X) = (g-1) \times$$
$$\left(\frac{1}{2^2 3^3}m(11m^2 - 18m - 18) + \frac{1}{2^2}(m+2)(3m-2) + \frac{1}{2^2 3}(m+3)(m+4)\right).$$

EXAMPLE 2.19. The filtration of $\mathcal{J}_3^6 X$ is given by $\mathcal{J}_3^6 X = \mathcal{S}^2 \supset \mathcal{S}^1 \supset \mathcal{S}^0 = \mathcal{J}_2^6 X$, and the associated exact sequences are $0 \to \mathcal{S}^1 \to \mathcal{J}_3^6 X \to \mathcal{K}_X^2 \to 0$ and

$$0 \to \mathcal{J}_2^6 X \to \mathcal{S}^1 \to \mathcal{K}_X \otimes \mathcal{J}_2^3 X \to 0.$$

Hence $h^0(\mathcal{J}_3^6 X) = 0$ and $h^0(\mathcal{J}_3^6 X) = h^0(\mathcal{J}_2^6 X) + h^0(\mathcal{K}_X^2) + h^0(\mathcal{K}_X \otimes \mathcal{J}_2^3 X)$. From the exact sequence $0 \to \mathcal{K}_X^3 \to \mathcal{J}_2^3 X \to \mathcal{K}_X^2 \to 0$ we obtain the exact sequence

$$0 \to \mathcal{K}_X^4 \to \mathcal{K}_X \otimes \mathcal{J}_2^3 X \to \mathcal{K}_X^3 \to 0$$

from which we conclude that

$$h^0(\mathcal{J}_3^6 X) = h^0(\mathcal{J}_2^6 X) + h^0(\mathcal{K}_X^2) + h^0(\mathcal{K}_X^3) + h^0(\mathcal{K}_X^4)$$
$$= h^0(\mathcal{K}_X^2) + 2(h^0(\mathcal{K}_X^3) + h^0(\mathcal{K}_X^4)) + h^0(\mathcal{K}_X^5) + h^0(\mathcal{K}_X^6)$$
$$= (3 + 2(5+7) + 9 + 11)(g-1) = 47(g-1).$$

Next we consider the problem of constructing an explicit basis for $H^0(\mathcal{J}_k^m X)$. First we recall the construction of a basis for $H^0(\mathcal{J}_1^1 X) = H^0(\mathcal{K}_X)$. The procedure of this construction works in any algebraically closed field and has been used toward resolving the *uniqueness problem for rational and meromorphic functions*. (The reader is referred to [An et al. 2003a; 2004; 2003b] for details.) Let z_0, z_1, z_2 be the homogeneous coordinates on $\mathbb{P}^2$. Then

$$d\left(\frac{z_i}{z_j}\right) = \frac{z_j dz_i - z_i dz_j}{z_j^2} = \frac{\begin{vmatrix} z_1 & z_2 \\ dz_1 & dz_2 \end{vmatrix}}{z_j^2} \tag{2.8}$$

is a well-defined rational 1-form on $\mathbb{P}^n$. Let $P(z_0, z_1, z_2)$ be a homogeneous polynomial of degree d and

$$X = \{[z_0, z_1, z_2] \in \mathbb{P}^2(\mathbb{C}) \mid P(z_0, z_1, z_2) = 0\}.$$

Then, by Euler's Theorem, for $[z_0, z_1, z_2] \in X$, we have

$$z_0 \frac{\partial P}{\partial z_0}(z_0, z_1, z_2) + z_1 \frac{\partial P}{\partial z_1}(z_0, z_1, z_2) + z_2 \frac{\partial P}{\partial z_2}(z_0, z_1, z_2) = 0.$$

The tangent space of X is defined by the equation $P(z_0, z_1, z_2) = 0$ and

$$dz_0 \frac{\partial P}{\partial z_0}(z_0, z_1, z_2) + dz_1 \frac{\partial P}{\partial z_1}(z_0, z_1, z_2) + dz_2 \frac{\partial P}{\partial z_2}(z_0, z_1, z_2) = 0.$$

These may be expressed as

$$z_0 \frac{\partial P}{\partial z_0}(z_0, z_1, z_2) + z_1 \frac{\partial P}{\partial z_1}(z_0, z_1, z_2) = -z_2 \frac{\partial P}{\partial z_2}(z_0, z_1, z_2),$$
$$dz_0 \frac{\partial P}{\partial z_0}(z_0, z_1, z_2) + dz_1 \frac{\partial P}{\partial z_1}(z_0, z_1, z_2) = -dz_2 \frac{\partial P}{\partial z_2}(z_0, z_1, z_2).$$

Then by Cramer's rule, we have on X

$$\frac{\partial P}{\partial z_0} = \frac{W(z_1, z_2)}{W(z_0, z_1)} \frac{\partial P}{\partial z_2}, \quad \frac{\partial P}{\partial z_1} = \frac{W(z_2, z_0)}{W(z_0, z_1)} \frac{\partial P}{\partial z_2}$$

provided that the *Wronskian* $W(z_0, z_1) = z_0 dz_1 - z_1 dz_0 \not\equiv 0$ on any component of X; that is, the defining homogeneous polynomial of X has no linear factor of the form $az_0 + bz_1$. Thus

$$\frac{W(z_1, z_2)}{\frac{\partial P}{\partial z_0}(z_0, z_1, z_2)} = \frac{W(z_2, z_0)}{\frac{\partial P}{\partial z_1}(z_0, z_1, z_2)} = \frac{W(z_0, z_1)}{\frac{\partial P}{\partial z_2}(z_0, z_1, z_2)} \tag{2.9}$$

is a globally well-defined rational 1-form on any component of $\pi^{-1}(X) \subset \mathbb{C}^3 \setminus \{0\}$, where $(\pi : \mathbb{C}^3 \setminus \{0\} \to \mathbb{P}^2(\mathbb{C})$ is the Hopf fibration), *provided that the expressions make sense* (*that is, the denominators are not identically zero when restricted to a component of $\psi^{-1}(X)$*). For our purpose, we also require that the form given by (2.9) *is not identically trivial when restricted to a component of $\pi^{-1}(X)$*. This is equivalent to the condition that *the Wronskians in the formula above are not identically zero; in other words, the defining homogeneous polynomial of X has no linear factor of the form $az_i + bz_j$ where $a, b \in \mathbb{C}$, $0 \le i, j \le 2$ and $i \ne j$.* If P, $\partial P/\partial z_0$, $\partial P/\partial z_1$, $\partial P/\partial z_2$ never vanish all at once (that is, X is smooth) then, at each point, one of the expressions in (2.9) is regular at the point. Hence so are the other expressions. This means that

$$\eta = \frac{\begin{vmatrix} z_1 & z_2 \\ dz_1 & dz_2 \end{vmatrix}}{\partial P/\partial z_0} \tag{2.10}$$

is regular on $\pi^{-1}(X)$. (Note that the form η is not well-defined on X unless $n = 3$; see (2.8)). The form

$$\omega = \frac{\begin{vmatrix} z_1 & z_2 \\ dz_1 & dz_2 \end{vmatrix} z_0^{n-1}}{z_0^2} \partial P / \partial z_0 = \frac{\begin{vmatrix} z_1 & z_2 \\ dz_1 & dz_2 \end{vmatrix}}{\partial P / \partial z_0} z_0^{n-3} = z_0^{n-3} \eta,$$

with $n = \deg P$, is a well-defined (again by (2.8)) rational 1-form on X. More-over, as η is regular on X, the 1-form ω is also regular if $n \geq 3$. If $n = 3$ then $\omega = \eta$ and if $n \geq 4$ then ω is regular and vanishes along the ample divisor $[z_0^{n-3}=0] \cap X$. Thus for any homogeneous polynomial $Q = Q(z_0, z_1, z_2)$ of degree $n - 3$, the 1-form

$$\frac{Q}{z_0^{n-3}} \omega = Q \eta$$

is regular on C and vanishes along $[Q=0]$. Note that the dimension of the vector space of homogeneous polynomials of degree $n-3$ (a basis is given by all possible monomials) is

$$\tfrac{1}{2}(n-1)(n-2) = \text{ genus of } X.$$

We summarize these observations:

PROPOSITION 2.20.   Let $X = \{[z_0, z_1, z_2] \in \mathbb{P}^2(\mathbb{C}) \mid P(z_0, z_1, z_2) = 0\}$ be a nonsingular curve of degree $d \geq 3$. If $d = 3$ then the space of regular 1-forms on X is $\{c\eta \mid c \in \mathbb{C}\}$, where η is defined by (2.2). If $d \geq 4$ take the set

$$\left\{Q_i \mid Q_i \text{ is a monomial of degree } d-3 \text{ for } 1 \leq i \leq \tfrac{1}{2}(d-1)(d-2)\right\}$$

as an ordered basis of homogeneous polynomials of degree $d - 3$. Then

$$\left\{\omega_i = Q_i \eta \mid 1 \leq i \leq \tfrac{1}{2}(d-1)(d-2)\right\}$$

is a basis of the space of regular 1-forms on X.

Using the preceding we may write down explicitly a basis for $H^0(\mathcal{J}_k^m X)$. We demonstrate via examples. For $d = 4$, $h^0(\mathcal{J}_2^2 X) = h^0(\mathcal{K}_X^2) + h^0(\mathcal{K}_X) = 6+3 = 9$ and, since the genus is 3, there are 3 linearly independent 1-forms $\omega_1, \omega_2, \omega_3$ which, as shown above, may be taken as

$$\omega_1 = \frac{z_0(z_0\, dz_1 - z_1\, dz_0)}{\partial P / \partial z_2}, \quad \omega_2 = \frac{z_1(z_0\, dz_1 - z_1\, dz_0)}{\partial P / \partial z_2}, \quad \omega_3 = \frac{z_2(z_0\, dz_1 - z_1\, dz_0)}{\partial P / \partial z_3}.$$

A basis for $H^0(\mathcal{J}_2^2 X)$ is given by

$$\omega_1^{\otimes 2}, \ \omega_2^{\otimes 2}, \ \omega_3^{\otimes 2}, \ \omega_1 \otimes \omega_2, \ \omega_1 \otimes \omega_3, \ \omega_2 \otimes \omega_3, \ \delta\omega_1, \ \delta\omega_2, \ \delta\omega_3,$$

where δ is the derivation defined in (1.9). The first six of these provide a basis of $H^0(\mathcal{K}_X^2)$ and the last three may be identified with a basis of $H^0(\mathcal{K}_X)$. For

$\mathcal{J}_2^3 X$ we have

$$
\begin{aligned}
h^0(\mathcal{J}_2^3 X) &= h^0(\mathcal{K}_X^2) + h^0(\mathcal{K}_X^3) \\
&= h^0\big(\mathcal{O}_X(2(d-3))\big) + h^0\big(\mathcal{O}_X(3(d-3))\big) \\
&= C_2^{2d-4} - C_2^{d-4} + C_2^{3d-7} - C_2^{2d-7}.
\end{aligned}
$$

In particular, for $d = 4$, $h^0(\mathcal{J}_2^3 X) = h^0(\mathcal{K}_X^2) + h^0(\mathcal{K}_X^3) = 6 + 10 = 16$. A basis for $H^0(\mathcal{J}_2^3 X)$ is given by the six elements (identified with a basis of $H^0(\mathcal{K}_X^2)$)

$$
\delta\omega_1^{\otimes 2}, \ \delta\omega_2^{\otimes 2}, \ \delta\omega_3^{\otimes 2}, \ \delta(\omega_1 \otimes \omega_2), \ \delta(\omega_1 \otimes \omega_3), \ \delta(\omega_2 \otimes \omega_3)
$$

and the 10 elements (a basis of $H^0(\mathcal{K}_X^3)$):

$$
\omega_1^{\otimes 3}, \ \omega_2^{\otimes 3}, \ \omega_3^{\otimes 3}, \ \omega_1 \otimes \omega_2 \otimes \omega_3,
$$
$$
\omega_1^{\otimes 2} \otimes \omega_2, \ \omega_1^{\otimes 2} \otimes \omega_3, \ \omega_2^{\otimes 2} \otimes \omega_1, \ \omega_2^{\otimes 2} \otimes \omega_3, \ \omega_3^{\otimes 2} \otimes \omega_1, \ \omega_3^{\otimes 2} \otimes \omega_2.
$$

3. Computation of Chern Classes
The Case of Surfaces

SUMMARY. *We exhibit here the explicit formulas due to* [Stoll and Wong 2002] (*see also* [Green and Griffiths 1980]) *for the Chern numbers of the projectivized parametrized jet bundles of a compact complex surface. The most important is the index formula given in Theorem 3.9:*

$$
\iota(\mathcal{J}_k^m X) = (\alpha_k c_1^2 - \beta_k c_2) m^{2k+1} + O(m^{2k})
$$

where $c_i = c_i(X), \alpha_k = \beta_k + \gamma_k$ *and*

$$
\beta_k = \frac{2}{(k!)^2 (2k+1)!} \sum_{i=1}^{k} \frac{1}{i^2}, \qquad \gamma_k = \frac{2}{(k!)^2 (2k+1)!} \sum_{i=1}^{k} \frac{1}{i} \sum_{j=1}^{i-1} \frac{1}{j}.
$$

This implies that $\alpha_k/\beta_k \to \infty$ *hence* $\alpha_k/\beta_k > c_2/c_1^2$ *for* k *sufficiently large provided that* $c_1^2 > 0$. *For example,* $c_2/c_1^2 = 11$ *for a smooth hypersurface of degree* $d = 5$ *and the explicit formula shows that* $\alpha_k/\beta_k > 11$ *for all* $k \geq 199$. (*See the table at the end of this section*). *The explicit formulas for* α_k *and* β_k *are crucial in the proof of the Kobayashi conjecture in Section 7.*

We now treat the case of a complex surface (complex dimension 2). The computations here are more complicated than those of Section 2 as we must deal with the second Chern number. The computation of the first Chern class is relatively easy since the *Whitney formula* is linear in this case; that is, if $0 \to \mathcal{S}' \to \mathcal{S} \to \mathcal{S}'' \to 0$ is exact, then $c_1(\mathcal{S}) = c_1(\mathcal{S}') + c_1(\mathcal{S}'')$. The Whitney formula for the second Chern classes on the other hand is nonlinear: $c_2(\mathcal{S}) = c_2(\mathcal{S}') + c_2(\mathcal{S}'') + c_1(\mathcal{S}') c_1(\mathcal{S}'')$. The (minor) nonlinearity may seem harmless at first but for filtrations the nonlinearity carries over at each step and the complexity increases rapidly. Thus the correct way to deal with the problem is not to calculate the second Chern class directly but to calculate the index $\iota(\mathcal{J}_k^m X) = c_1^2(\mathcal{J}_k^m X) - 2c_2(\mathcal{J}_k^m X)$ which does behave linearly, that is, $\iota(\mathcal{S}) = \iota(\mathcal{S}') + \iota(\mathcal{S}'')$. We then recover the second Chern class from the formula $c_2(\mathcal{J}_k^m X) = (c_1^2(\mathcal{J}_k^m X) - \iota(\mathcal{J}_k^m X))/2$. In order to compute the jet differentials we must first calculate the Chern classes and indices of the sheaves $\mathcal{S}_I = \bigodot^{i_1} T^* X \otimes \cdots \otimes \bigodot^{i_k} T^* X$ where $I = (i_1, \ldots, i_k)$. For details of the computations see [Stoll and Wong 2002].

By results from [Tsuji 1987; 1988; Maruyama 1981], the exterior, symmetric and tensor products of the cotangent sheaf of a manifold of general type are semistable in the sense of Mumford–Takemoto. For a coherent sheaf $\mathcal{S}$ on a variety of dimension n the *index of stability* relative to the canonical class is defined to be

$$\mu(\mathcal{S}) = \frac{c_1^{n-1}(\mathcal{S}) \, c_1(T^* X)}{(\mathrm{rk}\ \mathcal{S}) c_1^n(T^* X)}.$$

A sheaf $\mathcal{S}$ is said to be *semistable* in the sense of Mumford–Takemoto (relative to the canonical class) if $\mu(\mathcal{S}') \leq \mu(\mathcal{S})$ for all coherent subsheaves $\mathcal{S}'$ of $\mathcal{S}$. For a nonsemistable sheaf a subsheaf $\mathcal{S}'$ satisfying $\mu(\mathcal{S}') > \mu(\mathcal{S})$ is said to be a *destabilizing subsheaf*. In view of Tsuji's result it would seem reasonable to expect that the sheaves of jet differentials are also semistable. However using the explicit formulas for the Chern classes computed below we shall see that this is not the case. Tsuji's result is used in [Lu and Yau 1990] (see also [Lu 1991]) to show that a projective surface X satisfying the conditions that $\mathcal{K}_X$ is nef and $c_1^2(T^* X) - 2c_2(T^* X) > 0$ contains no rational nor elliptic curves. The instability of the jet differentials implies that the analogous result of Lu–Yau requires a different argument.

We list below some basic but very useful formulas (see [Wong 1999; Stoll and Wong 2002]):

LEMMA 3.1. *Let X be a nonsingular complex surface and E be a vector bundle of rank 2 over X. Then* rk $(\bigodot^m E) = m + 1$ *and*

$$c_1(\bigodot^m E) = \tfrac{1}{2} m(m+1) c_1(E),$$
$$c_2(\bigodot^m E) = \tfrac{1}{24} m(m^2 - 1)(3m + 2) c_1^2(E) + \tfrac{1}{6} m(m+1)(m+2) c_2(E).$$

Consequently the index is given by the formula:

$$\iota(\bigodot^m E) = \tfrac{1}{6} m(m+1)(2m+1) c_1^2(E) - \tfrac{1}{3} m(m+1)(m+2) c_2(E).$$

Moreover, if $c_1^2(E) \neq 0$ then

$$\delta_\infty(E) \overset{\mathrm{def}}{=} \lim_{m \to \infty} \frac{c_2(\bigodot^m E)}{c_1^2(\bigodot^m E)} = \frac{1}{2}.$$

Note that $\delta_\infty(E)$ is independent of $c_2(E)/c_1^2(E)$. The next formula gives the *Chern numbers for tensor products of different bundles.*

LEMMA 3.2. *Let E_i, $i = 1, \ldots, k$, be holomorphic vector bundles, of respective rank r_i, over a nonsingular complex surface X. Let $R = \prod_{l=1}^k r_l$. Then:*

(i) $c_1\left(\bigotimes_{i=1}^{k} E_i\right) = \sum_{i=1}^{k}(r_1 \ldots r_{i-1}r_{i+1}\ldots r_k)c_1(E_i) = Rr_i\sum_{i=1}^{k}\dfrac{c_1(E_i)}{r_i}.$

(ii) $c_2\left(\bigotimes_{i=1}^{k} E_i\right) = R\left(\sum_{i=1}^{k}\dfrac{c_2(E_i)}{r_i} + (R-1)\sum_{1\leq i<j\leq k}\dfrac{c_1(E_i)c_1(E_j)}{r_ir_j}\right)$

$$+\sum_{i=1}^{k}\left(\dfrac{\prod_{l=1,l\neq i}^{k}r_j}{2}\right)c_1^2(E_i).$$

In particular:

(iii) $c_2(E_1\otimes E_2) = \binom{r_2}{2}c_1^2(E_1) + (r_1r_2-1)c_1(E_1)c_1(E_2) + \binom{r_1}{2}c_1^2(E_2)$

$$+r_2c_2(E_1)+r_1c_2(E_2);$$

(iv) $c_2\left(\bigotimes_{i=1}^{3} E_i\right) = r_1r_2r_3\left(\sum_{i=1}^{3}\dfrac{c_2(E_i)}{r_i} + (r_1r_2r_3-1)\sum_{1\leq i<j\leq 3}\dfrac{c_1(E_i)c_1(E_j)}{r_ir_j}\right)$

$$+\sum_{i=1}^{3}\binom{r_1r_2r_3/r_i}{2}c_1^2(E_i);$$

and the index $\iota(E_1\otimes E_2) = c_1^2(E_1\otimes E_2) - 2c_2(E_1\otimes E_2)$ *is given by*

(v) $\iota(E_1\otimes E_2) = r_2c_1^2(E_1) + r_1c_1^2(E_2) + 2c_1(E_1)c_1(E_2) - 2r_2c_2(E_1) - 2r_1c_2(E_2).$

With the preceding formulas the computation of the *Chern numbers for* $J_k^m X$ can now be carried out by using the filtration given in Theorem 2.3, reducing the calculation to the Chern numbers of bundles of the form

$$S_I = \bigodot^{i_1} T^*X \otimes \cdots \otimes \bigodot^{i_k} T^*X,$$

where the indices $I = (i_1, \ldots, i_k)$ satisfy the condition $i_1 + 2i_2 + \cdots + ki_k = m$. More precisely, take

$$\mathcal{I}_{km} = \{I = (i_1, \ldots, i_k) \mid i_j \in \mathbb{N}\cup\{0\}, i_1 + 2i_2 + \cdots + ki_k = m\}$$

together with a fixed ordering of $\mathcal{I}_{km}$ (say, the lexicographical ordering). Then a brute force computation, applying Lemma 3.2 and Lemma 3.3 repeatedly yields the following formulas:

THEOREM 3.3. *Let* X *be a nonsingular complex surface and let* $S_{m-2i,i} = \bigodot^{m-2i} T^*X \otimes \bigodot^{i} T^*X$. *Denote by* $c_1 = c_1(T^*X), c_2 = c_2(T^*X)$. *Then*

$\text{rk}\,(S_{m-2i,i}) = (m-2i+1)(i+1) = (m+1)+(m-1)i-2i^2,$

$c_1(S_{m-2i,i}) = \frac{1}{2}(m-i)(m-2i+1)(i+1)c_1$

$\qquad\qquad = \frac{1}{2}m(m+1) + (m^2-2m-1)i - (3m-1)i^2 + 2i^3 c_1,$

$c_2(S_{m-2i,i}) = \frac{1}{24}\{m(m^2-1)(3m+2) + 2(3m^4-5m^3-3m^2+4m+1)i$

$\qquad\qquad +(3m^4-30m^3+12m^2+6m-7)i^2 - 2(9m^3-27m^2+5m+1)i^3$

$\qquad\qquad +(39m^2-42m+7)i^4 - 4(3m-1)i^5 + 4i^6\}c_1^2 + \frac{1}{6}b_{m-2i,i}c_2,$

where $c_i = c_i(T^*X)$. The index is given by

$$\iota(\mathcal{S}_{m-2i,i}) = \tfrac{1}{6}(a_{m-2i,i}c_1^2 - 2b_{m-2i,i}c_2),$$

where $a_{m-2i,i}$ and $b_{m-2i,i}$ are polynomials given by

$$a_{m-2i,i} = m(m+1)(2m+1) + (2m^3 - 6m^2 - 7m - 1)i - (9m^2 - 6m - 5)i^2$$
$$+ (14m-2)i^3 - 8i^4,$$

$$b_{m-2i,i} = m(m+1)(m+2) + (m^3 - 3m^2 - 8m - 2)i - (6m^2 - 3m - 7)i^2$$
$$+ (13m-1)i^3 - 10i^4.$$

The rank rk $\mathcal{J}_2^m X$ of the sheaf of $\mathcal{J}_2^m X$ is given by

$$\frac{1}{24}(m+1)(m+3)(m+5) = \frac{1}{2^3 3}(m^3 + 9m^2 + 23m + 15), \quad \text{if } m \text{ is odd,}$$

$$\frac{1}{24}(m+2)(m+3)(m+4) = \frac{1}{2^3 3}(m^3 + 9m^2 + 26m + 24), \quad \text{if } m \text{ is even;}$$

and the first Chern class of same sheaf, $c_1(\mathcal{J}_2^m X)$, is

$$\frac{(m+1)(m+3)(m+5)(3m+1)}{192}c_1 = \frac{3m^4 + 28m^3 + 78m^2 + 68m + 15}{2^6 3}c_1, \quad m \text{ odd,}$$

$$\frac{m(m+2)(m+4)(3m+10)}{192}c_1 = \frac{3m^4 + 28m^3 + 84m^2 + 80m}{2^6 3}c_1, \quad m \text{ even.}$$

The index of $\mathcal{J}_2^m X$ is given by

$$\iota(\mathcal{J}_2^m X) = c_1^2(\mathcal{J}_2^m X) - 2c_2(\mathcal{J}_2^m X) = a_m c_1^2 - b_m c_2,$$

where the coefficients a_m and b_m are polynomials in m given by

$$a_m = \begin{cases} \frac{1}{2^6 15}(7m^5 + 75m^4 + 270m^3 + 390m^2 + 203m + 15) & \text{if } m \text{ is odd,} \\ \frac{1}{2^6 15}(7m^5 + 75m^4 + 280m^3 + 420m^2 + 208m) & \text{if } m \text{ is even;} \end{cases}$$

$$b_m = \begin{cases} \frac{1}{2^6 15}(5m^5 + 75m^4 + 390m^3 + 810m^2 + 565m + 75) & \text{if } m \text{ is odd,} \\ \frac{1}{2^6 15}(5m^5 + 75m^4 + 400m^3 + 900m^2 + 720m) & \text{if } m \text{ is even.} \end{cases}$$

The formula for the index also yields the formula for $c_2(\mathcal{J}_2^m X)$:

$$c_2(\mathcal{J}_2^m X) = \frac{1}{2}\{c_1^2(\mathcal{J}_2^m X) - (a_m c_1^2 - b_m c_2)\} = \frac{1}{2}\{\lambda_m c_1^2 + b_m c_2\} \qquad (3.1)$$

where the coefficients a_m and b_m are given by Theorem 3.3, and the coefficient λ_m is given by

$$\lambda_m = \begin{cases} \left(\frac{1}{192}(m+1)(m+3)(3m^2 + 16m + 5)\right)^2 - a_m, & m \text{ odd,} \\ \left(\frac{1}{192}m(m+2)(m+4)(3m+10)\right)^2 - a_m, & m \text{ even.} \end{cases}$$

In particular:

COROLLARY 3.4. *Let X be a nonsingular complex surface and assume that $c_1^2(T^*X) > 0$. Then*

$$\delta(\mathcal{J}_2^m X) = \lim_{m \to \infty} \frac{c_2(\mathcal{J}_2^m X)}{c_1^2(\mathcal{J}_2^m X)} = \frac{1}{2}.$$

For simplicity, set $c_1 = c_1(T^*X)$, $c_2 = c_2(T^*X)$. For any sheaf $\mathcal{S}$, define

$$\iota(\mathcal{S}) = c_1^2(\mathcal{S}) - 2c_2(\mathcal{S}), \quad \mu(\mathcal{S}) = \frac{c_1(\mathcal{S}) c_1}{(\text{rk } \mathcal{S})c_1^2}, \quad \delta(\mathcal{S}) = \frac{c_2(\mathcal{S})}{c_1^2(\mathcal{S})}, \qquad (3.2)$$

provided that the denominators are not zero. Denote for simplicity $\delta = \delta(TX)$ $= c_2/c_1^2$. It is well known that $c_1^2 \leq 3c_2$ and $c_2 \leq 5c_1^2 + 36$ for a surface of general type with $c_1^2 > 0$ [Barth et al. 1984, p. 228]. Thus, for such surfaces, δ satisfies the estimate

$$\frac{1}{3} \leq \delta \leq 5 + \frac{36}{c_1^2} \leq 41. \qquad (3.3)$$

We give the precise numbers for a few special cases:

• $\mathcal{J}_2^2 X$. In this case $k = 2$, $m = 2$ and there are two weighted partitions (i_1, i_2) corresponding to the two solutions of $i_1 + 2i_2 = 2$ (Example 2.9), namely $I_1 = (2, 0)$ and $I_2 = (0, 1)$. The corresponding sheaves are $\mathcal{S}_{I_1} = \bigodot^2 T^*X$, $\mathcal{S}_{I_2} = T^*X$. The various invariants of these sheaves are as follows:

I	$\mathcal{S}$	rank	$c_1(\mathcal{S})$	$c_2(\mathcal{S})$	$\iota(\mathcal{S})$	$\mu(\mathcal{S})$	$\delta(\mathcal{S})$
$(2,0)$	$\bigodot^2 T^*X$	3	$3c_1$	$2c_1^2 + 4c_2$	$5c_1^2 - 8c_2$	1	$\frac{1}{9}(2 + 4\delta)$
$(0,1)$	T^*X	2	c_1	c_2	$c_1^2 - 2c_2$	$\frac{1}{2}$	δ
	$\mathcal{J}_2^2 X$	5	$4c_1$	$5c_1^2 + 5c_2$	$6c_1^2 - 10c_2$	$\frac{4}{5}$	$\frac{5}{16}(1 + \delta)$

The Chern numbers are calculated using Lemma 3.1 and Lemma 3.2. Note that $\bigodot^2 T^*X$ is a subsheaf of $\mathcal{J}_2^2 X$ (by Example 2.5, $0 \to \bigodot^2 T^*X \to \mathcal{J}_2^2 X \to T^*X \to 0$ is an exact sequence) with $\mu(\bigodot^2 T^*X) > \mu(\mathcal{J}_2^2 X)$. A subsheaf with such a property is said to be a *destabilizing subsheaf*. On the other hand T^*X is a quotient sheaf of $\mathcal{J}_2^2 X$ with $\mu(T^*X) < \mu(\mathcal{J}_2^2 X)$. A quotient sheaf with such a property is said to be a *destabilizing quotient sheaf*.

• $\mathcal{J}_2^3 X$. In this case $k = 2$, $m = 3$ and there are two weighted partitions $I_1 = (3, 0)$ and $I_2 = (1, 1)$ corresponding to the two solutions of $i_1 + 2i_2 = 3$.

I	$\mathcal{S}$	rk	$c_1(\mathcal{S})$	$c_2(\mathcal{S})$	$\iota(\mathcal{S})$	$\mu(\mathcal{S})$	$\delta(\mathcal{S})$
$(3,0)$	$\bigodot^3 T^*X$	4	$6c_1$	$11c_1^2 + 10c_2$	$14c_1^2 - 20c_2$	$\frac{3}{2}$	$\frac{1}{36}(11 + 10\delta)$
$(1,1)$	$\bigotimes^2 T^*X$	4	$4c_1$	$5c_1^2 + 4c_2$	$6c_1^2 - 8c_2$	1	$\frac{1}{16}(5 + \delta)$
	$\mathcal{J}_2^3 X$	8	$10c_1$	$40c_1^2 + 14c_2$	$20c_1^2 - 28c_2$	$\frac{5}{4}$	$\frac{1}{50}(20 + 7\delta)$

The sheaves $\bigodot^3 T^*X$ and $\bigotimes^2 T^*X$ are respectively a destabilizing subsheaf and a destabilizing quotient sheaf of $\mathcal{J}_2^3 X$. The sequence $0 \to \bigodot^3 T^*X \to \mathcal{J}_2^3 X \to \bigotimes^2 T^*X \to 0$ is exact, by Example 2.5.

- $\mathcal{J}_2^4 X$. In this case $k = 2$, $m = 4$ and there are 3 weighted partitions $I_1 = (4,0)$, $I_2 = (2,1)$ and $I_3 = (0,2)$ corresponding to the 3 solutions of $i_1 + 2i_2 = 4$.

I	$\mathcal{S}$	rk	$c_1(\mathcal{S})$	$c_2(\mathcal{S})$	$\iota(\mathcal{S})$	$\mu(\mathcal{S})$	$\delta(\mathcal{S})$
$(4,0)$	$\odot^4 T^*X$	5	$10c_1$	$35c_1^2 + 20c_2$	$30c_1^2 - 40c_2$	2	$\frac{7}{20} + \frac{1}{5}\delta$
$(2,1)$	$\odot^2 T^*X \otimes T^*X$	6	$9c_1$	$31c_1^2 + 11c_2$	$19c_1^2 - 22c_2$	$\frac{3}{2}$	$\frac{31}{81} + \frac{11}{81}\delta$
$(0,2)$	$\odot^2 T^*X$	3	$3c_1$	$2c_1^2 + 4c_2$	$5c_1^2 - 8c_2$	1	$\frac{2}{9} + \frac{4}{9}\delta$
	$\mathcal{J}_2^4 X$	14	$22c_1$	$215c_1^2 + 35c_2$	$54c_1^2 - 70c_2$	$\frac{11}{7}$	$\frac{215}{484} + \frac{35}{484}\delta$

The sheaves $\odot^4 T^*X$ and $\odot^2 T^*X$ are respectively a destabilizing subsheaf and a destabilizing quotient sheaf of $\mathcal{J}_2^4 X$. Note that $\odot^2 T^*X \otimes T^*X$ is neither a subsheaf nor a quotient sheaf of $\mathcal{J}_2^4 X$. We have two exact sequences:

$$0 \to \mathcal{F}_2^1 \to \mathcal{J}_2^4 X \to \odot^2 T^*X \to 0,$$
$$0 \to \odot^4 T^*X \to \mathcal{F}_2^1 \to \odot^2 T^*X \otimes T^*X \to 0.$$

- $\mathcal{J}_2^5 X$. In this case $k = 2$, $m = 5$ and there are 3 weighted partitions $I_1 = (5,0)$, $I_2 = (3,1)$ and $I_3 = (1,2)$ corresponding to the 3 solutions of $i_1 + 2i_2 5$.

I	$\mathcal{S}$	rk	$c_1(\mathcal{S})$	$c_2(\mathcal{S})$	$\iota(\mathcal{S})$	$\mu(\mathcal{S})$	$\delta(\mathcal{S})$
$(5,0)$	$\odot^5 T^*X$	6	$15c_1$	$85c_1^2 + 35c_2$	$55c_1^2 - 70c_2$	$\frac{5}{2}$	$\frac{17}{45} + \frac{7}{45}\delta$
$(3,1)$	$\odot^3 T^*X \otimes T^*X$	8	$16c_1$	$106c_1^2 + 24c_2$	$44c_1^2 - 48c_2$	2	$\frac{53}{128} + \frac{3}{32}\delta$
$(1,2)$	$T^*X \otimes \odot^2 T^*X$	6	$9c_1$	$31c_1^2 + 11c_2$	$19c_1^2 - 22c_2$	$\frac{3}{2}$	$\frac{31}{81} + \frac{11}{81}\delta$
	$\mathcal{J}_2^5 X$	20	$40c_1$	$741c_1^2 + 70c_2$	$118c_1^2 - 140c_2$	2	$\frac{741}{1600} + \frac{7}{160}\delta$

The sheaves $\odot^5 T^*X$ and $\odot^2 T^*X \otimes T^*X$ are respectively a destabilizing subsheaf and a destabilizing quotient sheaf of $\mathcal{J}_2^5 X$. Note that $\odot^3 T^*X \otimes T^*X$ is neither a subsheaf nor a quotient sheaf of $\mathcal{J}_2^5 X$. We have two exact sequences:

$$0 \to \mathcal{F}_2^1 \to \mathcal{J}_2^5 X \to T^*X \otimes \odot^2 T^*X \to 0,$$
$$0 \to \odot^5 T^*X \to \mathcal{F}_2^1 \to \odot^3 T^*X \otimes T^*X \to 0.$$

- $\mathcal{J}_2^6 X$. In this case $k = 2$, $m = 6$ and there are 4 weighted partitions $I_1 = (6,0)$, $I_2 = (4,1)$, $I_3 = (2,1)$ and $I_4 = (0,3)$ corresponding to the 3 solutions of $i_1 + 2i_2 = 6$.

I	$\mathcal{S}$	rk	$c_1(\mathcal{S})$	$c_2(\mathcal{S})$	$\iota(\mathcal{S})$	$\mu(\mathcal{S})$	$\delta(\mathcal{S})$
$(6,0)$	$\odot^6 T^*X$	7	$21c_1$	$175c_1^2 + 56c_2$	$91c_1^2 - 112c_2$	3	$\frac{25}{63} + \frac{8}{63}\delta$
$(4,1)$	$\odot^4 T^*X \otimes T^*X$	10	$25c_1$	$270c_1^2 + 45c_2$	$85c_1^2 - 90c_2$	$\frac{5}{2}$	$\frac{54}{125} + \frac{9}{125}\delta$
$(2,2)$	$\odot^2 T^*X \otimes \odot^2 T^*X$	9	$18c_1$	$138c_1^2 + 24c_2$	$48c_1^2 - 48c_2$	2	$\frac{23}{54} + \frac{2}{27}\delta$
$(0,3)$	$\odot^3 T^*X$	4	$6c_1$	$11c_1^2 + 10c_2$	$14c_1^2 - 20c_2$	$\frac{3}{2}$	$\frac{11}{36} + \frac{5}{18}\delta$
	$\mathcal{J}_2^6 X$	30	$70c_1$	$2331c_1^2 + 135c_2$	$238c_1^2 - 270c_2$	$\frac{7}{3}$	$\frac{333}{700} + \frac{27}{980}\delta$

We have three exact sequences:

$$0 \to \mathcal{F}_2^2 \to \mathcal{J}_2^6 X \to \bigodot^3 T^* X \to 0,$$
$$0 \to \mathcal{F}_2^1 \to \mathcal{F}_2^2 \to \bigodot^2 T^* X \otimes \bigodot^2 T^* X \to 0,$$
$$0 \to \bigodot^6 T^* X \to \mathcal{F}_2^1 \to \bigodot^4 T^* X \otimes \bigodot^2 T^* X \to 0.$$

The sheaves $\bigodot^6 T^* X$ and $\bigodot^3 T^* X$ are respectively a destabilizing subsheaf and a destabilizing quotient sheaf of $\mathcal{J}_2^6 X$.

REMARK 3.5. For each partition $I = (i_1, i_2)$ satisfying $i_1 + 2i_2 = m$ we associate the (nonweighted) sum $|I| = i_1 + i_2$. Let $I_{\max} = \max_I \{|I|\}$ and $I_{\min} = \max_I \{|I|\}$. Then the sheaf $\mathcal{S}_{I_{\max}}$ is a destabilizing subsheaf and the sheaf $\mathcal{S}_{I_{\min}}$ is a destabilizing quotient sheaf.

We now deal with the case of general k. We shall be content with asymptotic formulas as the general formulas become complicated since the general formula for sums of powers can only be given recursively. However the highest order term is quite simple; indeed, we have

$$\sum_{i=1}^m i^d = \frac{m^{d+1}}{d+1} + O(m^d). \tag{3.4}$$

Before dealing with the jet bundles $\mathcal{J}_k^m X$ we must first find the formulas for the sheaves $\mathcal{S}_I = \bigodot^{i_1} T^* X \otimes \cdots \otimes \bigodot^{i_k} T^* X$. This is easier due to the symmetry of the sheaves and we know, a priori, that the formulas can be expressed in terms of the symmetric functions in the exponents $i_1, \ldots, i_k$. For general k we introduce some notation for the j-th symmetric functions on k indices:

$$s_{0;k} = 1, \quad s_{1;k} = \sum_{p=1}^k i_p, \quad s_{2;k} = \sum_{1 \le p < q \le k} i_p i_q, \quad \ldots, \quad s_{k;k} = \prod_{p=1}^k i_p. \tag{3.5}$$

We have

$$\mu_k = \prod_{p=1}^k (i_p + 1) = \sum_{p=0}^k s_{p;k}. \tag{3.6}$$

Let $I = (i_1, \ldots, i_k)$ and $I' = (i_1, \ldots, i_{k-1})$, so that

$$\mathcal{S}_I = \bigodot^{i_1} T^* X \otimes \cdots \otimes \bigodot^{i_{k-1}} T^* X \otimes \bigodot^{i_k} T^* X = \mathcal{S}_{I'} \otimes \bigodot^{i_k} T^* X.$$

By Lemma 3.1, Lemma 3.2 and induction we obtain the following result, where we abbreviate $c_i = c_i(T^* X)$:

LEMMA 3.6. *Let X be a nonsingular complex surface and $\mathcal{S}_I = \mathcal{S}_{i_1, i_2, \ldots, i_k} = \bigodot^{i_1} T^* X \otimes \bigodot^{i_2} T^* X \otimes \cdots \otimes \bigodot^{i_k} T^* X$ where $i_1, i_2, \ldots, i_k$ are nonnegative integers.*

Then rk $\mathcal{S}_I = \mu_k$,

$$c_1(\mathcal{S}_I) = \frac{1}{2}\sum_{j=1}^{k} i_j \prod_{j=1}^{k}(i_j+1)c_1 = \tfrac{1}{2}s_{1;k}\mu_k c_1(T^*X) = \tfrac{1}{2}s_{1;k}\sum_{p=0}^{k} s_{p;k},$$

$$\iota(\mathcal{S}_I) = \tfrac{1}{6}\mu_k\big((2s_{1;k}^2+s_{1;k}-s_{2;k})c_1^2 - 2(s_{1;k}^2+2s_{1;k}-2s_{2;k})c_2\big),$$

$$c_2(\mathcal{S}_I) = \tfrac{1}{24}\mu_k s_{j;k}\big((3s_{1;k}^2\mu_k - 4s_{1;k}^2 - 2s_{1;k}+2s_{2;k})c_1^2 + 4(s_{1;k}^2+2s_{1;k}-2s_{2;k})c_2\big),$$

where $s_{j;k}, 1 \le j \le k$ *are the symmetric functions in* $i_1,\dots,i_k$ *as defined in* (3.5) *and* $\mu_k = \sum_{j=0}^{k} s_{j;k}$.

These formulas, together with the filtrations of Green–Griffiths, are now used to get the formulas for $\mathcal{J}_k^m X$. First we have the formula for the rank (the proof is similar to that of Theorem 2.15 though somewhat more complicated):

THEOREM 3.7. *For any positive integer* $k \ge 2$ *we have*

$$\mathrm{rk}\ \mathcal{J}_k^m X = \sum_{(i_1,\dots,i_k)\in\mathcal{I}_{k,m}} \prod_{j=1}^{k}(i_j+1) = A_k m^{2k-1} + O(m^{2k-2})$$

where the coefficient is given by

$$A_k = \frac{1}{\prod_{l=2}^{k} l^2(2l-2)(2l-1)} = \frac{1}{(k!)^2(2k-1)!}.$$

Next we derive the formulas for $c_1(\mathcal{J}_k^m X)$ from the formulas for $c_1(\mathcal{S}_I)$, for $I \in \mathcal{I}_{k,k}$. By Whitney's formula, we see that $c_1(\mathcal{J}_k^m X)$ is given by

$$c_1(\mathcal{J}_k^m X) = \sum_{i_k=0}^{[m/k]} \sum_{I'\in\mathcal{I}_{k-1,m-ki_k}} \big(c_1(\mathcal{S}_{I'})\,\mathrm{rk}\,\bigodot\nolimits^{i_k} T^*X + c_1(\bigodot\nolimits^{i_k} T^*X)\,\mathrm{rk}\,\mathcal{S}_{I'}\big), \quad (3.7)$$

where $i_1+\cdots+ki_k = m$ *and* $\mathcal{I}_{k-1,m-ki_k}$ *consists of all indices* $I' = (i_1,\dots,i_{k-1})$ *satisfying* $i_1 + 2i_2 + \cdots + (k-1)i_{k-1} = m - ki_k$. *We have already seen that*

$$c_1(\mathcal{J}_1^m X) = \big(\tfrac{1}{2}m^2 + O(m)\big)c_1,$$

$$c_1(\mathcal{J}_2^m X) = \big(\tfrac{1}{2^6}m^4 + O(m^3)\big)c_1,$$

where $c_1 = c_1(T^*X)$. For general k we have (using (3.7) and along the lines of the proof of Theorem 2.16):

THEOREM 3.8. *Let* X *be a nonsingular complex surface. Then, for any positive integer* $k \ge 2$,

$$c_1(\mathcal{J}_k^m X) = \big(B_k m^{2k} + O(m^{2k-1})\big)c_1,$$

where the coefficient B_k *is given by*

$$B_k = \frac{1}{(k!)^2(2k)!}\sum_{i=1}^{k}\frac{1}{i} = \frac{A_k}{2k}\sum_{i=1}^{k}\frac{1}{i}.$$

We now compute the *index of* $\mathcal{J}_k^m X$ for general k. As in the case of the first Chern number, the filtration theorem implies that

$$\iota(\mathcal{J}_k^m X) = \sum_{I \in \mathcal{I}_{k,m}} \iota(\mathcal{S}_I).$$

Since $\iota(\mathcal{S}_I) = (\mathrm{rk} \bigodot^{i_k} T^* X)\, \iota(\mathcal{S}_{I'}) + (\mathrm{rk}\, \mathcal{S}_{I'})\, \iota(\bigodot^{i_k} T^* X) + 2c_1(\mathcal{S}_{I'})\, c_1(\bigodot^{i_k} T^* X)$, where $I = (i_1, \ldots, i_k)$ and $I' = (i_1, \ldots, i_{k-1})$, we get

$$\iota(\mathcal{J}_k^m X) = \sum_{i_k=0}^{[m/k]} \Big((i_k + 1) \sum_{I'} \iota(\mathcal{S}_{I'}) + \iota(\bigodot^{i_k} T^* X) \sum_{I'} \mathrm{rk}\,(\mathcal{S}_{I'})$$
$$+ i_k(i_k + 1) \sum_{I'} c_1(\mathcal{S}_{I'}) \Big),$$

where we abbreviate $\sum_{I' \in \mathcal{I}_{k-1, m-ki_k}}$ by $\sum_{I'}$. Using the formulas for $\iota(\mathcal{S}_{I'})$ and $\mathrm{rk}\,(\mathcal{S}_{I'})$ obtained previously (Lemma 3.6) and induction we get:

THEOREM 3.9. *Let X be a nonsingular complex surface. For any positive integer $k \geq 2$,*

$$\iota(\mathcal{J}_k^m X) = (\alpha_k c_1^2 - \beta_k c_2) m^{2k+1} + O(m^{2k}),$$

where the coefficients α_k and β_k satisfy the respective recursive relations:

$$\alpha_k = \frac{\alpha_{k-1}}{2k^3(2k+1)} + \frac{B_{k-1}}{k^4(4k^2-1)} + \frac{A_{k-1}}{2k^5(k-1)(4k^2-1)},$$

$$\beta_k = \frac{\beta_{k-1}}{2k^3(2k+1)} + \frac{A_{k-1}}{2k^5(k-1)(4k^2-1)}$$

with $\alpha_1 = \beta_1 = \frac{1}{3}$ and A_i, B_i are the numbers given in Theorems 3.7 and 3.8 respectively. The coefficients are given explicitly as $\alpha_k = \beta_k + \gamma_k$, where $\gamma_1 = 0$ and for $k \geq 2$

$$\beta_k = \frac{2}{(k!)^2(2k+1)!} \sum_{i=1}^{k} \frac{1}{i^2}, \qquad \gamma_k = \frac{2}{(k!)^2(2k+1)!} \sum_{i=1}^{k} \frac{1}{i} \sum_{j=1}^{i-1} \frac{1}{j}.$$

COROLLARY 3.10. *With the assumptions and notations of Theorem 3.9,*

$$\lim_{k \to \infty} \frac{\alpha_k}{\beta_k} = \lim_{k \to \infty} \frac{\gamma_k}{\beta_k} = \infty.$$

Consequently if $c_1^2 > 0$ then $\iota(\mathcal{J}_k^m X) = cm^{2k+1} c_1^2 + O(m^{2k})$ for some positive constant c.

The asymptotic expansion for $c_2(\mathcal{J}_k^m X)$ now follows readily from Corollary 3.10 along with Theorems 3.8 and 3.9:

THEOREM 3.11. *Let X be a nonsingular complex surface. For any positive integer k,*

$$c_2(\mathcal{J}_k^m X) = \tfrac{1}{2}(c_1^2(\mathcal{J}_k^m X) - \iota(\mathcal{J}_k^m X)) = \tfrac{1}{2} c_1^2(\mathcal{J}_k^m X) = \tfrac{1}{2} A_k^2 c_1^2 m^{4k} + O(m^{4k-1}).$$

We tabulate the ratios α_k/β_k on the next page (they can be checked readily using Mathematica or Maple):

k	α_k	β_k	α_k/β_k
2	$\dfrac{7}{2^6 15}$	$\dfrac{5}{2^6 15}$	1.40000
3	$\dfrac{17}{2^7 3^6 7}$	$\dfrac{7}{2^7 3^6 5}$	1.73469
4	$\dfrac{83}{2^{16} 3^8 7}$	$\dfrac{41}{2^{16} 3^8 7}$	2.02439
5	$\dfrac{1717}{2^{17} 3^8 5^6 11}$	$\dfrac{479}{2^{17} 3^8 5^6 7}$	2.28108
6	$\dfrac{1927}{2^{21} 3^{11} 5^6 11\,13}$	$\dfrac{59}{2^{21} 3^{11} 5^6 11}$	2.51239
7	$\dfrac{726301}{2^{22} 3^{12} 5^7 7^6 11\,13}$	$\dfrac{266681}{2^{22} 3^{12} 5^7 7^6 11\,13}$	2.72348
8	$\dfrac{3144919}{2^{34} 3^{12} 5^7 7^6 11\,13\,17}$	$\dfrac{63397}{2^{34} 3^{12} 5^7 7^6 11\,13}$	2.91804
9	$\dfrac{2754581}{2^{35} 3^{20} 5^7 7^6 13\,17\,19}$	$\dfrac{514639}{2^{35} 3^{20} 5^7 7^6 11\,13\,17}$	3.09879
10	$\dfrac{2923673}{2^{39} 3^{21} 5^{10} 7^7 13\,17\,19}$	$\dfrac{178939}{2^{39} 3^{21} 5^9 7^7\,13\,17\,19}$	3.26779
11	$\dfrac{315566191}{2^{40} 3^{21} 5^{10} 7^7 11^6 17\,19\,23}$	$\dfrac{10410343}{2^{40} 3^{21} 5^9 7^7 11^6\,13\,17\,19}$	3.42666
12	$\dfrac{330851461}{2^{47} 3^{24} 5^{12} 7^7 11^6 17\,19\,23}$	$\dfrac{18500393}{2^{47} 3^{24} 5^{11} 7^7 11^6\,17\,19\,23}$	3.57670
197			10.9808
198			10.9987
199			11.0165
200			11.0345

4. Finsler Geometry of Projectivized Vector Bundles

SUMMARY. *Our use of projectivized jet bundles is initiated by the recognition that, for projectivized vector bundles, the algebraic geometric concept of ampleness is equivalent to the existence of a Finsler (not hermitian in general) metric with negative mixed holomorphic bisectional curvature. It is known, at least in the case of the tangent bundle that, even for Finsler metrics, negative holomorphic bisectional curvature implies hyperbolicity. We provide in this section some of the basic notions from Finsler geometry. For more details see* [Cao and Wong 2003; Chandler and Wong 2004] *and the references there.*

Many questions concerning a complex vector bundle E of rank greater than 1 may be reduced to problems about the tautological *line* bundle (or its dual) over the projectivization $\mathbb{P}(E)$. For example the algebraic geometric concept of ampleness (and the numerical effectiveness) of a holomorphic vector bundle

E may be interpreted in terms of Finsler geometry (see [Cao and Wong 2003], and also [Aikou 1995; 1998]; for general theory on Finsler geometry we refer to [Bao and Chern 1991; Bao et al. 1996; Abate and Patrizio 1994]). For the relationship with the *Monge–Ampère equation* see [Wong 1982]. We also provide some implications of this reformulation. For applications of the formulation using projectivized bundles to complex analysis see [Dethloff et al 1995a, b]. The dual of a vector bundle E will be denoted by E^*. For any positive integer k, denote by $\odot^k E$ the k-fold symmetric product. The dual vector bundle E^* is said to be *ample* if and only if the line bundle $\mathcal{L}_{\mathbb{P}(E)}$ is ample.

By a *Finsler metric* along the fibers of E we mean a function $h : E \to \mathbb{R}_{\geq 0}$ with the following properties:

(FM1) h is of class $\mathcal{C}^0$ on E and is of class $\mathcal{C}^\infty$ on $E \setminus \{\text{zero section}\}$.

(FM2) $h(z, \lambda v) = |\lambda| h(z, v)$ for all $\lambda \in \mathbb{C}$.

(FM3) $h(z, v) > 0$ on $E \setminus \{\text{zero section}\}$.

(FM4) For z and v fixed, the function $\eta_{z,v}(\lambda) = h^2(z, \lambda v)$ is smooth even at $\lambda = 0$.

(FM5) $h|_{E_z}$ is a *strictly pseudoconvex* function on $E_z \setminus \{0\}$ for all $z \in M$.

Denote by $\pi : TE \to E$ the projection and $\mathcal{V} = \ker \pi \subset TE$ the vertical sub-bundle. A Finsler metric F defines naturally a hermitian inner product on the vertical bundle $\mathcal{V} \subset TE$ by

$$\langle V, W \rangle_{\mathcal{V}} = \sum_{i,j=1}^{r} g_{i\bar{j}}(z, v) V^i \overline{W}^j, \quad g_{i\bar{j}}(z, v) = \frac{\partial^2 F^2(z, v)}{\partial v^i \partial \bar{v}^j} \tag{4.1}$$

for horizontal vector fields $V = \sum_i V^i \partial/\partial v^i, W = \sum_i W^i \partial/\partial v^i \in \mathcal{V}$ on E where $v_1, \ldots, v_r$ are the fiber coordinates. (The difference between a Finsler metric and a hermitian metric is that, for a hermitian metric, the components $(g_{i\bar{j}})$ of the hermitian inner product on the vertical bundle are independent of the fiber coordinates). The hermitian inner product defines uniquely a hermitian connection (known as the *Chern connection*) $\theta = (\theta_i^k)$ and the associate hermitian curvature $\Theta = (\Theta_i^k)$. If $(g_{i\bar{j}})$ comes from a hermitian metric then the curvature forms depend only on the base coordinates; however if it comes from a general Finsler metric then the curvature forms will have horizontal, vertical and mixed components:

$$\Theta_i^k = \sum_{\alpha,\beta=1}^{n} K_{i\alpha\bar{\beta}}^k dz^\alpha \wedge d\bar{z}^\beta + \sum_{j,l=1}^{r} \kappa_{ij\bar{l}}^k dv^j \wedge d\bar{v}^l + \sum_{\alpha=1}^{n} \sum_{l=1}^{r} \mu_{i\alpha\bar{l}}^k dz^\alpha \wedge d\bar{v}^l$$

$$+ \sum_{j=1}^{r} \sum_{\beta=1}^{n} \nu_{ij\bar{\beta}}^k dv^j \wedge d\bar{z}^\beta.$$

Denote by $P = \sum_{i=1}^{r} v^i \partial/\partial v^i$ the position vector field on E. The *mixed holomorphic bisectional curvature* of the Finsler metric is defined, for any nonzero vector field $X \in \Gamma(M, TM)$, to be

$$\langle \Theta(X, X)P, P \rangle_{\mathcal{V}} = \sum_{i,j,k=1}^{r} \sum_{\alpha,\beta=1}^{n} g_{k\bar{j}} K_{i\alpha\bar{\beta}}^{k} v^{i} \bar{v}^{j} X^{\alpha} \overline{X}^{\beta}, \qquad (4.2)$$

where the inner product is defined by (4.1). The following result can be found in [Cao and Wong 2003].

THEOREM 4.1. *Let E be a rank $r \geq 2$ holomorphic vector bundle over a compact complex manifold X. The following statements are equivalent:*

(1) E^* *is ample (resp. nef).*
(2) $\bigodot^{k} E^*$ *is ample (resp. nef) for some positive integer k.*
(3) *The dual $\mathcal{L}_{\mathbb{P}(\bigodot^{k} E)}$ of the tautological line bundle over the projectivized bundle $\mathbb{P}(\bigodot^{k} E)$ is ample (resp. nef) for some positive integer k.*
(4) *There exists a Finsler metric along the fibers of E with negative (resp. non-positive) mixed holomorphic bisectional curvature.*
(5) *For some positive integer k there exists a Finsler metric along the fibers of $\bigodot^{k} E$ with negative (resp. nonpositive) mixed holomorphic bisectional curvature.*

From the algebraic geometric point of view the key relationship between a vector bundle and its projectivization is the *Fundamental Theorem of Grothendieck* [Grothendieck 1958]:

THEOREM 4.2. *Let $p : E \to X$ be a holomorphic vector bundle of rank $r \geq 2$ over a complex manifold X of dimension n. Then for any analytic sheaf $\mathcal{S}$ on X and any $m \geq 1$,*

$$p_*^{i} \mathcal{L}_{\mathbb{P}(E)}^{m} \cong \begin{cases} \bigodot^{m} E^*, & \text{if } i = 0, \\ 0, & \text{if } i > 0, \end{cases}$$

where $p_*^{i} \mathcal{L}_{\mathbb{P}(E)}^{m}$ is the i-th direct image of $\mathcal{L}_{\mathbb{P}(E)}^{m}$. Consequently,

$$H^{i}(X, \bigodot^{m} E^* \otimes \mathcal{S}) \cong H^{i}(\mathbb{P}(E), \mathcal{L}_{\mathbb{P}(E)}^{m} \otimes p^*\mathcal{S}) \quad \text{for all } i \geq 0.$$

The theorem implies that the cohomology groups vanish beyond the dimension n of X although the dimension of $\mathbb{P}(E)$ is $n+r-1 > n$; moreover, $\chi(\bigodot^{m} E^* \otimes \mathcal{S}) = \chi(\mathcal{L}_{\mathbb{P}(E)}^{m} \otimes p^*\mathcal{S})$. For a vector bundle F over a smooth surface X, the *Chern character* and the *Todd class* are defined by

$$\begin{aligned} \mathrm{ch}(F) &= \mathrm{rk}(F) + c_1(F) + \tfrac{1}{2}(c_1^2(F) - 2c_2(F)), \\ \mathrm{td}(F) &= 1 + \tfrac{1}{2}c_1(F) + \tfrac{1}{12}(c_1^2(F) + c_2(F)). \end{aligned} \qquad (4.3)$$

The *Riemann–Roch formula* is

$$\chi(F) = \mathrm{ch}(F) \cdot \mathrm{td}(TX)[X] = \left(\tfrac{1}{2}(\iota(F) - c_1(F) c_1) + \tfrac{1}{12}\mathrm{rk}(F)(c_1^2 + c_2) \right)[X], \quad (4.4)$$

where $c_i = c_i(T^*X) = -c_i(TX)$. The notation $\omega[X]$ indicates the evaluation of a form of top degree on the *fundamental cycle* $[X]$, that is,

$$\omega[X] = \int_X \omega.$$

Assume rank $F = 2$ over a nonsingular complex surface X. Then, by Lemma 3.1,

$$\iota(\bigodot^m F) = \tfrac{1}{6}m(m+1)(2m+1)c_1^2(F) - \tfrac{1}{3}m(m+1)(m+2)c_2(F),$$

$$\mathrm{ch}(\bigodot^m F) = m+1 + \tfrac{1}{2}m(m+1)c_1(F) + \tfrac{1}{12}m(m+1)\big((2m+1)c_1^2(F)$$
$$- 2(m+2)c_2(F)\big),$$

$$\chi(\bigodot^m F) = \tfrac{1}{12}m(m+1)\big((2m+1)c_1^2(F) - 2(m+2)c_2(F)\big)$$
$$- \tfrac{1}{4}m(m+1)c_1(F)\,c_1 + \tfrac{1}{12}(m+1)(c_1^2 + c_2).$$

For example, taking $F = T^*X$,

$$\chi(\bigodot^m T^*X) = \tfrac{1}{12}(m+1)\big((2m^2 - 2m + 1)c_1^2 - (2m^2 + 4m - 1)c_2)\big);$$

in particular,

$$\chi(T^*X) = \tfrac{1}{6}(c_1^2 - 5c_2), \quad \chi(\bigodot^2 T^*X) = \tfrac{1}{4}(5c_1^2 - 15c_2).$$

In any case we have:

THEOREM 4.3. *Let $p : E \to X$ be a holomorphic vector bundle of rank $r = 2$ over a complex surface X. Then $\dim \mathbb{P}(E) = 3$ and for any positive integer m,*

$$\chi(\bigodot^m E^*) = \chi(\mathcal{L}_{\mathbb{P}(E)}^m) = \frac{m^3}{3!}(c_1^2(E^*) - c_2(E^*)) + O(m^2) = \frac{m^3}{3!}c_1^3(\mathcal{L}_{\mathbb{P}(E)}^m) + O(m^2).$$

Suppose that $h^2(\mathcal{L}_{\mathbb{P}(E)}^m)\,(= h^2(\bigodot^m E)) = O(m^2)$ and that $c_1^3(\mathcal{L}_{\mathbb{P}(E)}^m) > 0$ (equivalently, $c_1^2(E) - c_2(E) > 0$). The preceding theorem implies that E (or equivalently $\mathcal{L}_{\mathbb{P}(E)}$) is big, that is,

$$h^0(\mathcal{L}_{\mathbb{P}(E)}^m) = h^0(\bigodot^m E) \geq Cm^3$$

for some constant $C > 0$. Recall the following fact (from [Cao and Wong 2003] or [Kobayashi and Ochiai 1970], for example):

THEOREM 4.4. *Let E be a holomorphic vector bundle of rank $r \geq 2$ over a complex manifold X. Then the canonical bundles of X and $\mathbb{P}(E)$ are related by the formula*

$$\mathcal{K}_{\mathbb{P}(E)} \cong [p_E]^*(\mathcal{K}_X \otimes \det E^*) \otimes \mathcal{L}_{\mathbb{P}(E)}^{-r}$$

where $\mathcal{L}_{\mathbb{P}(E)}^{-r}$ is the dual of the r-fold tensor product of $\mathcal{L}_{\mathbb{P}(E)}$. In particular, we have

$$\mathcal{K}_{\mathbb{P}(TX)} \cong [p_{TX}]^*\mathcal{K}_X^2 \otimes \mathcal{L}_{\mathbb{P}(TX)}^{-n} \quad \text{and} \quad \mathcal{K}_{\mathbb{P}(T^*X)} \cong \mathcal{L}_{\mathbb{P}(T^*X)}^{-n}$$

where $n = \dim X$.

COROLLARY 4.5. *Let X be a complex manifold of dimension n.*

(i) *TX is ample (resp. nef) if and only if $\mathcal{K}_{\mathbb{P}(T^*X)}^{-1}$ is ample (resp. nef).*

(ii) *If $\mathcal{K}_X$ is nef then $\mathcal{K}_{\mathbb{P}(TX)} \otimes \mathcal{L}_{\mathbb{P}(TX)}^n$ is nef.*

(iii) *If T^*X is ample then $\mathcal{K}_{\mathbb{P}(TX)} \otimes \mathcal{L}_{\mathbb{P}(TX)}^n$ is nef and $\mathcal{K}_{\mathbb{P}(TX)} \otimes \mathcal{L}_{\mathbb{P}(TX)}^{n+1}$ is ample.*

We have the following *vanishing theorem* [Cao and Wong 2003] (for variants see [Chandler and Wong 2004]):

COROLLARY 4.6. *Let E be a nef holomorphic vector bundle of rank $r \geq 2$ over a compact complex manifold M of dimension n. Then*

$$H^i\big(X, \bigodot^m E \otimes \det E \otimes \mathcal{K}_X\big) = 0,$$
$$H^i\big(X, \bigodot^m (\bigotimes^k E) \otimes \det(\bigotimes^k E) \otimes \mathcal{K}_X\big) = 0,$$

for all $i, m, k \geq 1$. Consequently, if $E = TX$ then $H^i(X, \bigodot^m TX) = 0$ for all $i, m \geq 1$.

For a holomorphic line bundle $\mathcal{L}$ over a compact complex manifold Y with $h^0(\mathcal{L}^m) > 0$, m a positive integer, define a meromorphic map

$$\Phi_m = [\sigma_0, \dots, \sigma_N] : Y \to \mathbb{P}^N$$

where $\sigma_0, \dots, \sigma_N$ is a basis of $H^0(\mathcal{L}^m)$. The *Kodaira–Iitaka dimension* of $\mathcal{L}$ is defined to be

$$\kappa(\mathcal{L}) = \begin{cases} -\infty, & \text{if } h^0(\mathcal{L}^m) = 0 \text{ for all } m, \\ \max\{\dim \Phi_m(X) \mid h^0(\mathcal{L}^m) > 0\}, & \text{otherwise.} \end{cases}$$

The line bundle $\mathcal{L}$ is said to be *big* if $k(\mathcal{L}) = \dim Y$. This is equivalent to saying that, for $m \gg 0$

$$h^0(\mathcal{L}^m) \geq C m^{\dim Y}$$

for some positive constant C; in other words, the dimension of the space of sections $h^0(\mathcal{L}^m)$ has maximum possible growth rate. See [Chandler and Wong 2004] for a discussion of the differential geometric meaning of big bundles. Riemann–Roch asserts that if $c_1^{\dim Y}(\mathcal{L}) > 0$ the Euler characteristic is big:

$$\chi(\mathcal{L}^m) = \frac{c_1^{\dim Y}(\mathcal{L})}{(\dim Y)!} m^{\dim Y} + O(m^{\dim Y - 1}).$$

This, in general, is not enough to conclude that $\mathcal{L}$ is big. However, Corollary 4.6 implies that if T^*X is nef then the cohomology groups $H^i(X, T^*X) = 0$ for all $i \geq 1$. Hence T^*X is big if the Euler characteristic is big. In fact, for surfaces the weaker condition that $\mathcal{K}_X$ is nef suffices:

COROLLARY 4.7. *Suppose that the canonical bundle $\mathcal{K}_X$ of a nonsingular surface is nef and that $c_1^{\dim \mathbb{P}(TX)}(\mathcal{L}_{\mathbb{P}(TX)}) > 0$. Then $\mathcal{L}_{\mathbb{P}(TX)}$ is big.*

A vector bundle E of rank > 1 is said to be big if the line bundle $\mathcal{L}_{\mathbb{P}(E)}$ is big. By Theorem 4.3, for a surface X the condition $c_1^{\dim \mathbb{P}(TX)}(\mathcal{L}_{\mathbb{P}(TX)}) > 0$ is equivalent to the condition that $c_1^2(TX) - c_2(TX) = c_1^2(T^*X) - c_2(T^*X)$ is positive. Thus we may restate Corollary 4.7 as follows:

COROLLARY 4.8. *Let X be a nonsingular compact surface such that $c_1^2(T^*X) - c_2(T^*X) > 0$ and $\mathcal{K}_X$ is nef. Then T^*X is big.*

The preceding corollary implies the following theorem which may be viewed as an analogue, for surfaces, of the classical theorem that a curve of positive genus is hyperbolic [Lu and Yau 1990; Lu 1991; Dethloff et al. 1995a; 1995b]:

THEOREM 4.9. *Let X be a nonsingular surface such that $c_1^2(T^*X) - 2c_2(T^*X) > 0$ and $\mathcal{K}_X$ is nef. Then X is hyperbolic.*

We refer the readers to [Dethloff et al. 1995a; 1995b] for further information and refinements of the preceding theorem. The condition $c_1^2(T^*X) - c_2(T^*X) > 0$ is not satisfied by hypersurfaces in $\mathbb{P}^3$ which is the main reason that jet differentials are introduced. The computations in the previous section will provide conditions (on the Chern numbers $c_1^2(T^*X)$ and $c_2(T^*X)$) under which the sheaves of jet differentials $\mathcal{J}_k^m X$ must be big.

5. Weighted Projective Spaces and Projectivized Jet Bundles

SUMMARY. *The fibers of the k-jet bundles $\mathbb{P}(J^k X)$ are special types of weighted projective spaces. We collect some of the known facts of these spaces in this section. The main point is that these spaces are, in general, not smooth but with very mild singularities and we show that the usual theory of fiber integration for smooth manifolds extends to $\mathbb{P}(J^k X)$. This will be used in later sections.*

We follow the approach of the previous section by reducing questions concerning k-jet differentials to questions about the line bundle over the projectivization $\mathbb{P}(J^k X)$. Since $J^k X$ is only a $\mathbb{C}^*$-bundle rather than a vector bundle the fibers of the projectivized bundle $\mathbb{P}(J^k X)$ is not the usual projective space but a special type of weighted projective space. We give below a brief account concerning these spaces; see [Beltrametti and Robbiano 1986; Dolgachev 1982; Dimca 1992] for more detailed discussions and further references. The general theory of the projectivization of coherent sheaves can be found in [Banica and Stanasila 1976].

Consider $\mathbb{C}^{r+1}$ together with a vector $Q = (q_0, \ldots, q_r)$ of positive integers. The space $\mathbb{C}^{r+1}$ is then denoted $(\mathbb{C}^{r+1}, Q)$ and we say that each coordinate z_i, $0 \leq i \leq r$, has *weight* (or *degree*) q_i. A $\mathbb{C}^*$-action is defined on $(\mathbb{C}^{r+1}, Q)$ by

$$\lambda.(z_0, \ldots, z_r) = (\lambda^{q_0} z_0, \ldots, \lambda^{q_r} z_r) \quad \text{for } \lambda \in \mathbb{C}^*.$$

The quotient space $\mathbb{P}(Q) = (\mathbb{C}^{r+1}, Q)/\mathbb{C}^*$ is the *weighted projective space* of type Q. The equivalence class of an element $(z_0, \ldots, z_r)$ is denoted by $[z_0, \ldots, z_r]_Q$.

For $Q = (1, \ldots, 1) = \mathbf{1}$, $\mathbb{P}(Q) = \mathbb{P}_r$ is the usual complex projective space of dimension r and an element of $\mathbb{P}_r$ is denoted simply by $[z_0, \ldots, z_r]$. The case $r = 1$ is special as it can be shown that $\mathbb{P}(q_0, q_1) \cong \mathbb{P}^1$ for any tuple (q_0, q_1). This is not so if $r \geq 2$. For a tuple Q define a map $\psi_Q : (\mathbb{C}^{r+1}, \mathbf{1}) \to (\mathbb{C}^{r+1}, Q)$ by

$$\psi_Q(z_0, \ldots, z_r) = (z_0^{q_0}, \ldots, z_r^{q_r}).$$

It is easily seen that ρ_Q is compatible with the respective $\mathbb{C}^*$-actions and hence descends to a well-defined morphism:

$$[\psi_Q] : \mathbb{P}_r \to \mathbb{P}(Q), \quad [\psi_Q]([z_0, \ldots, z_r]) = [z_0^{q_0}, \ldots, z_r^{q_r}]_Q. \tag{5.1}$$

The weighted projective space can also be described as follows. Denote by Θ_{q_i} the group consisting of all q_i-th roots of unity. The group $\Theta_Q = \bigoplus_{i=0}^r \Theta_{q_i}$ acts on $\mathbb{P}_r$ by coordinatewise multiplication:

$$(\theta_0, \ldots, \theta_r).[z_0, \ldots, z_r] = [\theta_0 z_0, \ldots, \theta_r z_r], \quad \theta_i \in \Theta_{q_i},$$

and the quotient space is denoted by $\mathbb{P}_r / \Theta_Q$. The next result is easily verified [Dimca 1992]:

THEOREM 5.1. *The weighted projective space $\mathbb{P}(Q)$ is isomorphic to the quotient $\mathbb{P}_r / \Theta_Q$. In particular, $\mathbb{P}(Q)$ is irreducible and normal (the singularities are cyclic quotients and hence rational).*

Given a tuple Q we assign the degree (or weight) q_i to the variable z_i $(i = 1, \ldots, q)$ and denote by $S_Q(m)$ the space of homogeneous polynomials of degree m. In other words, a polynomial P is in $S(Q)(m)$ if and only if $P(\lambda \cdot (z_0, \ldots, z_r)) = \lambda^m P(z_0, \ldots, z_r)$. We may express such a polynomial explicitly as

$$P = \sum_{(i_0, \ldots, i_r) \in \mathcal{I}_{Q,m}} a_{i_0 \ldots i_r} z_0^{i_0} \ldots z_r^{i_r},$$

where the index set $\mathcal{I}_{Q,m}$ is defined by

$$\mathcal{I}_{Q,m} = \left\{ (i_0, \ldots, i_r) \, \middle| \, \sum_{j=0}^r q_j i_j = m, \, i_l \in \mathbb{N} \cup \{0\} \right\}.$$

The sheaf $\mathcal{O}_{\mathbb{P}(Q)}(m)$, $m \in \mathbb{N}$, is by definition the sheaf over $\mathbb{P}(Q)$ whose global regular sections are precisely the elements of $S_Q(m)$, i.e., $H^0(\mathbb{P}(Q), \mathcal{O}_{\mathbb{P}(Q)}(m)) = S_Q(m)$. For a negative integer $-m$ the sheaf $\mathcal{O}_{\mathbb{P}(Q)}(-m)$ is defined to be the dual of $\mathcal{O}_{\mathbb{P}(Q)}(m)$ and $\mathcal{O}_{\mathbb{P}(Q)}(0)$ is the structure sheaf $\mathcal{O}_{\mathbb{P}(Q)}$ of $\mathbb{P}(Q)$. Here are some basic properties of these sheaves (see [Beltrametti and Robbiano 1986]):

THEOREM 5.2. *Let $Q = (q_0, \ldots, q_r)$ be an $r + 1$-tuple of positive integers.*

(i) *For any for any $m \in \mathbb{Z}$, the line sheaf $\mathcal{O}_{\mathbb{P}(Q)}(m)$ is a reflexive coherent sheaf.*

(ii) *$\mathcal{O}_{\mathbb{P}(Q)}(m)$ is locally free if m is divisible by each q_i (hence by the least common multiple).*

(iii) *Let m_Q be the least common multiple of $\{q_0, \ldots, q_r\}$. Then $\mathcal{O}_{\mathbb{P}(Q)}(m_Q)$ is ample.*

(iv) *There exists an integer n_0 depending only on Q such that $\mathcal{O}_{\mathbb{P}(Q)}(nm_Q)$ is very ample for all $n \geq n_0$.*

(v) $\mathcal{O}_{\mathbb{P}(Q)}(\alpha m_Q) \otimes \mathcal{O}_{\mathbb{P}(Q)}(\beta) \cong \mathcal{O}_{\mathbb{P}(Q)}(\alpha m_Q + \beta)$ *for any* $\alpha, \beta \in \mathbb{Z}$.

For $Q = 1$ the assertions of the preceding theorem reduce to well-known properties of the usual twisted structure sheaves of the projective space. For any subset $J \subset \{0, 1, \dots, r\}$ denote by m_J the least common multiple of $\{q_j, j \in J\}$ and define

$$m(Q) = -\sum_{i=0}^{q} q_i + \frac{1}{r} \sum_{i=2}^{r+1} \frac{\sum_{\#J=i} m_J}{\binom{r-1}{i-2}},$$

where $\#J$ is the number of elements in the set J. It is known that we may take $n_0 = m(Q) + 1$ in assertion (iv) above. In general the line sheaf $\mathcal{O}_{\mathbb{P}(Q)}(m)$ is not invertible if m is not an integral multiple of m_Q. It can be shown that for $Q = (1, 1, 2)$ the sheaf $\mathcal{O}_{\mathbb{P}(Q)}(1)$ is not invertible and hence, neither is $\mathcal{O}_{\mathbb{P}(Q)}(1) \otimes \mathcal{O}_{\mathbb{P}(Q)}(1)$. On the other hand, by part (ii) of the preceding theorem we know that $\mathcal{O}_{\mathbb{P}(Q)}(2)$ is invertible, thus $\mathcal{O}_{\mathbb{P}(Q)}(1) \otimes \mathcal{O}_{\mathbb{P}(Q)}(1) \not\cong \mathcal{O}_{\mathbb{P}(Q)}(2)$. The following theorem on the cohomologies of the sheaf $\mathcal{O}_{\mathbb{P}}(Q)(p)$ is similar to the case of standard projective space (see [Beltrametti and Robbiano 1986] or [Dolgachev 1982]):

THEOREM 5.3. *If $Q = (q_0, \dots, q_r)$ is an $(r+1)$-tuple of positive integers then for $p \in \mathbb{Z}$,*

$$H^i(\mathbb{P}(Q), \mathcal{O}_{\mathbb{P}}(Q)(p)) = \begin{cases} \{0\}, & i \neq 0, r \\ S_Q(p), & i = 0, \\ S(Q)(-p - |Q|), & i = r, \end{cases}$$

where $|Q| = q_0 + \cdots + q_r$.

The cohomology group $H^i(\mathbb{P}(Q), \mathcal{O}_{\mathbb{P}(Q)}(p))$ vanishes provided that $i \neq 0, r$. Let $Q = (q_0, \dots, q_r)$ be a $(r+1)$-tuple of positive integers and define, for $k = 1, \dots, r$,

$$l_{Q,k} = \operatorname{lcm}\left\{ \frac{q_{i_0} \cdots q_{i_k}}{\gcd(q_0, \dots, q_{i_k})} \,\middle|\, 0 \leq i_0 < \cdots < i_k \leq r \right\}.$$

For integral cohomology we have:

THEOREM 5.4. *Let Q be an $(r+1)$-tuple of positive integers. Then*

$$H^i(\mathbb{P}(Q); \mathbb{Z}) \cong \begin{cases} \mathbb{Z}, & \text{if } i \text{ is even,} \\ 0, & \text{if } i \text{ is odd.} \end{cases}$$

Further, take $[\psi_Q] : \mathbb{P}^r \to \mathbb{P}(Q)$ as the quotient map defined by (5.1). Then the diagram

$$
\begin{array}{ccc}
H^{2k}(\mathbb{P}(Q); \mathbb{Z}) & \xrightarrow{\;[\psi_Q]^*\;} & H^{2k}(\mathbb{P}^r; \mathbb{Z}) \\
\cong \Big\downarrow & & \Big\downarrow \cong \\
\mathbb{Z} & \xrightarrow{\;\;l_{Q,k}\;\;} & \mathbb{Z}
\end{array}
$$

commutes, where the lower map is given by multiplication by the number $l_{Q,k}$.

Note that the number $l_{Q,r}$ is precisely the number of preimages of a point in $\mathbb{P}(Q)$ under the quotient map $[\psi_Q]$ (see (5.1)). The proof of the preceding Theorem for $k = r$ is easy; the reader is referred to [Kawasaki 1973] for the general case.

Let $Q = (q_0, q_1, \ldots, q_r)$, $r \geq 1$, be an $(r+1)$-tuple of positive integers. The tuple Q is said to be *reduced* if the greatest common divisor (gcd) of $(q_0, q_1, \ldots, q_r)$ is 1. In general, if the gcd is d, the tuple

$$Q_{\text{red}} = Q/d = (q_0/d, \ldots, q_r/d)$$

is called the *reduction* of Q. Let

$$\begin{aligned}
d_0 &= \gcd(q_1, \ldots, q_r), \\
d_i &= \gcd(q_0, q_1, \ldots, q_{i-1}, q_{i+1}, \ldots, q_r), \quad 1 \leq i \leq r-1, \\
d_r &= \gcd(q_0, \ldots, q_{r-1})
\end{aligned}$$

and define

$$\begin{aligned}
a_0 &= \text{lcm}(d_1, \ldots, d_r), \\
a_i &= \text{lcm}(d_0, d_1, \ldots, d_{i-1}, d_{i+1}, \ldots, d_r), \quad 1 \leq i \leq r-1, \\
a_r &= \text{lcm}(d_0, \ldots, d_{r-1}),
\end{aligned}$$

where "lcm" is short for "least common multiple". Define the normalization of Q by

$$Q_{\text{norm}} = (q_0/a_0, \ldots, q_r/a_r).$$

A tuple Q is said to be *normalized* if $Q = Q_{\text{norm}}$.

THEOREM 5.5. *Let Q be a normalized $(r+1)$-tuple of positive integers. Then the Picard group $\text{Pic}(\mathbb{P}(Q))$ and the divisor class group $\text{Cl}(\mathbb{P}(Q))$ are both isomorphic to $\mathbb{Z}$, and are generated, respectively, by*

$$\left[\mathcal{L}_{\mathbb{P}(Q)}^{m_Q} = \mathcal{O}_{\mathbb{P}(Q)}(m_Q) \right] \quad and \quad \left[\mathcal{L}_{\mathbb{P}(Q)} = \mathcal{O}_{\mathbb{P}(Q)}(1) \right].$$

Note that the generators of the two groups are different in general. For the standard projective space we have $m_Q = 1$ and so the generators are the same. For the k-jet bundles the fibers of their projectivization are weighted projective spaces with $m_Q = k!$, so we shall only be concerned with the case where $n, k \geq 1$ are positive integers and

$$Q = ((\underbrace{1, \ldots, 1}_{n}), (\underbrace{2, \ldots, 2}_{n}), \ldots, (\underbrace{k, \ldots, k}_{n})),$$

which is normalized. In this case we shall write $\mathbb{P}_{n,k}$ for $\mathbb{P}(Q)$. Note that $r = \dim \mathbb{P}_{n,k} = nk - 1$; the least common multiple of Q is

$$m_Q = k! \quad \text{and} \quad l_{Q,r} = (k!)^n. \tag{5.2}$$

Define a positive function

$$\rho_Q(z_0, \ldots, z_r) = \sum_{i=0}^{r} |z_i|^{2/q_i} \tag{5.3}$$

on $(\mathbb{C}^{r+1}, Q) \setminus \{0\}$. Then

$$\rho_Q(\lambda^{q_0} z_0, \ldots, \lambda^{q_r} z_r) = |\lambda|^2 \sum_{i=0}^{r} |z_i|^{2/q_i} = |\lambda|^2 \rho_Q(z_0, \ldots, z_r)$$

and

$$\psi^*(\rho_Q)(z_0, \ldots, z_r) = \sum_{i=0}^{r} |z_i^{q_i}|^{2/q_i} = \sum_{i=0}^{r} |z_i|^2 = \rho_1(z_0, \ldots, z_r)$$

is the standard Euclidean norm function on $(\mathbb{C}^{r+1}, 1)$. The function ρ_Q is not differentiable along $Z = \bigcup \{[z_{q_i} = 0], \, q_i \neq 1\}$. However, on $\mathbb{C}^{r+1} \setminus Z$, we deduce from the above that

$$\partial\bar{\partial} \log \rho_Q(\lambda^{q_0} z_0, \ldots, \lambda^{q_r} z_r) = \partial\bar{\partial} \log \rho_Q(z_0, \ldots, z_r)$$

and that

$$\psi_Q^*(\partial\bar{\partial} \log \rho_Q) = \partial\bar{\partial} \log \rho_1.$$

The first identity shows that $\partial\bar{\partial} \log \rho_Q$ is invariant under the $\mathbb{C}^*$-action hence descends to a well-defined $(1, 1)$-form ω_Q on $\mathbb{P}(Q) \setminus \pi_Q(Z)$. The second identity says that $\psi_Q^*(\omega_Q)$ is the Fubini–Study metric ω_{FS} on the standard projective space $\mathbb{P}^r \setminus \pi(Z)$ (hence actually extends smoothly across $\pi(Z)$). The Fubini–Study metric $[\omega_{FS}]$ is the first Chern form of $\mathcal{O}_{\mathbb{P}^r}(1)$ which is the (positive) generator of $\mathrm{Pic}\,\mathbb{P}^r = \mathrm{Cl}\,\mathbb{P}^r$. Hence $[\omega_{FS}]$ is the positive generator of $H^2(\mathbb{P}^r, \mathbb{Z})$. Theorem 5.4 implies that $[l_{Q,1}\omega_Q]$ is the generator of $H^2(\mathbb{P}(Q), \mathbb{Z})$.

Consider the function $\left(\sum_{i=0}^{r} |z_i|^{2\kappa}\right)^{1/\kappa}$, for κ a positive integer, defined on $\mathbb{C}^{r+1}$. It clearly satisfies

$$\left(\sum_{i=0}^{r} |\lambda z_i|^{2\kappa}\right)^{1/\kappa} = |\lambda|^2 \left(\sum_{i=0}^{r} |z_i|^{2\kappa}\right)^{1/\kappa};$$

hence is a metric along the fibers of the tautological line bundle over $\mathbb{P}^r$. Moreover, the form

$$\partial\bar{\partial} \log \left(\sum_{i=0}^{r} |z_i|^{2\kappa}\right)^{1/\kappa}$$

descends to a well-defined form on the standard projective space $\mathbb{P}^r$, indeed a Chern form, denoted by η_κ, for the hyperplane bundle of $\mathbb{P}^r$; moreover it is cohomologous to the Fubini–Study form. With this we may define an alternative to ρ_Q,

$$\tau_Q(z_0, \ldots, z_r) = \left(\sum_{i=0}^{r} |z_i|^{2\kappa/q_i}\right)^{1/\kappa}, \quad \kappa = \prod_{i=0}^{r} q_i. \tag{5.4}$$

It is of class $\mathcal{C}^\infty$ on $\mathbb{C}^{r+1} \setminus \{0\}$. Just like ρ_Q, the function τ_Q satisfies

$$\tau_Q(\lambda^{q_0} z_0, \ldots, \lambda^{q_r} z_r) = \left(\sum_{i=0}^r |\lambda|^{2\kappa} |z_i|^{2\kappa/q_i} \right)^{1/\kappa} = |\lambda|^2 \left(\sum_{i=0}^r |z_i|^{2\kappa/q_i} \right)^{1/\kappa}$$

and

$$(\psi^* \tau_Q)(z_0, \ldots, z_r) = \tau_Q(z_0^{q_0}, \ldots, z_r^{q_r}) = \left(\sum_{i=0}^r |z_i|^{2\kappa} \right)^{1/\kappa}.$$

These equalities imply that $\partial\bar\partial \log \tau_Q$ descends to a well-defined form γ_Q on $\mathbb{P}(Q)$ with the property that $\psi_Q^* \gamma_Q = \eta_k$, and consequently is cohomologous to ω_Q.

Let $\pi : J^k X \to X$ be the parametrized k-jet bundle of a complex manifold X and denote by $p : \mathbb{P}(J^k X) \to X$ and $pr : p^* J^k X \to \mathbb{P}(J^k X)$ the corresponding projection maps. The following diagram is commutative:

$$
\begin{array}{ccc}
p^* J^k X & \xrightarrow{\ p_* \ } & J^k X \\
\downarrow {\scriptstyle pr} & & \downarrow {\scriptstyle \pi} \\
\mathbb{P}(J^k X) & \xrightarrow{\ p \ } & X
\end{array}
$$

and the tautological subsheaf of $p^* J^k X$ is the line sheaf defined by

$$\{ ([\xi], \eta) \in p^* J^k X \mid [\xi] \in \mathbb{P}(J^k X),\ p([\xi]) = \pi(\eta) = x,\ [\eta] = [\xi] \}$$

where, for ξ (resp. η) in $J^k X$, its equivalence class in $\mathbb{P}(J^k X)$ is denoted by $[\xi]$ (resp. $[\eta]$). The "hyperplane sheaf", denoted $\mathcal{L} = \mathcal{L}_k$, is defined to be the dual of the tautological line sheaf. The fiber $\mathbb{P}(J_x^k X)$ over a point $x \in X$ is the weighted projective space of type $Q = ((1, \ldots, 1); \ldots; (k, \ldots, k))$ and the restriction of $\mathcal{L}_k$ to $\mathbb{P}(J_x^k X)$ is the line sheaf $\mathcal{O}_{\mathbb{P}(Q)}(1)$ as defined in Theorem 5.2. The next result follows readily from Theorem 5.2:

THEOREM 5.6. *Let X be a complex manifold.*

(i) *For any $m \in \mathbb{Z}, \mathcal{L}^m_{\mathbb{P}(J^k X)}$ is a reflexive coherent sheaf.*

(ii) *$\mathcal{L}^{k!}_{\mathbb{P}(J^k X)}$ is the generator of $\mathrm{Pic}(\mathbb{P}(J^k X))$, that is, $\mathcal{L}^m_{\mathbb{P}(J^k X)}$ is locally free if m is divisible by $k!$.*

(iii) *For any $\alpha, \beta \in \mathbb{Z}$, $\mathcal{L}^{k!\alpha}_{\mathbb{P}(J^k X)} \otimes \mathcal{L}^\beta_{\mathbb{P}(J^k X)} \cong \mathcal{L}^{k!\alpha+\beta}_{\mathbb{P}(J^k X)}$.*

The Chern class of the bundle $\mathcal{L}^{k!}_{\mathbb{P}(J^k X)}$ is $k! \omega_Q = l_{Q,1} \omega_Q$, where ω_Q is constructed after Theorem 5.5. By (5.3) the function $\rho_Q^{k!}$ is a Finsler metric along the fibers of $\mathcal{L}^{k!}_{\mathbb{P}(J^k X)}$. The same is of course also true if we use γ_Q and τ_Q instead. Just as in the case of projectivized vector bundles we still have the identification of the spaces $\mathcal{L}^{-1}_{\mathbb{P}(J^k X)} \setminus \{0\}$ with $J^k X \setminus \{0\}$, which is compatible with the respective $\mathbb{C}^*$ action. Thus, as in the case of vector bundles, we conclude that a metric along the fibers of $\mathcal{L}^{-1}_{\mathbb{P}(J^k X)}$ is identified with a Finsler metric along the fibers of

$J^k X$. As $J^k X$, in general, is only a $\mathbb{C}^*$-bundle and not a vector bundle, we see that Finsler geometry is indispensable.

Note that, for $k \geq 2$, the sheaf $\mathcal{L}_{\mathbb{P}(J^k X)}$ is not locally free and, in general,

$$\mathcal{L}^a_{\mathbb{P}(J^k X)} \otimes \mathcal{L}^b_{\mathbb{P}(J^k X)}(a) \not\cong \mathcal{L}^{a+b}_{\mathbb{P}(J^k X)}.$$

Hence some of the proofs of the results that are valid for projectivized vector bundles require modifications. Basically, things work well if we use integer multiples of $k!$ (that is, $\mathcal{J}^{mk!}_k X$); for example, Grothendieck's Theorem (Theorem 4.2) remains valid:

THEOREM 5.7. *Let X be a complex manifold and $p : \mathbb{P}(J^k X) \to X$ the k-th parametrized jet bundle and let $\mathcal{S}$ be an analytic sheaf on X. For any $m \geq 1$,*

$$p^i_* \mathcal{L}^{mk!}_{\mathbb{P}(J^k X)} \cong \begin{cases} \mathcal{J}^{mk!}_k X, & \text{if } i = 0, \\ 0, & \text{if } i > 0, \end{cases}$$

where $p^i_* \mathcal{L}^{mk!}_{\mathbb{P}(J^k X)}$ is the i-th direct image of $\mathcal{L}^{mk!}_{\mathbb{P}(J^k X)}$. Consequently, we have

$$H^i(X, \mathcal{J}^{mk!}_k X \otimes \mathcal{S}) \cong H^i(\mathbb{P}(J^k X), \mathcal{L}^{mk!}_{\mathbb{P}(J^k X)} \otimes p^* \mathcal{S})$$

for all i.

In the case of vector bundles, Theorem 4.3 provides a relation between the Chern numbers of the bundle and that of the line bundle over the projectivization. Theorem 4.3 may be proved directly via fiber integration. Although the projectivized k-jet bundles are not smooth for $k \geq 2$ this correspondence is still valid. These technicalities are needed when we deal with problem of degeneration; as we shall see in Sections 6 and 7, under the condition that $\mathcal{L}^{k!}_k$ is big, k-jets of holomorphic maps into X are *algebraically degenerate*, that is, the images are contained in some (special type of) subvarieties of $\mathbb{P}(J^k X)$ which may be very singular. In order to calculate the Euler characteristic of $\mathcal{L}^{k!}_k X$ of these subvarieties it is necessary to compute the intersection numbers, as usual, via Chern classes and this is best handled by going down, via fiber integration, to the base variety X which is nonsingular. We take this opportunity to formulate a criterion for certain type of singular spaces on which fiber integration works well. The purpose here is not to exhibit the most general results but results general enough for our purpose. First we recall some basic facts concerning fiber integration. Let P and X be complex manifolds and $p : P \to X$ be a holomorphic surjection. The map p is said to be *regular* at a point $y \in P$ if the Jacobian of p at y is of maximal rank. The set of regular points is an open subset of P and p is said to be *regular* if every point of P is a regular point. The following statements concerning fiber integration are well-known (see [Stoll 1965], for example):

THEOREM 5.8. *Let P and X be connected complex manifolds of dimension N and n respectively. Let $p : P \to X$ be a regular holomorphic surjection. Let r, s be integers with $r, s \geq N - n = q$. Then for any (r, s)-form ω of class $\mathcal{C}^k$ on*

*P that is integrable along the fibers of p, the fiber integral $p_*\omega$ is a well defined
$(r-q, s-q)$ form of class $\mathcal{C}^k$ on X. Moreover:*

(i) *For any $(N-r, N-s)$-form on X such that $\omega \wedge p^*\eta$ is integrable on P, we
have*

$$\int_P \omega \wedge p^*\eta = \int_X p_*\omega \wedge \eta.$$

(ii) *If ω is of class $\mathcal{C}^1$ and if ω and $d\omega$ are integrable along the fibers p then
$dp_*\omega = p_*d\omega, \partial p_*\omega = p_*\partial\omega$ and $\bar{\partial}p_*\omega = p_*\bar{\partial}\omega$.*

(iii) *If ω is nonnegative and integrable along the fibers of p then $p_*\omega$ is also
nonnegative.*

(iv) *Suppose that Y is another connected complex manifold of dimension n' with
a regular holomorphic surjection $\pi : X \to Y$. Assume that ω is a (r, s)-form
such that $r, s \geq q+q'$ where $q = N-n$ and $q' = n-n'$. If ω is integrable along
the fibers of p and $p_*\omega$ is integrable along the fibers of π then $\pi_* p_* \omega = (\pi \circ p)_* \omega$.*

If ω is a form of bidegree (r, s) so that either $r < q$ or $s < q$, where q is the fiber
dimension, then we set $p_*\omega = 0$. If $p : P \to X$ is a holomorphic fiber bundle
with smooth fiber S, then p is a regular surjection and the preceding Theorem is
applicable. Consider now P, an irreducible complex space of complex dimension
N, with a holomorphic *surjection* $p : P \to X$ where X is *nonsingular* and of
complex dimension n. The map p is said to be *regular* if there exists a connected
complex manifold $\tilde{P}$ of the same dimension as P and a surjective morphism
$\tau : \tilde{P} \to P$ such that the composite map $\tilde{p} = p \circ \tau : \tilde{P} \to X$ is regular. Let
$U \subset P$ be an open set and $\iota : U \to V \subset \mathbb{C}^{N'}$ a local embedding, where V is
an open set of $\mathbb{C}^{N'}$ for some N'. If η is a differential form on V then $\iota^*\eta$ is a
differential form on U. Conversely, a differential form ω on U is of the form $\iota^*\eta$
for some embedding $\iota : U \to V$ and some differential form η on V. Suppose that
ω is a differential form on P of bidegree (r, s); hence $\tau^*\omega$ is a differential form
on $\tilde{P}$ of bidegree (r, s). If either r or s is less than the fiber dimension $q = N-n$
then $p_*\omega$ is defined to be zero. For the case $r, s \geq q$ and assuming that $\tau^*\omega$ is
integrable along the fibers of $\tilde{p}$ (for example, this is the case if ω is integrable
along the fibers of p), we are in the nonsingular situation; hence $\tilde{p}_*\tau^*\omega$ is defined.
The pushforward $p_*\omega$ is naturally defined by

$$p_*\omega \overset{\text{def}}{=} \tilde{p}_*\tau^*\omega. \tag{5.5}$$

From this definition it is clear (since $\tilde{p} = p \circ \tau$) that the basic properties of fiber
integrals remain valid in the more general situation:

THEOREM 5.9. *Let $\tilde{\tau} : \tilde{P} \to P, p : P \to X$ and $\tilde{p} : \tilde{P} \to X$ be as above and
let ω be a form of bidegree (r, s) on P with $r, s \geq N-n$ where $n = \dim X, N =
\dim P = \dim \tilde{P}$. Then:*

(i) *If $\tau^*\omega$ is integrable along the fibers $\tilde{P}_y = \tilde{p}^{-1}(x)$ for almost all $x \in X$ then
for any $(N-r, N-s)$-form on X such that $\omega \wedge p^*\eta$ is integrable on P and*

$\tau^* \omega \wedge \tilde{p}^* \eta$ is integrable on $\tilde{P}$,

$$\int_P \omega \wedge p^* \eta = \int_X p_* \omega \wedge \eta = \int_{\tilde{P}} \tilde{p}_* \tau^* \omega \wedge \tau^* \eta.$$

(ii) If ω is of class $\mathcal{C}^1$ and if $\tau^* \omega$ and $\tau^* d\omega$ are integrable along the fibers $\tilde{P}_x$ for all $x \in X$ then $dp_* \omega = p_* d\omega$, $\partial p_* \omega = p_* \partial \omega$ and $\overline{\partial} p_* \omega = p_* \overline{\partial} \omega$.

(iii) If $\tau^* \omega$ is integrable along $\tilde{M}_x$ for all $x \in X$ then $p_* \omega$ is a form of type $(p - N + n, q - N + n)$ on X.

(iv) If ω is a continuous nonnegative form and $\tau^* \omega$ is integrable along $\tilde{P}_x$ for all $x \in X$ then $p_* \omega$ is also nonnegative.

The converse of part (iv) is not true in general.

The next theorem shows that the preceding theorem is applicable to the projectivized k-jet bundles (we refer the readers to [Stoll and Wong 2002] for details).

THEOREM 5.10. *Let X be a complex manifold of complex dimension n and let $p : P = \mathbb{P}(J^k X) \to X$ be the projectivized k-jet bundle of X. Then there exists a complex manifold $\tilde{P}$ of the same dimension as P and a surjective finite morphism $\tilde{\tau} : \tilde{P} \to P$ such that $\tilde{p} = p \circ \tau : \tilde{P} \to X$ is a regular holomorphic surjection. Moreover, $\tilde{P}$ can be chosen so that each of the fibers of $\tilde{p}$ is the complex projective space $\mathbb{P}^q$ where $q = nk - 1$.*

A similar argument (see [Stoll and Wong 2002]) shows that in general we have:

THEOREM 5.11. *Let X be a connected complex manifold of complex dimension n. Suppose that P is an irreducible complex space for which there exists a holomorphic surjective morphism $p : P \to X$ that is locally trivial; that is, for any $x \in X$ there exists an open neighborhood V of X, a complex space Y and a biholomorphic map $\alpha_V : p^{-1} V \to V \times Y$ such that the diagram*

$$
\begin{array}{ccc}
p^{-1}(V) & \xrightarrow{\alpha_V} & V \times Y \\
{\scriptstyle p}\downarrow & \cong & \downarrow{\scriptstyle p_V} \\
V & === & V
\end{array}
$$

commutes, where p_V is the projection onto the first factor. Then there exists a complex manifold $\tilde{P}$ of the same dimension as P and a surjective morphism $\tau : \tilde{P} \to P$ such that $\tilde{p} = p \circ \tau : \tilde{P} \to X$ is a regular holomorphic surjection.

Next we extend the definition of *pushforward* of forms to subvarieties of a complex space P with a projection map $p : P \to X$ satisfying the local triviality condition of the preceding theorem. In general the pushforwards exist only as currents. Suppose that $Y \subset P$ is an irreducible subvariety of dimension ν of P and assume that $p|_Y : Y \to X$ is surjective. Let $\Sigma \subset Y$ be the set of singular points of Y; so the set $S_1 = \{z \in X \mid (p|_Y)^{-1}(z) \subset \Sigma\}$ is a subvariety of codimension at least one in X. Note that

$$p|_{Y \setminus \Sigma} : Y \setminus \Sigma \to X \setminus S_1$$

is surjective, hence generically regular; that is, there exists a subvariety $S_2 \subset X$ of codimension at least 1 such that

$$p|_{Y \setminus (\Sigma \cup (p|_Y)^{-1}(S_2))} : Y_1 = Y \setminus (\Sigma \cup (p|_Y)^{-1}(S_2)) \to X_1 = X \setminus (S_1 \cup S_2)$$

is a regular surjection. Let ω be a smooth (r, s)-form on Y, $r, s \geq N - n =$ generic fiber dimension of $p|_{Y_1}$, which is integrable along the fibers of $p|_{Y_1}$. Then $(p|_{Y_1})_* \omega$ is a $(p - N + n, q - N + n)$-form on X_1. Meanwhile, the pushforward $(p|_Y)_* \omega$ exists as a current on X, that is,

$$(p|_Y)_* \omega((p|_Y)^* \phi) \overset{\text{def}}{=} \omega(\phi) = \int_Y (p|_Y)^* \phi \wedge \omega \qquad (5.6)$$

for any $(N-r, N-s)$-form ϕ with compact support on X. Clearly, $(p|_Y)_* \omega|_{X_1} = (p|_Y)_* \omega$. Note that as a current the pushforward commutes with exterior differentiation, that is, $d(p|_Y)_* \omega = p_* d\omega$, $\partial(p|_Y)_* \omega = (p|_Y)_* \partial \omega$ and $\overline{\partial}(p|_Y)_* \omega = (p|_Y)_* \overline{\partial} \omega$. Also, by definition, the pushforward preserves nonnegativity.

The *Riemann–Roch formulas for jet differentials* follow from those of the bundles $\bigodot^{i_1} T^* X \otimes \cdots \otimes \bigodot^{i_k} T^* X$, given below (see [Stoll and Wong 2002] for details):

THEOREM 5.12. *Let X be a smooth compact complex surface. Set $I = (i_1, \ldots, i_k)$ and $\mathcal{S}_I = \bigodot^{i_1} T^* X \otimes \cdots \otimes \bigodot^{i_k} T^* X$, where each i_j is a nonnegative integer. Then*

$$\chi(X; \mathcal{S}_I) = \tfrac{1}{12} \mu_k (2 s_{1;k}^2 - 2 s_{1;k} - s_{2;k} + 1) c_1^2(T^* X)$$
$$- \tfrac{1}{12} \mu_k (2 s_{1;k}^2 + 4 s_{1;k} - 4 s_{2;k} - 1) c_2(T^* X),$$

where $s_{j;k}$, for $1 \leq j \leq k$, is the degree-j symmetric function in $i_1, \ldots, i_k$ and $\mu_k = \sum_{j=0}^{k} s_{j;k}$ is as in (3.6).

Given an exact sequence of coherent sheaves $0 \to E_1 \to E_2 \to E_3 \to 0$ the ranks, the first Chern classes, the Chern characters, the indices and the Euler characteristics are additive in the sense that $\text{rk}\, E_2 = \text{rk}\, E_1 + \text{rk}\, E_3$, $c_1(E_2) = c_1(E_1) + c_1(E_3)$, $\iota(E_2) = \iota(E_1) + \iota(E_3)$, $ch(E_2) = ch(E_1) + ch(E_3)$ and $\chi(X; E_2) = \chi(X; E_1) + \chi(X; E_3)$. The Euler characteristic of $\mathcal{J}_k^{k!} X$ is given thus:

THEOREM 5.13. *Let X be a nonsingular surface. We have, for $m \gg k$,*

$$\chi(\mathcal{J}_k^{k!m} X) = \tfrac{1}{2} \iota(\mathcal{J}_k^{k!m} X) + O(m^{2k}) = \tfrac{1}{2} (k!)^{2k+1} (\alpha_k c_1^2 - \beta_k c_2) \, m^{2k+1} + O(m^{2k}),$$

where α_k and β_k are constants given in Theorem 3.9.

EXAMPLE 5.14. We record below explicit formulas for the sheaves that occur in the preceding computations:

$\mathcal{S}$	$\mathrm{ch}(\mathcal{S})$	$\chi(\mathcal{S})$
T^*X	$2 + c_1 + \frac{1}{2}(c_1^2 - 2c_2)$	$\frac{1}{6}(c_1^2 - 5c_2)$
$\bigodot^2 T^*X$	$3 + 3c_1 + \frac{1}{2}(5c_1^2 - 8c_2)$	$\frac{1}{4}(5c_1^2 - 15c_2)$
$\bigodot^3 T^*X$	$4 + 6c_1 + 7c_1^2 - 10c_2$	$\frac{1}{3}(13c_1^2 - 29c_2)$
$\bigodot^4 T^*X$	$5 + 10c_1 + 15c_1^2 - 20c_2$	$\frac{1}{12}(125c_1^2 - 235c_2)$
$\bigodot^5 T^*X$	$6 + 15c_1 + \frac{1}{2}(55c_1^2 - 70c_2)$	$\frac{1}{2}(41c_1^2 - 84c_2)$
$\bigodot^6 T^*X$	$7 + 21c_1 + \frac{1}{2}(91c_1^2 - 112c_2)$	$\frac{1}{12}(427c_1^2 - 665c_2)$
$\bigodot^7 T^*X$	$7 + 21c_1 + \frac{1}{2}(91c_1^2 - 112c_2)$	$\frac{1}{3}(170c_1^2 - 250c_2)$
$T^*X \otimes T^*X$	$4 + 4c_1 + 3c_1^2 - 4c_2$	$\frac{1}{3}(4c_1^2 - 11c_2)$
$(\bigodot^2 T^*X) \otimes T^*X$	$6 + 9c_1 + \frac{1}{2}(19c_1^2 - 22c_2)$	$\frac{1}{2}(11c_1^2 - 21c_2)$
$(\bigodot^3 T^*X) \otimes T^*X$	$8 + 16c_1 + 22c_1^2 - 24c_2$	$\frac{1}{3}(44c_1^2 - 70c_2)$
$(\bigodot^2 T^*X) \otimes (\bigodot^2 T^*X)$	$9 + 18c_1 + 24c_1^2 - 24c_2$	$\frac{1}{4}(63c_1^2 - 93c_2)$
$(\bigodot^4 T^*X) \otimes T^*X$	$10 + 25c_1 + \frac{1}{2}(85c_1^2 - 90c_2)$	$\frac{1}{6}(185c_1^2 - 265c_2)$
$(\bigodot^3 T^*X) \otimes (\bigodot^2 T^*X)$	$12 + 30c_1 + 49c_1^2 - 46c_2$	$35c_1^2 - 45c_2$
$(\bigodot^5 T^*X) \otimes T^*X$	$12 + 36c_1 + 73c_1^2 - 76c_2$	$56c_1^2 - 75c_2$
$\mathcal{J}_2^2 X$	$5 + 4c_1 + 3c_1^2 - 5c_2$	$\frac{1}{12}(17c_1^2 - 55c_2)$
$\mathcal{J}_2^3 X$	$8 + 10c_1 + 10c_1^2 - 14c_2$	$\frac{1}{4}(23c_1^2 - 53c_2)$
$\mathcal{J}_2^4 X$	$14 + 22c_1 + 27c_1^2 - 35c_2$	$\frac{1}{6}(103c_1^2 - 207c_2)$
$\mathcal{J}_2^5 X$	$20 + 40c_1 + 59c_1^2 - 70c_2$	$\frac{1}{3}(122c_1^2 - 205c_2)$
$\mathcal{J}_2^6 X$	$30 + 70c_1 + 119c_1^2 - 135c_2$	$\frac{1}{2}(173c_1^2 - 265c_2)$
$\mathcal{J}_2^7 X$	$40 + 110c_1 + 214c_1^2 - 200c_2$	$\frac{1}{3}(487c_1^2 - 590c_2)$

Although the space $\mathbb{P}(J^k X)$ is not smooth, the following Riemann–Roch Theorem is still valid, by Theorems 5.7 and 5.13:

THEOREM 5.15. *Let X be a nonsingular surface and $p : \mathbb{P}(J^k X) \to X$ the k-jet bundle. Then*

$$
\begin{aligned}
\chi(\mathcal{L}^m_{\mathbb{P}(J^k X)}) &= \mathrm{ch}(\mathcal{L}^m) . \mathrm{td}(\mathbb{P}(J^k X))[\mathbb{P}(J^k X)] \\
&= \mathrm{ch}(\mathcal{L}^m) . \mathrm{td}(T_p) . p^* \mathrm{td}(X)[\mathbb{P}(J^k X)] \\
&= p_*\big(\mathrm{ch}(\mathcal{L}^m) . \mathrm{td}(T_p)\big) . \mathrm{td}(X)[X],
\end{aligned}
$$

where T_p is the relative tangent sheaf of the projection $p : \mathbb{P}(J^k X) \to X$, that is, the restriction of T_p to each fiber of p is the tangent sheaf of the weighted projective space $\mathbb{P}(Q)$.

On $\mathbb{P}(J^k X)$ we have

$$\text{ch}(\mathcal{L}^m) = \sum_{i=0}^{2k+1} \frac{c_1^i(\mathcal{L}^m)}{i!} = \sum_{i=0}^{2k+1} \frac{c_1^i(\mathcal{L})}{i!} m^i$$

which implies that

$$\chi(\mathcal{L}^m) = \frac{c_1^{2k+1}(\mathcal{L})}{(2k+1)!} m^{2k+1} + O(m^{2k}).$$

Theorems 5.13 and 5.15 imply:

COROLLARY 5.16. *Let* X *be a nonsingular surface and* $p : \mathbb{P}(J^k X) \to X$ *the* k-*jet bundle. Then*

$$p_* \left(\frac{c_1^{2k+1}(\mathcal{L})}{(2k+1)!} \right) = \frac{1}{2} \iota(\mathcal{J}_k X) = \frac{1}{2}(\alpha_k c_1^2 - \beta_k c_2)$$

where α_k *and* β_k *are constants given in Theorem 3.9.*

6. The Lemma of Logarithmic Derivatives and the Schwarz Lemma

SUMMARY. *In this section we use Nevanlinna Theory to show (Corollary 6.2) that if* ω *is a holomorphic* k-*jet differential of weight* m *vanishing on an effective divisor of a projective manifold* X *then* $f^* \omega \equiv 0$ *for any holomorphic map* $f : \mathbb{C} \to X$. *(For our application in Section 7 it is enough to assume that the divisor is a hyperplane section.) This implies (see Theorem 6.4) that if* $f : \mathbb{C} \to X$ *is an algebraically nondegenerate holomorphic map then the irreducible component of the base locus containing* $[j^k f]$ *is of codimension at most* $(n-1)k$; *equivalently the dimension is at least* $n + k - 1, n = \dim X$. *This result is crucial in the proof of our main result in Section 7. We must point out that the method of this section works only for the parametrized jet bundles but not the full jet bundles. (Otherwise we could have avoided the complicated computations of the Chern numbers of the parametrized jet bundles; computing the Chern numbers of the full jet bundles, as honest vector bundles, is much simpler!) The idea of the proof is relatively standard from the point of view of Nevanlinna Theory. The main step is to construct, using a standard algebraic geometric argument, a Finsler metric of logarithmic type, reducing the problem to a situation in which the Lemma of Logarithmic Derivatives is applicable. If* $f^* \omega \not\equiv 0$ *this lemma implies that the integral of* $\log |f^* \omega|$ *is small. On the other hand, the first Crofton formula in Nevanlinna Theory asserts that the integral of* $\log |f^* \omega|$ *(as the counting function of the zeros of* $f^* \omega$ *by the Poincaré–Lelong formula) is not small. This contradiction establishes Theorem 6.1 and Corollary 6.2. Theorem 6.4 and Corollary 6.5 then follow from the Schwarz Lemma via a reparametrization argument often used in Nevanlinna Theory. The main point is that a reparametrization does* *not change, as a set, the (algebraic closure of the) image of the map f but does change the image of $f^*\omega$ if ω is a k-jet differential and $k \geq 2$.*

The *classical Schwarz Lemma* in one complex variable asserts that a holomorphic map $f : \mathbb{C} \to X$ is constant for a compact Riemann surface X of genus ≥ 2; that is, there are at least 2 independent regular 1-forms on X. Further, there is a noncompact version; namely, let X be a compact Riemann surface and let D be a finite number of points in X. Then a holomorphic map $f : \mathbb{C} \to X \setminus D$ is constant if the logarithmic genus is at least 2, that is, there are at least 2 independent 1-forms regular on $X \setminus D$ and with no worse than logarithmic singularity at each of the points in D. There are of course many proofs of this classical result, one of which is to find a nontrivial holomorphic 1-form (or, logarithmic 1-form) on X such that $f^*\omega \equiv 0$ for any entire holomorphic map $f : \mathbb{C} \to X$. This is not so difficult to do because $g \geq 2$ implies that T^*X is ample (and a priori, spanned). The main difficulty of proving the preceding comes from the fact that a big bundle is not necessary spanned. A coherent sheaf $\mathcal{S}$ is said to be *spanned* (by global regular sections) if, for every $v \in \mathcal{S}_x$ there is a global regular section $\sigma \in H^0(\mathcal{S})$ such that $\sigma(x) = v$. However, it is easily seen that a coherent sheaf $\mathcal{S}$ is spanned by global rational sections. For example, the complex projective space has no global regular 1-form. Hence it cannot span any of the fibers of $T^*\mathbb{P}^n$. However take any point $x \in \mathbb{P}^n$, assuming without loss of generality that $x = [x_0, \ldots, x_n]$ with $x_0 \neq 0$, then $T_x^*\mathbb{P}^n$ is spanned by dt_i, $i = 1, \ldots, n$, where $t_i = x_i/x_0$. Now dt_i is a global rational one-form since t_i is a global rational function; in fact

$$dt_i = \frac{x_0 dx_i - x_i dx_0}{x_0^2}$$

has a pole of order 2 along the "hyperplane at infinity", $[x_0 = 0]$. This shows that $T^*\mathbb{P}^n$ is spanned by global rational one-forms. In fact we can do better, namely, we may replace dt_i by $d\log t_i$

$$d\log t_i = \frac{dt_i}{t_i} = \frac{dx_i}{x_i} - \frac{dx_0}{x_0}.$$

A simple argument shows that there is a finite set of logarithmic one-forms $\{dL_i/L_i\}$ where each L_i is a rational function which span $T^*\mathbb{P}^n$ at every point. The mild singularity can be dealt with using the classical Lemma of Logarithmic Derivatives in Nevanlinna Theory and a weak form of the analytic Bézout Theorem known as Crofton's Formula.

It is not hard to see that the preceding procedure can be extended to deal with jet differentials. The details are given in the next theorem. The most convenient way to get to the Schwarz Lemma is via Nevanlinna Theory. First we recall some standard terminology. The characteristic function of a map $f : \mathbb{C} \to X$ is

$$T_f(r) = \int_0^r \frac{1}{t} \int_{\Delta_t} f^* c_1(H),$$

where H is a hyperplane section in X and the characteristic function of a nontrivial holomorphic function $F : \mathbb{C} \to \mathbb{C}$ is

$$T_F(r) = \int_0^{2\pi} \log^+ |F(re^{\sqrt{-1}\theta})| \frac{d\theta}{2\pi}.$$

Note that $\omega(j^k f)$ is a holomorphic function if f is a holomorphic map and ω is a k-jet differential of weight m.

THEOREM 6.1 (LEMMA OF LOGARITHMIC DERIVATIVES). *Let X be a projective variety and let (i) D be an effective divisor with simple normal crossings, or (ii) D be the trivial divisor in X (that is, the support of D is empty or equivalently, the line bundle associated to D is trivial). Let $f : \mathbb{C} \to X$ be an algebraically nondegenerate holomorphic map and $\omega \in H^0(X, \mathcal{J}_k^m X(\log D))$ (resp. $H^0(X, \mathcal{J}_k^m X)$ in case (ii)) a jet differential such that $\omega \circ j^k f$ is not identically zero. Then*

$$T_{\omega \circ j^k f}(r) = \int_0^{2\pi} \log^+ |\omega(j^k f(re^{\sqrt{-1}\theta}))| \frac{d\theta}{2\pi} \leq O(\log T_f(r)) + O(\log r).$$

PROOF. We claim that there exist a finite number of rational functions $t_1, \ldots, t_q$ on X such that

(†)   *the logarithmic jet differentials $\{(d^{(j)} t_i / t_i)^{m/j} \mid 1 \leq i \leq q, 1 \leq j \leq k\}$ span the fibers of $\mathcal{J}_k^m X(\log D)$ (resp. $\mathcal{J}_k^m X$) over every point of X.*

Note that rational jet differentials span the fibers of $\mathcal{J}_k^m X(\log D)$ (resp. $\mathcal{J}_k^m X$); the claim here is that this can be achieved by those of logarithmic type. Without loss of generality we may assume that D is ample; otherwise we may replace D by $D + D'$ so that $D + D'$ is ample. (This is so because a section of $\mathcal{J}_k^m X(\log D)$ is a priori a section of $\mathcal{J}_k^m X(\log(D + D'))$.) Observe that if s is a function that is holomorphic on a neighborhood U such that $[s = 0] = D \cap U$ then $[s^\tau = 0] = \tau D \cap U$ for any rational number τ. Thus $\delta^{(j)}(\log s^\tau) = \tau \delta^{(j)}(\log s)$ is still a jet differential with logarithmic singularity along $D \cap U$ so the multiplicity causes no problem. This implies that we may assume without loss of generality that D is very ample (after perhaps replacing D with τD for some τ for which τD is very ample).

Let $u \in H^0(X, [D])$ be a section such that $D = [u = 0]$. At a point $x \in D$ choose a section $v_1 \in H^0(X, [D])$ so that $E_1 = [v_1 = 0]$ is smooth, $D + E_1$ is of *simple normal crossings* and v_1 is nonvanishing at x. (This is possible because the line bundle $[D]$ is very ample.) The rational function $t_1 = u_1 / v_1$ is regular on the affine open neighborhood $X \setminus E_1$ of x and $(X \setminus E_1) \cap [t_1 = 0] = (X \setminus E_1) \cap D$. Choose rational functions $t_2 = u_2 / v_2, \ldots, t_n = u_n / v_n$ where u_i and v_i are sections of a very ample bundle $\mathcal{L}$ so that $t_2, \ldots, t_n$ are regular at x, the divisors $D_i = [u_i = 0], E_i = [v_i = 0]$ are smooth and the divisor $D + D_2 + \cdots + D_n + E_1 + \cdots + E_n$ is of simple normal crossings. Further, since the bundles involved are very ample the sections can be chosen so that $dt_1 \wedge \cdots \wedge dt_n$ is nonvanishing at x; the complete system of sections provides an embedding. Hence at each point there are $n+1$ sections with the property that n of the quotients of these $n+1$ sections form a local coordinate system on some open neighborhood U_x of x. This implies that (†) is satisfied over U_x. Since D is compact it is covered by a finite number of such open neighborhoods, say $U_1, \ldots, U_p$, and a finite number of rational functions (constructed as above for each U_i) on X so that (†) is satisfied on $\bigcup_{1 \leq i \leq p} U_i$. Moreover, there exist relatively compact open subsets U_i' of U_i ($1 \leq i \leq p$) such that $\bigcup_{1 \leq i \leq p} U_i'$ still covers D.

Next we consider a point x in the compact set $X \setminus \bigcap_{1 \leq i \leq p} U_i'$. Repeating the procedure as above we may find rational functions $s_1 = a_1/b_1, \ldots, s_n = a_n/b_n$ where a_i and b_i are sections of some very ample line bundle so that $s_1, \ldots, s_n$ form a holomorphic local coordinate system on some open neighborhood V_x of x. Thus (†) is satisfied on V_x by the rational functions $s_1, \ldots, s_n$. Note that we must also choose these sections so that the divisor $H = [s_1 \ \ldots \ s_n = 0]$ together with those divisors (finite in number), which have been already constructed above, is still a divisor with simple normal crossings (this is possible by the very ampleness of the line bundle $\mathcal{L}$.) Since $X \setminus \bigcap_{1 \leq i \leq p} U_i'$ is compact, it is covered by a finite number of such coordinate neighborhoods. The coordinates are rational functions and finite in number and by construction it is clear that the condition (†) is satisfied on $X \setminus \bigcap_{1 \leq i \leq p} U_i'$. Since $\bigcup_{1 \leq i \leq p} U_i$ together with $X \setminus \bigcap_{1 \leq i \leq p} U_i'$ covers X, the condition (†) is satisfied on X. If D is the trivial divisor, then it is enough to use only the second part of the construction above and again (†) is verified with $J_k^m X(\log D) = J_k^m X$. To obtain the estimate of the theorem observe that the function $\rho : J^k X(-\log D) \to [0, \infty]$ defined by

$$\rho(\xi) = \sum_{i=1}^{q} \sum_{j=1}^{k} \left| (d^{(j)} t_i/t_i)^{m/j}(\xi) \right|^2, \quad \xi \in J^k X(-\log D), \qquad (6.1)$$

$\{t_i\}$ being the family of rational functions satisfying condition (†), is continuous in the extended sense; it is continuous in the usual sense outside the fibers over the divisor E (the sum of the divisors associated to the rational functions $\{t_i\}$; note that E contains D). Over the fiber of each point $x \in X - E$, $\left| (d^{(j)} t_i/t_i)^{m/j}(\xi) \right|^2$ is finite for $\xi \in J^k X(-\log D)_x$, thus ρ is not identically infinite. Moreover, since

$$\{ (d^{(j)} t_i/t_i)^{m/j} \mid 1 \leq i \leq q, 1 \leq j \leq k \}$$

span the fiber of $J_k^m X(\log D)$ over every point of X, ρ is strictly positive (possibly $+\infty$) outside the zero section of $J^k X(-\log D)$. The quotient

$$|\omega|^2/\rho : J^k X(-\log D) \to [0, \infty]$$

does not take on the extended value ∞ when restricted to $J^k X(-\log D) \setminus \{\text{zero section}\}$ because, as we have just observed, ρ is nonvanishing (although it does blow up along the fibers over E so that the reciprocal $1/\rho$ is zero there) and the singularity of $|\omega|$ is no worse than that of ρ since the singularity of ω occurs only along D (which is contained in E) and is of log type. Thus the restriction to $J_k X(-\log D) \setminus \{\text{zero section}\}$,

$$|\omega|^2/\rho : J^k X(-\log D) \setminus \{\text{zero section}\} \to [0, \infty),$$

is a continuous nonnegative function. Moreover, $|\omega|$ and ρ have the same homogeneity,

$$|\omega(\lambda.\xi)|^2 = |\lambda|^{2m}|\omega(\lambda.\xi)|^2 \quad \text{and} \quad \rho(\lambda.\xi) = |\lambda|^{2m}\rho(\xi),$$

for all $\lambda \in \mathbb{C}^*$ and $\xi \in J^k X(-\log D)$; therefore $|\omega|^2/\rho$ descends to a well-defined function on $\mathbb{P}(E_{k,D}) = (J^k X(-\log D) \setminus \{\text{zero section}\})/\mathbb{C}^*$, that is,

$$|\omega|^2/\rho : \mathbb{P}(E_{k,D}) \to [0, \infty)$$

is a well-defined continuous function and so, by compactness, there exists a constant c with the property that $|\omega|^2 \le c\rho$. This implies that

$$\begin{aligned}
T_{\omega \circ j^k f}(r) &= \int_0^{2\pi} \log^+ |\omega(j^k f(re^{\sqrt{-1}\theta}))| \frac{d\theta}{2\pi} \\
&\le \int_0^{2\pi} \log^+ |\rho(j^k f(re^{\sqrt{-1}\theta}))| \frac{d\theta}{2\pi} + O(1).
\end{aligned}$$

Since t_i is a rational function on X, the function

$$(d^{(j)} t_i/t_i)^{m/j} (j^k f) = \left((t_i \circ f)^{(j)}/t_i \circ f \right)^{m/j}$$

(m is divisible by $k!$) is meromorphic on $\mathbb{C}$ and so, by the definition of ρ,

$$\log^+ |\rho(j^k f)| \le O\left(\max_{1 \le i \le q, 1 \le j \le k} \log^+ |(t_i \circ f)^{(j)}/t_i \circ f| \right) + O(1).$$

Now by the classical lemma of logarithmic derivatives for meromorphic functions,

$$\int_0^{2\pi} \log^+ |(t_i \circ f)^{(j)}/t_i \circ f| \frac{d\theta}{2\pi} \cdot\le\cdot O(\log r) + O(\log T_{t_i \circ f}(r)),$$

where $\cdot\le\cdot$ indicates that the estimate holds outside a set of finite Lebesgue measure in $\mathbb{R}_+$. Since t_i is a rational function,

$$\log T_{t_i \circ f}(r) \le O(\log T_f(r)) + O(\log r)$$

and we arrive at the estimate

$$\int_0^{2\pi} \log^+ |\rho(j^k f(re^{\sqrt{-1}\theta}))| \frac{\theta}{2\pi} \le O\left(\int_0^{2\pi} \log^+ |(t_i \circ f)^{(j)}/t_i \circ f| \frac{d\theta}{2\pi} \right) + O(1)$$

$$\cdot\le\cdot O(\log T_f(r)) + O(\log r).$$

This implies that $T_{\omega \circ j^k f}(r) \cdot\le\cdot O(\log T_f(r)) + O(\log r)$, as claimed. $\square$

We obtain as a consequence the following Schwarz type lemma for logarithmic jet differentials.

COROLLARY 6.2. *Let X be a projective variety and D be an effective divisor (possibly the trivial divisor) with simple normal crossings. Let $f : \mathbb{C} \to X \setminus D$ be a holomorphic map. Then*

$$\omega(j^k f) \equiv 0 \quad \text{for all } \omega \in H^0(X, \mathcal{J}_k^m X(\log D) \otimes [-H]),$$

where H is a generic hyperplane section (and hence any hyperplane section).

PROOF. If f is constant the corollary holds trivially, so we may assume that f is nonconstant. Now suppose that $\omega \circ j^k f \not\equiv 0$. Since f is nonconstant, we may assume without loss of generality that $\log r = o(T_f(r))$ (after perhaps replacing f with $f \circ \phi$, where ϕ is a transcendental function on $\mathbb{C}$). By Theorem 6.1, we have

$$\int_0^{2\pi} \log^+ |\omega \circ j^k f| \, \frac{d\theta}{2\pi} = T_{\omega \circ j^k f}(r) \quad \cdot \leq \cdot \quad O\big(\log(r T_f(r))\big).$$

On the other hand, since ω vanishes on H and H is generic, we obtain via Jensen's Formula ,

$$T_f(r) \leq N_f(H; r) + O\big(\log(r T_f(r))\big)$$

$$= \int_0^{2\pi} \log |\omega \circ j^k f| \, \frac{d\theta}{2\pi} + O\big(\log\big(r T_f(r)\big)\big),$$

which, together with the preceding estimate, implies that

$$T_f(r) \leq O\big(\log(r T_f(r))\big).$$

This is impossible; hence we must have $\omega \circ j^k f \equiv 0$. If $H_1 = [s_1 = 0]$ is any hyperplane section then it is linearly equivalent to a generic hyperplane section $H = [s=0]$. If ω vanishes along H' then $(s/s_1)\omega$ vanishes along H. The preceding discussion implies that $(s/s_1)\omega(j^k f) \equiv 0$. Further, this implies that $\omega(j^k f) \equiv 0$ as we may choose a generic section H so that the image of f is not entirely contained in H.   □

Interpreting this corollary via Grothendieck's isomorphism we may restate the result in terms of sections of $\mathcal{L}_{\mathbb{P}(J^k X)}^m|_Y \otimes p|_Y^*[-D]$ on the projectivized bundle:

COROLLARY 6.3. *Let $Y \subset \mathbb{P}(J^k X)$ be a subvariety and suppose that there exists a nontrivial section*

$$\sigma \in H^0\big(Y, \mathcal{L}_{\mathbb{P}(J^k X)}^m|_Y \otimes p|_Y^*[-D]\big),$$

where D is an ample divisor in X and $p : \mathbb{P}(J^k X) \to X$ is the projection map. If the image of the lifting $[j^k f] : \mathbb{C} \to \mathbb{P}(J^k X)$ of a holomorphic curve $f : \mathbb{C} \to X$ is contained in Y, then $\sigma([j^k f]) \equiv 0$.

Theorem 6.1 and Corollaries 6.2 and 6.3 tell us about the *base locus* $B_k^m(D)$ of the line sheaves $\mathcal{L}_k^m \otimes p^*[-D]$, where we write for simplicity $\mathcal{L}_k^m = \mathcal{L}_{\mathbb{P}(J^k X)}^m$ and

D is an ample divisor in X; by that we mean the (geometric) intersection of all possible sections of powers of $\mathcal{L}_k$:

$$B_k^m(D) = \bigcap_{\sigma \in H^0(\mathbb{P}(J^k X), \mathcal{L}_k^m)} [\sigma = 0]. \tag{6.2}$$

Indeed, Corollary 6.3 implies that the image of the (projectivized k-jet) $[j^k f]$: $\mathbb{C} \to \mathbb{P}(J^k X)$ of a nonconstant holomorphic map $f : \mathbb{C} \to X$ must be contained in $B_k^m(D)$ for all $m \in \mathbb{N}$ and $D \in \mathcal{A}$ = the cone of all ample divisors; that is, the image $[j^k f](\mathbb{C})$ is contained in

$$B_k(\mathcal{L}_k) = \bigcap_{m \in \mathbb{N}} \bigcap_{D \in \mathcal{A}} B_k^m(D), \tag{6.3}$$

which is a subvariety of $\mathbb{P}(J^k X)$. Moreover, the image $[j^k f](\mathbb{C})$, being a connected set, must be contained in an irreducible component of $B_k(\mathcal{L})$. If f is algebraically nondegenerate then $\dim \overline{f(\mathbb{C})} = \dim X = n$. Since $p_*\overline{[j^k f(\mathbb{C})]} = \overline{f(\mathbb{C})}$ and $\overline{[j^k f(\mathbb{C})]} \subset B_k(\mathcal{L})$ (where $p : \mathbb{P}(J^k X) \to X$ is the projection) we conclude that *the dimension of the base locus is at least $n = \dim X$ if f is algebraically nondegenerate.* We shall show that the dimension is actually higher, for $k \geq 2$, by considering a reparametrization of the curve f.

Define

$$\mathcal{A} = \{\phi \mid \phi : \mathbb{C} \to \mathbb{C} \text{ is a nonconstant holomorphic map}\},$$
$$\mathcal{A}_{\zeta_0} = \{\phi \in \mathcal{A} \mid \phi(\zeta_0) = \zeta_0, \phi'(\zeta_0) \neq 0\},$$
$$\mathcal{A}_{\zeta_0, \zeta_1} = \{\phi \in \mathcal{A} \mid \phi(\zeta_0) = \zeta_1, \phi'(\zeta_0) \neq 0\}.$$

By a *reparametrization* of f we mean the composite map $f \circ \phi : \mathbb{C} \to X$, where $\phi \in \mathcal{A}$. It is clear that, as a set, the algebraic closure of the image of f is invariant by reparametrization. Moreover, since a reparametrization is again a curve in X, the Schwarz Lemma implies that its k-jet is contained in the base locus $B_k(\mathcal{L})$. As remarked earlier, if f is algebraically nondegenerate the dimension of the base locus is at least n.

The first order jet of a reparametrization is given by

$$j^1(f \circ \phi) = (f(\zeta), f'(\phi)\phi').$$

Thus, if $\phi \in \mathcal{A}_0$ (that is, $\phi(0) = 0$), then

$$j^1(f \circ \phi)(0) = \big(f(\phi(0)), f'(\phi(0))\phi'(0)\big) = \big(f(0), f'(0)\phi'(0)\big),$$

which implies that the projectivization satisfies

$$[j^1(f \circ \phi)(0)] = [f'(0)\phi'(0)] = [f'(0)] = [j^1 f(0)];$$

that is, the fiber $\mathbb{P}_{f(\zeta_0)}(J^1 X)$ is invariant by $\phi \in \mathcal{A}_{\zeta_0}$.

Assume from here on that the map f is algebraically nondegenerate. For $k \geq 2$ we have

$$j^2(f(\phi)) = \big(f(\phi),\, f'(\phi)\phi',\, f'(\phi)\phi'' + f''(\phi)(\phi')^2\big).$$

Moreover, $\phi \in \mathcal{A}_0$ implies

$$j^2(f \circ \phi)(0) = \big((f \circ \phi)(0),\, (f \circ \phi)'(0),\, (f \circ \phi)''(0)\big)$$
$$= \big(f(0),\, s_\phi f'(0),\, t_\phi f'(0) + s_\phi^2 f''(0)\big),$$

where $s_\phi = \phi'(0)$ and $t_\phi = \phi''(0)$. We are free to prescribe the complex numbers s_ϕ and t_ϕ. The bundle $\mathbb{P}(J^2 X)$ is algebraic and locally trivial, hence locally algebraically trivial (as a $\mathbb{C}^*$-bundle). In particular, we have a $\mathbb{C}^*$-isomorphism $J^2_{f(0)} X \cong \mathbb{C}^n \oplus \mathbb{C}^n$, where $\lambda(\boldsymbol{z}, \boldsymbol{w}) = (\lambda \boldsymbol{z}, \lambda^2 \boldsymbol{w})$ for $(\boldsymbol{z}, \boldsymbol{w}) \in \mathbb{C}^n \oplus \mathbb{C}^n$. The Jacobian matrix of the map

$$(s_\phi, t_\phi) \mapsto (s_\phi f'(0), t_\phi f'(0) + s_\phi^2 f''(0))$$

is given by

$$\begin{pmatrix} \dfrac{\partial(f \circ \phi)'(0)}{\partial s_\phi} & \dfrac{\partial(f \circ \phi)''(0)}{\partial s_\phi} \\[2mm] \dfrac{\partial(f \circ \phi)'(0)}{\partial t_\phi} & \dfrac{\partial(f \circ \phi)''(0)}{\partial t_\phi} \end{pmatrix} = \begin{pmatrix} f'(0) & 2s_\phi f''(0) \\ 0 & f'(0) \end{pmatrix}.$$

It is clear that the rank is 2 if $f'(0) \neq 0$ (which we may assume without loss of generality because $f' \not\equiv 0$ so $f'(\zeta) \neq 0$ for generic ζ). Thus, as ϕ varies through the space $\mathcal{A}_0$, $j^2(f \circ \phi)(0)$ sweeps out a complex 2-dimensional set in the fiber $J^2_{f(0)} X$ over the point $f(0) \in X$, and the projectivization is a set of dimension at least 1 in $\mathbb{P}(J^2_{f(0)} X)$. If f is algebraically nondegenerate, the algebraic closure of $[j^2 f(\mathbb{C})]$ is of dimension $n = \dim X$, as remarked earlier. The preceding argument shows that

$$\bigcup_{\phi \in \mathcal{A}} \overline{[j^2(f \circ \phi)(\mathbb{C})]}$$

is of dimension at least $n+1$. By Schwarz's Lemma the set $\bigcup_{\phi \in \mathcal{A}} \overline{[j^2(f \circ \phi)(\mathbb{C})]}$ is contained in the base locus $B_2(\mathcal{L})$ thus $\dim B_2(\mathcal{L}) \geq n+1$. Since $\dim \mathbb{P}(J^2 X) = n(2+1) - 1$, the codimension of $B_2(\mathcal{L})$ in $\mathbb{P}(J^2 X)$ is at most $3n - 1 - (n+1) = 2(n-1)$.

For $k = 3$ we get

$$j^3(f(\phi)) =$$
$$\big(f(\phi),\, f'(\phi)\phi',\, f'(\phi)\phi'' + f''(\phi)(\phi')^2,\, f'(\phi)\phi''' + 3f''(\phi)\phi'\phi'' + f'''(\phi)(\phi')^3\big).$$

Hence, for $\phi \in \mathcal{A}_0$,

$$j^3(f \circ \phi)(0) = \big(f(0),\, s_\phi f'(0),\, t_\phi f'(0) + s_\phi^2 f''(0),\, u_\phi f'(0) + 3s_\phi t_\phi f''(0) + s_\phi^3 f'''(0)\big),$$

where $s_\phi = \phi'(0)$, $t_\phi = \phi''(0)$, and $u_\phi = \phi'''(0)$. The Jacobian matrix of the map

$$(s_\phi, t_\phi, u_\phi) \mapsto \big(s_\phi f'(0),\, t_\phi f'(0) + s_\phi^2 f''(0),\, u_\phi f'(0) + 3s_\phi t_\phi f''(0) + s_\phi^3 f'''(0)\big)$$

is

$$
\begin{pmatrix}
\dfrac{\partial(f\circ\phi)'(0)}{\partial s_\phi} & \dfrac{\partial(f\circ\phi)''(0)}{\partial s_\phi} & \dfrac{\partial(f\circ\phi)'''(0)}{\partial s_\phi} \\[2mm]
\dfrac{\partial(f\circ\phi)'(0)}{\partial t_\phi} & \dfrac{\partial(f\circ\phi)''(0)}{\partial t_\phi} & \dfrac{\partial(f\circ\phi)'''(0)}{\partial t_\phi} \\[2mm]
\dfrac{\partial(f\circ\phi)'(0)}{\partial u_\phi} & \dfrac{\partial(f\circ\phi)''(0)}{\partial u_\phi} & \dfrac{\partial(f\circ\phi)'''(0)}{\partial u_\phi}
\end{pmatrix}
$$

$$
= \begin{pmatrix}
f'(0) & 2s_\phi f''(0) & 3t_\phi f''(0)+3s_\phi^2 f'''(0) \\
0 & f'(0) & 3t_\phi f''(0) \\
0 & 0 & f'(0)
\end{pmatrix}.
$$

It is clear that the rank is 3 if $f'(0)\neq 0$ (which we may assume without loss of generality). Thus, as ϕ varies through the space $\mathcal{A}_0$, $j^3(f\circ\phi)(0)$ sweeps out a complex 3-dimensional set in the fiber $J^3_{f(0)}X$ over the point $f(0)\in X$ and the projectivization is a set of dimension at least 2 in $\mathbb{P}(J^3_{f(0)}X)$. If f is algebraically nondegenerate then the set $\bigcup_{\phi\in\mathcal{A}}\overline{[j^3(f\circ\phi)(\mathbb{C})]}$ is of dimension at least $n+2$ in $\mathbb{P}(J^3X)$. By Schwarz's Lemma this same set is contained in the base locus $B_3(\mathcal{L})$ thus $\dim B_3(\mathcal{L})\geq n+2$. Since $\dim\mathbb{P}(J^3X)=n(3+1)-1$, the codimension of $B_3(\mathcal{L})$ in $\mathbb{P}(J^3X)$ is at most $4n-1-(n+2)=3(n-1)$.

The case for general k is argued in a similar fashion. Define polynomials P_{ij}, $1\leq j\leq i$, by setting $P_{1,1}=\phi'$, $P_{2,1}=\phi''$, $P_{2,2}=(\phi')^2$ and, for $i\geq 3$,

$$P_{i,1}=\phi^{(i)},$$
$$P_{i,2}=P_{i-1,1}+P'_{i-1,2},\cdots$$
$$P_{i-1,i-1}=P_{i-1,i-2}+P'_{i-1,i-1},$$
$$P_{i,i}=(\phi')^i.$$

In particular, $P_{i,1}$ is the only polynomial involving $\phi^{(i)}$; each $P_{i,j}$, for $j\geq 2$, involves only derivatives of ϕ of order less than i. We get, by induction:

$$
(f\circ\phi)^{(i)}=\sum_{j=1}^{i} f^{(j)}(\phi)P_{i,j}=f'(\phi)\phi^{(i)}+\sum_{j=2}^{i} f^{(j)}(\phi)P_{i,j}.
$$

Thus the k-th jet $j^k(f\circ\phi)$ is given by

$$
\left(f'(\phi)\phi',\cdots,f'(\phi)\phi^{(i)}+\sum_{j=2}^{i} f^{(j)}(\phi)P_{i,j},\cdots,f^{(k)}(\phi)\phi^{(k)}+\sum_{j=2}^{k} f^{(j)}(\phi)P_{k,j}\right),
$$

and we have k parameters $s_{\phi,i}=\phi^{(i)}(0)$, $i=1,\ldots,k$. The Jacobian matrix of the map (with $\phi(0)=0$)

$$
(s_{\phi,1},\ldots,s_{\phi,k})\mapsto \left(s_{\phi,1}f'(0),\, s_{\phi,2}f'(0)+s_{\phi,1}^2 f''(0),\ldots,\, s_{\phi,k}f'(0)+\sum_{j=2}^{k} f^{(j)}(0)P_{k,j}\right)
$$

is given by the $k \times nk$ matrix

$$
\begin{pmatrix}
f'(0) & 2s_\phi f''(0) & 3t_\phi f''(0) + 3s_\phi^2 f'''(0) & \cdots & & \cdot \\
0 & f'(0) & 3t_\phi f''(0) & \cdots & \cdot & \cdot \\
0 & 0 & f'(0) & \cdots & \cdot & \cdot \\
\cdot & & & 0 & \cdots & \cdot & \cdot \\
\cdot & \cdot & & & \cdots & f'(0) & \cdot \\
0 & 0 & & \cdot & & \cdots & 0 & f'(0)
\end{pmatrix}.
$$

It is clear that the rank is k if $f'(0) \neq 0$ (which we may assume without loss of generality). Thus, as ϕ varies through the space $\mathcal{A}_0$, $j^k(f \circ \phi)(0)$ sweeps out a complex k-dimensional set in the fiber $J^k_{f(0)}X$ over the point $f(0) \in X$, and the projectivization is a set of dimension at least $k-1$ in $\mathbb{P}(J^k_{f(0)}X)$. If f is algebraically nondegenerate then the set $\bigcup_{\phi \in \mathcal{A}} \overline{[j^k(f \circ \phi)(\mathbb{C})]}$ is of dimension at least $n+k-1$ in $\mathbb{P}(J^k X)$. By Schwarz's Lemma this same set is contained in the base locus $B_k(\mathcal{L})$ thus $\dim B_k(\mathcal{L}) \geq n+k-1$. Since $\dim \mathbb{P}(J^k X) = n(k+1) - 1$, the codimension of $B_k(\mathcal{L})$ in $\mathbb{P}(J^k X)$ is at most $(k+1)n - 1 - (n+k-1) = k(n-1)$. This completes the proof of the following Theorem:

THEOREM 6.4. *Let X be a connected compact manifold of dimension n and let $\mathcal{L}_k$ be the dual of the tautological line bundle over $\mathbb{P}(J^k X)$, $k \geq 2$. Suppose that $f : \mathbb{C} \to X$ is an algebraically nondegenerate holomorphic map. Then the irreducible component of the base locus containing $[j^k f]$ is of codimension at most $(n-1)k$; equivalently the dimension is at least $n+k-1$.*

COROLLARY 6.5. *Let X be a connected projective manifold of complex dimension n and let $\mathcal{L}_k$ be the dual of the tautological line bundle over $\mathbb{P}(J^k X)$. If the dimension of the base locus $B_k(\mathcal{L}_k) \leq n+k-2$ then every holomorphic map $f : \mathbb{C} \to X$ is algebraically degenerate.*

7. Surfaces of General Type

SUMMARY. *In this section we shall show that every holomorphic map $f : \mathbb{C} \to X$ is algebraically degenerate, where X is a minimal surface of general type such that $p_g(X) > 0$ and $\operatorname{Pic} X \cong \mathbb{Z}$. These conditions, together with the explicit calculations in Section 3, imply that $J_k X$ is big (equivalently, the line bundle $\mathcal{L}_k$ over $\mathbb{P}(J^k X)$ is big) for $k \gg 0$. The Schwarz Lemma of the preceding section implies that the image of the lifting $[j^k f] : \mathbb{C} \to \mathbb{P}(J^k X)$ is contained in the base locus $B_k(\mathcal{L}_k)$ (see (6.3)). (Note that the dimension of $\mathbb{P}(J^k X)$ is $2k+1$.) Moreover, if f is algebraically nondegenerate, $\dim B_k(\mathcal{L}_k) \geq k+1$.*

On the other hand, we show (Theorem 7.20) that the base locus is at most of dimension k. This contradiction establishes the theorem. The result in Theorem 7.20 is obtained by a cutting procedure (each cut lowers the dimension of the base locus by one) pioneered by Lu and Yau and extended by Dethloff–Schumacher–Wong (in which the condition $\operatorname{Pic} X \cong \mathbb{Z}$ was first introduced).

The starting point in the process is the explicit formulas obtained by Stoll and Wong in Theorem 3.9 and Corollary 3.10, namely, the index $\iota(\mathcal{J}_k^m X) = \chi(\mathcal{L}_k^m) + O(m^{2k}) = (\alpha_k c_1^2 - \beta_k c_2)m^{2k+1} + O(m^{2k})$ (here $c_i = c_i(X)$) is very big; indeed we have, $\lim_{k\to\infty} \alpha_k/\beta_k = \infty$. Consequently if $c_1^2 > 0$, which is the case if X is minimal, then $\chi(\mathcal{L}_k^m) = cm^{2k+1}c_1^2 + O(m^{2k})$ for some positive constant c (as, eventually, $\alpha_k/\beta_k > c_2/c_1^2$). If the base locus Y_1 were of codimension one (which we show that there is no loss of generality in assuming that it is irreducible) then for $k \gg 0$, $\chi(\mathcal{L}_k|_{Y_1})$ is still big and Schwarz Lemma implies that the base locus must be of codimension 2. The computation is based on the intersection formulas obtained in Lemma 7.15 (requiring the assumption $\text{Pic } X \cong \mathbb{Z}$) and Theorem 7.16. The cutting procedure can be repeated and, as to be expected, each time with a loss which can be explicitly estimated using the intersection formulas. These losses are compensated by taking a larger k. In the proof of Theorem 7.20 we show that, after k cuts, the Euler characteristic is bounded below by

$$\mu_k \left(\delta_k c_1^2 - \left(\sum_{i=1}^{k} \frac{1}{i^2} \right) c_2 \right),$$

where μ_k is a positive integer and

$$\delta_k = \left(\sum_{i=1}^{k} \frac{1}{i^2} + \sum_{i=2}^{k} \frac{1}{i} \sum_{j=1}^{i-1} \frac{1}{j} \right) - \frac{1}{4} \left(\sum_{i=1}^{k} \frac{1}{i} \right)^2 + \frac{(k+1)}{4(k!)^2} \left(\sum_{i=1}^{k} \frac{1}{i} \right)^2.$$

It remains to show that

$$\frac{\delta_k}{\sum_{i=1}^{k} \frac{1}{i^2}} > \frac{c_2}{c_1^2}$$

for k sufficiently large. A little bit of combinatorics shows that

$$\lim_{k\to\infty} \frac{\delta_k}{\sum_{i=1}^{k} \frac{1}{i^2}} = \infty$$

(compare the proof of Corollary 3.10). This completes the proof of our main result. Indeed, for a hypersurface of degree $d \geq 5$ in $\mathbb{P}^3$, our colleague B. Hu checked, using Maple, that $k \geq 2283$ is sufficient. This, together with a result of Xu implies that a generic hypersurface of degree $d \geq 5$ in $\mathbb{P}^3$ is hyperbolic.

We recall first some well-known results on manifolds of general type. The following result can be found in [Barth et al. 1984]:

THEOREM 7.1.   Let X be a minimal surface of general type. The following Chern-number inequalities hold:

  (i) $c_1^2(T^*X)[X] > 0$.
  (ii) $c_2(T^*X)[X] > 0$.
  (iii) $c_1^2(T^*X)[X] \leq 3c_2(T^*X)[X]$.
  (iv) $5c_1^2(T^*X)[X] - c_2(T^*X[X]) + 36 \geq 0$ if $c_1^2(T^*X)[X]$ is even.
  (v) $5c_1^2(T^*X)[X] - c_2(T^*X)[X] + 30 \geq 0$ if $c_1^2(T^*X)[X]$ is odd.

Let L_0 be a nef line bundle on a variety X of complex dimension n. A coherent sheaf E over X is said to be *semistable* (or *semistable in the sense of Mumford–Takemoto*) with respect to L_0 if $c_1(E).c_1^{n-1}(L_0) \geq 0$ and if, for any coherent subsheaf $\mathcal{S}$ of E with $1 \leq \mathrm{rk}\, \mathcal{S} < \mathrm{rk}\, E$, we have $\mu_{\mathcal{S},L_0} \leq \mu_{E,L_0}$, where

$$\mu_{\mathcal{S},L_0} \overset{\mathrm{def}}{=} \frac{c_1(\mathcal{S}).c_1^{n-1}(L_0)}{\mathrm{rk}\,\mathcal{S}}[X] \quad \text{and} \quad \mu_{E,L_0} \overset{\mathrm{def}}{=} \frac{c_1(E).c_1^{n-1}(L_0)}{\mathrm{rk}\,E}[X]. \qquad (7.1)$$

It is said to be *stable* if the inequality is strict, that is, $\mu_{\mathcal{S},L_0} < \mu_{E,L_0}$.

The number $\mu_{\mathcal{S},L_0}$ shall be referred to as *the normalized degree relative to L_0*. We shall write $\mu_{\mathcal{S}}$ for $\mu_{\mathcal{S},L_0}$ if L_0 is the canonical bundle. If X is of general type then (see [Maruyama 1981] in the case of surfaces and [Tsuji 1987, 1988] for general dimensions):

THEOREM 7.2. *Let X be a smooth variety of general type. Then the bundles $\bigotimes^m T^*X$, $\bigodot^m T^*X$ are semistable with respect to the canonical bundle K_X.*

Recall from Section 2 that for a vector bundle E of rank r,

$$\mathrm{rk}\, \bigodot^m E = \frac{(m+r-1)!}{(r-1)!\,m!}, \quad c_1(\bigodot^m E) = \frac{(m+r-1)!}{r!\,(m-1)!}c_1(E).$$

Thus, for surfaces of general type, we have

$$\mu_{\bigodot^m T^*X} = \tfrac{1}{2}mc_1^2(T^*X)[X]$$

with respect to the canonical bundle. More generally:

THEOREM 7.3. *Let X be a surface of general type. If D is a divisor in X such that $H^0(X, \mathcal{S}_I \otimes [-D]) \neq 0$ where $\mathcal{S}_I = (\bigodot^{i_1} T^*X \otimes \cdots \otimes \bigodot^{i_k} T^*X)$ and $I = (i_1, \ldots, i_k)$ is a k-tuple of positive integers satisfying $m = i_1 + 2i_2 + \cdots + ki_k$, then*

$$\mu_{[D]} \leq \mu_{\mathcal{S}_I} = \frac{\sum_{j=1}^{k} i_j}{2}c_1^2(T^*X)[X] \leq \tfrac{1}{2}mc_1^2(T^*X)[X],$$

where $[D]$ is the line bundle associated to the divisor D.

The examples at the end of Section 2 show that the sheaves of k-jet differentials are not semistable unless $k = 1$. However we do have (by Theorems 3.7 and 3.8):

THEOREM 7.4. *Let X be a surface of general type. Then*

$$\mu_{\mathcal{J}_k^m X} = \frac{\sum_I c_1(\mathcal{S}_I)\,c_1(T^*X)}{\sum_I \mathrm{rk}\,\mathcal{S}_I} = \sum_I \frac{\mathrm{rk}\,\mathcal{S}_I}{\sum_I \mathrm{rk}\,\mathcal{S}_I}\mu_{\mathcal{S}_I} \leq \frac{m}{2}c_1^2(T^*X),$$

and equality holds if and only if $k = 1$; moreover, asymptotically,

$$\mu_{\mathcal{J}_k^m X} = \left(\frac{\sum_{i=1}^{k}\frac{1}{i}}{2k}m + O(1)\right)c_1^2(T^*X).$$

A coherent sheaf E is said to be *Euler semistable* if for any coherent subsheaf $\mathcal{S}$ of E with $1 \leq \mathrm{rk}\,\mathcal{S} < \mathrm{rk}\,E$, we have

$$\frac{\chi(\mathcal{S})}{\mathrm{rk}\,\mathcal{S}} \leq \frac{\chi(E)}{\mathrm{rk}\,E} \tag{7.2}$$

It is said to be *Euler stable* if the inequality is strict.

There is a concept of semistability due to Gieseker–Maruyama (see [Okonek et al. 1980]) for coherent sheaves on $\mathbb{P}^n$ in terms of the Euler characteristic that differs from the concept introduced here.

EXAMPLE 7.5. From the exact sequence

$$0 \to \bigodot^2 T^*X \to \mathcal{J}_2^2 X \to T^*X \to 0,$$

we get, via the table on page 163,

$$\chi(\mathcal{J}_2^2 X) = \chi(T^*X) + \chi(\bigodot^2 T^*X) = \tfrac{1}{6}(c_1^2 - 5c_2) + \tfrac{1}{4}(5c_1^2 - 15c_2) = \tfrac{1}{12}(17c_1^2 - 55c_2).$$

Theorem 6.1 yields $c_1^2 - 3c_2 \leq 0$, which implies that

$$\chi(T^*X) = \frac{c_1^2 - 5c_2}{6} < 0.$$

Thus $\chi(\mathcal{J}_2^2 X) < \chi(\bigodot^2 T^*X)$, that is, $\mathcal{J}_2^2 X$ is not semistable in the sense of (7.2).

Recall that the index of each of the sheaves $\mathcal{S}_I$ and $\mathcal{J}_k^m X$ of a surface X is of the form $ac_1^2(T^*X) + bc_2(T^*X)$. Thus the ratio $\gamma(X) = \gamma(T^*X) = c_2(T^*X)/c_1^2(T^*X)$ is an important invariant. More generally, we define

$$\gamma(\mathcal{S}) = \frac{c_2(\mathcal{S})}{c_1^2(\mathcal{S})}, \tag{7.3}$$

provided that $c_1^2(\mathcal{S}) \neq 0$.

Let X be a smooth hypersurface in $\mathbb{P}^3$. Then

$$c_1 = c_1(TX) = -c_1(T^*X) = d - 4,$$
$$c_2 = c_2(TX) = c_2(T^*X) = d^2 - 4d + 6.$$

Hence the ratio of $c_1^2(T^*X)$ and $c_2(T^*X)$ is given by

$$\gamma_d(\mathcal{J}_1 X) = \gamma_d(T^*X) = \frac{c_2(T^*X)}{c_1^2(T^*X)} = \frac{d^2 - 4d + 6}{(d-4)^2} = 1 + \frac{4d - 10}{(d-4)^2}, \tag{7.4}$$

provided that $d \neq 4$. Note that $\gamma_\infty(T^*X) = \lim_{d\to\infty} \gamma_d(T^*X) = 1$. Table A on the next page shows the first few values of $\gamma_g = \gamma_d(\mathcal{J}_1 X)$.

Recall from Theorem 5.12 that

$$\chi(X; \bigodot^m T^*X) = \tfrac{1}{12}(m+1)\big((2m^2 - 2m + 1)c_1^2 - (2m^2 + 4m - 1)c_2\big)$$
$$= \tfrac{1}{12}(m+1)\big((2m^2 - 2m + 1)(d-4)^2 - (2m^2 + 4m - 1)(d^2 - 4d + 6)\big)$$
$$= \tfrac{1}{3}(5 - 2d)3m^3 + O(m^2).$$

d	γ_d	d	γ_d	d	γ_d
5	11	12	$\frac{51}{32} \sim 1.5937$	19	$\frac{97}{75} \sim 1.2934$
6	$\frac{9}{2} = 4.5$	13	$\frac{14}{27} \sim 1.5175$	20	$\frac{163}{128} \sim 1.2735$
7	3	14	$\frac{73}{50} = 1.46$	21	$\frac{363}{289} \sim 1.2561$
8	$\frac{19}{8} = 2.375$	15	$\frac{171}{121} \sim 1.4132$	22	$\frac{67}{54} \sim 1.2408$
9	$\frac{51}{25} = 2.04$	16	$\frac{11}{8} = 1.375$	23	$\frac{443}{361} \sim 1.2244$
10	$\frac{11}{6} = 1.8\bar{3}$	17	$\frac{227}{169} \sim 1.3432$	24	$\frac{243}{200} = 1.215$
11	$\frac{83}{49} \sim 1.6939$	18	$\frac{129}{98} \sim 1.3164$	25	$\frac{59}{49} \sim 1.2041$

Table A. Values of $\gamma_d(\mathcal{J}_1 X)$ as a function of d.

It is clear that $\chi(X; \bigodot^m T^* X) < 0$ for all $m \geq 1$ if $d \geq 3$. If $d \geq 5$ it is well-known that $H^0(X, \bigodot^m T^* X) = 0$, whence the following nonvanishing theorem:

THEOREM 7.6. *Let X be a smooth hypersurface of degree $d \geq 5$ in $\mathbb{P}^3$. Then*

$$\dim H^1(X, \bigodot^m T^* X) \geq \dim H^1(X, \bigodot^m T^* X) - \dim H^2(X, \bigodot^m T^* X)$$
$$= \tfrac{1}{6}(m+1)\big(2(2d-5)m^2 - (3d^2 - 16d + 28)m - (d^2 - 6d + 11)\big)$$
$$= \tfrac{1}{3}(2d-5)3m^3 + O(m^2)$$

for all $m \gg 0$.

Next we consider the case of 2-jets. We have, by Riemann–Roch:

$$\chi(\mathcal{J}_2^m X) = \frac{1}{2}\big(\iota(\mathcal{J}_2^m X) - c_1(\mathcal{J}_2^m X) \cdot c_1\big) + \frac{1}{12}(\text{rk } \mathcal{J}_2^m X)(c_1^2 + c_2).$$

(Here $c_1 = c_1(T^* X), c_2 = c_2(T^* X)$ and, using the formulas for $c_1(\mathcal{J}_2^m X)$, rk $\mathcal{J}_2^m X$ and $\iota(\mathcal{J}_2^m X)$ in Theorem 3.3 we get:

$$\chi(\mathcal{J}_2^m X) = \frac{1}{2^7 3^2 5}(p_m c_1^2 - q_m c_2)$$

with

$$p_m = \begin{cases} 21m^5 + 180m^4 + 410m^3 + 180m^2 + 49m + 120, & \text{if } m \text{ is odd,} \\ 21m^5 + 180m^4 + 420m^3 + 180m^2 - 56m + 480, & \text{if } m \text{ is even;} \end{cases}$$

$$q_m = \begin{cases} 15m^5 + 225m^4 + 1150m^3 + 2250m^2 + 1235m - 75, & \text{if } m \text{ is odd,} \\ 15m^5 + 225m^4 + 1180m^3 + 2520m^2 + 1640m - 480, & \text{if } m \text{ is even.} \end{cases}$$

The index $\chi(\mathcal{J}_2^m X)$ is positive if and only if $p_m/q_m > c_2/c_1^2$, and taking the limit as $m \to \infty$ yields the inequality $c_2/c_1^2 \leq \tfrac{7}{5}$. For a smooth hypersurface of degree d in $\mathbb{P}^3$ the ratio $c_2/c_1^2 = 1 + \big((4d-10)/(d-4)^2\big)$ and we arrive at the inequality

$$\frac{4d-10}{(d-4)^2} \leq \frac{2}{5},$$

which is equivalent to the inequality $0 \leq d^2 - 18d + 41 = (d-9)^2 - 40$. We deduce:

THEOREM 7.7. *Let X be a smooth hypersurface in $\mathbb{P}^3$. Then $\chi(\mathcal{J}_2^m X)$ is big if and only if $d = \deg X \geq 16$.*

We use the terminology that the Euler characteristic is *big* if and only if there is a constant $c > 0$ such that

$$\chi(\mathcal{J}_2^m X) \geq cm^5 + O(m^4)$$

for all $m \gg 0$. In order to lower the degree in the preceding theorem we must use jet differentials of higher order. We see from Table A on page 177 that the ratio c_2/c_1^2 of a hypersurface of degree $d \geq 5$ in $\mathbb{P}^3$ is bounded above by 11. By Theorem 3.7,

$$\iota(\mathcal{J}_k^m X) = (\alpha_k c_1^2 - \beta_k c_2)m^{2k+1} + O(m^{2k})$$

thus the index is positive if and only if

$$\frac{\alpha_k}{\beta_k} > \frac{c_2}{c_1^2}.$$

In the table on page 148 we see that the ratio α_k/β_k crosses the threshold 11 as k increases from 198 to 199. Putting this together with Theorem 5.13, we get:

THEOREM 7.8. *Let X be a generic smooth hypersurface of degree $d \geq 5$ in $\mathbb{P}^3$. For each $k \geq 199$,*

$$\chi(\mathcal{J}_k^m X) \geq cm^5 + O(m^4)$$

for all $m \gg k$.

For a minimal surface of general type, Theorem 7.1 implies that

$$\frac{1}{3} \leq \gamma(X) = \frac{c_2(X)}{c_1^2(X)} \leq \begin{cases} 5 + 36c_1^{-2} \leq 41 & \text{if } c_1^2 \text{ is even,} \\ 5 + 30c_1^{-2} \leq 34 & \text{if } c_1^2 \text{ is odd.} \end{cases}$$

The ratio α_k/β_k was shown to tend to ∞ as $k \to \infty$. Thus Theorem 7.8 extends to any minimal surface of general type:

THEOREM 7.9. *Let X be a smooth minimal surface of general type. Then*

$$\chi(\mathcal{J}_k^m X) \geq cm^5 + O(m^4) \quad \text{for all } m \gg k \gg 0.$$

In [Green and Griffiths 1980] we find the following result:

THEOREM 7.10. *Let X be a smooth surface of general type. If $i_1 + \cdots + i_k$ is even then a nontrivial section of the bundle $\bigodot^{i_1} TX \otimes \cdots \otimes \bigodot^{i_k} TX \otimes \mathcal{K}^{(i_1 + \cdots + i_k)/2}$ is nonvanishing.*

Using this, Green and Griffiths deduced the following vanishing Theorem. We include their argument here, with minor modifications.

THEOREM 7.11. *Let X be a smooth surface of general type. Assume that the canonical bundle $\mathcal{K}_X$ admits a nontrivial section. Then $H^2(X, \mathcal{J}_k^m X) = 0$ for all $k \geq 1$ and $m > 2k$.*

PROOF. Let σ be a nontrivial section of $\mathcal{K}_X$, so that we have an exact sequence:

$$0 \to \mathcal{S}_I \otimes \mathcal{K}^{(/i_1+\cdots+i_k)2-1} \overset{\otimes\sigma}{\to} \mathcal{S}_I \otimes \mathcal{K}^{(/i_1+\cdots+i_k)2} \to \mathcal{S}_I \otimes \mathcal{K}^{(/i_1+\cdots+i_k)2}\big|_D \to 0$$

where $D = [\sigma=0]$, $\mathcal{S}_I = \bigodot^{i_1} TX \otimes \cdots \otimes \bigodot^{i_k} TX$ and $i_1+\cdots+i_k$ is even. Hence,

$$0 \to H^0(X, \mathcal{S}_I \otimes \mathcal{K}^{(/i_1+\cdots+i_k)2-1}) \overset{\otimes\sigma}{\to} H^0(X, \mathcal{S}_I \otimes \mathcal{K}^{(/i_1+\cdots+i_k)2})$$

is exact. By Theorem 7.10 the image of the map $\otimes\sigma$ is 0; hence

$$H^0(X, \mathcal{S}_I \otimes \mathcal{K}^{(/i_1+\cdots+i_k)2-1}) = 0.$$

The argument applies also to the exact sequence:

$$0 \to \mathcal{S}_I \otimes \mathcal{K}^{(/i_1+\cdots+i_k)2-l} \overset{\otimes\sigma}{\to} \mathcal{S}_I \otimes \mathcal{K}^{(/i_1+\cdots+i_k)2=l+1} \to \mathcal{S}_I \otimes \mathcal{K}^{(/i_1+\cdots+i_k)2}\big|_D \to 0$$

for any $l \geq 1$ and we conclude via induction that

$$H^0(X, \mathcal{S}_I \otimes \mathcal{K}^q) = 0$$

for all $q < (i_1 + \cdots + i_k)/2$. If $i_1 + \cdots + i_k$ is odd then taking $i_{k+1} = 1$ we have

$$H^0(X, \mathcal{S}_{i_1,\ldots,i_k,i_{k+1}} \otimes \mathcal{K}^{q+1}) = 0$$

provided that $q+1 < (i_1 + \cdots + i_k + 1)/2$ (equivalently $q < (i_1 + \cdots + i_k - 1)/2$). Suppose that $H^0(X, \mathcal{S}_I \otimes \mathcal{K}^q) \neq 0$. Then there exists a nontrivial section ρ of $H^0(X, \mathcal{S}_I \otimes \mathcal{K}^q)$ and we obtain a nontrivial section $\rho \otimes \sigma$ of $\mathcal{S}_{i_1,\ldots,i_k,i_{k+1}} \otimes \mathcal{K}^{q+1}$. This shows that:

$$H^0(X, \mathcal{S}_I \otimes \mathcal{K}^q) = \begin{cases} 0, & \text{for all } q < \frac{1}{2}(i_1 + \cdots + i_k - 1) \text{ if } i_1 + \cdots + i_k \text{ is odd}, \\ 0 & \text{for all } q < \frac{1}{2}(i_1 + \cdots + i_k) \text{ if } i_1 + \cdots + i_k \text{ is even}. \end{cases}$$

By Serre duality,

$$H^2(X, \mathcal{S}_I \otimes \mathcal{K}^{1-q}) = \begin{cases} 0, & \text{for all } q < \frac{1}{2}(i_1 + \cdots + i_k - 1) \text{ if } i_1 + \cdots + i_k \text{ is odd}, \\ 0 & \text{for all } q < \frac{1}{2}(i_1 + \cdots + i_k) \text{ if } i_1 + \cdots + i_k \text{ is even}, \end{cases}$$

where $\mathcal{S}_I = \bigodot^{i_1} T^*X \otimes \cdots \otimes \bigodot^{i_k} T^*X$. If $|I| = i_1 + \cdots + i_k \geq 3$ then we may take $q = 1$ in the formulas above. Thus we have: $H^2(X, \mathcal{S}_I) = 0$, if $|I| \geq 3$. Note $\mathcal{J}_k^m X$ admits a composition series by $\mathcal{S}_I$ satisfying the condition $\sum_{j=1}^{k} j i_j = m$. Thus $H^2(X, \mathcal{J}_k^m X) = 0$ if each of these $\mathcal{S}_I$ satisfies the condition $|I| \geq 3$. If $k = 2$ we have:

$$i_1 + 2i_2 = m \iff i_2 = (m - i_1)/2 \iff i_1 + i_2 = (m + i_1)/2.$$

Thus $i_1 + i_2 \geq 3$ if and only if $m \geq 6 - i_1$. Since $i_1 \geq 0$ we conclude that $m \geq 6$ implies $i_1 + i_2 \geq 3$. If $k = 3$ then

$$i_1 + 2i_2 + 3i_3 = m \iff (i_1+i_3) + 2(i_2+i_3) = m \iff i_2 + i_3 = \tfrac{1}{2}(m - i_1 - i_3)$$
$$\iff i_1 + i_2 + i_3 = \tfrac{1}{2}(m + i_1 - i_3).$$

Thus $i_1 + i_2 + i_3 \geq 3$ if and only if $m \geq 6 - i_1 + i_3 \geq 6 + i_3$. Since i_3 is at most $[m/3]$ we conclude that $i_1 + i_2 + i_3 \geq 3$ if $m \geq 9$. The case of general k can be established by an induction argument. $\square$

For our purpose only the following weaker result is needed:

THEOREM 7.12. *Let E be a holomorphic vector bundle of rank $r \geq 2$ over a nonsingular projective surface X. Assume that*

(i) $\mathcal{K}_X$ *is nef and not the trivial bundle;*
(ii) Pic $X \cong \mathbb{Z}$;
(iii) *det E^* is nef;*
(iv) *there exists a positive integer s with the property that there is a nontrivial global regular section ρ of $(\mathcal{K}_X \otimes \det E^*)^s$ such that the zero divisor $[\rho = 0]$ is smooth.*

Then $H^i(X, \bigodot^m E^*) = 0$ for all $i \geq 2$ and for m sufficiently large.

The canonical bundle $\mathcal{K}_X$ of a minimal surface X of general type is nef. If Pic $X \cong \mathbb{Z}$ then $\mathcal{K}_X$ is ample, so $\mathcal{K}_X \otimes \det \left(\bigodot^{i_1} T^*X \otimes \cdots \otimes \bigodot^{i_k} T^*X \right)$ is ample for any nonnegative integers $i_1, \ldots, i_k$. Hence:

COROLLARY 7.13. *Let X be a nonsingular minimal surface of general type. Assume that Pic $X \cong \mathbb{Z}$ and $p_g(X) > 0$. Let $I = (i_1, \ldots, i_k)$ be a k-tuple of nonnegative integers. Then $H^2 \left(X, \bigodot^{i_1} T^*X \otimes \cdots \otimes \bigodot^{i_k} T^*X \right) = 0$ if $i_1 + \cdots + i_k$ is sufficiently large; consequently, $H^2(X, \mathcal{J}_k^m X) = 0$ if $m \gg k$.*

COROLLARY 7.14. *Let X be a nonsingular minimal surface of general type with Pic $X \cong \mathbb{Z}$. Then*
$$h^0(X, \mathcal{J}_k^{k!m} X) \geq cm^{2k+1} + O(m^{2k})$$
for some positive constant c; that is, $\mathcal{J}_k^{k!} X$ is big.

A good source for the general theory of vanishing theorems is [Esnault and Viehweg 1992].

Next we deal with the question of algebraic degeneration of holomorphic maps and hyperbolicity of surfaces of general type. The condition that $\mathcal{J}_k^{k!} X$ is big implies that $\mathcal{J}_k^{k!} X \otimes [-D]$ is big for any ample divisor D on X. We may write $D = a_0 D_0$ for $a_0 > 0$, with D_0 as the positive generator of Pic X. The Schwarz Lemma for jet differentials implies that the image of $[j^k f]$ is contained in the zero set of all k-jet differentials vanishing along an ample divisor. Thus we may assume that $[j^k f](\mathbb{C})$ is contained in an effective irreducible divisor in $\mathbb{P}(J^k X)$ and is the zero set of a section
$$\sigma \in H^0 \left(\mathbb{P}(J^k X), \mathcal{L}_k^{k!m_k} \otimes p^*[-\nu_k D_0] \right), \tag{7.5}$$
where we abbreviate $\mathcal{L}_k = \mathcal{L}_{\mathbb{P}(J^k X)}$ (note that Pic $\mathbb{P}(J^k X) \cong \mathbb{Z}\langle \mathcal{L}_k^{k!} \rangle \oplus$ Pic X). Our aim is to show that the restriction $\mathcal{L}_k^{k!}|_{[\sigma = 0]}$ is big. First we need a lemma:

LEMMA 7.15.   *Let X be a nonsingular minimal surface of general type with* Pic $X \cong \mathbb{Z}$. *Suppose that* $H^0(\mathcal{J}_k^{k!m} X \otimes [-D]) \neq 0$ *for some divisor D in X. Then for all $m \gg 0$,*

$$c_1([D]) \leq \frac{B_k}{A_k} k! \, m c_1(T^*X) + O(1)$$

where A_k and B_k are the constants defined in Theorems 3.7 and 3.8.

PROOF.   The assumption that Pic $X \cong \mathbb{Z}$ implies that $c_1(\mathcal{J}_k^{k!} X \otimes [-D]) = q c_1$, where $c_1 = c_1(T^*X)$ and $q \in \mathbb{Q}$. Let σ be a nontrivial section of $\mathcal{J}_k^{k!} X \otimes [-D]$. The Poincaré–Lelong formula implies that

$$0 = \int_X dd^c \log \|\sigma\|^2 \wedge c_1 \geq \int_{[\sigma=0]} c_1 - \int_X c_1(\mathcal{J}_k^{k!m} X \otimes [-D]) \wedge c_1$$

$$= \int_{[\sigma=0]} c_1 - q \int_X c_1^2,$$

implying $q > 0$. On the other hand, the usual formula for Chern classes yields

$$0 < c_1(\mathcal{J}_k^{k!m} X \otimes [-D]) = c_1(\mathcal{J}_k^{k!m} X) - (\mathrm{rk}\, \mathcal{J}_k^{k!m} X) c_1([D]).$$

By the asymptotic formula in Section 3, we have

$$c_1(\mathcal{J}_k^{k!m} X) = B_k (k!m)^{2k} c_1 + O((k!m)^{2k-1});$$

hence the preceding inequality may be written as

$$A_k c_1([D])(k!m)^{2k-1} = (\mathrm{rk}\, \mathcal{J}_k^{k!m} X) c_1([D]) < B_k (k!m)^{2k} c_1 + O((k!m)^{2k-1}),$$

where A_k and B_k are the constants defined in Theorems 3.7 and 3.8. Thus we get the estimate

$$c_1([D]) \leq \frac{B_k}{A_k} k! \, m c_1 + O(1). \qquad \square$$

THEOREM 7.16.   *Let X be a smooth surface and $\mathcal{L}_k$ be the "hyperplane line sheaf" over $\mathbb{P}(J^k X)$. Then*

$$p_* c_1^{2k+1}(\mathcal{L}_k^{k!}) = (2k+1)! \chi(\mathcal{L}_k^{k!}) = \tfrac{1}{2}(2k+1)!(k!)^{2k+1}(\alpha_k c_1^2 - \beta_k c_2)$$

$$= (k!)^{2k-1} \left(\sum_{i=1}^{k} \frac{1}{i^2} + \sum_{i=2}^{k} \frac{1}{i} \sum_{j=1}^{i-1} \frac{1}{j} \right) c_1^2 - \left(\sum_{i=1}^{k} \frac{1}{i^2} \right) c_2,$$

$$p_* \big(c_1^{2k}(\mathcal{L}_k^{k!}) p^* c_1 \big) = \tfrac{1}{2}(k!)^{2k}(2k!) B_k c_1^2 = \left(\frac{(k!)^{2k-2}}{2} \sum_{i=1}^{k} \frac{1}{i} \right) c_1^2,$$

$$p_* \big(c_1^{2k-1}(\mathcal{L}_k^{k!}) p^* c_1^2 \big) = (k!)^{2k-1}(2k-1)! A_k c_1^2 = (k!)^{2k-3} c_1^2.$$

PROOF.   Let E be a coherent sheaf of rank r and L be a line bundle. Then

$$c_k(E \otimes L) = \sum_{i=0}^{k} \frac{(r-i)!}{(k-i)!(r-k)!} c_i(E) c_1(L)^{k-i}.$$

For a surface we have only two Chern classes, $c_1(E \otimes L) = rc_1(L) + c_1(E)$ and
$c_2(E \otimes L) = \frac{1}{2}r(r-1)c_1^2(L) + (r-1)c_1(E)c_1(L) + c_2(E)$. From this we get

$$\iota(E \otimes L) = r^2 c_1^2(L) + 2rc_1(L)c_1(E) + c_1^2(E) - r(r-1)c_1^2(L)$$
$$- 2(r-1)c_1(L)c_1(E) - 2c_2(E)$$
$$= c_1^2(E) - 2c_2(E) + rc_1^2(L) + 2c_1(L)c_1(E)$$
$$= \iota(E) + rc_1^2(L) + 2c_1(L)c_1(E)$$

and the Euler characteristic (with $c_i = c_i(T^*X)$):

$$\chi(E \otimes L) = \frac{1}{2}\big(\iota(E \otimes L) - c_1(E \otimes L)c_1\big) + \frac{1}{12}\mathrm{rk}(E \otimes L)(c_1^2 + c_2)$$
$$= \frac{1}{2}\big(\iota(E) + rc_1^2(L) + 2c_1(L)c_1(E) - (rc_1(L) + c_1(E))c_1\big) + \frac{1}{12}r(c_1^2 + c_2)$$
$$= \chi(E) + \frac{1}{2}\big(rc_1^2(L) + 2c_1(L)c_1(E) - rc_1(L)c_1\big).$$

For the sheaf of jet differentials we have the asymptotic expansions

$$c_1(\mathcal{J}_k^m X) = B_k m^{2k} c_1 + O(m^{2k-1}),$$
$$\mathrm{rk}\, \mathcal{J}_k^m X = A_k m^{2k-1} + O(m^{2k-2})$$
$$\chi(\mathcal{J}_k^m X) = \chi(\mathcal{L}^m) = \frac{1}{2}(\alpha_k c_1^2 - \beta_k c_2)m^{2k+1} + O(m^{2k})';$$

hence

$$c_1(\mathcal{J}_k^{k!m} X) = (k!)^{2k} B_k m^{2k} c_1 + O(m^{2k-1}),$$
$$\mathrm{rk}\, \mathcal{J}_k^{k!m} X = (k!)^{2k-1} A_k m^{2k-1} + O(m^{2k-2})$$
$$\chi(\mathcal{J}_k^{k!m} X) = \chi(\mathcal{L}^{k!m}) = \frac{1}{2}(k!)^{2k+1}(\alpha_k c_1^2 - \beta_k c_2)m^{2k+1} + O(m^{2k}).$$

We get from these the asymptotic expansion for $\chi(\mathcal{J}_k^{k!m} X \otimes L^m)$:

$$\chi(\mathcal{J}_k^{k!m} X \otimes L^m) = \chi(\mathcal{J}_k^{k!m} X) + \frac{1}{2}m\big(m(\mathrm{rk}\, \mathcal{J}_k^{k!m} X)c_1^2(L) + c_1(L)c_1(\mathcal{J}_k^{k!m} X)$$
$$- (\mathrm{rk}\, \mathcal{J}_k^{k!m} X)c_1(L)c_1\big)$$
$$= \frac{1}{2}\big((k!)^{2k+1}(\alpha_k c_1^2 - \beta_k c_2)$$
$$+ (k!)^{2k-1} A_k c_1^2(L) + (k!)^{2k} B_k c_1(L)c_1\big)m^{2k+1} + O(m^{2k}).$$

If $c_1(L) = \lambda c_1$ then

$$\chi((\mathcal{L}_k^{k!} \otimes p^*L)^m)$$
$$= \chi((\mathcal{J}_k^{k!m} X \otimes L^m)$$
$$= \frac{1}{2}\big((k!)^{2k+1}(\alpha_k c_1^2 - \beta_k c_2) + (\lambda^2 (k!)^{2k-1} A_k + \lambda(k!)^{2k} B_k)c_1^2\big)m^{2k+1} + O(m^{2k})$$
$$= \chi(\mathcal{J}_k^{k!m} X) + \frac{1}{2}\lambda^2(k!)^{2k-1} A_k c_1^2 m^{2k+1} + \frac{1}{2}\lambda(k!)^{2k} B_k c_1^2 m^{2k+1} + O(m^{2k}).$$

Since $c_1^i(p^*L) = 0$ for all $i \geq 3$, we have

$$c_1^{2k+1}(\mathcal{L}_k^{k!} \otimes p^*L)$$
$$= (c_1(\mathcal{L}_k^{k!}) + c_1(p^*L))^{2k+1}$$
$$= c_1^{2k+1}(\mathcal{L}_k^{k!}) + (2k+1)c_1^{2k}(\mathcal{L}_k^{k!})c_1(p^*L) + k(2k+1)c_1^{2k-1}(\mathcal{L}_k^{k!})c_1^2(p^*L),$$

and we get, *up to* $O(m^{2k})$,

$$\chi((\mathcal{L}_k^{k!} \otimes p^*L)^m)$$

$$= \frac{c_1^{2k+1}(\mathcal{L}_k^{k!} \otimes p^*L)}{(2k+1)!}m^{2k+1}$$

$$= \frac{c_1^{2k+1}(\mathcal{L}_k^{k!}) + (2k+1)c_1^{2k}(\mathcal{L}_k^{k!})c_1(p^*L) + k(2k+1)c_1^{2k-1}(\mathcal{L}_k^{k!})c_1^2(p^*L)}{(2k+1)!}m^{2k+1}$$

$$= \frac{c_1^{2k+1}(\mathcal{L}_k^{k!})}{(2k+1)!}m^{2k+1} + \frac{c_1^{2k}(\mathcal{L}_k^{k!})c_1(p^*L)}{(2k)!}m^{2k+1} + \frac{1}{2}\frac{c_1^{2k-1}(\mathcal{L}_k^{k!})c_1^2(p^*L)}{(2k-1)!}m^{2k+1}$$

$$= \frac{c_1^{2k+1}(\mathcal{L}_k^{k!})}{(2k+1)!}m^{2k+1} + \lambda\frac{c_1^{2k}(\mathcal{L}_k^{k!})p^*c_1}{(2k)!}m^{2k+1} + \lambda^2\frac{1}{2}\frac{c_1^{2k-1}(\mathcal{L}_k^{k!})p^*c_1^2}{(2k-1)!}m^{2k+1}.$$

Comparing the two expressions for $\chi((\mathcal{L}_k^{k!} \otimes p^*L)^m)$ we deduce that

$$p_*c_1^{2k+1}(\mathcal{L}_k^{k!}) = (2k+1)!\chi(\mathcal{L}_k^{k!}),$$
$$p_*c_1^{2k}(\mathcal{L}_k^{k!})p^*c_1 = \tfrac{1}{2}(k!)^{2k}(2k!)B_kc_1^2,$$
$$p_*c_1^{2k-1}(\mathcal{L}_k^{k!})p^*c_1^2 = (k!)^{2k-1}(2k-1)!A_kc_1^2.$$

The theorem follows from these by substituting the asymptotic expansions for $\chi(\mathcal{L}_k^{k!})$, A_k and B_k into the expressions above. $\qquad\square$

As a means toward understanding the general case we treat the special case of 2-jets and 3-jets (for the case of $\mathbb{P}(TX)$, that is, 1-jets, see [Miyaoka 1977; Lu and Yau 1990; Lu 1991; Dethloff et al. 1995b]). For 2-jets the intersection formulas in Lemma 7.15 and Theorem 7.16 read as:

$$\begin{aligned}
c_1([D]) &\le \tfrac{3}{4}mc_1,\\
p_*c_1^5(\mathcal{L}_2^2) &= 14c_1^2 - 10c_2,\\
p_*c_1^4(\mathcal{L}_2^2)p^*c_1 &= 3c_1^2,\\
p_*c_1^3(\mathcal{L}_2^2)p^*c_1^2 &= 2c_1^2.
\end{aligned} \qquad (7.6)$$

We shall use these formulas to deal with holomorphic maps from the complex plane into a minimal surface X of general type satisfying the conditions that $\mathrm{Pic}\,X \cong \mathbb{Z}$ and $\mathcal{K}_X$ is effective and nontrivial (for example X is a hypersurface in $\mathbb{P}^3$ of degree $d \ge 5$). *The condition was first introduced in* [Dethloff et al. 1995b] *and is crucial in the rest of this article.* We shall use the following terminology. An irreducible subvariety Y in $\mathbb{P}(J^kX)$ is said to be *horizontal* if $p(Y) = X$, where $p : \mathbb{P}(J^kX) \to X$ is the projection; otherwise it is said to be *vertical*. A variety is said to be horizontal (resp. vertical) if every irreducible component is horizontal (resp. vertical). A subvariety Y may be decomposed as $Y = Y^{\mathrm{hor}} + Y^{\mathrm{ver}}$, where Y^{hor} and Y^{ver} consist respectively of the horizontal and vertical components. Note that $Y^{\mathrm{ver}} = (p^{-1}C) \cap Y$, where C is a subvariety of X; indeed $C = p(Y^{\mathrm{ver}})$. We shall need a lemma:

LEMMA 7.17. *Let X be a surface such that $p_g(X) > 0$ and Pic $X \cong \mathbb{Z}$ with ample generator $[D_0]$. There exist positive integers m and a and a nontrivial section $\sigma \in H^0\big(\mathbb{P}(J^k X), \mathcal{L}_k^{k!m} \otimes p^*[-aD_0]\big)$ such that $[\sigma = 0]^{\mathrm{hor}}$ is reduced and irreducible, that is, there exists exactly one horizontal component with multiplicity 1.*

For the proof of the case $k = 1$, see [Dethloff et al. 1995b, Lemmas 3.5 and 3.6]. The proof depends only on the assumption Pic $X \cong \mathbb{Z}$, which implies that Pic $\mathbb{P}(J^1 X) \cong \mathbb{Z} \oplus \mathbb{Z}$. This is of course also valid for Pic $\mathbb{P}(J^k X)$ for any k. Indeed the proof (with $J^1 X$ replaced by $J^k X$) is word for word the same.

THEOREM 7.18. *Let X be a minimal surface of general type with effective ample canonical bundle such that Pic $X \cong \mathbb{Z}$, $p_g(X) > 0$, and*

$$17c_1^2(T^* X) - 16c_2(T^* X) > 0.$$

(This is satisfied if X is a hypersurface of degree $d \geq 70$.) Then every holomorphic map $f : \mathbb{C} \to X$ is algebraically degenerate.

PROOF. We start with the weaker assumption $7c_1^2(T^* X) - 5c_2(T^* X) > 0$. (By Theorem 7.7, this is satisfied for smooth hypersurfaces in $\mathbb{P}^3$ if and only if $\deg X \geq 16$.) Under this assumption the sheaf $\mathcal{J}_2^2 X$ is big. This implies that, for any ample divisor D in X there is a section $0 \not\equiv \sigma_1 \in H^0(\mathcal{L}_2^{2m} \otimes p^*[-aK])$ provided that $m \gg 0$ where $a > 0$ and K is the canonical divisor. By Schwarz Lemma (Corollary 6.3) the image of $[j^2 f]$ (as f is algebraically nondegenerate) is contained in the horizontal component of $[\sigma_1 = 0]$. By Lemma 7.17 we may assume that the horizontal component of $[\sigma_1 = 0]$ is irreducible. The vertical component of $[\sigma_1 = 0]$ must be of the form $p^*(bK)$ for some $b \geq 0$ which admits a section s^b. Replacing σ_1 with $\sigma_1 \otimes s^{-b} \in H^0(\mathcal{L}_2^{2m} \otimes p^*[-(a-b)K])$, $Y_1 = [\sigma_1 \otimes s^{-b} = 0]$ is horizontal, irreducible and contains the image of $[j^2 f]$. Since $\dim \mathbb{P}(J^2 X) = 5$ the dimension of Y_1 is 4. As remarked earlier we may assume that $a_1 = a - b \geq 0$. We get from the first and third intersection formulas of (7.6):

$$c_1^4(\mathcal{L}_2^2|_{Y_1}) = c_1^4(\mathcal{L}_2^2) \cdot (c_1(\mathcal{L}_2^{2m_1}) - a_1 p^* c_1) \geq m_1\big(c_1^5(\mathcal{L}_2^2) - a_1 c_1^4(\mathcal{L}_2^2) \cdot p^* c_1\big)$$

$$\geq m_1\big((14c_1^2 - 10c_2) - \tfrac{9}{4}c_1^2\big) = \frac{m_1}{2^2}(47c_1^2 - 40c_2) > 0.$$

(For a hypersurface of degree d in $\mathbb{P}^n$ we have $c_1^2 = (d-4)^2$, $c_2 = d^2 - 4d + 6$. Thus, for $d = 16$, $47c_1^2 = 6768$ and $40c_2 = 7920$, so $\chi(\mathcal{L}_2^2|_Y) < 0$; however,

$$47c_1^2 - 40c_2 = 47(d^2 - 8d + 16) - 40(d^2 - 4d + 6) = 7d(d - 30) - 2(3d - 256)$$

is positive if and only if $d \geq 40$.) We claim that $\mathcal{L}_2^2|_{Y_1}$ is big. It suffices to show that $H^2(\mathcal{L}_2^{2m_1} \otimes [-Y_1]) = 0$ for $m \gg 0$. To see this consider the exact sequence

$$0 \to \mathcal{L}_2^{2m_1} \otimes [-Y_1] \overset{\otimes \sigma}{\to} \mathcal{L}_2^{2m_1} \to \mathcal{L}_2^{2m_1}|_{Y_1} \to 0$$

and the induced exact sequence

$$\cdots \to H^2(\mathcal{L}_2^{2m_1} \otimes [-Y_1]) \overset{\otimes \sigma}{\to} H^2(\mathcal{L}_2^{2m_1}) \to H^2(\mathcal{L}_2^{2m_1}|_{Y_1}) \to 0.$$

The vanishing of $H^2(\mathcal{L}_2^{2m_1}|_{Y_1})$ for $m_1 \gg 0$ follows from the vanishing of $H^2(\mathcal{L}_2^{2m})$. By Schwarz's Lemma, the image of $[j^k f]$ is contained in the zero set of any nontrivial section $\sigma_2 \in H^0(Y_1, \mathcal{L}_2^{2m_1}|_{Y_1} \otimes p^*[-a_2 K])$, $a_2 > 0$ and $m_2 \gg 0$. Since Y_1 is irreducible $Y_2 = [\sigma_2 = 0] \cap Y_1$ is of codimension 2 (so $\dim Y_2 = 3$) in $\mathbb{P}(J^2 X)$ where $\sigma_2 \in H^0(\mathcal{L}_2^{2m_2}|_{Y_1} \otimes [-a_2 D])$. By Schwarz's Lemma the reparametrized k-jets $\{[j^k(f \circ \phi)]\}$ is contained in Y_2.

We may assume that Y_2 is irreducible. Otherwise $Y_2 = \sum_{i=1}^n Y_{2,i}$, where $n \geq 2$ and each $Y_{2,i}$, is irreducible and hence effective. We have $\bigotimes_{i=1}^n [Y_{2,i}] = [Y_2] = \mathcal{L}_k^{k!m_2} \otimes p^*[-a_2 K]|_{Y_1}$ (we use the notation $[Z]$ to denote the line bundle associated to a divisor Z). The image $[j^k f](\mathbb{P}^n)$ is contained in Y_{2,i_0} for some $1 \leq i_0 \leq n$. Let s_i be the (regular) section such that $[s_i = 0] = Y_{2,i}$ (an effective divisor in Y_1); then we have an exact sequence

$$0 \to [Y_{2,i_0}] \overset{\rho_{i_0}}{\to} \mathcal{L}_k^{k!m_2} \otimes p^*[-a_2 K]|_{Y_1} \to \mathcal{L}_k^{k!m_2} \otimes p^*[-a_2 K]|_{Y_{2,i_0}} \to 0.$$

In particular, we have an injection

$$0 \to [Y_{2,i_0}] \overset{\rho_{i_0}}{\to} \mathcal{L}_k^{k!m_2} \otimes p^*[-a_2 K]|_{Y_1},$$

where the map ρ_{i_0} is defined by multiplication with the section $\bigotimes_{i=1, i \neq i_0}^n s_i$. In other words we may consider each $[Y_{2,i_0}]$ as a subsheaf of $\mathcal{L}_k^{k!m_2} \otimes p^*[-a_2 K]|_{Y_1}$ hence a section of $[Y_{2,i_0}]$ is identified also as a section of $\mathcal{L}_k^{k!m_2} \otimes p^*[-a_2 K]|_{Y_1}$. The Schwarz Lemma applies and we conclude that $s_{i_0}([j^k f]) \equiv 0$ for each i. Thus we may assume that Y_2 is irreducible by replacing Y_2 with Y_{2,i_0}.

We now repeat the previous calculation for Y_1 to Y_2 using again the intersection formulas listed above; we get

$$c_1^3(\mathcal{L}_2^2|_{Y_2}) = c_1^3(\mathcal{L}_2^2) \cdot (c_1(\mathcal{L}_2^{2m_1}) - a_1 p^* c_1) \cdot (c_1(\mathcal{L}_2^{2m_2}) - a_2 p^* c_1)$$
$$\geq \left(m_1 m_2 c_1^5(\mathcal{L}_2^2) - (a_1 m_2 + m_1 a_2) c_1^4(\mathcal{L}_2^2) \cdot p^* c_1 + a_1 a_2 c_1^3(\mathcal{L}_2^2) \cdot p^* c_1^2 \right)$$
$$= m_1 m_2 \left(c_1^5(\mathcal{L}_2^2) - (l_1 + l_2) c_1^4(\mathcal{L}_2^2) \cdot p^* c_1 + l_1 l_2 c_1^3(\mathcal{L}_2^2) \cdot p^* c_1^2 \right)$$
$$= m_1 m_2 \left((14 c_1^2 - 10 c_2) - 3(l_1 + l_2) c_1^2 + 2 l_1 l_2 c_1^2 \right),$$

where $0 \leq l_i = a_i/m_i \leq \frac{3}{4}$, for $i = 1, 2$. Elementary calculus shows that the function $14 - 3(l_1 + l_2) + 2 l_1 l_2$ achieves its minimum value $14 - 3(\frac{3}{4} + \frac{3}{4}) + 2(\frac{3}{4})^2 = \frac{85}{8}$ at $l_1 = l_2 = \frac{3}{4}$; thus we get

$$c_1^3(\mathcal{L}_2^2|_{Y_2}) \geq m_1 m_2 (\tfrac{85}{8} c_1^2 - 10 c_2) = \frac{5 m_1 m_2}{2^3} (17 c_1^2 - 16 c_2) > 0.$$

This shows that $\mathcal{L}_2^2|_{Y_2}$ is big and the image of $[j^2 f]$ is contained in

$$Y_3 = Y_2 \cap [\sigma_3 = 0]$$

(where the intersection is taken over all global sections σ_3 of $\mathcal{L}_2^{2m}|_{Y_2}$ vanishing on an ample divisor), which is of dimension 2. By Corollary 6.5 the dimension of the base locus is at least 3 if f is algebraically nondegenerate. Thus f must be algebraically degenerate (and if X contains no rational or elliptic curve then X is hyperbolic). □

Note that the intersection procedure was applied twice. For a smooth hypersurface X in $\mathbb{P}^3$, condition (iii) is satisfied if and only degree of $X \geq 70$ (this is easily checked from the formulas $c_1^2 = (d-4)^2$, $c_2 = d^2 - 4d + 6$). This can be improved if we use 3-jets. For 3-jets the intersection formulas of Lemma 7.15 and Theorem 7.16 are given explicitly as follows:

$$c_1([D]) \leq \frac{11m}{3!}c_1,$$

$$p_* c_1^7(\mathcal{L}_3^{3!}) = \frac{7!(3!)^7}{2}\left(\frac{17}{2^7 3^6 7}c_1^2 - \frac{7}{2^7 3^6 5}c_2\right) = (3!)^3(85c_1^2 - 49c_2),$$

$$p_* c_1^6(\mathcal{L}_3^{3!})p^* c_1 = \frac{(3!)^3 11}{2}c_1^2,$$

$$p_* c_1^5(\mathcal{L}_3^{3!})p^* c_1^2 = (3!)^3 c_1^2.$$

THEOREM 7.19. *Let X be a minimal surface with* $\operatorname{Pic} X \cong \mathbb{Z}$, $p_g(X) > 0$, *and*

$$389c_1^2(T^*X) - 294c_2(T^*X) > 0. \tag{$*$}$$

Then every holomorphic map $f : \mathbb{C} \to X$ is algebraically degenerate.

PROOF. The sheaf $\mathcal{J}_3 X$ is big if and only if degree $d \geq 11$. As in the case of 2-jets we know that the image of $[j^3 f]$ is contained in $Y_1 = [\sigma_1 = 0]$ for some $\sigma_1 \in H^0(\mathcal{L}_2^{2m_1} \otimes p^*[-a_1 K])$. Since $\dim \mathbb{P}(J^3 X) = 7$ the dimension of Y_1 is 6. From the intersection formulas listed above we get

$$c_1^6(\mathcal{L}_2^2|_{Y_1}) = c_1^6(\mathcal{L}_2^{3!}) \cdot (c_1(\mathcal{L}_2^{2m_1}) - a_1 p^* c_1) \geq m_1\left(c_1^7(\mathcal{L}_2^{3!}) - a_1 c_1^6(\mathcal{L}_2^{3!}) \cdot p^* c_1\right)$$
$$\geq (3!)^3 m_1\left((85c_1^2 - 49c_2) - \tfrac{1}{12}11^2 c_1^2\right) = (3!)^3 \tfrac{1}{12}m_1(899c_1^2 - 588c_2) > 0.$$

For a smooth hypersurface in $\mathbb{P}^3$, $899c_1^2 - 588c_2 > 0$ if and only if $d \geq 13$.

Continuing as in the case of 2-jets, we see that the image of $[j^3 f]$ is contained in the zero set of any nontrivial section $\sigma_2 \in H^0(Y_1, \mathcal{L}_2^{(3!)m_1}|_{Y_1} \otimes p^*[-a_2 K])$, $a_2 > 0$ and $m_2 \gg 0$. The dimension of $Y_2 = [\sigma_2 = 0] \cap Y_1$ is 5. By Schwarz's Lemma the reparametrized 3-jet $\{[j^3(f \circ \phi)]\}$ is contained in Y_2. As in the case of 2-jets we may assume that Y_2 is irreducible. We now repeat the previous calculation using the intersection formulas above:

$$c_1^5(\mathcal{L}_3^{3!}|_{Y_2}) = c_1^5(\mathcal{L}_3^{3!}) \cdot (c_1(\mathcal{L}_3^{(3!)m_1}) - a_1 p^* c_1) \cdot (c_1(\mathcal{L}_3^{(3!)m_2}) - a_2 p^* c_1)$$
$$\geq \left(m_1 m_2 c_1^7(\mathcal{L}_3^{3!}) - (a_1 m_2 + m_1 a_2)c_1^6(\mathcal{L}_3^{3!}) \cdot p^* c_1 + a_1 a_2 c_1^5(\mathcal{L}_3^{3!}) \cdot p^* c_1^2\right)$$
$$= m_1 m_2\left(c_1^7(\mathcal{L}_3^{3!}) - (l_1 + l_2)c_1^6(\mathcal{L}_3^{3!}) \cdot p^* c_1 + l_1 l_2 c_1^5(\mathcal{L}_3^{3!}) \cdot p^* c_1^2\right)$$
$$= (3!)^3 m_1 m_2\left((85c_1^2 - 49c_2) - \tfrac{11}{2}(l_1 + l_2)c_1^2 + l_1 l_2 c_1^2\right)$$

where $0 \leq l_i = a_i/m_i \leq \frac{11}{6}$ for $i = 1, 2$. Elementary calculus shows that the function $85 - \frac{11}{2}(l_1 + l_2) + l_1 l_2$ achieves its minimum value at $l_1 = l_2 = \frac{11}{6}$; thus we have

$$c_1^5(\mathcal{L}_3^{3!}|_{Y_2}) \geq (3!)^3 m_1 m_2 \left(\tfrac{2455}{36} c_1^2 - 49 c_2\right) > 0.$$

For a smooth hypersurface in $\mathbb{P}^3$ this occurs if and only if $d \geq 18$.

Now the image of $[j^3 f]$ is contained in $Y_3 = [\sigma_3 = 0] \cap Y_2$ and has dimension 4. Moreover, an argument identical to the case of Y_2 shows that we may assume Y_3 irreducible. Continuing with the procedure we get

$$c_1^4(\mathcal{L}_3^{3!}|_{Y_3}) = c_1^4(\mathcal{L}_3^{3!}) \prod_{i=1}^{3} \left(c_1(\mathcal{L}_3^{(3!)m_i}) - a_i p^* c_1\right).$$

Expanding the right-hand side above yields (note that $p^* c_1^3 \equiv 0$ because the dimension of the base space is 2 hence $c_1^3 = c_1^3(X) \equiv 0$)

$$m_1 m_2 m_3 c_1^7(\mathcal{L}_3^{3!}) - (a_1 m_2 m_3 + m_1 a_2 m_3 + m_1 m_2 a_3) c_1^6(\mathcal{L}_3^{3!}) . p^* c_1$$
$$+ (a_1 a_2 m_3 + a_1 m_2 a_3 + m_1 a_2 a_3) c_1^5(\mathcal{L}_3^{3!}) . p^* c_1^2,$$

so we have

$$c_1^4(\mathcal{L}_3^{3!}|_{Y_3}) \geq \big(m_1 m_2 m_3 c_1^7(\mathcal{L}_3^{3!}) - (a_1 m_2 m_3 + m_1 a_2 m_3 + m_1 m_2 a_3) c_1^6(\mathcal{L}_3^{3!}) . p^* c_1$$
$$+ (a_1 a_2 m_3 + a_1 m_2 a_3 + m_1 a_2 a_3) c_1^5(\mathcal{L}_3^{3!}) . p^* c_1^2\big)$$
$$= m_1 m_2 m_3 \big(c_1^7(\mathcal{L}_3^{3!}) - (l_1 + l_2 + l_3) c_1^6(\mathcal{L}_3^{3!}) . p^* c_1$$
$$+ (l_1 l_2 + l_2 l_3 + l_3 l_1) c_1^5(\mathcal{L}_3^{3!}) . p^* c_1^2\big)$$
$$= (3!)^3 m_1 m_2 m_3 \big((85 c_1^2 - 49 c_2) - \tfrac{11}{2}(l_1 + l_2 + l_3) c_1^2$$
$$+ (l_1 l_2 + l_2 l_3 + l_3 l_1) c_1^2\big),$$

where $0 \leq l_i = a_i / m_i \leq \tfrac{11}{6}$ for $i = 1, 2, 3$. Elementary calculus shows that the function $85 - \tfrac{11}{2}(l_1 + l_2 + l_3) + (l_1 l_2 + l_2 l_3 + l_3 l_1)$ achieves its minimum value at $l_1 = l_2 = l_3 = \tfrac{11}{6}$; thus we get

$$c_1^4(\mathcal{L}_3^{3!}|_{Y_3}) \geq \frac{(3!)^3}{6} m_1 m_2 m_3 (389 c_1^2 - 294 c_2) > 0.$$

For hypersurfaces in $\mathbb{P}^3$ this happens if and only if $d \geq 20$. Thus the image of $[j^3 f]$ is contained in a subvariety $Y_4 = Y_3 \cap [\sigma_4 = 0]$ which is of dimension 3. By Corollary 6.5 the map f must be algebraically degenerate.   □

Note that the intersection procedure was applied three times. In order to remove condition $(*)$ in Theorem 7.19 we must use very high order jets, and if we use k-jets then it is necessary to carry out the intersection procedure k times. The preceding proof underscores the importance of the explicit formulas obtained in Section 3.

THEOREM 7.20.   *Let X be a smooth minimal surface of general type with $p_g(X) > 0$ and $\mathrm{Pic}\, X \cong \mathbb{Z}$. Then every holomorphic map $f : \mathbb{C} \to X$ is algebraically degenerate. If, in addition, the surface X contains no rational nor elliptic curve then X is hyperbolic.*

PROOF. As remarked earlier, we have to work with the k-jet bundles for k sufficiently large. In the case of 2-jets the cutting procedure was applied twice and for 3-jets, 3 times. Now we have to do this k-times, each time making sure (by using the explicit formulas of section 3) that the bundle is still big.

The assumption implies that $\mathcal{L}_k^{k!}$ is big for $k \gg 0$ hence there exists $m_1 \gg k$ and $a_1 > 0$ such that $h^0(\mathbb{P}(J^k X), \mathcal{L}_k^{k!m_1} \otimes p^*[-a_1 K]) > 0$ where K is the canonical divisor. As in the proof of Theorem 7.18 (and Theorem 7.19) we may, by Lemma 7.17, assume that there exists $\sigma_1 \in H^0(\mathbb{P}(J^k X), \mathcal{L}_k^{k!m_1} \otimes p^*[-a_1 K])$ such that $Y_1 = [\sigma_1 = 0]$ is horizontal and irreducible. This implies that codim $Y_1 = 1$ (equivalently, $\dim Y_i = \dim \mathbb{P}(J^k X) - 1 = 2k + 1 - 1 = 2k$).

By the Schwarz Lemma of the preceding section, we conclude that the image of $[j^k f]$ is contained in Y_1. The proof of Theorem 7.19 shows that $\mathcal{L}_{\mathbb{P}(J^k X)}|_{Y_1}$ is still big and so there exists $\sigma_2 \in H^0(Y_1, \mathcal{L}_k^{k!m_2} \otimes p^*[-a_2 K]), m_2, a_2 > 0$ and (because Y_1 is irreducible) that $Y_2 = [\sigma_2 = 0]$ is of codimension 2 in $\mathbb{P}(J^k X)$. Schwarz's Lemma implies that the image of $[j^k f]$ is contained in Y_2. As was shown earlier, we may assume that Y_2 is irreducible. A calculation similar to that of Theorem 7.19 shows that $\mathcal{L}_k^{k!}|_{Y_2}$ is still big (see the calculation below).

The process can be continued k times, resulting in a sequence of reduced and irreducible horizontal subvarieties,

$$\mathbb{P}(J^k X) = Y_0 \supset Y_1 \supset Y_2 \supset \cdots \supset Y_k \supset [j^k f](\mathbb{C}),$$

where codim $Y_i = i$ (equivalently, $\dim Y_i = 2k + 1 - i$ as $\dim \mathbb{P}(J^k X) = 2k + 1$), and each of the subvarieties is the zero set of a section σ_i:

$$Y_i = [\sigma_i = 0], \ \sigma_i \in H^0\left(Y_{i-1}, \mathcal{L}_k^{k!m_i} \otimes p^*[-a_i K]\right) \quad \text{for } 1 \le i \le k.$$

We claim that $\mathcal{L}_k^{k!}|_{Y_i}$ is big for $1 \le i \le k$, by a calculation (to be carried out below) analogous to that in Theorems 7.19 and 7.20.

Assuming this for the moment, we see that there exists a nontrivial section $\sigma_{k+1} \in H^0\left(Y_k, \mathcal{L}_k^{k!m_k} \otimes p^*[-a_{k+1} K]\right)$ and $[j^k f](\mathbb{C})$ is contained in an irreducible component of $[\sigma_{k+1} = 0] \cap Y_k$. Since Y_k is irreducible this component, denoted Y_{k+1}, is of codimension $k+1$ (equivalently, $\dim Y_{k+1} = 2k + 1 - (k+1) = k$). This however contradicts Corollary 6.5 that the component containing all the reparametrization $[j^k (f \circ \phi)](\mathbb{C})$ must be of codimension at most k (equivalently, dimension at least $k+1$) if f is algebraically nondegenerate. Thus the map f must be algebraically degenerate. Since the image of an algebraically degenerate map must be contained in a rational or an elliptic curve in X, we conclude readily that X is hyperbolic if it contains no rational nor elliptic curve.

It remains to verify the claim by carrying out the computations for k-jets — more precisely, computations for the Chern numbers $c_1^{2k+1-\lambda}(\mathcal{L}_k^{k!}|_{Y_\lambda})$, for $1 \le \lambda \le k$ — using Theorem 7.16 the intersection formulas obtained in Lemma 7.15:

$$c_1^{2k+1-\lambda}(\mathcal{L}_k^{k!}|_{Y_\lambda})$$

$$= c_1^{2k+1-\lambda}(\mathcal{L}_k^{k!}) \prod_{i=1}^{\lambda} \left(m_i c_1(\mathcal{L}_k^{k!}) - a_i c_1 \right)$$

$$= \left(\prod_{i=1}^{\lambda} m_i \right) c_1^{2k+1}(\mathcal{L}_k^{k!}) - \left(\sum_{i=1}^{\lambda} a_i \prod_{1 \le j \ne i \le \lambda} m_j \right) c_1^{2k}(\mathcal{L}_k^{k!} X) c_1$$

$$+ \left(\sum_{1 \le i < j \le l} a_i a_j \prod_{1 \le q \ne i,j \le \lambda} m_q c_1^{2k-1}(\mathcal{L}_k^{k!}) \right) c_1^2$$

$$= (k!)^{2k-3} \left(\prod_{i=1}^{\lambda} m_i \right) \left((k!)^2 \left(\sum_{i=1}^{k} \frac{1}{i^2} + \sum_{i=2}^{k} \frac{1}{i} \sum_{j=1}^{i-1} \frac{1}{j} \right) c_1^2 - (k!)^2 \left(\sum_{i=1}^{k} \frac{1}{i^2} \right) c_2 \right.$$

$$\left. - \left(\sum_{i=1}^{\lambda} l_i \right) \left(\frac{k!}{2} \sum_{i=1}^{k} \frac{1}{i} \right) c_1^2 + \left(\sum_{1 \le i < j \le \lambda} l_i l_j \right) c_1^2 \right)$$

for $1 \le l_j = a_j/m_j \le (B_k/A_k) k!$ and $1 \le \lambda \le k$. The coefficient of c_1^2 is

$$D_{k,\lambda} = (k!)^2 \left(\sum_{i=1}^{k} \frac{1}{i^2} + \sum_{i=2}^{k} \frac{1}{i} \sum_{j=1}^{i-1} \frac{1}{j} \right) - \left(\sum_{i=1}^{\lambda} l_i \right) \left(\frac{k!}{2} \sum_{i=1}^{k} \frac{1}{i} \right) + \sum_{1 \le i < j \le \lambda} l_i l_j.$$

The minimum occurs at

$$l_j = \frac{B_k}{A_k} k! = \frac{(k-1)!}{2} \sum_{i=1}^{k} \frac{1}{i}$$

for all $1 \le j \le \lambda \le k$. By the intersection formulas in Lemma 7.15 and Theorem 7.16, we have:

$$D_{k,\lambda} \ge (k!)^2 \left(\sum_{i=1}^{k} \frac{1}{i^2} + \sum_{i=2}^{k} \frac{1}{i} \sum_{j=1}^{i-1} \frac{1}{j} \right) - \lambda \frac{(k-1)! \, k!}{4} \left(\sum_{i=1}^{k} \frac{1}{i} \right)^2 + \frac{\lambda(\lambda+1)}{4k} \left(\sum_{i=1}^{k} \frac{1}{i} \right)^2.$$

It is clear that the worst case occurs for $\lambda = k$, namely, $D_{k,\lambda} \ge D_{k,k}$, and that

$$D_{k,k} \ge (k!)^2 \left(\sum_{i=1}^{k} \frac{1}{i^2} + \sum_{i=2}^{k} \frac{1}{i} \sum_{j=1}^{i-1} \frac{1}{j} \right) - \frac{(k!)^2}{4} \left(\sum_{i=1}^{k} \frac{1}{i} \right)^2 + \frac{(k!)^2(k+1)}{4(k!)^2} \left(\sum_{i=1}^{k} \frac{1}{i} \right)^2.$$

In other words, denoting the expression on the right-hand side above by $(k!)^2 \delta_k$, we have

$$c_1^{2k+1-\lambda}(\mathcal{L}_k^{k!}|_{Y_\lambda}) = c_1^{2k+1-\lambda}(\mathcal{L}_k^{k!}) \prod_{i=1}^{\lambda} \left(m_i c_1(\mathcal{L}_k^{k!} X) - a_i c_1 \right)$$

$$\ge (k!)^{2k-3} \left(\prod_{i=1}^{\lambda} m_i \right) \left((k!)^2 \delta_k c_1^2 - (k!)^2 \left(\sum_{i=1}^{k} \frac{1}{i^2} \right) c_2 \right)$$

$$= (k!)^{2k-1} \left(\prod_{i=1}^{\lambda} m_i \right) \left(\delta_k c_1^2 - \left(\sum_{i=1}^{k} \frac{1}{i^2} \right) c_2 \right)$$

for $1 \leq \lambda \leq k$. It remains to show that

$$\delta_k c_1^2 - \left(\sum_{i=1}^{k} \frac{1}{i^2} \right) c_2 > 0$$

or, equivalently,

$$\frac{\delta_k}{\sum_{i=1}^{k} 1/i^2} > \frac{c_2}{c_1^2} \qquad (7.7)$$

for k sufficiently large. We claim that

$$\lim_{k \to \infty} \frac{\delta_k}{\sum_{i=1}^{k} 1/i^2} = \infty, \qquad (7.8)$$

where

$$\delta_k = \left(\sum_{i=1}^{k} \frac{1}{i^2} + \sum_{i=2}^{k} \frac{1}{i} \sum_{j=1}^{i-1} \frac{1}{j} \right) - \frac{1}{4} \left(\sum_{i=1}^{k} \frac{1}{i} \right)^2 + \frac{(k+1)}{4(k!)^2} \left(\sum_{i=1}^{k} \frac{1}{i} \right)^2.$$

Observe that

$$\sum_{i=2}^{k} \frac{1}{i} \sum_{j=1}^{i-1} \frac{1}{j} = \sum_{1 \leq i < j \leq k} \frac{1}{ij} \quad \text{and} \quad \left(\sum_{i=1}^{k} \frac{1}{i} \right)^2 = \sum_{i=1}^{k} \frac{1}{i^2} + 2 \sum_{1 \leq i < j \leq k} \frac{1}{ij};$$

hence

$$\delta_k \geq \frac{1}{2} \sum_{1 \leq i < j \leq k} \frac{1}{ij} - \frac{3}{4} \sum_{i=1}^{k} \frac{1}{i^2},$$

and the ratio satisfies

$$\frac{\delta_k}{\sum_{i=1}^{k} 1/i^2} \geq \frac{1}{2} \frac{\sum_{1 \leq i < j \leq k} 1/(ij)}{\sum_{i=1}^{k} 1/i^2} - \frac{3}{4}.$$

Since

$$\sum_{1 \leq i < j \leq k} \frac{1}{ij} = \sum_{i=2}^{k} \frac{1}{i} \sum_{j=1}^{i-1} \frac{1}{j},$$

we must show that

$$\lim_{k \to \infty} \frac{\sum_{i=2}^{k} 1/i \sum_{j=1}^{i-1} 1/j}{\sum_{i=1}^{k} 1/i^2} = \infty,$$

just as in the limit in Corollary 3.10. But this is clear, because

$$\lim_{k \to \infty} \sum_{i=2}^{k} \frac{1}{i} \sum_{j=1}^{i-1} \frac{1}{j} \geq \lim_{k \to \infty} \sum_{i=2}^{k} (i-1) \frac{1}{i^2} = \infty,$$

whereas $\lim_{k \to \infty} \sum_{i=1}^{k} 1/i^2 < \infty$. Thus (7.7) is verified.

We remark that $c_2/c_1^2 = 11$ for a smooth hypersurface of degree $d = 5$ in $\mathbb{P}^3$. Thus, by (7.7), it is enough to choose k so that

$$\frac{\delta_k}{\sum_{i=1}^{k} 1/i^2} > 11.$$

With the aid of a computer, we found that this occurs at $k = 2283$ (for $k = 2282$ the ratio on the left above is approximately 10.9998). By Theorem 7.1,

$$\begin{cases} 5c_1^2 - c_2 + 36 \geq 0, & \text{if } c_1^2 \text{ is even,} \\ 5c_1^2 - c_2 + 30 \geq 0, & \text{if } c_1^2 \text{ is odd,} \end{cases}$$

which implies that

$$\begin{cases} 23 \geq 5 + (36/c_1^2) \geq c_2/c_1^2, & \text{if } c_1^2 \text{ is even,} \\ 35 \geq 5 + (30/c_1^2) \geq c_2/c_1^2, & \text{if } c_1^2 \text{ is odd.} \end{cases}$$

Thus, by (7.7), we need k so that the ratio $\delta_k/\sum_{i=1}^{k} 1/i^2$ is > 23 if c_1^2 is even and > 35 if it is odd. We did not find the explicit k satisfying these conditions; this would take a lot of time, even for the computer. However we do know from (7.8) that k exists. This shows that $c_1^{2k+1-\lambda}(\mathcal{L}_k^{k!}|_{Y_\lambda}) > 0$ hence

$$\sum_{i=0}^{2} (-1)^i H^i(Y_\lambda, \mathcal{L}_k^{k!}|_{Y_\lambda}) = \chi(\mathcal{L}_k^{k!}|_{Y_\lambda}) > 0.$$

To show that $\mathcal{L}_k^{k!}|_{Y_\lambda}$ is big it is sufficient to show that $H^2(Y_\lambda, \mathcal{L}_k^{k!}|_{Y_\lambda}) = 0$ for $0 \leq \lambda \leq k$. This is done as in Theorems 7.18 and 7.19 by considering the exact sequences

$$0 \to \mathcal{L}_k^{k!m_\lambda}|_{Y_{\lambda-1}} \otimes [-Y_\lambda] \xrightarrow{\otimes \sigma_\lambda} \mathcal{L}_k^{k!m_\lambda}|_{Y_{\lambda-1}} \to \mathcal{L}_k^{k!m_\lambda}|_{Y_\lambda} \to 0$$

and the induced exact sequence

$$\cdots \to H^2\big(\mathcal{L}_k^{k!m_\lambda}|_{Y_{\lambda-1}} \otimes [-Y_\lambda]\big) \to H^2(\mathcal{L}_k^{k!m_\lambda}|_{Y_{\lambda-1}}) \to H^2(\mathcal{L}_k^{k!m_\lambda}|_{Y_\lambda}) \to 0.$$

By induction $H^2(\mathcal{L}_k^{k!m_\lambda}|_{Y_{\lambda-1}}) = 0$ for $m_\lambda \gg 0$ and the exact sequence above implies the vanishing of $H^2(\mathcal{L}_k^{k!m_\lambda}|_{Y_\lambda})$. This completes the proof of the theorem. $\qquad\square$

COROLLARY 7.21. *A generic hypersurface surface of degree $d \geq 5$ in $\mathbb{P}^3$ is hyperbolic.*

PROOF. The assumptions of Theorem 7.19 are satisfied by a generic hypersurface of degree $d \geq 5$ in $\mathbb{P}^3$. Thus the image of a holomorphic map $f : \mathbb{C} \to X$ is contained in a curve, necessarily rational or elliptic curve. By a theorem of Xu [1994] a generic hypersurface surface of degree $d \geq 5$ in $\mathbb{P}^3$ contains no rational nor elliptic curve. Hence f must be a constant. $\qquad\square$

The generic condition in Xu means that the statement holds for all curves outside a countable union of Zariski closed sets. A variety X satisfying the condition that every holomorphic curve $f : \mathbb{C} \to X$ is constant is usually referred to as Brody hyperbolic. In general Kobayashi hyperbolic implies Brody hyperbolic. For compact varieties Brody hyperbolic is equivalent to Kobayashi hyperbolic but for open varieties this is not the case. As a consequence of Corollary 7.21 we have:

COROLLARY 7.22. *There exists a curve C of degree $d = 5$ in $\mathbb{P}^2$ such that $\mathbb{P}^2 \setminus C$ is Kobayashi hyperbolic.*

PROOF. It is well-known that the complement of 5 lines, in general position, in $\mathbb{P}^2$ is Kobayashi hyperbolic. By a Theorem of Zaidenberg [1989] any sufficiently small (in the sense of the classical topology, rather than the Zariski topology) deformation of a Brody hyperbolic manifold is Brody hyperbolic. Thus, for any curve C of degree 5 in a sufficiently small open (in the classical topology) neighborhood U, of 5 lines in general position, the complement $\mathbb{P}^2 \setminus C$ is Brody hyperbolic. Let $\bigcup Z_i$ be a countable union of Zariski closed sets in the space of surfaces of degree 5 in $\mathbb{P}^3$ such that any surface $S \notin \cup Z_i$ is hyperbolic. Embed $\mathbb{P}^2$ in $\mathbb{P}^3$ as a linear subspace. Any curve $C \in \mathcal{C} = \{S \cap \mathbb{P}^2 \mid S \notin \cup Z_i\}$ is a curve of degree 5 and is hyperbolic. It is clear that $\mathcal{C} \cap U$ is nonempty. Thus there exists a hyperbolic curve C of degree 5 in $\mathbb{P}^2$ such that $\mathbb{P}^2 \setminus C$ is Brody hyperbolic. It is well-known that this implies that $\mathbb{P}^2 \setminus C$ is Kobayashi hyperbolic. $\square$

Acknowledgements

The authors would like to express their gratitude to Professors B. Hu, J.-G. Cao, the editor and the referees for many helpful suggestions. We would also like to thank one of the referees for pointing out the work of Ehresmann [1952] on jet bundles.

References

[Abate and Patrizio 1994] M. Abate and G. Patrizio, *Finsler metrics: a global approach*, Lecture Notes in Math. **1591**, Springer, Berlin, 1994.

[Aikou 1995] T. Aikou, "Complex manifolds modelled on a complex Minkowski space", *J. Math. Kyoto Uni.* **35** (1995), 83–101.

[Aikou 1998] T. Aikou, "A partial connection on complex Finsler bundles and its applications", *Illinois J. Math.* **42** (1998), 481–492.

[Alexander 1988] J. Alexander, "Singularités imposables en position génerale aux hypersurfaces de $\mathbf{P}^n$", *Compositio Math.* **68** (1988), 305–354.

[Alexander and Hirschowitz 1992a] J. Alexander and A. Hirschowitz, "Un lemme d'Horace différentiel: application aux singularités hyperquartiques de $\mathbf{P}^5$", *J. Algebraic Geom.* **1** (1992), 411–426.

[Alexander and Hirschowitz 1992b] J. Alexander and A. Hirschowitz, "La métho-de d'Horace eclatée: application a l'interpolation en degré quatre." *Invent. Math.* **107** (1992), 585–602.

[Alexander and Hirschowitz 1995] J. Alexander and A. Hirschowitz, "Polynomial interpolation in several variables." *J. Algebraic Geom.* **4** (1995), 201–222.

[An et al. 2003a] T.-T.-H. An, J. T.-Y. Wang and P.-M. Wong, "Unique range sets and uniqueness polynomials in positive characteristic", *Acta Arith.* **109** (2003), 259–280.

[An et al. 2003b] T.-T.-H. An, J. T.-Y. Wang and P.-M. Wong, "Unique range sets and Uniqueness polynomials in positive characteristic II", preprint, 2003.

[An et al. 2004] T.-T.-H. An, J. T.-Y. Wang and P.-M. Wong, "Strong uniqueness polynomials: the complex case", *Complex Variables* **9** (2004), 25–54.

[Banica and Stanasila 1976] C. Banica and O. Stanasila, *Algebraic methods in the global theory of complex spaces*, Wiley, New York, 1976. Translation of the original Romanian book of the same title published in 1974.

[Besse 1978] A.L. Besse, *Manifolds all of whose geodesics are closed*, Ergebnisse der Math. (n.F.) **93**, Springer, Berlin, 1978.

[Bloch 1926] A. Bloch, "Sur les systemes de fonctions uniformes satisfaisant a l'équation d'une varieté algebraique dont l'irregularité dépasse la dimension", *J. de Math.* **5** (1926), 19–66.

[Bogomolov 1977] F. Bogomolov, "Families of curves on a surface of general type", *Sov. Math. Dokl.* **18** (1977), 1294–1297.

[Bogomolov 1979] F. Bogomolov, "Holomorphic tensors and vector bundles on projective varieties", *Math. USSR Izv.* **13** (1979), 499–555.

[Bao and Chern 1991] D. Bao and S.-S. Chern, *On a notable connection in Finsler geometry*, *Houston J. Math.* **19** (1995), 135–180.

[Bao et al. 1996] D. Bao, S.-S. Chern and Z. Shen (editors), *Finsler geometry*, Contemporary Math. **196**, Amer. Math. Soc., Providence, 1996.

[Barth et al. 1984] W. Barth, C. Peters and A. Van de Ven, *Compact complex surfaces* Ergebnisse der Math. (3) **4**, Springer, Berlin, (1984).

[Beltrametti and Robbiano 1986] M. Beltrametti and L. Robbiano, "Introduction to the theory of weighted projective spaces", *Exposition. Math.* **4** (1986), 111–162.

[Cao and Wong 2003] J.-K. Cao and P.-M. Wong, "Finsler geometry of projectivized vector bundles", *J. Math. Kyoto Univ.* **43** (2003), 369–410.

[Chandler 1995] K. Chandler, "Geometry of dots and ropes", *Trans. Amer. Math. Soc.* **347** (1995), 767–784.

[Chandler 1998a] K. Chandler, "A brief proof of a maximal rank theorem for general points in projective space", *Trans. Amer. Math. Soc.* **353** (1998), 1907–1920.

[Chandler 1998b] K. Chandler, "Higher infinitesimal neighborhoods", *J. of Algebra* **205** (1998), 460–479.

[Chandler 2002] K. Chandler, "Linear systems of cubics singular at general points in projective space", *Compositio Math.* **134** (2002), 269–282.

[Chandler 2003] K. Chandler, "The geometric interpretation of Froberg–Iarrobino conjectures on the Hilbert functions of fat points", to appear in *J. Algebra*.

[Chandler and Wong 2004] K. Chandler and P.-M. Wong, "On the holomorphic sectional and bisectional curvatures in complex Finsler geometry", *Periodica Mathematica Hungarica* **48** (2004), 1–31.

[Dimca 1992] A. Dimca, *Singularities and topology of hypersurfaces*, Universitext, Springer, Berlin, 1992.

[Dolgachev 1982] I. Dolgachev, "Weighted projective varieties", pp. 34–71 in "Group actions and vector fields" (Vancouver, 1981), edited by J. B. Carrell, Lecture Notes in Math. **956**, Springer, New York, 1982.

[Dethloff et al. 1995a] G. Dethloff, G. Schumacher and P.-M. Wong, "Hyperbolicity of the complements of plane algebraic curves", Amer. J. Math. **117** (1995), 573–599.

[Dethloff et al. 1995b] G. Dethloff, G. Schumacher and P.-M. Wong, "Hyperbolicity of the complements of plane algebraic curves: the three components case", Duke Math. J. **78** (1995), 193–212.

[Ehresmann 1952] C. Ehresmann, "Introduction à la théorie des structures infinitésimales et des pseudo-groupes de Lie", pp. 97–110 in *Géométrie différentielle* (Strasbourg, 1953), Colloques internationaux du Centre National de la Recherche Scientifique **52**, Paris, CNRS, 1953.

[Esnault and Viehweg 1992] H. Esnault and E. Viehweg, *Lectures on Vanishing Theorems*, DMV Seminar **20**, Birkhäuser, Basel, 1992.

[Faltings 1991] G. Faltings, "Diophantine approximation on abelian varieties", *Ann. Math.* **133** (1991), 549–576.

[Fujimoto 2001] H. Fujimoto, "A family of hyperbolic hpersurfaces in the complex projective space", *Complex Variables* **43** (2001), 273–283.

[Fulton and Harris 1991] W. Fulton and J. Harris, *Representation Theory*, Grad. Texts in Math. **129**, Springer, New York, 1991.

[Grant 1986] C. G. Grant, "Entire holomorphic curves in surfaces", *Duke Math. J.* **53** (1986), 345–358.

[Grant 1989] C. G. Grant, "Hyperbolicity of surfaces modular rational and elliptic Curves", *Pacific J. Math.* **139** (1989), 241–249.

[Grauert 1989] H. Grauert, "Jetmetriken und hyperbolische Geometrie", *Math. Z.* **200** (1989), 149–168.

[Griffiths 1969] P. Griffiths, "Hermitian differential geometry, Chern classes, and positive vector bundles", pp. 185–251 in *Global Analysis*, edited by D. C. Spencer and S. Iyanaga, University of Tokyo Press and Princeton University Press (1969).

[Green and Griffiths 1980] M. Green and P. Griffiths, "Two applications of algebraic geometry to entire holomorphic mappings", pp. 41–74 in *The Chern Symposium* (Berkeley, 1979), Springer, New York, 1980.

[Grothendieck 1958] A. Grothendieck, "La theorie des classes de Chern", *Bull. Soc. math. France* **86**, *137–154* (1958).

[Hardy and Wright 1970] G. H. Hardy and E. M. Wright, "An introduction to the theory of numbers", 3rd ed., Clarendon Press, Oxford, 1970.

[Hartshorne 1968] R. Hartshorne, "Ample vector bundles", *Publ. IHES* **29** (1968), 319–350.

[Hirzebruch 1966] F. Hirzebruch, "Topological methods in algebraic geometry", Grundlehren der Math. Wiss. **131**, Springer, Berlin, 1966.

[Iarrobino and Kanev 1999] A. Iarrobino and V. Kanev, *Power sums, Gorenstein algebras and determinantal loci*, Lectures Notes in Math. **1721** Springer, Berlin, 1999.

[Jung 1995] E.K. Jung, "Holomorphic Curves in Projective Varieties", Ph.D. thesis, Univ. Notre Dame, Notre Dame (IN), 1995.

[Kawasaki 1973] T. Kawasaki, "Cohomology of twisted projective spaces and lens complexes", *Math. Ann.* **206**, *243–248* (1973).

[Kobayashi 1970] S. Kobayashi, "Hyperbolic manifolds and holomorphic mappings", Dekker, New York, 1970.

[Kobayashi and Ochiai 1970] S. Kobayashi and T. Ochiai, "On complex manifolds with positive tangent bundles", *J. Math. Soc. Japan* **22** (1970), 499–525.

[Lang 1987] S. Lang, "Introduction to Complex Hyperbolic Spaces", Springer, New York, 1987.

[Lu 1991] S. Lu, "On meromorphic maps into varieties of log-general type", pp. 305–333 in *Several complex variables and complex geometry*, edited by Eric Bedford et al., Proc. Symp. Amer. Math. Soc. **52**, Amer. Math. Soc., Providence (RI), 1991.

[Lu and Yau 1990] S. Lu and S.T. Yau, "Holomorphic curves in surfaces of general type", *Proc. Natl. Acad. Sci. USA* **87** (1990), 80–82.

[Maruyama 1981] M. Maruyama, "The theorem of Grauert–Mulich–Splindler", *Math. Ann.* **255** (1981), 317–333.

[Miyaoka 1977] Y. Miyaoka, "On the Chern numbers of surfaces of general type", *Invent. Math.* **42** (1977), 225–237.

[Nadel 1989] A. Nadel, "Hyperbolic surfaces in $\mathbb{P}^3$", *Duke Math. J.* **58** (1989), 749–771.

[Okonek et al. 1980] C. Okonek, M. Schneider and H. Spindler, *Vector bundles on complex projective spaces*, Progress in Math. **3**, Birkhäuser, Boston, 1980.

[Royden 1971] H. Royden, "Integral points of $\mathbf{P}^n \setminus \{2n+1$ hyperplanes in general position$\}$", *Invent. Math.* **106** (1991), 195–216.

[Ru and Wong 1991] M. Ru and P.-M. Wong, *Integral points of $\mathbf{P}^n \setminus \{2n+1$ hyperplanes in general position$\}$*, *Invent. Math.* **106** (1991), 195–216.

[Sakai 1980] F. Sakai, "Semi-stable curves on algebraic surfaces", *Math. Ann.* **254** (1980), 89–120.

[Stoll 1965] W. Stoll, *Value distribution of holomorphic maps into compact complex manifolds*, Lectures Notes in Math. **135**, Springer, Berlin, 1965.

[Stoll and Wong 1994] W. Stoll and P.-M. Wong, "Second Main Theorem of Nevanlinna Theory for non-equidimensional meromorphic maps", *Amer. J. Math.* **116** (1994), 1031–1071.

[Stoll and Wong 2002] W. Stoll and P.-M. Wong, "On holomorphic jet bundles", preprint, 2002.

[Tsuji 1987] H. Tsuji, "An inequality of Chern numbers for open algebraic varieties", *Math. Ann.* **277** (1987), 483–487.

[Tsuji 1988] H. Tsuji, "Stability of tangent bundles of minimal algebraic varieties", *Topology* **22** (1988), 429–441.

[Wong 1982] P.-M. Wong, "Geometry of the complex homogeneous Monge–Ampère equation", *Invent. Math.* **67** (1982), 261–274.

[Wong 1989] P.-M. Wong, "On the Second Main Theorem of Nevanlinna Theory", *Amer. J. of Math.* **111** (1989), 549–583.

[Wong 1993a] P.-M. Wong, "Holomorphic curves in spaces of constant curvature", pp. 201–223 in *Complex geometry*, edited by G. Komatsu and Y. Sakane, Lecture Notes in Pure and Applied Math. **143**, Dekker, New York, 1993.

[Wong 1993b] P.-M. Wong, "Recent results in hyperbolic geometry and diophantine geometry", pp. 120–135 in *Proc. International Symposium on Holomorphic Mappings, Diophantine Geometry and Related Topics*, edited by J. Noguchi, RIMS, Kyoto, 1993.

[Wong 1999] P.-M. Wong, *Topics on higher dimensional Nevanlinna theory and intersection theory*, Six lectures at the Center of Theoretical Science, National Tsing-Hua University, 1999.

[Xu 1994] G. Xu, "Subvarieties of general hypersurfaces in projective space", *J. Diff. Geom.* **39** (1994), 139–172.

[Zaidenberg 1989] M.G. Zaidenberg, "Stability of hyperbolic embeddedness and construction of examples", Math. USSR Sbornik **63** (1989), 351–361.

KAREN CHANDLER
DEPARTMENT OF MATHEMATICS
UNIVERSITY OF NOTRE DAME
NOTRE DAME, IN 46556
UNITED STATES
chandler.6@nd.edu

PIT-MANN WONG
DEPARTMENT OF MATHEMATICS
UNIVERSITY OF NOTRE DAME
NOTRE DAME
IN 46556
UNITED STATES
pmwong@nd.edu

Riemann–Finsler Geometry
MSRI Publications
Volume **50**, 2004

Ricci and Flag Curvatures
in Finsler Geometry

DAVID BAO AND COLLEEN ROBLES

CONTENTS

Mathematics Subject Classification: 53B40, 53C60.

Keywords: Finsler metric, flag curvature, Ricci curvature, Einstein metric, Zermelo navigation, Randers space, Riemannian metric, sectional curvature.

Bao's research was supported in part by R. Uomini and the S.-S. Chern Foundation for Mathematical Research. Robles' research was supported in part by the UBC Graduate Fellowship Program.

Introduction

It is our goal in this article to present a current and uniform treatment of flag and Ricci curvatures in Finsler geometry, highlighting recent developments. (The flag curvature is a natural extension of the Riemannian sectional curvature to Finsler manifolds.) Of particular interest are the Einstein metrics, constant Ricci curvature metrics and, as a special case, constant flag curvature metrics. Our understanding of Einstein spaces is inchoate. Much insight may be gained by considering the examples that have recently proliferated in the literature. This motivates us to discuss many of these metrics.

Happily, the theory is developing as well. The Einstein and constant flag curvature metrics of spaces of Randers type, a fecund class of Finsler spaces, are now properly understood. Enlightenment comes from being able to identify the class as solutions to Zermelo's problem of navigation, a perspective that allows a very apt characterisation of the Einstein spaces. When specialised to flag curvature, the navigation description yields a complete classification of the constant flag curvature Randers metrics.

We hope to bring out the rich variety of behaviour displayed by these metrics. For example, Finsler metrics of constant flag curvature exhibit qualities not found in their constant sectional curvature Riemannian counterparts.

- Beltrami's theorem guarantees that a Riemannian metric is projectively flat if and only if it has constant sectional curvature. On the other hand, there are many Finsler metrics of constant flag curvature which are *not* projectively flat. See Section 3.2.3 and [Shen 2004].
- Every Riemannian metric of constant sectional curvature K is locally isometric to a round sphere, Euclidean space, or a hyperbolic space, depending on K. Hence, for each K, there is only one Riemannian standard model, up to isometry. By contrast, on S^n, $\mathbb{R}^n$, and the unit ball B^n, there are numerous nonisometric Randers metrics of constant flag curvature K. In fact, isometry classes of Randers type standard models make up a moduli space $\mathcal{M}_K$ whose dimension is linear in n. See the table on page 242 and [Bryant 2002].

A roadmap. This article is written with a variety of readers in mind, ranging from the geometric neophyte to the Finsler aficionado. We anticipate that these users will approach the manuscript with distinct aims. The outline below is intended to help readers navigate the article efficiently.

Section 1 introduces Finsler metrics and their curvatures, as well as tools and constructions that are endemic to non-Riemannian Finsler geometry. The reader conversant with Finsler metrics might only skim this section to glean our notation and conventions.

In Section 2 we develop a useful characterisation of the Einstein spaces among a ubiquitous class of Finsler metrics. This description generalises a characterisation of constant flag curvature Randers metrics ([Bao–Robles 2003] and

[Matsumoto–Shimada 2002]) to the Einstein realm. The resulting conditions form a tensorial, coupled system of nonlinear second order partial differential equations, whose unknowns consist of Riemannian metrics a and 1-forms b. These equations provide a substantial step forward in computational efficiency over the defining Einstein criterion (which stipulates that the 'average flag curvature' is to be a function of position only). Indeed, study of these equations has led to the construction of the Finslerian Poincaré metric ([Okada 1983] and [Bao et al. 2000]), as well as the S^3 metric [Bao–Shen 2002]. However, while the characterisation improves the computational accessibility of Einstein metrics, it does little to advance our understanding of their geometry. It is in the following section that we pursue this geometric insight.

The *sine qua non* here is Shen's observation that Randers metrics may be identified with solutions to Zermelo's problem of navigation on Riemannian manifolds. This navigation structure establishes a bijection between Randers spaces $(M, F = \alpha + \beta)$ and pairs (h, W) of Riemannian metrics h and vector fields W on the manifold M. From this perspective, the characterisation of Section 2 is parlayed into a breviloquent geometric description of Einstein metrics. Explicitly, the Randers metric F with navigation data (h, W) is Einstein if and only if h is Einstein and W is an infinitesimal homothety of h. (In particular, these h and W *solve* the system of partial differential equations in Section 2.) The transparent nature of the navigation description immediately yields a Schur lemma for the Ricci scalar, together with a certain rigidity in three dimensions.

The variety of examples in the article may be categorised as follows.

- Metrics in their defining form: Sections 1.1.1, 1.2.1, 2.1.1, 2.3.2, 3.1.2
- Solutions to Zermelo navigation: Sections 1.1.1, 3.1.1, 3.1.2, 3.2.3
- Of constant flag curvature: Sections 1.2.1, 2.3.2, 3.1.2, 3.2.3, 4.1.1
- Einstein but not of const. flag curvature: Sections 4.1.1, 4.1.2, 4.2.3, 4.3.3
- Ricci-flat Berwald: Sections 3.1.1 (locally Minkowski), 4.3.3 (not loc. Mink.)

In Section 4, the emphasis is on Einstein metrics of nonconstant flag curvature, especially those on compact boundaryless manifolds. The spaces studied include Finsler surfaces with Ricci scalar a function on M alone (the scalar is *a priori* a function on the tangent bundle), as well as non-Riemannian Ricci-constant solutions of Zermelo navigation on Cartesian products and Kähler–Einstein manifolds. Section 5 discusses open problems.

1. Flag and Ricci Curvatures

1.1. Finsler metrics

1.1.1. Definition and examples. A Finsler metric is a continuous function

$$F : TM \to [0, \infty)$$

with the following properties:

(i) *Regularity*: F is smooth on uniformized in favor of : $TM \smallsetminus 0 := \{(x, y) \in TM : y \neq 0\}$.

(ii) *Positive homogeneity*: $F(x, cy) = cF(x, y)$ for all $c > 0$.

(iii) *Strong convexity*: the fundamental tensor

$$g_{ij}(x, y) := \left(\tfrac{1}{2}F^2\right)_{y^i y^j}$$

is positive definite for all $(x, y) \in TM \smallsetminus 0$. Here the subscript $_{y^i}$ denotes partial differentiation by y^i.

Strong convexity implies that $\{y \in T_x M : F(x, y) \leqslant 1\}$ is a strictly convex set, but not vice versa; see [Bao et al. 2000].

The function F for a Riemannian metric a is $F(x, y) := \sqrt{a_{ij}(x) y^i y^j}$. In this case, one finds that $g_{ij} := \left(\tfrac{1}{2}F^2\right)_{y^i y^j}$ is simply a_{ij}. Thus the fundamental tensor for general Finsler metrics may be thought of as a direction-dependent Riemannian metric. This viewpoint is treated more carefully in Section 1.1.2.

Many calculations in Finsler geometry are simplified, or magically facilitated, by *Euler's theorem* for homogeneous functions:

Let ϕ be a real valued function on $\mathbb{R}^n$, differentiable at all $y \neq 0$. The following two statements are equivalent.

- $\phi(cy) = c^r \phi(y)$ *for all $c > 0$ (positive homogeneity of degree r).*
- $y^i \phi_{y^i} = r\phi$; *that is, the radial derivative of ϕ is r times ϕ.*

(See, for example, [Bao et al. 2000] for a proof.) This theorem, for instance, lets us invert the defining relation of the fundamental tensor given above to get

$$F^2(x, y) = g_{ij}(x, y) y^i y^j.$$

Consequently, strong convexity implies that F must be positive at all $y \neq 0$. The converse, however, is false; positivity does not in general imply strong convexity. This is because while $g_{ij}(x, y) y^i y^j = F^2(x, y)$ may be positive for $y \neq 0$, the quadratic $g_{ij}(x, y) \tilde{y}^i \tilde{y}^j$ could still be $\leqslant 0$ for some nonzero $\tilde{y}$.

Here are some 2-dimensional examples. Being in two dimensions, we revert to the common notation of denoting position coordinates by x, y rather than x^1, x^2, and components of tangent vectors by u, v rather than y^1, y^2.

EXAMPLE (QUARTIC METRIC). Let

$$F(x, y; u, v) := (u^4 + v^4)^{1/4}.$$

Positivity is manifest. However, $\det(g_{ij}) = 3u^2 v^2 / (u^4 + v^4)$ along the tangent vector $u\partial_x + v\partial_y$ based at the point (x, y). Thus (g_{ij}) fails to be a positive definite matrix when u or v vanishes. For instance, if $u = 0$ but $v \neq 0$, we have $g_{ij}(x, y; 0, v) = \left(\begin{smallmatrix} 1 & 0 \\ 0 & 0 \end{smallmatrix}\right)$, which is not positive definite. It is shown in [Bao et al. 2000] that F can be regularised to restore strong convexity.   ◇

We shall see in Section 2.1.2 that, surprisingly, positivity of F does imply strong convexity for Randers metrics.

Next, consider a surface S given by the graph of a smooth function $f(x, y)$. Parametrise S via $(x, y) \mapsto (x, y, f)$. By a slight abuse of notation, set $\partial_x := (1, 0, f_x)$ and $\partial_y := (0, 1, f_y)$, and denote the natural dual of this basis by dx, dy. The Euclidean metric of $\mathbb{R}^3$ induces a Riemannian metric on S:

$$h := \left(1 + f_x^2\right) dx \otimes dx + f_x f_y (dx \otimes dy + dy \otimes dx) + \left(1 + f_y^2\right) dy \otimes dy.$$

If $Y := u\partial_x + v\partial_y$ is an arbitrary tangent vector on S, we have

$$|Y|^2 := h(Y, Y) = u^2 + v^2 + (uf_x + vf_y)^2.$$

We note for later use that the contravariant description of $df = f_x\, dx + f_y\, dy$ is the vector field

$$(df)^\sharp = \frac{1}{1 + f_x^2 + f_y^2} (f_x \partial_x + f_y \partial_y), \quad \text{with} \quad |(df)^\sharp|^2 = \frac{f_x^2 + f_y^2}{1 + f_x^2 + f_y^2}.$$

EXAMPLE (METRIC FROM ZERMELO NAVIGATION). If we assume that gravity acts perpendicular to S, a person's weight does not affect his motion along the surface. Now introduce a wind $W = W^x\partial_x + W^y\partial_y$ blowing tangentially to S. The norm function F that measures travel time on S can be derived using a procedure (Section 3.1) due to Zermelo and generalised by Shen. With

$$|W|^2 = (W^x)^2 + (W^y)^2 + (W^x f_x + W^y f_y)^2,$$
$$h(W, Y) = uW^x + vW^y + (uf_x + vf_y)(W^x f_x + W^y f_y),$$

and $\lambda := 1 - |W|^2$, the formula for F reads

$$F(x, y; u, v) = \frac{\sqrt{h(W, Y)^2 + |Y|^2\lambda}}{\lambda} - \frac{h(W, Y)}{\lambda}.$$

This Zermelo navigation metric F is strongly convex if and only if $|W| < 1$. The unit circle of h in each tangent plane represents the destinations reachable in one unit of time when there is no wind. It will be explained (Section 3.1) that the effect of the wind is to take this unit circle and translate it rigidly by the amount W. The resulting figure is off-centered and represents the locus of unit time destinations under windy conditions, namely the indicatrix of F. Since the latter lacks central symmetry, F could not possibly be Riemannian; the above formula makes explicit this fact. $\diamondsuit$

EXAMPLE (MATSUMOTO'S SLOPE-OF-A-MOUNTAIN METRIC). Take the same surface S, but without the wind. View S as the slope of a mountain resting on level ground, with gravity pointing down instead of perpendicular to S. A person who can walk with speed c on level ground navigates this hillside S along a path that makes an angle θ with the steepest downhill direction. The acceleration of gravity (of magnitude g), being perpendicular to level ground, has a component of magnitude $g_{\parallel} = g\sqrt{(f_x^2 + f_y^2)/(1 + f_x^2 + f_y^2)}$ along the steepest downhill direction. The hiker then experiences an acceleration $g_{\parallel} \cos\theta$ along her path, and compensates against the $g_{\parallel} \sin\theta$ which tries to drag her off-course. Under suitable assumptions about frictional forces, the acceleration $g_{\parallel} \cos\theta$ *rapidly* effects a terminal addition $\frac{1}{2}g_{\parallel}\cos\theta$ to the pace c generated by her leg muscles. In other words, her speed is effectively of the form $c + a\cos\theta$, where a is independent of θ. Thus the locus of unit time destinations is a limaçon. The unit circle of h, instead of undergoing a rigid translation as in Zermelo navigation, has now experienced a direction-dependent deformation. The norm function F with this limaçon as indicatrix measures travel time on S. It was worked out by Matsumoto, after being inspired by a letter from P. Finsler, and reads

$$\frac{|Y|^2}{c|Y| - (g/2)(uf_x + vf_y)};$$

see [Matsumoto 1989] and [Antonelli et al. 1993].

For simplicity, specialise to the case $c = g/2$. Multiplication by c then converts this norm function to

$$F(x, y; u, v) := \frac{|Y|^2}{|Y| - (uf_x + vf_y)} = |Y|\,\varphi\left(\frac{(df)(Y)}{|Y|}\right),$$

with $\varphi(s) := 1/(1-s)$. We see from [Shen 2004] in this volume that metrics of the type $\alpha\varphi(\beta/\alpha)$ are strongly convex whenever the function $\varphi(s)$ satisfies $\varphi(s) > 0$, $\varphi(s) - s\varphi'(s) > 0$, and $\varphi''(s) \geqslant 0$. For the φ at hand, this is equivalent to $(df)(Y) < \frac{1}{2}|Y|$, which is in turn equivalent to $|(df)^{\sharp}| < \frac{1}{2}$. (In one direction, set $Y = (df)^{\sharp}$; the converse follows from a Cauchy–Schwarz inequality.) Using the formula for $|(df)^{\sharp}|^2$ presented earlier, this criterion produces

$$f_x^2 + f_y^2 < \tfrac{1}{3}.$$

Whenever this holds, F defines a Finsler metric. Such is the case for $f(x, y) := \frac{1}{2}x$ but not for $f(x, y) := x$, even though the surface S is an inclined plane in both instances. As for the elliptic paraboloid given by the graph of $f(x, y) := 100 - x^2 - y^2$, we have strong convexity only in a circular vicinity of the hilltop. $\diamond$

The functions F in these two examples are not absolutely homogeneous (and therefore non-Riemannian) because at any given juncture, the speed with which one could move forward typically depends on the direction of travel. Our discussion also raises a tantalising question: if the wind were blowing on the slope of a mountain, would the indicatrix of the resulting F be a rigid translate of the limaçon?

1.1.2. The pulled-back bundle and the fundamental tensor. The matrix g_{ij} involved in the definition of strong convexity is known as the fundamental tensor, and has a geometric meaning, which is most transparent through the use of the pulled-back bundle introduced by Chern.

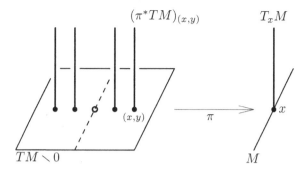

Figure 1. The pulled-back tangent bundle π^*TM is a vector bundle over the "parameter space" $TM \setminus 0$. The fiber over any point (x,y) is a copy of T_xM. The dotted part is the deleted zero section.

The g_{ij} depend on both x and $y \in T_xM$. Over each fixed point $(x,y) \in TM \setminus 0$, the bundle π^*TM provides a copy of T_xM. Endow this T_xM with the symmetric bilinear form

$$g := g_{ij}(x,y)\, dx^i \otimes dx^j.$$

Since the Finsler metric is strongly convex, this bilinear form is positive definite, which renders it an inner product. Thus, every Finsler metric endows the fibres of the pulled-back bundle with a Riemannian metric. Two facts stand out:

(1) The fundamental tensor $g_{ij}(x,y)$ is invariant under $y \mapsto \lambda y$ for all $\lambda > 0$. This invariance directly follows from the hypothesis that F is positively homogeneous of degree 1 in y. Thus, over the points $\{(x, \lambda y) : \lambda > 0\}$ in the parameter space $TM \setminus 0$, not only the fibres are identical, but the inner products are too.

(2) That g_{ij} arises as the y-Hessian of $\frac{1}{2}F^2$ imposes a stringent symmetry condition. Namely, $(g_{ij})_{y^k}$ must be totally symmetric in its three indices i, j, k. For this reason, not every Riemannian metric on the fibres of the pulled-back bundle comes from a Finsler metric.

Property (1) is a redundancy that will be given a geometrical interpretation below. We will also explain why the symmetry criterion (2) is an integrability condition in disguise.

Let's switch our perspective from the pulled-back bundle π^*TM to the tangent bundle TM itself. Recall that a Riemannian metric on M is a smooth assignment of inner products, one for each tangent space T_xM. By contrast, a Finsler metric F gives rise to a family of inner products $g_{ij}(x,y)\, dx^i \otimes dx^j$ on each tangent space T_xM. Item (1) above means that there is exactly one inner product for each *direction*. This sphere's worth of inner products on each tangent space has to satisfy the symmetry condition described in (?)

The converse is also true. Suppose we are given a family of inner products $g_{ij}(x,y)$ on each tangent space T_xM, smoothly dependent on x and nonzero

$y \in T_x M$, invariant under positive rescaling of y (that is, maps $y \mapsto \lambda y$, for $\lambda > 0$) and such that $(g_{ij})_{y^k}$ is totally symmetric in i, j, and k. We construct a Finsler function as follows:

$$F(x, y) := \sqrt{g_{pq}(x, y) y^p y^q}.$$

This F is smooth on the entire slit tangent bundle $TM \smallsetminus 0$. It is positively homogeneous of degree 1 in y because $g_{pq}(x, y)$ is invariant under positive rescaling in y. Also, with the total symmetry of $(g_{ij})_{y^k}$, we have

$$(F^2)_{y^i y^j} = (g_{pq} y^p y^q)_{y^i y^j} = ((g_{pq})_{y^i} y^p y^q + g_{iq} y^q + g_{pi} y^p)_{y^j}$$

$$= (g_{iq})_{y^j} y^q + g_{ij} + (g_{pi})_{y^j} y^p + g_{ji} = 2 g_{ij}.$$

Thus F is strongly convex because, for each nonzero y, the matrix $g_{ij}(x, y)$ is positive definite.

In the calculation above, the quantity $(g_{pq})_{y^i} y^p$ on the first line and the terms $(g_{iq})_{y^j} y^q$ and $(g_{pi})_{y^j} y^p$ on the second line are all zero because of the hypothesis that $(g_{rs})_{y^t}$ is totally symmetric in r, s, t. For instance,

$$(g_{pq})_{y^i} y^p = (g_{iq})_{y^p} y^p = 0,$$

where the last equality follows from Euler's theorem, and the assumption that $g_{iq}(x, y)$ is positive homogeneous of degree 0 in y. The symmetry hypothesis therefore plays the role of an integrability condition.

We conclude that *the concept of a Finsler metric on M is equivalent to an assignment of a sphere's worth of inner products $g_{ij}(x, y)$ on each tangent space $T_x M$, such that $(g_{ij})_{y^k}$ is totally symmetric in i, j, k and appropriate smoothness holds. Similarly, a Finsler metric on M is also equivalent to a smooth Riemannian metric on the fibres of the pulled-back bundle $\pi^* TM$, satisfying the above redundancy (1) and integrability condition (2).*

The pulled-back bundle $\pi^* TM$ and its natural dual $\pi^* T^* M$ each contains an important *global* section. They are

$$\ell := \frac{y^i}{F(x, y)} \partial_{x^i},$$

the *distinguished section*, and

$$\omega := F_{y^i} dx^i,$$

the *Hilbert form*. As another application of the integrability condition, we differentiate the statement $F^2 = g_{ij} y^i y^j$ to find

$$F_{y^i} = \frac{g_{y_i}}{F} = g_{ij} \ell^j =: \ell_i, \quad \text{with} \quad {}^g y_i := g_{ij} y^j.$$

This readily gives

$$\ell^i \ell_i = 1 \quad \text{and} \quad \ell_{i;j} = g_{ij} - \ell_i \ell_j,$$

where the semicolon abbreviates the differential operator $F \partial_{y^i}$.

1.1.3. Geodesic spray coefficients and the Chern connection. We have seen in Section 1.1.1 that if F is the Finsler function of a Riemannian metric, its fundamental tensor has no y-dependence. The converse follows from the fact that $g_{ij}y^iy^j$ reconstructs F^2 for us, thanks to Euler's theorem. Thus, the y derivative of g_{ij} measures the extent to which F fails to be Riemannian. More formally, we define the *Cartan tensor*

$$A_{ijk} := \tfrac{1}{2}F(g_{ij})_{y^k} = \tfrac{1}{4}F(F^2)_{y^iy^jy^k},$$

which is totally symmetric in all its indices. As an illustration of Euler's theorem, note that since g is homogeneous of degree zero in y, we have $y^iA_{ijk} = 0$.

Besides the Cartan tensor, we can also associate to g its formal Christoffel symbols of the second kind ,

$$\gamma^i{}_{jk} := \tfrac{1}{2}g^{is}(g_{sj,x^k} - g_{jk,x^s} + g_{ks,x^j}),$$

and the *geodesic spray coefficients*

$$G^i := \tfrac{1}{2}\gamma^i{}_{jk}y^jy^k.$$

The latter are so named because the (constant speed) geodesics of F are the solutions of the differential equation $\ddot{x}^i + \dot{x}^j\dot{x}^k\gamma^i{}_{jk}(x,\dot{x}) = 0$, which may be abbreviated as $\ddot{x}^i + 2G^i(x,\dot{x}) = 0$.

Caution: the G^i defined here is equal to *half* the G^i in [Bao et al. 2000].

Covariant differentiation of local sections of the pulled-back bundle π^*TM requires a connection, for which there are many name-brand ones. All of them have their genesis in the *nonlinear connection*

$$N^i{}_j := (G^i)_{y^j}.$$

This $N^i{}_j$ is a connection in the Ehresmann sense, because it specifies a distribution of *horizontal* vectors on the manifold $TM \smallsetminus 0$, with basis

$$\frac{\delta}{\delta x^j} := \partial_{x^j} - N^i{}_j\partial_{y^i}.$$

As another application of Euler's theorem, note that $N^i{}_jy^j = 2G^i$, since G^i is homogeneous of degree 2 in y. We also digress to observe that the Finsler function F is constant along such horizontal vector fields:

$$F_{|j} := \frac{\delta}{\delta x^j}F = 0.$$

The key lies in the following sketch of a computation:

$$N^i{}_j\ell_i = (G^i\ell_i)_{y^j} - \frac{1}{F}G^i\ell_{i;j} = F_{x^j},$$

in which establishing $(\ell^iG_i)_{y^j} = \tfrac{1}{2}(F_{x^j} + y^kF_{x^ky^j})$ and its companion statement $-(1/F)G^i\ell_{i;j} = \tfrac{1}{2}(F_{x^j} - y^kF_{x^ky^j})$ takes up the bulk of the work.

Using A, γ, and N, we can now state the formula of the *Chern connection* in natural coordinates [Bao et al. 2000]:

$$\Gamma^i{}_{jk} = \gamma^i{}_{jk} - \frac{1}{F} g^{is} (A_{sjt} N^t{}_k - A_{jkt} N^t{}_s + A_{kst} N^t{}_j),$$

with associated connection 1-forms $\omega_j{}^i := \Gamma^i{}_{jk} \, dx^k$. These $\omega_j{}^i$ represent the unique, torsion-free ($\Gamma^i{}_{kj} = \Gamma^i{}_{jk}$) connection which is almost g-compatible:

$$dg_{ij} - g_{kj}\omega_i{}^k - g_{ik}\omega_j{}^k = \frac{2}{F} A_{ijk} (dy^k + N^k{}_l dx^l).$$

As yet another application of Euler's Theorem, we shall show that the nonlinear connection N can be recovered from the Chern connection Γ as follows:

$$\Gamma^i{}_{jk} y^k = N^i{}_j.$$

A crucial ingredient in the derivation is $(\gamma^i{}_{st})_{y^j} y^s y^t = 2(g^{ip})_{y^j} G_p$. Indeed,

$$(\gamma^i{}_{st})_{y^j} y^s y^t = (g^{ip} \gamma_{pst})_{y^j} y^s y^t$$

$$= (g^{ip})_{y^j} \gamma_{pst} y^s y^t + g^{ip} \left(\left(\frac{1}{F} A_{psj} \right)_{x^t} - \left(\frac{1}{F} A_{stj} \right)_{x^p} + \left(\frac{1}{F} A_{tpj} \right)_{x^s} \right) y^s y^t$$

$$= 2(g^{ip})_{y^j} G_p + 0.$$

Whence, with the help of $A_{ijk} y^k = 0$ and $N^t{}_k y^k = 2G^t$, we have

$$\Gamma^i{}_{jk} y^k = \gamma^i{}_{jk} y^k - \frac{2}{F} g^{is} A_{stj} G^t = \gamma^i{}_{jk} y^k - g^{is} (g_{st})_{y^j} G^t$$

$$= \gamma^i{}_{jk} y^k + (g^{is})_{y^j} g_{st} G^t$$

$$= \gamma^i{}_{jk} y^k + \tfrac{1}{2} (\gamma^i{}_{st})_{y^j} y^s y^t$$

$$= (\tfrac{1}{2} \gamma^i{}_{st} y^s y^t)_{y^j} = (G^i)_{y^j} = N^i{}_j,$$

as claimed.

It is now possible to covariantly differentiate sections of $\pi^* TM$ (and its tensor products) along the horizontal vector fields $\delta/\delta x^k$ of the manifold $TM \setminus 0$. For example,

$$T^i{}_{j|k} := \frac{\delta}{\delta x^k} T^i{}_j + T^s{}_j \Gamma^i{}_{sk} - T^i{}_s \Gamma^s{}_{jk}.$$

If F arises from a Riemannian metric a, then $A = \frac{1}{2} F(a_{ij})_{y^k} = 0$ because a has no y-dependence. In that case, Γ is given by the Christoffel symbols of a. If the tensor T also has no y-dependence, $T^i{}_{j|k}$ reduces to the familiar covariant derivative in Riemannian geometry.

Let's return to the Finsler setting. Using the symmetry $\Gamma^i{}_{sk} = \Gamma^i{}_{ks}$ and the recovery property $y^s \Gamma^i{}_{ks} = N^i{}_k$, we have

$$y^i{}_{|k} = \frac{\delta}{\delta x^k} y^i + y^s \Gamma^i{}_{ks} = 0, \quad \text{hence} \quad \ell^i{}_{|k} = 0.$$

Also, the covariant derivative of the Cartan tensor A along the special horizontal vector field $\ell^s \, \delta/\delta x^s$ gives the *Landsberg tensor*, which makes frequent appearances in Finsler geometry:

$$\dot{A}_{ijk} := A_{ijk|s} \ell^s.$$

Note that $\dot{A}$ is totally symmetric, and its contraction with y vanishes.

The Chern connection gives rise to two curvature tensors:

$$R_j{}^i{}_{kl} = \frac{\delta}{\delta x^k} \Gamma^i{}_{jl} - \frac{\delta}{\delta x^l} \Gamma^i{}_{jk} + \Gamma^i{}_{sk} \Gamma^s{}_{jl} - \Gamma^i{}_{sl} \Gamma^s{}_{jk},$$

$$P_j{}^i{}_{kl} = -F \frac{\partial}{\partial y^l} \Gamma^i{}_{jk} \qquad \text{(note the symmetry: } P_j{}^i{}_{kl} = P_k{}^i{}_{jl}),$$

both invariant under positive rescaling in y. In the case of Riemannian metrics, Γ reduces to the standard Levi-Civita (Christoffel) connection, which is independent of y; hence $\frac{\delta}{\delta x} \Gamma$ becomes $\frac{\partial}{\partial x} \Gamma$. The curvature R is then the usual Riemann tensor, and P is zero.

In Finsler geometry, there are many Bianchi identities. A leisurely account of their derivation can be found in [Bao et al. 2000].

1.2. Flag curvature. This is a generalisation of the sectional curvature of Riemannian geometry. Alternatively, flag curvatures can be treated as Jacobi endomorphisms [Foulon 2002]. The flag curvature has also led to a pinching (sphere) theorem for Finsler metrics; see [Rademacher 2004] in this volume.

1.2.1. The flag curvature versus the sectional curvature. Installing a flag on a Finsler manifold (M, F) implies choosing

○ a basepoint $x \in M$ at which the flag will be planted,
○ a flagpole given by a nonzero $y \in T_x M$, and
○ an edge $V \in T_x M$ transverse to the flagpole.

See Figure 2. Note that the flagpole $y \neq 0$ singles out an inner product

$$g_y := g_{ij}(x, y) \, dx^i \otimes dx^j$$

from among the sphere's worth of inner products described in Section 1.1.2. This g_y allows us to measure the angle between V and y. It also enables us to calculate the area of the parallelogram formed by V and $\ell := y/F(x, y)$.

The flag curvature is defined as

$$K(x, y, V) := \frac{V^i \, (y^j \, R_{jikl} \, y^l) \, V^k}{g_y(y, y) \, g_y(V, V) - g_y(y, V)^2},$$

where the index i on $R_j{}^i{}_{kl}$ has been lowered by g_y. When the Finsler function F comes from a Riemannian metric, g_y is simply the Riemannian metric, R_{jikl} is the usual Riemann tensor, and $K(x, y, V)$ reduces to the familiar sectional curvature of the 2-plane spanned by $\{y, V\}$.

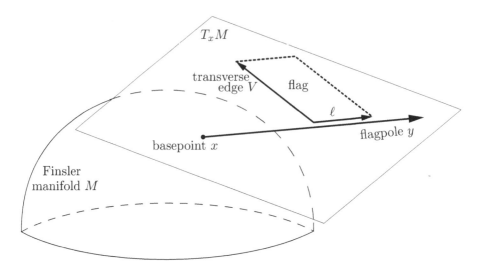

Figure 2. A typical flag, based at the point x on a Finsler manifold M. The flagpole is y, and the "cloth" part of the flag is $\ell \wedge V$. The entire flag lies in the tangent space $T_x M$.

Since $g_y(y, y) = F^2(x, y)$, the flag curvature can be reexpressed as

$$K(x, y, V) = \frac{V^i (\ell^j R_{jikl} \ell^l) V^k}{g_y(V, V) - g_y(\ell, V)^2} .$$

The denominator here is the area-squared of the parallelogram formed by V and the g_y unit vector $\ell = y/F$.

The tensor $R_{ik} := \ell^j R_{jikl} \ell^l$ is called the *predecessor* of the flag curvature. It is proved in [Bao et al. 2000] that $R_{ki} = R_{ik}$.

We also note that $\ell^i R_{ik} = 0 = R_{ik} \ell^k$. The second equality is immediate because $R_j{}^i{}_{kl}$ is manifestly skew-symmetric (Section 1.1.3) in k and l. The first equality follows from the symmetry of R_{ik}.

One can use $y^i R_{ik} = 0 = R_{ik} y^k$ to show that if $\{y, V_1\}$ and $\{y, V_2\}$ have the same span, then $K(x, y, V_1) = K(x, y, V_2)$. In other words, K depends on x, y, and $\operatorname{span}\{y, V\}$.

A Finsler metric is *of scalar curvature* if $K(x, y, V)$ does not depend on V, that is, if $R_{ik} V^i V^k = K(x, y)(g_{ik} - \ell_i \ell_k) V^i V^k$. This says that two symmetric bilinear forms generate the same quadratic form. A polarisation identity then tells us that the bilinear forms in question must be equal. So, Finsler metrics of scalar curvature are described by

$$R_{ik} = K(x, y)(g_{ik} - \ell_i \ell_k),$$

where $\ell_i = F_{y^i} = {}^g y_i / F$ (see Section 1.1.2).

EXAMPLE (NUMATA METRICS).   Numata [1978] has shown that the Finsler metrics

$$F(x,y) = \sqrt{q_{ij}(y)y^i y^j} + b_i(x)y^i,$$

where q is positive definite and b is *closed*, are of scalar curvature. Note that the first term of F is a locally *Minkowski* norm, Riemannian only when the q_{ij} are constant.

Consider the case that $q_{ij} = \delta_{ij}$ and $b = df$, where f is a smooth function on $\mathbb{R}^n$. If necessary, scale f so that the open set $M := \{x \in \mathbb{R}^n : \sqrt{\delta^{ij} f_{x^i} f_{x^j}} < 1\}$ is nonempty. Then a straightforward calculation reveals that F is of scalar curvature on M with

$$K(x,y) = \frac{3}{4} \frac{1}{F^4} (f_{x^i x^j} y^i y^j)^2 - \frac{1}{2} \frac{1}{F^3} (f_{x^i x^j x^k} y^i y^j y^k).$$

The computation in [Bao et al. 2000] utilises the spray curvature and Berwald's formula, to be discussed in Section 1.2.3. The Numata metrics are projectively flat. For Finsler metrics of (nonconstant) scalar curvature but not projectively flat, see [Shen 2004] in this volume.   ◇

The flag curvature is an important geometric invariant because its sign governs the growth of Jacobi fields, which in turn gives qualitative information about short geodesic rays with close initial data. See [Bao et al. 2000]. To bring out the essential difference between the Finslerian and Riemannian settings, we consider the case of surfaces. There, once the basepoint x and the flagpole y are chosen, span$\{y, V\}$ is equal to the tangent plane $T_x M$ for all transverse edges V. Thus every Finsler surface is of scalar curvature $K(x,y)$.

REMARK.   It is evident from the Numata metrics that the sign of $K(x,y)$ can depend on the direction of the flagpole $y \in T_x M$. By contrast, when the surface is Riemannian, this $K(x,y)$ reduces to the usual Gaussian curvature $K(x)$, which does not depend on y. The implication of this difference is profound when we survey the immediate vicinity of any fixed x. If the landscape is Riemannian, the sign of $K(x)$ creates only one type of geometry near x: hyperbolic, flat, or spherical. If the landscape is Finslerian, the sign of $K(x,y)$ can depend on the direction y of our line of sight, making it possible to encounter all three geometries during the survey!

If K is a constant (namely, it depends neither on V, nor y, nor x), the Finsler metric F is said to be *of constant flag curvature*. The tensorial criterion is $R_{ik} = K(g_{ik} - \ell_i \ell_k)$, with K constant. For later purposes, we rewrite it as

$$F^2 R^i{}_k = K(F^2 \delta^i{}_k - y^i {}^g y_k),$$

where ${}^g y_k := g_{ks} y^s$ (see Section 1.1.2).

EXAMPLE (BRYANT'S METRICS).   Bryant discovered an interesting 2-parameter family of projectively flat Finsler metrics on the sphere S^2, with constant flag curvature $K = 1$. Here we single out one metric from this family for presentation; see [Bryant 1997] for the geometry behind the construction.

Parametrise the hemispheres of S^2 via the map $(x, y) \mapsto \left(x, y, s\sqrt{1-x^2-y^2}\right)$, with $s = \pm 1$. Denote tangent vectors by $(u, v) = u\partial_x + v\partial_y \in T_{(x,y)}S^2$, and introduce the following abbreviations:

$$r^2 := x^2 + y^2, \quad P^2 := 1 - r^2, \quad B := 2r^2 - 1;$$
$$R^2 := u^2 + v^2, \quad C := xu + yv;$$
$$a := (1+B^2)\left((P^2 R^2 + C^2) + B(P^2 R^2 - C^2)\right) + 8(1+B)C^2 P^4;$$
$$b := (1+B^2)\left((P^2 R^2 - C^2) - B(P^2 R^2 + C^2)\right) - 8(0+B)C^2 P^2.$$

We emphasise that in b, the very last term contains $C^2 P^2$ and not $C^2 P^4$. The formula for the Finsler function is then

$$F(x, y; u, v) = \frac{1}{1+B^2}\left(\frac{1}{P}\sqrt{\frac{a + \sqrt{a^2 + b^2}}{2}} + 2C\right).$$

Note that a is a quadratic and $a^2 + b^2$ is a *quartic*. All the geodesics of F are arcs of great circles with Finslerian length 2π. As a comparison, the corresponding Finsler function for the standard Riemannian metric on S^2 is simply $(1/P)\sqrt{P^2 R^2 + C^2}$. For additional discussions, see [Bao et al. 2000; Sabau 2003; Shen 2004]. ◇

In dimension greater than two, a Schur lemma says that if K does not depend on V and y, it must be constant. This was proved in [del Riego 1973], then in [Matsumoto 1986]; see also [Berwald 1947].

In two dimensions, K can be a function of position x only without being constant. All Riemannian surfaces with nonconstant Gaussian curvature belong to this category; for non-Riemannian examples, see Section 4.1.1.

1.2.2. Rapscák's identity. We now prepare to relate the flag curvature to one of Berwald's spray curvatures. Suppose F and $\mathcal{F}$ are two arbitrary Finsler metrics, with geodesic spray coefficients G^i and $\mathcal{G}^i$. We think of $\mathcal{F}$ as a "background" metric, and use a colon to denote horizontal covariant differentiation with respect to the Chern connection of $\mathcal{F}$. (Note that $\mathcal{F}_{:j}$ vanishes by Section 1.1.3, but not $F_{:j}$.) Finally, let g^{ij} denote the inverse of the fundamental tensor of F, and let the subscript 0 abbreviate contraction with y: for instance, $F_{:0} = F_{:j}y^j$.

Then Rapcsák's identity [Rapcsák 1961] reads

$$G^i = \mathcal{G}^i + y^i\left(\frac{1}{2F}F_{:0}\right) + \tfrac{1}{2}Fg^{ij}\left((F_{:0})_{y^j} - 2F_{:j}\right).$$

Since Finsler metrics generally involve square roots, another form of Rapcsák's identity is more user-friendly:

$$G^i = \mathcal{G}^i + \tfrac{1}{4}g^{ij}\left((F^2_{:0})_{y^j} - 2F^2_{:j}\right).$$

The derivation of the identity involves three key steps, in which the basic fact $FF_{y^i} = g_{ij}y^j$ will be used repeatedly without mention.

We first verify that $2G_i = (FF_{x^k})_{y^i}y^k - FF_{x^i}$.

$$2G_i = \gamma_{ijk}y^j y^k = \tfrac{1}{2}(g_{ij,x^k} - g_{jk,x^i} + g_{ki,x^j})y^j y^k$$
$$= (g_{ij}y^j)_{x^k}y^k - \tfrac{1}{2}(g_{jk}y^j y^k)_{x^i} = (\tfrac{1}{2}F^2)_{y^i x^k}y^k - (\tfrac{1}{2}F^2)_{x^i}, \qquad (\dagger)$$

where $g_{rs} := (\tfrac{1}{2}F^2)_{y^r y^s}$, and the last displayed equality follows from two applications of Euler's theorem.

Next, observe that $FF_{x^r} = FF_{;r} + \mathcal{N}^j{}_r\,{}^g y_j$, where ${}^g y_j := g_{js}y^s$ and $\mathcal{N}$ is the nonlinear connection of $\mathcal{F}$. Indeed, horizontal differentiation with respect to $\mathcal{F}$ is given by $(\cdots)_{;r} = (\cdots)_{x^r} - \mathcal{N}^j{}_r(\cdots)_{y^j}$. Thus

$$FF_{x^r} = F(F_{;r} + \mathcal{N}^j{}_r F_{y^j}) = FF_{;r} + \mathcal{N}^j{}_r(g_{js}y^s). \qquad (\ddagger)$$

Here we have chosen to reexpress F_{x^r} using the nonlinear connection of $\mathcal{F}$ rather than that of F, thereby opening the door for $\mathcal{G}$ to enter the picture.

Finally, we substitute equality $(\ddagger)$ into the purpose of $(\dagger)$, getting

$$2G_i = F_{y^i}F_{;k}y^k + F(F_{;k})_{y^i}y^k - FF_{;i} + (\mathcal{N}^j{}_k)_{y^i}\,{}^g y_j y^k + \mathcal{N}^j{}_k({}^g y_j)_{y^i}y^k - \mathcal{N}^j{}_i\,{}^g y_j.$$

Note that $(F_{;k})_{y^i}y^k = (F_{;0})_{y^i} - F_{;i}$. Also, since $\mathcal{N}^j{}_k = (\mathcal{G}^j)_{y^k}$, Euler's theorem gives

$$(\mathcal{N}^j{}_k)_{y^i}\,{}^g y_j y^k = (\mathcal{G}^j)_{y^i}\,{}^g y_j = \mathcal{N}^j{}_i\,{}^g y_j,$$
$$\mathcal{N}^j{}_k({}^g y_j)_{y^i}y^k = (2\mathcal{G}^j)\left((\tfrac{1}{2}F^2)_{y^j y^s y^i}y^s + g_{js}\delta^s{}_i\right) = (2\mathcal{G}^j)(0 + g_{ji}).$$

These two statements constitute the heart of the entire derivation. After using them to simplify the above expression for $2G_i$, and relabelling i as r, we have

$$2G_r = 2\mathcal{G}^j g_{jr} + \frac{1}{F}\,{}^g y_r F_{;0} + F\left((F_{;0})_{y^r} - 2F_{;r}\right).$$

Contracting with $\tfrac{1}{2}g^{ir}$ yields Rapcsák's identity in its original form. The user-friendly version follows without much trouble.

1.2.3. Spray curvatures and Berwald's formulae.

The definition of the flag curvature through a connection (for example, Chern's) has theoretical appeal, but is not practical if one wants to compute it. Even for relatively simple Finsler metrics, the machine computation of any name-brand connection is already a daunting and often insurmountable task, let alone the curvature. This is where Berwald's spray curvatures come to the rescue. They originate from his study of systems of coupled second order differential equations, and are defined entirely in terms of the geodesic spray coefficients [Berwald 1929].

$$K^i{}_k := 2(G^i)_{x^k} - (G^i)_{y^j}(G^j)_{y^k} - y^j(G^i)_{x^j y^k} + 2G^j(G^i)_{y^j y^k},$$
$$G_j{}^i{}_{kl} := (G^i)_{y^j y^k y^l}.$$

These spray curvatures are related to the predecessor of the flag curvature, and to the P curvature of Chern's, in the following manner:

$$F^2 R^i{}_k = y^j R_j{}^i{}_{kl} y^l = K^i{}_k,$$

$$\frac{1}{F} P_j{}^i{}_{kl} = -G_j{}^i{}_{kl} + (\dot{A}^i{}_{jk})_{y^l}.$$

The statement about P follows from the fact that the Berwald connection $(G)_{yy}$ can be obtained from the Chern connection Γ by adding $\dot{A}$. This, together with a companion formula, is discussed in the reference [Bao et al. 2000] (whose G^i is *twice* the G^i in the present article). Explicitly,

$$\dot{A}^i{}_{jk} = (G^i)_{y^j y^k} - \Gamma^i{}_{jk} \quad \text{and} \quad \dot{A}_{ijk} = -\tfrac{1}{2} {}^g y_s (G^s)_{y^i y^j y^k},$$

where ${}^g y_s := g_{st} y^t$. The key that helps establish the first claim is the realisation that $G^i = \tfrac{1}{2} \Gamma^i{}_{pq} y^p y^q$, which holds because contracting A with y gives zero. When calculating the y-Hessian of G^i, we need the latter part of Section 1.1.3 and the Bianchi identity $\ell^j P_j{}^i{}_{kl} = -\dot{A}^i{}_{kl}$. As for the second claim, here is a sketch of the derivation:

o Start with $\dot{A}^s{}_{jk} = (G^s)_{y^j y^k} - \Gamma^s{}_{jk}$, apply ∂_{y^i}, contract with ${}^g y_s$, and use $P = -F \partial_y \Gamma$ (Section 1.1.3). We get ${}^g y_s (\dot{A}^s{}_{jk})_{y^i} = {}^g y_s (G^s)_{y^i y^j y^k} + \ell^s P_{jski}$.

o With the two Bianchi identities (3.4.8) and (3.4.9) of [Bao et al. 2000], it can be shown that $\ell^s P_{jski} = -\ell^s P_{sjki} = \dot{A}_{jki}$.

o The term ${}^g y_s (\dot{A}^s{}_{jk})_{y^i}$ on the left is equal to $\ell_s (\dot{A}^s{}_{jk})_{;i}$, which in turn $= (\ell_s \dot{A}^s{}_{jk})_{;i} - \ell_{s;i} \dot{A}^s{}_{jk} = 0 - (g_{si} - \ell_i \ell_s) \dot{A}^s{}_{jk} = -\dot{A}_{ijk}$.

o These manoeuvres produce $\dot{A}_{ijk} = -\tfrac{1}{2} {}^g y_s (G^s)_{y^i y^j y^k}$ as stated.

We are now in a position to express the Chern curvatures in terms of the Berwald spray curvatures. Note that applying ∂_{y^l} to $\dot{A}^i{}_{jk} = (G^i)_{y^j y^k} - \Gamma^i{}_{jk}$ immediately yields the statement involving P. The derivation of the formula $F^2 R^i{}_k = K^i{}_k$ makes frequent use of the relationship

$$y^k \Gamma^i{}_{jk} = N^i{}_j, \qquad (*)$$

from Section 1.1.3. We first demonstrate that

$$y^r \frac{\delta T}{\delta x^s} = \frac{\delta}{\delta x^s} (y^r T) + N^r{}_s T. \qquad (**)$$

This arises from $0 = y^r{}_{|s} = \frac{\delta}{\delta x^s} y^r + y^k \Gamma^r{}_{ks}$; see Section 1.1.3. Property $(*)$ gives $y^k \Gamma^r{}_{ks} = y^k \Gamma^r{}_{sk} = N^r{}_s$. Thus $\frac{\delta}{\delta x^s} y^r = -N^r{}_s$, and $(**)$ follows from the product rule.

Using $(**)$ and $(*)$, we ascertain that

$$F^2 R^i{}_k = y^j \left(\frac{\delta}{\delta x^k} N^i{}_j - \frac{\delta}{\delta x^j} N^i{}_k \right). \qquad (***)$$

Indeed, $F^2 R^i{}_k$ is obtained by contracting with $y^j y^l$ the explicit formula for $R_j{}^i{}_{kl}$ (Section 1.1.3). After several uses of $(*)$, we get the intermediate expression

$y^j(y^l\,\delta_{x^k}\Gamma^i{}_{jl}) - y^l(y^j\,\delta_{x^l}\Gamma^i{}_{jk}) + y^j\,\Gamma^i{}_{hk}\,N^h{}_j - N^i{}_h\,N^h{}_k$. Now apply $(**)$ to the first two sets of parentheses, and use $(*)$ again. The result simplifies to $(***)$.

Finally, recall from Section 1.1.3 that $N^i{}_j := (G^i)_{y^j}$, so Euler's theorem gives $y^j N^i{}_j = 2G^i$. Consequently,

$$
y^j\,\frac{\delta}{\delta x^k}\,N^i{}_j = \frac{\delta}{\delta x^k}(2G^i) + N^i{}_j\,N^j{}_k \qquad \text{by } (**)
$$

$$
= (2G^i)_{x^k} - N^j{}_k\,(2G^i)_{y^j} + N^i{}_j\,N^j{}_k = (2G^i)_{x^k} - (G^i)_{y^j}\,(G^j)_{y^k},
$$

$$
-y^j\,\frac{\delta}{\delta x^j}\,N^i{}_k = -y^j(N^i{}_k)_{x^j} + (y^j N^h{}_j)(N^i{}_k)_{y^h} = -y^j(G^i)_{y^k x^j} + 2G^j(G^i)_{y^k y^j}.
$$

Summing these two conclusions gives Berwald's formula for $K^i{}_k$. (For ease of exposition, we shall refer to $K^i{}_k$ simply as the *spray curvature*.)

We now return to the setting of two arbitrary Finsler metrics F and $\mathcal{F}$, as discussed in the previous subsection. Denote their respective spray curvatures by $K^i{}_k$ and $\mathcal{K}^i{}_k$. According to Rapcsák's identity, the geodesic coefficients of F and $\mathcal{F}$ are related by $G^i = \mathcal{G}^i + \zeta^i$. Inspired by Shibata–Kitayama, we now show that substituting this decomposition into Berwald's formula for $K^i{}_k$ allows us to rewrite the latter in a split and covariantised form

$$
K^i{}_k = \mathcal{K}^i{}_k + 2(\zeta^i)_{:k} - (\zeta^i)_{y^j}\,(\zeta^j)_{y^k} - y^j(\zeta^i{}_{:j})_{y^k} + 2\zeta^j(\zeta^i)_{y^j y^k} + 3\zeta^j\dot{\mathcal{A}}^i{}_{jk},
$$

where the colon refers to horizontal covariant differentiation with respect to the Chern connection of $\mathcal{F}$, and $\dot{\mathcal{A}}$ is associated to $\mathcal{F}$ as well.

REMARKS. 1. Had we used the Berwald connection (which equals the Chern connection plus $\dot{\mathcal{A}}$), that $3\zeta\dot{\mathcal{A}}$ term in the above formula would have been absorbed away.

2. If the background metric were Riemannian, $\mathcal{F}^2 = a_{ij}(x)y^i y^j$, then $\mathcal{A}_{ijk} = \frac{1}{2}\mathcal{F}(a_{ij})_{y^k}$ would be zero because a is independent of y. Hence $\dot{\mathcal{A}} = 0$ as well. Since $\mathcal{A} = 0$, the Chern connection is given by the usual Christoffel symbols of a. Also, the Chern and Berwald connections coincide for Riemannian metrics because they differ merely by $\dot{\mathcal{A}}$.

Here is a sketch of the derivation of the split and covariantised formula. First we replace G by $\mathcal{G} + \zeta$ in Berwald's original formula, obtaining

$$
K^i{}_k = \mathcal{K}^i{}_k + 2(\zeta^i)_{x^k} - (\zeta^i)_{y^j}\,(\zeta^j)_{y^k} - y^j(\zeta^i)_{x^j y^k} + 2\zeta^j(\zeta^i)_{y^j y^k}
$$

$$
- \mathcal{N}^i{}_j(\zeta^j)_{y^k} - (\zeta^i)_{y^j}\mathcal{N}^j{}_k + 2\mathcal{G}^j(\zeta^i)_{y^j y^k} + 2\zeta^j(\mathcal{G}^i)_{y^j y^k}. \quad (\dagger)
$$

Next, let $\tilde{\Gamma}$ denote the Chern connection of $\mathcal{F}$. Then horizontal covariant differentiation of ζ is

$$
\zeta^i{}_{:j} = \big((\zeta^i)_{x^j} - \mathcal{N}^l{}_j(\zeta^i)_{y^l}\big) + \zeta^l\tilde{\Gamma}^i{}_{lj},
$$

from which we solve for $(\zeta^i)_{x^j}$ in terms of $\zeta^i{}_{:j}$. The result is used to replace ("covariantise") all the x-derivatives in (†). The outcome reads

$$
\begin{aligned}
K^i{}_k = {}& \mathcal{K}^i{}_k + 2(\zeta^i)_{:k} - (\zeta^i)_{y^j}(\zeta^j)_{y^k} - y^j(\zeta^i{}_{:j})_{y^k} + 2\zeta^j(\zeta^i)_{y^j y^k} \\
& - 2\zeta^l \tilde{\Gamma}^i{}_{lk} + 2(\zeta^i)_{y^l}\mathcal{N}^l{}_k + y^j(\zeta^l)_{y^k}\tilde{\Gamma}^i{}_{lj} + y^j \zeta^l (\tilde{\Gamma}^i{}_{lj})_{y^k} \\
& - y^j(\mathcal{N}^l{}_j)_{y^k}(\zeta^i)_{y^l} - y^j \mathcal{N}^l{}_j(\zeta^i)_{y^l y^k} - \mathcal{N}^i{}_j(\zeta^j)_{y^k} - (\zeta^i)_{y^j}\mathcal{N}^j{}_k \\
& + 2\mathcal{G}^j(\zeta^i)_{y^j y^k} + 2\zeta^j(\mathcal{G}^i)_{y^j y^k} .
\end{aligned}
$$

It remains to check that the last ten terms on the above right-hand side actually coalesce into the single term $3\zeta^j \dot{\mathcal{A}}^i{}_{jk}$. To that end, homogeneity and Euler's theorem enable us to make the substitutions

$$
\begin{aligned}
y^j(\mathcal{N}^l{}_j)_{y^k} &= (\mathcal{N}^l{}_j y^j)_{y^k} - \mathcal{N}^l{}_k = 2(\mathcal{G}^l)_{y^k} - \mathcal{N}^l{}_k = \mathcal{N}^l{}_k, \\
y^j(\tilde{\Gamma}^i{}_{lj})_{y^k} &= (\tilde{\Gamma}^i{}_{lj} y^j)_{y^k} - \tilde{\Gamma}^i{}_{lk} = (\mathcal{N}^i{}_l)_{y^k} - \tilde{\Gamma}^i{}_{lk}, \\
y^j \tilde{\Gamma}^i{}_{lj} &= y^j \tilde{\Gamma}^i{}_{jl} = \mathcal{N}^i{}_j, \\
y^j \mathcal{N}^l{}_j &= 2\mathcal{G}^l .
\end{aligned}
$$

After some cancellations, those final ten terms consolidate into

$$
3\zeta^j y^l(\tilde{\Gamma}^i{}_{jl})_{y^k} = 3\zeta^j\big((\mathcal{G}^i)_{y^j y^k} - \tilde{\Gamma}^i{}_{jk}\big) = 3\zeta^j \dot{\mathcal{A}}^i{}_{jk}.
$$

1.3. Ricci curvature. The importance of Ricci curvatures (defined below, in Section 1.3.1) can be seen from the following Bonnet–Myers theorem:

Let (M, F) be a forward-complete connected Finsler manifold of dimension n. Suppose its Ricci curvature has the uniform positive lower bound

$$
\mathcal{R}ic \geqslant (n-1)\lambda > 0;
$$

equivalently, $y^i y^j \, \mathrm{Ric}_{ij}(x, y) \geqslant (n-1)\lambda F^2(x, y)$, with $\lambda > 0$. Then:

(i) *Every geodesic of length at least $\pi/\sqrt{\lambda}$ contains conjugate points.*
(ii) *The diameter of M is at most $\pi/\sqrt{\lambda}$.*
(iii) *M is in fact compact.*
(iv) *The fundamental group $\pi(M, x)$ is finite.*

The Riemannian version of this result is one of the most useful comparison theorems in differential geometry; see [Cheeger–Ebin 1975]. It was first extended to Finsler manifolds in [Auslander 1955]. See [Bao et al. 2000] for a leisurely exposition and references.

1.3.1. Ricci scalar and Ricci tensor. Our geometric definition of the Ricci curvature begins with $K(x, y, V) = V^i R_{ik} V^k / \big(g_y(V, V) - g_y(\ell, V)^2\big)$, a formula for the flag curvature (Section 1.2.1). If, with respect to g_y, the transverse edge V has unit length and is orthogonal to the flagpole y, that formula simplifies to

$$
K(x, y, V) = V^i R_{ik} V^k.
$$

Using g_y to measure angles and length, we take any collection of $n-1$ orthonormal transverse edges $\{e_\nu : \nu = 1, \ldots, n-1\}$ perpendicular to the flagpole. They give rise to $n-1$ flags whose flag curvatures are $K(x, y, e_\nu) = (e_\nu)^i R_{ik}(e_\nu)^k$. The inclusion of $e_n := \ell = y/F$ completes our collection into a g_y orthonormal basis $\mathcal{B}$ for T_xM. Note that $K(x, y, e_\nu)$ is simply $R_{\nu\nu}$ (no sum), the (ν, ν) component of the tensor R_{ik} with respect to $\mathcal{B}$. Also, as mentioned in Section 1.2.1, $\ell^i R_{ik}\ell^k = 0$. Thus $R_{nn} = 0$ with respect to the orthonormal basis $\mathcal{B}$.

Define, geometrically, the *Ricci scalar* $\mathcal{R}ic(x, y)$ as the sum of those $n-1$ flag curvatures $K(x, y, e_\nu)$. Then

$$\mathcal{R}ic(x,y) := \sum_{\nu=1}^{n-1} R_{\nu\nu} = \sum_{a=1}^{n} R_{aa} = R^a{}_a = R^i{}_i = \frac{1}{F^2}(y^j R_j{}^i{}_{il} y^l) = \frac{1}{F^2} K^i{}_i,$$

where the last equality follows from Section 1.2.3.

REMARKS. 1. The indices on R are to be manipulated by the fundamental tensor, and the latter is the Kronecker delta in the g_y orthonormal basis $\mathcal{B}$. Thus each component R_{aa} has the same numerical value as $R^a{}_a$ (no sum).
2. The fact that R_{ik} is a tensor ensures that its trace is independent of the basis used to carry that out. Hence $R^a{}_a = R^i{}_i$. Consequently, the definition of the Ricci scalar is independent of the choice of those $n-1$ orthonormal transverse edges.
3. The invariance of $R_j{}^i{}_{kl}$ under positive rescaling in y makes clear that $\mathcal{R}ic(x, y)$ has the same property. It is therefore meant to be a function on the projectivised sphere bundle of (M, F), but could just as well live on the slit tangent bundle $TM \setminus 0$. In any case, being a function justifies the name *scalar*.

We obtain the *Ricci tensor* from the Ricci scalar as follows:

$$\mathrm{Ric}_{ij} := (\tfrac{1}{2}F^2 \mathcal{R}ic)_{y^i y^j} = \tfrac{1}{2}(y^k R_k{}^s{}_{sl} y^l)_{y^i y^j}.$$

This definition, due to Akbar-Zadeh, is motivated by the fact that, when F arises from any Riemannian metric a, the curvature tensor depends on x alone and the y-Hessian in question reduces to the familiar expression ${}^aR_i{}^s{}_{sj}$, which is ${}^a\mathrm{Ric}_{ij}$.

The Ricci tensor has the same geometrical content as the Ricci scalar. It can be shown that

$$\mathcal{R}ic = \ell^i \ell^k \mathrm{Ric}_{ik},$$

$$\mathrm{Ric}_{ik} = g_{ik} \mathcal{R}ic + \tfrac{3}{4}(\ell_i \mathcal{R}ic_{;k} + \ell_k \mathcal{R}ic_{;i}) + \tfrac{1}{4}(\mathcal{R}ic_{;i;k} + \mathcal{R}ic_{;k;i}),$$

where the semicolon means $F\partial_y$. See [Bao et al. 2000].

1.3.2. Einstein metrics. We defined the Ricci scalar $\mathcal{R}ic$ as the sum of $n-1$ appropriately chosen flag curvatures. We showed in the last section that this sum depends only on the position x and the flagpole y, not on the specific $n-1$ flags with transverse edges orthogonal to y. Thus it is legitimate to think of $\mathcal{R}ic$ as $n-1$ times the average flag curvature at x in the direction y. In Riemannian geometry this is the average sectional curvature among sections spanned by y and a vector orthogonal to y; so generically the result again depends on both x and y.

Using the above perspective, it would seem quite remarkable if the said average does not depend on the flagpole y. Finsler metrics F with such a property, namely $\mathcal{R}ic = (n-1)K(x)$ for some function K on M, are called *Einstein metrics*. This nomenclature is due to Akbar-Zadeh. The reciprocal relationship (Section 1.3.1) between the Ricci scalar and the Ricci tensor tells us that

$$\mathcal{R}ic = (n-1)K(x) \quad \Longleftrightarrow \quad \text{Ric}_{ij} = (n-1)K(x)\,g_{ij}.$$

Going one step further, if that average does not depend on the location x either, F is said to be *Ricci-constant*; in this case, the function K is constant.

REMARKS. 1. Every Riemannian surface is Einstein (because $\mathcal{R}ic$ equals the familiar Gaussian curvature $K(x)$), but not necessarily Ricci-constant. On the other hand, Finsler surfaces are typically not Einstein, with counterexamples provided by the Numata metrics in Section 1.2.1.

2. In dimension at least 3, a Schur type lemma ensures that every Riemannian Einstein metric is necessarily Ricci-constant. The proof uses the second Bianchi identity for the Riemann curvature tensor.

3. It is not known at the moment whether such a Schur lemma holds for Finsler Einstein metrics in general. However, if we restrict our Finsler metrics to those of Randers type, then there is indeed a Schur lemma for dim$M \geqslant 3$; see [Robles 2003] and Section 3.3.1.

A good number of non-Riemannian Einstein metrics and Ricci-constant metrics are presented in Section 4.

It follows immediately from our geometric definition of $\mathcal{R}ic$ that every Finsler metric of constant flag curvature K must be Einstein with constant Ricci scalar $(n-1)K$. As a consistency check, we derive the same fact from the constant flag curvature criterion in Section 1.2.1, $R^i{}_k = K(\delta^i{}_k - y^i\,g_{yk}/F^2)$. Indeed, tracing on i and k, and noting that $y^i\,g_{yi} = y^iy^j\,g_{ij} = F^2$, we get $\mathcal{R}ic = R^i{}_i = (n-1)K$.

1.3.3. Known rigidity results and topological obstructions. We summarise a few basic results on Riemannian Einstein manifolds. References for this material include [Besse 1987; LeBrun–Wang 1999].

In two dimensions: Every 2-dimensional manifold M admits a complete Riemannian metric of constant (Gaussian) curvature, which is therefore Einstein. The construction involves gluing together manifolds with boundary, and is due to Thurston; see [Besse 1987] for a summary. If M is compact, there is a variational approach. In [Berger 1971], existence was established by starting with any Riemannian metric on M and deforming it conformally to one with constant curvature. For an exposition of the hard analysis behind this so-called (Melvyn) Berger problem, see [Aubin 1998].

In three dimensions: For 3-dimensional Riemannian manifolds (M, h), the Weyl conformal curvature tensor is automatically zero. This has the immediate consequence that h is Einstein if and only if it is of constant sectional curvature. Though such rigidity extends to Finsler metrics of Randers type (Section 3.3.2), it is not known whether the same holds for arbitrary Finsler metrics. The said rigidity in the Riemannian setting precludes some topological manifolds from admitting Einstein metrics. Take, for example, $M = S^2 \times S^1$. Were M to admit an Einstein metric h, the latter would perforce be of constant sectional curvature. Now M is compact, so h is complete and Hopf's classification of Riemannian space forms implies that the universal cover of M is either compact or contractible. This is a contradiction because the universal cover of M is $S^2 \times \mathbb{R}$.

In four dimensions: The basic result is the Hitchin–Thorpe Inequality [Hitchin 1974; Thorpe 1969; LeBrun 1999]: *If a smooth compact oriented 4-dimensional manifold M admits an Einstein metric, then*

$$\chi(M) \geqslant \tfrac{3}{2} |\tau(M)|.$$

Here $\chi(M)$ is the Euler characteristic, $\tau(M) = \tfrac{1}{3} p_1(M)$ is the signature, and $p_1(M)$ is the first Pontryagin number. The key lies in the following formulae peculiar to four dimensions [Besse 1987]:

$$\tau(M) = \frac{1}{12\pi^2} \int_M (|W^+|^2 - |W^-|^2) \sqrt{h} \, dx,$$

$$\chi(M) = \frac{1}{8\pi^2} \int_M (|S|^2 + |W^+|^2 + |W^-|^2) \sqrt{h} \, dx,$$

where S and W are respectively the scalar curvature part and the Weyl part of the Riemann curvature tensor of h. In the second formula, the fact that h is Einstein has already been used to zero out an otherwise negative term from the integrand. The complex projective space $\mathbb{C}P^2$ has the Fubini–Study metric, which is Einstein; there, $\tau = 1$ and $\chi = 3$. On the other hand, the connected sum of 4 or more copies of $\mathbb{C}P^2$ fails the inequality and hence cannot admit any Einstein metric. Finally, the said inequality is not sufficient. See [LeBrun 1999] for compact simply connected 4-manifolds which satisfy $\chi > \tfrac{3}{2} |\tau|$, but which do not admit Einstein metrics.

In three dimensions, rigidity equips us with all the tools available to space forms. In particular, there are well-understood universal models. Without this structure in dimensions at least 4, the analysis of Einstein metrics becomes considerably more difficult. The saving grace in four dimensions may be attributed to the fact [Singer–Thorpe 1969] that a 4-manifold is Einstein if and only if its curvature operator (as a self-adjoint linear operator on 2-forms) commutes with the Hodge star operator. This is the property which leads us to the Hitchin–Thorpe Inequality. The tools above do not apply in dimensions greater than four, where there are no known topological obstructions.

2. Randers Metrics in Their Defining Form

2.1. Basics

2.1.1. Definition and examples. Randers metrics were introduced by Randers [1941] in the context of general relativity, and later named by Ingarden [1957]. In the positive definite category, they are Finsler spaces built from

- a Riemannian metric $a := a_{ij} \, dx^i \otimes dx^j$, and
- a 1-form $b := b_i \, dx^i$, with equivalent description $b^\sharp := b^i \, \partial_{x^i}$,

both living globally on the smooth n-dimensional manifold M. The Finsler function of a Randers metric has the simple form $F = \alpha + \beta$, where

$$\alpha(x,y) := \sqrt{a_{ij}(x)\,y^i y^j}\,, \quad \beta(x,y) := b_i(x)y^i.$$

Generic Randers metrics are only positively homogeneous. No Randers metric can satisfy absolute homogeneity $F(x, cy) = |c|\,F(x,y)$ unless $b = 0$, in which case it is Riemannian.

EXAMPLES. The Zermelo navigation metric (page 201) is a Randers metric with defining data

$$a_{ij} = \frac{\lambda h_{ij} + W_i W_j}{\lambda^2} \quad \text{and} \quad b_i = \frac{-W_i}{\lambda}, \quad \text{where } \lambda := 1 - h(W, W).$$

Matsumoto's slope-of-a-mountain metric (page 201) is *not* of Randers type. This is because it has the form $F = \alpha^2/(\alpha - \beta)$, where α comes from the Riemannian metric on the graph of a certain function f, and $\beta = df$.

Of the examples in Section 1.2.1, a subclass of the Numata metrics—those with constant q_{ij} (which serves as our a_{ij}) and closed 1-forms b—is Randers, while Bryant's metrics are manifestly not Randers.   ◇

2.1.2. Criterion for strong convexity. In order that $F = \alpha + \beta : TM \to \mathbb{R}$ be a Finsler function, it must be nonnegative, regular, positively homogeneous, and strongly convex (Section 1.1.1). Regularity and positive homogeneity can be established by inspection. Strong convexity concerns the positive definiteness of the fundamental tensor, which for Randers metrics is

$$g_{ij} = \frac{F}{\alpha}\left(a_{ij} - \frac{{}^a y_i \, {}^a y_j}{\alpha \; \alpha}\right) + \left(\frac{{}^a y_i}{\alpha} + b_i\right)\left(\frac{{}^a y_j}{\alpha} + b_j\right), \quad \text{where } {}^a y_i := a_{ij}\,y^j.$$

It turns out that the following three criteria are equivalent:

(1) The a-norm $\|b\|$ of b is strictly less than 1 on M.
(2) $F(x,y)$ is positive for all $y \neq 0$.
(3) The fundamental tensor $g_{ij}(x,y)$ is positive definite at all $y \neq 0$.

Proof: (1) $\Longrightarrow$ (2). Suppose $\|b\| < 1$. A Cauchy–Schwarz type argument gives

$$\pm\beta \leqslant |\beta| = |b_i y^i| \leqslant \|b\|\,\|y\| < 1 \cdot \sqrt{a_{ij}\,y^i y^j} = \alpha.$$

In particular, $F = \alpha + \beta$ is positive.

(2) $\Longrightarrow$ (3). (As stressed in Section 1.1.1, this is false for general Finsler metrics.) Suppose $F = \alpha + \beta$ is positive. Then so is $F_t := \alpha + t\beta$, where $|t| \leqslant 1$. Let g_t denote the fundamental tensor of F_t. Starting with the cited formula for the fundamental tensor of Randers metrics, a standard matrix identity gives

$$\det(g_t) = \left(\frac{F_t}{\alpha}\right)^{n+1} \det(a).$$

For a leisurely treatment, see [Bao et al. 2000]. This tells us that g_t has positive determinant, hence none of its eigenvalues can vanish. These eigenvalues depend continuously on t. At $t = 0$, they are all positive because $g_0 = a$ is Riemannian. If any eigenvalue were to become nonpositive, it would have to go through zero at some t, in which case $\det(g_t)$ could not possibly remain positive. Thus all eigenvalues stay positive; in particular, $g = g_1$ is positive definite.

(3) $\Longrightarrow$ (2). As in Section 1.1.1, this follows from $F^2(x,y) = g_{ij}(x,y)y^iy^j$.

(2) $\Longrightarrow$ (1). Suppose F is positive. Then $F(x, -b(x)) = \|b\|(1 - \|b\|)$ forces $\|b\| < 1$ wherever $b(x) \neq 0$. At points where $b(x)$ vanishes, the said inequality certainly holds.

2.1.3. Explicit formula of the spray curvature. Let $^aK^i{}_k$ denote the spray curvature tensor of the Riemannian metric a. Then the spray curvature tensor $K^i{}_k$ of the Randers metric $F(x,y) := \alpha + \beta = \sqrt{a_{ij}(x)y^iy^j} + b_i(x)y^i$ can be expressed in terms of $^aK^i{}_j$ and the quantities

$$\mathrm{lie}_{ij} := b_{i|j} + b_{j|i}, \quad \mathrm{curl}_{ij} := b_{i|j} - b_{j|i}, \qquad \theta_j := b^i\,\mathrm{curl}_{ij},$$

$$^ay_i := a_{ij}y^j, \qquad\qquad \xi := \tfrac{1}{2}\mathrm{lie}_{00} - \alpha\theta_0, \quad \xi_{|0} := \tfrac{1}{2}\mathrm{lie}_{00|0} - \alpha\theta_{0|0},$$

through the use of Berwald's formula (Section 1.2.3) in a split and covariantised form. When applying Section 1.2.3, set the background metric $\mathcal{F}$ to be the Riemannian a, with Christoffel symbols $^a\gamma^i{}_{jk}$, and let $_|$ instead of $_:$ denote the corresponding covariant differentiation. We also need the fact, derived in [Bao et al. 2000], that $G^i = {}^aG^i + \zeta^i$ with $2\zeta^i = (y^i/F)\xi + \alpha\,\mathrm{curl}^i{}_0$. The resulting formula for the spray curvature (also independently obtained by Shen) reads

$$K^i{}_k = {}^aK^i{}_k + \text{yy-Coeff } y^i\,{}^ay_k + \text{yb-Coeff } y^ib_k + \text{δ-Coeff } \delta^i{}_k$$

$$+ \tfrac{1}{4}\,\mathrm{curl}^i{}_j\,\mathrm{curl}^j{}_0\,{}^ay_k - \tfrac{1}{4}\alpha^2\,\mathrm{curl}^i{}_j\,\mathrm{curl}^j{}_k + \tfrac{3}{4}\,\mathrm{curl}^i{}_0\,\mathrm{curl}_{k0}$$

$$+ \tfrac{1}{4}(\alpha^2/F)y^i\theta_j\,\mathrm{curl}^j{}_k - \tfrac{3}{4}(1/F)y^i\theta_0\,\mathrm{curl}_{k0}$$

$$+ \tfrac{1}{2}(\alpha/F)y^i\,\mathrm{curl}^j{}_0\,\mathrm{lie}_{jk} - \tfrac{1}{4}(\alpha/F)y^i\,\mathrm{lie}_{j0}\,\mathrm{curl}^j{}_k$$

$$+ \alpha\,\mathrm{curl}^i{}_{0|k} - \tfrac{1}{2}\alpha\,\mathrm{curl}^i{}_{k|0} - \tfrac{1}{2}(1/\alpha)\,\mathrm{curl}^i{}_{0|0}\,{}^ay_k$$

$$+ \tfrac{1}{2}(\alpha/F)y^i\theta_{k|0} - (\alpha/F)y^i\theta_{0|k} + \tfrac{1}{2}(1/F)y^i\,\mathrm{lie}_{00|k} - \tfrac{1}{2}(1/F)y^i\,\mathrm{lie}_{k0|0}.$$

The three suppressed coefficients are

$$\text{yy-Coeff} := \left(\alpha/(2F^2) - 1/(4F)\right) \text{curl}^j{}_0 \theta_j - \left(1/(2F^2) + 1/(4F\alpha)\right) \text{curl}^j{}_0 \text{lie}_{j0}$$
$$+ \tfrac{1}{2}\theta_{0|0}/(F\alpha) - \tfrac{3}{4}\xi^2/(F^3\alpha) + \tfrac{1}{2}\xi_{|0}/(F^2\alpha),$$

$$\text{yb-Coeff} := \tfrac{1}{2}(\alpha^2/F^2)\,\text{curl}^j{}_0\theta_j - \tfrac{1}{2}(\alpha/F^2)\,\text{curl}^j{}_0\,\text{lie}_{j0} - \tfrac{3}{4}(1/F^3)\xi^2 + \tfrac{1}{2}(1/F^2)\xi_{|0},$$

$$\delta\text{-Coeff} := -\tfrac{1}{2}(\alpha^2/F)\,\text{curl}^j{}_0\theta_j + \tfrac{1}{2}(\alpha/F)\,\text{curl}^j{}_0\,\text{lie}_{j0} + \tfrac{3}{4}(1/F^2)\xi^2 - \tfrac{1}{2}(1/F)\xi_{|0}.$$

REMARK. Covariant differentiation with respect to the Riemannian metric a, indicated by our vertical slash, can be lifted horizontally to $TM \smallsetminus 0$, using the nonlinear connection and the Christoffel symbols of a. The section y of π^*TM then satisfies $y^i{}_{|k} = 0$; see Section 1.1.3. So, in the above expressions, we can interpret the subscript 0 as contraction with y either before or after the vertical slash has been carried out, with no difference in the outcome.

2.2. Characterising Einstein–Randers metrics.

In this section we derive necessary and sufficient conditions on a and b for the Randers metric to be Einstein. Recall that F is Einstein with Ricci scalar $\mathcal{R}ic(x)$ if and only if $K^i{}_i = \mathcal{R}ic(x)F^2$ (Section 1.3.2). We begin by assuming that this equality holds, and deduce the necessary conditions for the metric to be Einstein. Then we show that these necessary conditions are also sufficient.

Compute $K^i{}_i$ by tracing the expression for $K^i{}_k$ in Section 2.1.3 to arrive at

$$0 = K^i{}_i - F^2\,\mathcal{R}ic(x)$$

$$= {}^a\text{Ric}_{00} + \alpha\,\text{curl}^i{}_{0|i} + \tfrac{1}{2}(n-1)\frac{\alpha}{F}\theta_{0|0} - \tfrac{1}{4}(n-1)\frac{1}{F}\text{lie}_{00|0}$$

$$+ \tfrac{1}{2}(n-1)\frac{\alpha}{F}\,\text{curl}^i{}_0\,\text{lie}_{i0} - \tfrac{1}{2}(n-1)\frac{\alpha^2}{F}\theta^i\,\text{curl}_{i0} + \tfrac{1}{2}\,\text{curl}^i{}_0\,\text{curl}_{i0}$$

$$+ \tfrac{1}{4}\alpha^2\,\text{curl}^{ij}\,\text{curl}_{ij} + \tfrac{3}{16}(n-1)\frac{1}{F^2}(\text{lie}_{00})^2 - \tfrac{3}{4}(n-1)\frac{\alpha}{F^2}\,\text{lie}_{00}\,\theta_0$$

$$+ \tfrac{3}{4}(n-1)\frac{\alpha^2}{F^2}(\theta_0)^2 - F^2\,\mathcal{R}ic(x).$$

Here, we have used the fact that ${}^aK^i{}_k$, the spray curvature of the Riemannian metric a, is related to the latter's Riemann tensor via ${}^aK^i{}_k = y^j\,{}^aR_j{}^i{}_{kl}\,y^l$, as shown in Section 1.2.3. Hence ${}^aK^i{}_i = y^j\,{}^aR_j{}^i{}_{il}\,y^l = y^j\,{}^a\text{Ric}_{jl}\,y^l = {}^a\text{Ric}_{00}$.

Multiplying this displayed equation by F^2 removes y from the denominators. The criterion for a Randers metric to be Einstein then takes the form

$$\text{Rat} + \alpha\text{Irrat} = 0, \quad \text{where} \quad \alpha := \sqrt{a_{ij}(x)\,y^i y^j}.$$

Here Rat and Irrat are homogeneous polynomials in y, of degree 4 and 3 respectively, whose coefficients are functions of x. Their formulae are given below.

As observed by Crampin, the displayed equation becomes $\text{Rat} - \alpha\text{Irrat} = 0$ if we replace y by $-y$. The two equations then effect $\text{Rat} = 0$ and $\alpha\text{Irrat} = 0$.

Being homogeneous in y, Irrat certainly vanishes at $y = 0$. At nonzero y, we have $\alpha > 0$ because a_{ij} is positive-definite. Hence Irrat $= 0$.

LEMMA 1. *Let $F(x,y) := \sqrt{a_{ij}(x)y^i y^j} + b_i(x)y^i$ be a Randers metric with positive definite (i.e., Riemannian) a_{ij}. Then F is Einstein if and only if* Rat $= 0$ *and* Irrat $= 0$.

The formulae for Rat and Irrat are

$$
\begin{aligned}
\text{Rat} = {} & (\alpha^2 + \beta^2)\,{}^a\text{Ric}_{00} + 2\alpha^2\beta\,\text{curl}^i{}_{0|i} + \tfrac{1}{2}(\alpha^2 + \beta^2)\,\text{curl}^i{}_0\,\text{curl}_{i0} \\
& + \tfrac{1}{4}\alpha^2(\alpha^2 + \beta^2)\,\text{curl}^{ij}\,\text{curl}_{ij} - (\alpha^4 + 6\alpha^2\beta^2 + \beta^4)\,\mathcal{R}ic(x) \\
& + \tfrac{1}{2}(n-1)\big(\alpha^2\theta_{0|0} - \tfrac{1}{2}\beta\,\text{lie}_{00|0} + \alpha^2\,\text{curl}^i{}_0\,\text{lie}_{i0} \\
& \hspace{3cm} - \alpha^2\beta\theta^i\,\text{curl}_{i0} + \tfrac{3}{8}(\text{lie}_{00})^2 + \tfrac{3}{2}\alpha^2(\theta_0)^2\big),
\end{aligned}
$$

$$
\begin{aligned}
\text{Irrat} = {} & 2\beta\,{}^a\text{Ric}_{00} + (\alpha^2 + \beta^2)\,\text{curl}^i{}_{0|i} + \beta\,\text{curl}^i{}_0\,\text{curl}_{i0} \\
& + \tfrac{1}{2}\alpha^2\beta\,\text{curl}^{ij}\,\text{curl}_{ij} - 4\beta(\alpha^2 + \beta^2)\,\mathcal{R}ic(x) \\
& + \tfrac{1}{2}(n-1)\big(\beta\theta_{0|0} - \tfrac{1}{2}\,\text{lie}_{00|0} + \beta\,\text{curl}^i{}_0\,\text{lie}_{i0} - \alpha^2\theta^i\,\text{curl}_{i0} - \tfrac{3}{2}\,\text{lie}_{00}\,\theta_0\big).
\end{aligned}
$$

From these two expressions we will derive the preliminary form of three necessary and sufficient conditions for a Randers metric to be Einstein.

2.2.1. Preliminary form of the characterisation. Assume F is Einstein, so that Rat $= 0$ and Irrat $= 0$. For convenience abbreviate $\mathcal{R}ic(x)$ by $\mathcal{R}ic$. Then the weaker statement Rat $- \beta\,$Irrat $= 0$ certainly holds, and reads

$$
\begin{aligned}
0 = {} & (\alpha^2 - \beta^2)\big({}^a\text{Ric}_{00} + \beta\,\text{curl}^i{}_{0|i} + \tfrac{1}{2}\,\text{curl}^i{}_0\,\text{curl}_{i0} + \tfrac{1}{4}\alpha^2\,\text{curl}^{ij}\,\text{curl}_{ij} - (\alpha^2 + 3\beta^2)\,\mathcal{R}ic \\
& \hspace{3cm} + \tfrac{1}{2}(n-1)(\text{curl}^i{}_0\,\text{lie}_{i0} + \tfrac{3}{2}(\theta_0)^2 + \theta_{0|0})\big) \\
& + \tfrac{3}{16}(n-1)\big(\text{lie}_{00} + 2\beta\theta_0\big)^2.
\end{aligned}
$$

Fix x. Considering the right-hand side as a polynomial in y, we see that $\alpha^2 - \beta^2$ divides $(\text{lie}_{00} + 2\beta\theta_0)^2$. The polynomial $\alpha^2 - \beta^2$ is irreducible, because if it were to factor — necessarily into two linear terms — its zero set would contain a hyperplane, contradicting the strong convexity condition ($\|b\| < 1$), which requires that it be positive at all $y \neq 0$ (Section 2.1.2).

Being irreducible, $\alpha^2 - \beta^2$ must divide not just the square but $\text{lie}_{00} + 2\beta\theta_0$ itself. Thus there exists a scalar function $\sigma(x)$ on M such that

$$
\text{lie}_{00} + 2\beta\theta_0 = \sigma(x)(\alpha^2 - \beta^2).
$$

This is our *Basic Equation*, the first necessary condition for a Randers metric to be Einstein. Differentiating with respect to y^i and y^k gives an equivalent version:

$$
\text{lie}_{ik} + b_i\theta_k + b_k\theta_i = \sigma(x)(a_{ik} - b_ib_k).
$$

To recover the original version, just contract this with y^iy^k. The Basic Equation is equivalent to the statement that the S-curvature of the Randers metric F is given by $S = \tfrac{1}{4}(n+1)\sigma(x)F$; see [Chen–Shen 2003].

Now return to the expression for $0 = \mathrm{Rat} - \beta\,\mathrm{Irrat}$. We use the Basic Equation to replace $\mathrm{lie}_{00} + 2\beta\theta_0$ with $\sigma(x)(\alpha^2 - \beta^2)$, and then divide off by a uniform factor of $\alpha^2 - \beta^2$. The result reads

$$^a\mathrm{Ric}_{00} = (\alpha^2 + 3\beta^2)\,\mathcal{R}ic - \beta\,\mathrm{curl}^j{}_{0|j} - \tfrac{1}{4}\alpha^2\,\mathrm{curl}^{hj}\,\mathrm{curl}_{hj} - \tfrac{1}{2}\,\mathrm{curl}^j{}_0\,\mathrm{curl}_{j0}$$
$$- \tfrac{1}{2}(n-1)\big(\tfrac{3}{8}\sigma^2(x)(\alpha^2 - \beta^2) + \mathrm{curl}^j{}_0\,\mathrm{lie}_{j0} + \tfrac{3}{2}(\theta_0)^2 + \theta_{0|0}\big).$$

This is the *Ricci Curvature Equation*, so named because it describes the Ricci tensor of a. We obtain the indexed version by differentiating with respect to y^i and y^k, and making use of the symmetry $^a\mathrm{Ric}_{ik} = {}^a\mathrm{Ric}_{ki}$.

$$^a\mathrm{Ric}_{ik} = (a_{ik} + 3b_i b_k)\,\mathcal{R}ic - \tfrac{1}{2}(b_i\,\mathrm{curl}^j{}_{k|j} + b_k\,\mathrm{curl}^j{}_{i|j})$$
$$- \tfrac{1}{4}a_{ik}\,\mathrm{curl}^{hj}\,\mathrm{curl}_{hj} - \tfrac{1}{2}\,\mathrm{curl}^j{}_i\,\mathrm{curl}_{jk}$$
$$- \tfrac{1}{2}(n-1)\big(\tfrac{3}{8}\sigma^2(x)(a_{ik} - b_i b_k) + \tfrac{1}{2}(\mathrm{curl}^j{}_i\,\mathrm{lie}_{jk} + \mathrm{curl}^j{}_k\,\mathrm{lie}_{ji})$$
$$+ \tfrac{3}{2}\theta_i\theta_k + \tfrac{1}{2}(\theta_{i|k} + \theta_{k|i})\big).$$

From the Basic and Ricci Curvature Equations we derive the final character-ising condition, which we call the E_{23} *Equation* (the number 23 being of some chronological significance in our research notes). Two pieces of information from the Basic Equation are required. To reduce clutter, abbreviate $\sigma(x)$ as σ. First, differentiate to obtain

$$\mathrm{lie}_{00|0} = \sigma_{|0}(\alpha^2 - \beta^2) - \mathrm{lie}_{00}(\sigma\beta + \theta_0) - 2\beta\theta_{0|0}.$$

Next, contract the indexed form of the Basic Equation with $y^i\,\mathrm{curl}^k{}_0$ to get

$$\mathrm{curl}^j{}_0\,\mathrm{lie}_{j0} = -\beta\theta^j\,\mathrm{curl}_{j0} - (\theta_0)^2 - \sigma\beta\theta_0.$$

Return to the equation $0 = \mathrm{Irrat}$. Replace the term $^a\mathrm{Ric}_{00}$ by the right-hand side of the Ricci Curvature Equation. Then, wherever possible, insert the expressions for lie_{00}, $\mathrm{lie}_{00|0}$ and $\mathrm{curl}^j{}_0\,\mathrm{lie}_{j0}$ given by the Basic Equation. After dividing off a factor of $\alpha^2 - \beta^2$, we obtain the E_{23} Equation:

$$\mathrm{curl}^j{}_{0|j} = 2\,\mathcal{R}ic\,\beta + (n-1)\big(\tfrac{1}{8}\sigma^2\beta + \tfrac{1}{2}\sigma\theta_0 + \tfrac{1}{2}\theta^j\,\mathrm{curl}_{j0} + \tfrac{1}{4}\sigma_{|0}\big). \qquad (\mathrm{E}_{23})$$

Again, differentiating by y^i produces the indexed version

$$\mathrm{curl}^j{}_{i|j} = 2\,\mathcal{R}ic\,b_i + (n-1)\big(\tfrac{1}{8}\sigma^2 b_i + \tfrac{1}{2}\sigma\theta_i + \tfrac{1}{2}\theta^j\,\mathrm{curl}_{ji} + \tfrac{1}{4}\sigma_{|i}\big).$$

The Basic, Ricci Curvature and E_{23} Equations are all necessary conditions for the Randers metric F to be Einstein. Together, they are also sufficient. In view of Lemma 1 (page 221), we can demonstrate this by showing that they imply $\mathrm{Rat} = 0 = \mathrm{Irrat}$.

Recall that we deduced the E_{23} Equation from $\mathrm{Irrat} = 0$ by

○ expressing $^a\mathrm{Ric}_{00}$ via the Ricci Curvature Equation,
○ computing lie_{00}, $\mathrm{lie}_{00|0}$ and $\mathrm{curl}^j{}_0\,\mathrm{lie}_{j0}$ with the Basic Equation, and
○ dividing by a uniform factor of $\alpha^2 - \beta^2$.

Reversing these three algebraic steps allows us to recover Irrat $= 0$ from the E_{23} Equation.

Likewise, the Ricci Curvature Equation came from Rat $- \beta$ Irrat $= 0$ by

◦ using the Basic Equation to replace $\mathrm{lie}_{00} + 2\beta\theta_0$ with $\sigma(\alpha^2 - \beta^2)$, and
◦ dividing by $\alpha^2 - \beta^2$.

Again, reversing the two steps above will give us Rat $- \beta$ Irrat $= 0$, whence Rat $= 0$ because Irrat $= 0$.

To summarise, *the Basic Equation, the Ricci Curvature Equation and the E_{23} Equation characterise strongly convex Einstein Randers metrics.*

In the next section we will refine the three characterising equations by showing that σ must be constant.

2.2.2. Constancy of the S-curvature. In the previous section we commented that the S-curvature of any Randers metric F satisfying the Basic Equation is given by $S = \frac{1}{4}(n+1)\sigma(x)F$ [Chen–Shen 2003]. The S-curvature is positively homogeneous of degree 1 in y. In Section 1 we demonstrated a strong preference for working with objects that are positively homogeneous of degree zero in y. That is because such objects naturally live on the projectivised sphere bundle SM as well as the larger slit tangent bundle $TM \smallsetminus 0$. The compact parameter space provided by the sphere bundle is generally better suited for global and analytic considerations. So the object we are really interested in is not S, but S/F, which is homogeneous of degree zero in y. In this context, when we say that the S-curvature of any Randers metric satisfying the Basic Equation is *isotropic*, we mean that the quotient $\frac{S}{F}$ is a function of x alone. Similarly, when $\sigma(x)$ is constant, F is said to be a metric of *constant S-curvature*.

The following lemma plays a crucial role in establishing the constancy of the S-curvature for Einstein Randers metrics.

LEMMA 2. *The covariant derivative of the tensor* curl *associated to any Randers metric is given by*

$$\mathrm{curl}_{ij|k} = -2b^s \,{}^aR_{ksij} + \mathrm{lie}_{ik|j} - \mathrm{lie}_{kj|i}.$$

PROOF. Using Ricci identities and the definition of lie_{ij} we have

$$b_{i|j|k} - b_{i|k|j} = b^s \,{}^aR_{isjk},$$
$$b_{i|k|j} + b_{k|i|j} = \mathrm{lie}_{ik|j},$$
$$-b_{k|i|j} + b_{k|j|i} = -b^s \,{}^aR_{ksij},$$
$$-b_{k|j|i} - b_{j|k|i} = -\mathrm{lie}_{kj|i},$$
$$b_{j|k|i} - b_{j|i|k} = b^s \,{}^aR_{jski}.$$

Summing these five equalities and applying the first Bianchi identity produces the desired formula.   □

PROPOSITION 3. *Let F be a strongly convex Randers metric on a connected manifold, satisfying the Basic Equation (with σ a function of x) and the Ricci Curvature Equation. Then F is of constant S-curvature (i.e., σ is constant) if and only if the E_{23} Equation holds.*

In practice, the E_{23} Equation has proved to be remarkably useful. Proposition 3 shows us that such efficacy is attributable to the constancy of the S-curvature. Also, since strongly convex Einstein Randers metrics satisfy the Basic, Ricci Curvature and E_{23} Equations, the following corollary is immediate.

COROLLARY 4. *Any strongly convex Einstein Randers metric on a connected manifold is necessarily of constant S-curvature.*

PROOF OF PROPOSITION. The key is to compute a formula for the tensor $\mathrm{curl}^i{}_{0|i}$. Lemma 2 plays a pivotal role. We first contract that lemma with $a^{ik}y^j$ to obtain

$$\mathrm{curl}^i{}_{0|i} = 2b^{i\,a}\mathrm{Ric}_{i0} + \mathrm{lie}^i{}_{i|0} - \mathrm{lie}^i{}_{0|i}, \qquad (*)$$

a preliminary formula around which all further analysis is centered. Another contracted version of Lemma 2 will be needed when we calculate a certain term in $2b^{i\,a}\mathrm{Ric}_{i0}$ and $-\mathrm{lie}^i{}_{0|i}$. Before plunging into details, here is an outline:

If σ is constant, we use the Basic and Ricci Curvature Equations to finish calculating the right-hand side of $(*)$. A third contracted version of Lemma 2 will come into play. The outcome is none other than the E_{23} Equation.

Conversely, if the E_{23} Equation is presumed to hold, we immediately get one formula for $\mathrm{curl}^i{}_{0|i}$. We use the Basic, Ricci Curvature, and E_{23} Equations to finish calculating the right-hand side of $(*)$, thereby deducing a second formula for $\mathrm{curl}^i{}_{0|i}$. A comparison of the two then tells us that σ is constant.

Now for the calculations. We first reexpress the terms in the right-hand side of $(*)$. The last two terms are handled using the Basic Equation:

$$\mathrm{lie}^i{}_{i|0} = (n - \|b\|^2)\sigma_{|0} - \sigma(1 - \|b\|^2)(\sigma\beta + \theta_0),$$

$$\mathrm{lie}^i{}_{0|i} = \sigma_{|0} - \beta b^i \sigma_{|i} - \tfrac{1}{2}\sigma^2(n - 2\|b\|^2 + 1)\beta + \tfrac{1}{2}\sigma(2\|b\|^2 - n)\theta_0$$
$$+ \tfrac{1}{2}\beta\theta_i\theta^i + \tfrac{1}{2}\theta^i\,\mathrm{curl}_{i0} - \beta\theta^i{}_{|i} - b^i\theta_{0|i}.$$

The remaining term, $2b^{i\,a}\mathrm{Ric}_{i0}$, is handled by the Ricci Curvature Equation:

$$2b^{i\,a}\mathrm{Ric}_{i0} = \theta^i\,\mathrm{curl}_{i0} - (n-1)\left(\tfrac{1}{4}\|b\|^2\sigma_{|0} + \tfrac{1}{2}b^i(\theta_{i|0} + \theta_{0|i})\right)$$
$$+ \beta\left(2(1 + \|b\|^2)\,\mathcal{R}ic - \tfrac{1}{2}\,\mathrm{curl}^{ij}\,\mathrm{curl}_{ij} - (n-1)\left(\tfrac{1}{8}\sigma^2(3 - \|b\|^2) + \tfrac{1}{4}b^i\sigma_{|i}\right)\right).$$

Next we compute the quantities $b^i\theta_{i|0}$, $b^i\theta_{0|i}$, and $\theta^i{}_{|i}$ that occur in these formulae. We will use without explicit mention the equalities

$$b_{i|j} = \tfrac{1}{2}(\mathrm{lie}_{ij} + \mathrm{curl}_{ij}) \qquad \text{and} \qquad b_{i|j}\,\mathrm{curl}^{ij} = \tfrac{1}{2}\,\mathrm{curl}_{ij}\,\mathrm{curl}^{ij}.$$

For $b^i\theta_{i|0}$, notice that $b^i\theta_i = b^ib^j\,\mathrm{curl}_{ij} = 0$, because curl_{ij} is skew-symmetric. Differentiating $b^i\theta_i = 0$ and using the Basic Equation gives

$$b^i\theta_{i|0} = \tfrac{1}{2}\theta_i\theta^i\beta - \tfrac{1}{2}\sigma\theta_0 - \tfrac{1}{2}\theta^i\,\mathrm{curl}_{i0}. \qquad (**)$$

To compute $b^i \theta_{0|i}$, expand it as $\frac{1}{2} b^i (\text{lie}_{ik} - \text{curl}_{ik}) \, \text{curl}^k{}_0 + b^i b^k \, \text{curl}_{i0|k}$. By Lemma 2, $b^i b^k \, \text{curl}_{i0|k} = b^i b^k (-2 b^j \, {}^a R_{ijk0} + \text{lie}_{ik|0} + \text{lie}_{i0|k})$. However, $b^i b^j \, {}^a R_{ijk0}$ vanishes because ${}^a R$ is skew-symmetric in the first two indices. Hence

$$b^i \theta_{0|i} = \tfrac{1}{2} b^i (\text{lie}_{ik} - \text{curl}_{ik}) \, \text{curl}^k{}_0 + b^i b^k (\text{lie}_{ik|0} + \text{lie}_{i0|k}).$$

Our calculation of $b^i \theta_{0|i}$ can now be completed in three steps as follows.

○ Use the Basic Equation to remove all occurrences of the tensor lie and its covariant derivatives.
○ Replace the $b^i \theta_{i|0}$ term, which resurfaces twice, by the right-hand side of (∗∗).
○ Simplification leads to an expression of the form $(1 - \|b\|^2)(\cdots)$ for the quantity $(1 - \|b\|^2) b^i \theta_{0|i}$. Strong convexity (Section 2.1.2) allows us to divide both sides by $1 - \|b\|^2$ to obtain

$$b^i \theta_{0|i} = \tfrac{1}{2} \theta_i \theta^i \beta + \tfrac{1}{2} \sigma \theta_0 - \tfrac{1}{2} \theta^i \, \text{curl}_{i0} + \|b\|^2 \sigma_{|0} - b^i \sigma_{|i} \beta.$$

The last term of interest, $\theta^i{}_{|i}$, is computed separately for each direction of the proof. First, if we assume that F is of constant S-curvature (σ is constant), the Ricci Curvature Equation simplifies. Carrying out an appropriate trace on Lemma 2, followed by contracting with b and rearranging, we find that

$$\theta^i{}_{|i} = \tfrac{1}{2} \text{curl}_{ij} \, \text{curl}^{ij} - b^i (2 b^j \, {}^a \text{Ric}_{ij} + \text{lie}^j{}_{j|i} - \text{lie}^j{}_{i|j}).$$

Now replace the Ricci tensor by the right-hand side of the Ricci Curvature Equation, and use the Basic Equation (with σ constant) to remove all occurrences of lie and its covariant derivatives. After simplifying we find that each term contains a factor of $1 + \|b\|^2$ (not the $1 - \|b\|^2$ occurring previously). Dividing out by that factor gives

$$\theta^i{}_{|i} = \tfrac{1}{2} \text{curl}_{ij} \, \text{curl}^{ij} - \left(2 \, \mathcal{R}ic + \tfrac{1}{8} (n-1) \sigma^2 \right) \|b\|^2 + \tfrac{1}{2} (n-1) \theta_i \theta^i. \qquad (\dagger)$$

Conversely, assume that the E_{23} Equation holds. Without the constancy of σ it is not possible to compute $\theta^i{}_{|i}$ via Lemma 2. Happily, the hypothesised E_{23} Equation saves the day:

$$\theta^i{}_{|i} = (b_i \, \text{curl}^{ij})_{|j} = \tfrac{1}{2} \text{curl}^{ij} \, \text{curl}_{ij} - b^i \, \text{curl}^j{}_{i|j}$$
$$= \tfrac{1}{2} \text{curl}_{ij} \, \text{curl}^{ij} - \left(2 \, \mathcal{R}ic + \tfrac{1}{8} (n-1) \sigma^2 \right) \|b\|^2 + \tfrac{1}{2} (n-1) \theta_i \theta^i - \tfrac{1}{4} (n-1) b^i \sigma_{|i}. \; (\ddagger)$$

REMARK. This is the only place where the E_{23} Equation gets used in the proof.

We are now ready to complete the proof. Consider the expressions for $2 b^i \, {}^a \text{Ric}_{i0}$, $\text{lie}^i{}_{i|0}$, and $-\text{lie}^i{}_{0|i}$ found on the previous page. By (∗), the sum of the three is $\text{curl}^i{}_{0|i}$. Now substitute into this sum the formulae for $b^i \theta_{i|0}$, $b^i \theta_{0|i}$, and $\theta^i{}_{|i}$ just found, and simplify.

Under the hypothesis that F is of constant S-curvature, using (†) for the value of $\theta^i{}_{|i}$, we get

$$\text{curl}^i{}_{0|i} = 2 \, \mathcal{R}ic \, \beta + (n-1) \left(\tfrac{1}{8} \sigma^2 \beta + \tfrac{1}{2} \sigma \theta_0 + \tfrac{1}{2} \theta^i \, \text{curl}_{i0} \right).$$

This is the E_{23} Equation (with σ constant, i.e. $\sigma_{|i} = 0$).

If instead we assume that F satisfies the E_{23} Equation, and use ($\ddagger$) as the value of $\theta^i{}_{|i}$, we obtain

$$\mathrm{curl}^i{}_{0|i} = 2\,\mathcal{R}ic\,\beta + (n-1)\left(\tfrac{1}{8}\sigma^2\beta + \tfrac{1}{2}\sigma\theta_0 + \tfrac{1}{2}\theta^i\,\mathrm{curl}_{i0} + \sigma_{|0} - \tfrac{3}{4}\|b\|^2\sigma_{|0}\right).$$

Comparing this formula for $\mathrm{curl}^i{}_{0|i}$ with the one given by the E_{23} Equation indicates that $\tfrac{3}{4}(1-\|b\|^2)\sigma_{|0} = 0$. Since $\|b\| < 1$, we must have $\sigma_{|0} = 0$; equivalently, all covariant derivatives $\sigma_{|i}$ vanish. But σ is a function of x, so all its partial derivatives are zero. Therefore σ is constant on the connected M. □

2.2.3. Final characterisation of Einstein–Randers metrics.
In Section 2.2.1 we showed that strongly convex Einstein Randers metrics are characterised by the preliminary form of the Basic, Ricci Curvature and E_{23} Equations. The constancy of σ, established in Corollary 4, can now be used to refine these conditions to their final form.

Let's begin with the final form of the *Basic Equation*. Since the equation involves no derivatives of σ, it undergoes little cosmetic alteration. We simply write σ instead of $\sigma(x)$ to emphasise the constancy of the function:

$$\mathrm{lie}_{00} + 2\beta\theta_0 = \sigma(\alpha^2 - \beta^2).$$

Equivalently, the indexed form reads

$$\mathrm{lie}_{ik} + b_i\theta_k + b_k\theta_i = \sigma(a_{ik} - b_ib_k).$$

The final form of the *Ricci Curvature Equation* is derived in two steps. First, use the Basic Equation above to remove the tensor lie and its covariant derivatives from the preliminary expression in Section 2.2.1. Then replace the $\mathrm{curl}^j{}_{0|j}$ term with the formula given by the E_{23} Equation in Section 2.2.1. Keep in mind that the covariant derivatives $\sigma_{|i}$ vanish because σ is constant. After simplifying, the result is

$$\begin{aligned}
{}^a\mathrm{Ric}_{00} = (\alpha^2 + \beta^2)\,\mathcal{R}ic(x) &- \tfrac{1}{4}\alpha^2\,\mathrm{curl}^{hj}\,\mathrm{curl}_{hj} - \tfrac{1}{2}\mathrm{curl}^j{}_0\,\mathrm{curl}_{j0} \\
&- (n-1)\left(\tfrac{1}{16}\sigma^2(3\alpha^2 - \beta^2) + \tfrac{1}{4}(\theta_0)^2 + \tfrac{1}{2}\theta_{0|0}\right).
\end{aligned}$$

Differentiating by y^i and y^k and applying the symmetry of ${}^a\mathrm{Ric}_{ik}$ produces the indexed version

$$\begin{aligned}
{}^a\mathrm{Ric}_{ik} = (a_{ik} + b_ib_k)\,\mathcal{R}ic(x) &- \tfrac{1}{4}a_{ik}\,\mathrm{curl}^{hj}\,\mathrm{curl}_{hj} - \tfrac{1}{2}\mathrm{curl}^j{}_i\,\mathrm{curl}_{jk} \\
&- (n-1)\left(\tfrac{1}{16}\sigma^2(3a_{ik} - b_ib_k) + \tfrac{1}{4}\theta_i\theta_k + \tfrac{1}{4}(\theta_{i|k} + \theta_{k|i})\right).
\end{aligned}$$

The constancy of σ updates the E_{23} *Equation* to

$$\mathrm{curl}^j{}_{0|j} = 2\,\mathcal{R}ic(x)\,\beta + (n-1)\left(\tfrac{1}{8}\sigma^2\beta + \tfrac{1}{2}\sigma\theta_0 + \tfrac{1}{2}\theta^j\,\mathrm{curl}_{j0}\right),$$

or

$$\mathrm{curl}^j{}_{i|j} = 2\,\mathcal{R}ic(x)\,b_i + (n-1)\left(\tfrac{1}{8}\sigma^2b_i + \tfrac{1}{2}\sigma\theta_i + \tfrac{1}{2}\theta^j\,\mathrm{curl}_{ji}\right).$$

REMARK. The final forms of the Basic, E_{23} and Ricci Curvature Equations are equivalent to the preliminary forms. In the case of the Basic and E_{23} Equations, this follows immediately from the constancy of σ. As for the Ricci Curvature Equation, its final form was deduced from the preliminary form by replacing the terms lie_{00}, $\text{lie}_{00|0}$, $\text{curl}^j{}_0 \text{lie}_{j0}$ and $\text{curl}^j{}_{0|j}$ with the expressions given by the Basic and E_{23} Equations. Reversing this algebraic substitution resurrects the preliminary form of the Ricci Curvature Equation.

We saw in Section 2.2.1 that the preliminary forms characterise strongly convex Einstein metrics. Therefore the final forms of the Basic, Ricci Curvature and E_{23} Equations are necessary and sufficient conditions for the metric to be Einstein. Moreover, Proposition 3 assures us that, with σ constant, the Basic and Ricci Curvature Equations alone do the trick.

THEOREM 5 (EINSTEIN CHARACTERISATION). *Let $F = \alpha + \beta$ be a strongly convex Randers metric on a smooth manifold M of dimension $n \geqslant 2$, with $\alpha^2 = a_{ij}(x)y^i y^j$ and $\beta = b_i(x)y^i$. Then (M, F) is Einstein with Ricci scalar $\mathcal{R}ic(x)$ if and only if the Basic Equation*

$$\text{lie}_{ik} + b_i\,\theta_k + b_k\,\theta_i = \sigma(a_{ik} - b_i b_k)$$

and the Ricci Curvature Equation

$$
\begin{aligned}
{}^a\text{Ric}_{ik} = {}&(a_{ik} + b_i b_k)\,\mathcal{R}ic(x) - \tfrac{1}{4}a_{ik}\,\text{curl}^{hj}\,\text{curl}_{hj} - \tfrac{1}{2}\,\text{curl}^j{}_i\,\text{curl}_{jk} \\
&-(n-1)\big(\tfrac{1}{16}\sigma^2(3a_{ik} - b_i b_k) + \tfrac{1}{4}\theta_i\theta_k + \tfrac{1}{4}(\theta_{i|k} + \theta_{k|i})\big)
\end{aligned}
$$

are satisfied for some constant σ.

Tracing the Basic Equation tells us that σ, besides being related to the S-curvature (Section 2.2.2), also has the geometrically significant value

$$\sigma = \frac{2\,\text{div}\,b^\sharp}{n - \|b\|^2},$$

where $\text{div}\,b^\sharp := b^i{}_{|i}$ is the divergence of the vector field $b^\sharp := b^i \partial_{x^i}$.

REMARK. The Basic Equation, Ricci Curvature Equation, and E_{23} Equation are tensorial equations, and highly nonlinear due to the presence of ${}^a\text{Ric}_{ik}$. They constitute a coupled system of second order partial differential equations.

   Their redeeming feature is being polynomial in the tangent space coordinates y^i, whereas the original Einstein criterion is not (unless $b = 0$). This greatly reduces computational complexity. While testing Randers metrics to see whether they satisfy the Einstein criterion $K^i{}_i = F^2\,\mathcal{R}ic(x)$, we encountered cases in which the software Maple was unable to complete computations of $K^i{}_i$. So, for those examples we could not directly verify the Einstein criterion. Maple was able, however, to work efficiently with the three *indirect* characterising equations.

2.3. Characterising constant flag curvature Randers metrics

2.3.1. The result. Recall from Section 1.3.1 that the Ricci scalar $\mathcal{R}ic$ is the sum of $n-1$ appropriately chosen flag curvatures. Thus, Finsler metrics of constant flag curvature K necessarily have constant Ricci scalar $(n-1)K$, and are therefore Einstein. By Corollary 4, they must have constant S-curvature.

Computationally, the equation $K^i{}_k = KF^2(\delta^i{}_k - \ell^i \ell_k)$ characterizing constant flag curvature is even more challenging than the Einstein equation, which already gives Maple trouble (see end of previous section). The need for machine-friendly characterisation equations is acute.

Partly motivated by this, Randers metrics of constant flag curvature have been characterised in [Bao–Robles 2003]. The same conclusion was simultaneously obtained in [Matsumoto–Shimada 2002], albeit by a different method. The result is similar to that described in Theorem 5.

THEOREM 6 (CONSTANT FLAG CURVATURE CHARACTERISATION). *Let $F = \alpha + \beta$ be a strongly convex Randers metric on a smooth manifold M of dimension $n \geqslant 2$, with $\alpha^2 = a_{ij}(x)y^i y^j$ and $\beta = b_i(x)y^i$. Then (M, F) is of constant flag curvature K if and only if there exists a constant σ such that the Basic Equation*

$$\mathrm{lie}_{ik} + b_i\,\theta_k + b_k\,\theta_i = \sigma\,(a_{ik} - b_i b_k)$$

holds and the Riemann tensor of a satisfies the Curvature Equation

$$
\begin{aligned}
{}^a R_{hijk} = {} & \xi\,(a_{ij}\,a_{hk} - a_{ik}\,a_{hj}) - \tfrac{1}{4}a_{ij}\,\mathrm{curl}^t{}_h\,\mathrm{curl}_{tk} + \tfrac{1}{4}a_{ik}\,\mathrm{curl}^t{}_h\,\mathrm{curl}_{tj} \\
& + \tfrac{1}{4}a_{hj}\,\mathrm{curl}^t{}_i\,\mathrm{curl}_{tk} - \tfrac{1}{4}a_{hk}\,\mathrm{curl}^t{}_i\,\mathrm{curl}_{tj} \\
& - \tfrac{1}{4}\mathrm{curl}_{ij}\,\mathrm{curl}_{hk} + \tfrac{1}{4}\mathrm{curl}_{ik}\,\mathrm{curl}_{hj} + \tfrac{1}{2}\mathrm{curl}_{hi}\,\mathrm{curl}_{jk},
\end{aligned}
$$

with $\xi := (K - \tfrac{3}{16}\sigma^2) + (K + \tfrac{1}{16}\sigma^2)\|b\|^2 - \tfrac{1}{4}\theta^i\theta_i$.

2.3.2. Utility. Theorem 6 provides an *indirect* but efficient way for checking whether a given Randers metric is of constant flag curvature.

As we mentioned in Section 2.2.3, tracing the Basic Equation reveals that the constant σ, whenever it exists, is equal to $2\,\mathrm{div}\,b^\sharp/(n - \|b\|^2)$. Hence, this quotient is one of the first items we must compute. If the answer is not a constant, it is pointless to proceed any further.

If the computed value for σ is constant, surviving the Basic Equation constitutes the next checkpoint. After that, we solve the Curvature Equation for K, and see whether it is constant.

EXAMPLE (FINSLERIAN POINCARÉ DISC). This metric is implicit in [Okada 1983], and is extensively discussed in [Bao et al. 2000]. Let r and θ denote polar coordinates on the open disc of radius 2 in $\mathbb{R}^2$. The Randers metric in question is defined by the following Riemannian metric a and 1-form b:

$$
a = \frac{dr \otimes dr + r^2 d\theta \otimes d\theta}{(1 - \tfrac{1}{4}r^2)^2}, \qquad b = \frac{r\,dr}{(1 + \tfrac{1}{4}r^2)(1 - \tfrac{1}{4}r^2)}.
$$

Note that a is the Riemannian Poincaré model of constant sectional curvature -1, and $b = d \log\big((4+r^2)/(4-r^2)\big)$ is exact.

This metric is interesting because its geodesic trajectories agree with those of the Riemannian Poincaré model a. However, as shown in [Bao et al. 2000], the travel time from the boundary to the center is finite ($\log 2$ seconds), while the return trip takes infinite time! We summarise below three key steps in ascertaining that our Randers metric has constant flag curvature $K = -\frac{1}{4}$.

○ The value of the S-curvature σ is computed to be 2.
○ Since b is exact, it is closed and curl $= 0$. In particular, $\theta = 0$ and $\text{lie}_{ik} = 2b_{i|k}$. The Basic Equation is shown to hold with $\sigma = 2$.
○ Since curl $= 0$ and a has constant curvature -1, the Curvature Equation reduces to $(K + \frac{1}{4})(1 + \|b\|^2) = 0$, which gives $K = -\frac{1}{4}$.   ◇

Another byproduct of Theorem 6 is the *corrected* Yasuda–Shimada theorem, proved in [Bao–Robles 2003; Matsumoto–Shimada 2002], and discussed near the end of [Bao et al. 2003]. That theorem characterises, within the family of Randers metrics satisfying $\theta = 0$, those that have constant flag curvature K. (Those with $K > 0$ and $\theta = 0$ were *classified* in [Bejancu–Farran 2002; 2003].)

Lastly, Theorem 6 provides an important link in the complete classification of constant flag curvature Randers metrics; see [Bao et al. 2003].

2.3.3. Comparing with the Einstein case.

In Sections 1.3.2 and 2.3.1, we pointed out that Finsler metrics of constant flag curvature are necessarily Einstein. In particular, Randers metrics characterised by Theorem 6 should satisfy the criteria stipulated in Theorem 5. This is indeed the case. The two Basic Equations are identical; and tracing the Curvature Equation of Theorem 6 produces the Ricci Curvature Equation of Theorem 5.

In [Bao–Robles 2003], the characterisation result includes a third condition, called the CC(23) Equation, which gives a formula for the covariant derivative $\text{curl}_{ij|k}$ of curl. We excluded this equation from our statement of Theorem 6 because, like the E$_{23}$ Equation, it is automatically satisfied whenever the Basic and Curvature Equations hold with constant σ.

PROPOSITION 7. *Suppose F is a strongly convex Randers metric on a connected manifold, satisfying the preliminary form of the Basic and Curvature Equations described in [Bao–Robles 2003]. Then F is of constant S-curvature (σ is constant) if and only if the CC(23) Equation of [Bao–Robles 2003] holds.*

We omit the proof because it is structurally similar to the one we gave for Proposition 3. For *constant* σ, the *CC(23) Equation* reads:

$$\text{curl}_{ij|k} = a_{ik}\left((2K + \tfrac{1}{8}\sigma^2)b_j + \tfrac{1}{2}\text{curl}^h{}_j\theta_h + \tfrac{1}{2}\sigma\theta_j\right)$$
$$- a_{jk}\left((2K + \tfrac{1}{8}\sigma^2)b_i + \tfrac{1}{2}\text{curl}^h{}_i\theta_h + \tfrac{1}{2}\sigma\theta_i\right).$$

Tracing the CC(23) Equation on its first and third indices gives the E$_{23}$ Equation (with σ constant) in Section 2.2.1.

3. Randers Metrics Through Zermelo Navigation

3.1. Zermelo navigation. Zermelo [1931] posed and answered the following question (see also [Carathéodory 1999]): Consider a ship sailing on the open sea in calm waters. Suppose a mild breeze comes up. How must the ship be steered in order to reach a given destination in the shortest time?

Zermelo assumed that the open sea was $\mathbb{R}^2$ with the flat/Euclidean metric. Recently, Shen generalised the problem to the setting where the sea is an arbitrary Riemannian manifold (M, h). Shen [2002] finds that, when the wind is time-independent, the paths of shortest time are the geodesics of a Randers metric. This will be established in Section 3.1.1. For the remainder of this section, we develop some intuition by considering the problem on the infinitesimal scale.

Given any Riemannian metric h on a differentiable manifold M, denote the corresponding norm-squared of tangent vectors $y \in T_x M$ by

$$|y|^2 := h_{ij} y^i y^j = h(y, y).$$

Think of $|y|$ as measuring the *time* it takes, using an engine with a fixed power output, to travel from the base-point of the vector y to its tip. Note the symmetry property $|-y| = |y|$.

The unit tangent sphere in each $T_x M$ consists of all those tangent vectors u such that $|u| = 1$. Now introduce a vector field W such that $|W| < 1$, the spatial velocity vector of our mild wind on the Riemannian landscape (M, h). Before W sets in, a journey from the base to the tip of any u would take 1 unit of time, say, 1 second. The effect of the wind is to cause the journey to veer off course (or merely off target if u is collinear with W). Within the same 1 second, we traverse not u but the resultant $v = u + W$ instead.

As an example, suppose $|W| = \frac{1}{2}$. If u points along W (that is, $u = 2W$), then $v = \frac{3}{2}u$. Alternatively, if u points opposite to W ($u = -2W$), then $v = \frac{1}{2}u$. In these two scenarios, $|v|$ equals $\frac{3}{2}$ and $\frac{1}{2}$ instead of 1. So, with the wind present, our Riemannian metric h no longer gives the travel time along vectors. This prompts the introduction of a function F on the tangent bundle TM, to keep track of the travel time needed to traverse tangent vectors y under windy conditions. For all those resultants $v = u + W$ mentioned above, we have $F(v) = 1$. In other words, within each tangent space $T_x M$, the unit sphere of F is simply the W-translate of the unit sphere of h. Since this W-translate is no longer centrally symmetric about the origin 0 of $T_x M$, the Finsler function F cannot be Riemannian.

Given any Finsler manifold (M, F), the *indicatrix* in each tangent space is $S_x(F) := \{y \in T_x M : F(x, y) = 1\}$. The indicatrices of h and the Randers metric F with navigation data (h, W) are related by a rigid translation: $S_x(F) = S_x(h) + W_x$. In particular, the Randers indicatrix is simply an ellipse centered at the tip of W_x.

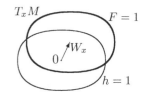

In the following section we will algebraically derive an expression for F, showing that it is a Randers metric. Then we demonstrate that the paths of shortest time are indeed the geodesics of this F.

3.1.1. Algebraic and calculus-of-variations aspects. We return to the earlier discussion and consider those $u \in T_x M$ with $|u| = 1$; equivalently, $h(u, u) = 1$. Into this, we substitute $u = v - W$ and then $h(v, W) = |v||W|\cos\theta$. Introducing the abbreviation $\lambda := 1 - |W|^2$, we have

$$|v|^2 - (2|W|\cos\theta)\,|v| - \lambda = 0.$$

Since $|W| < 1$, the resultant v is never zero, hence $|v| > 0$. This leads to $|v| = |W|\cos\theta + \sqrt{|W|^2\cos^2\theta + \lambda}$, which we abbreviate as $p+q$. Since $F(v) = 1$, we see that

$$F(v) = 1 = |v|\frac{1}{q+p} = |v|\frac{q-p}{q^2 - p^2} = \frac{\sqrt{[h(W, v)]^2 + |v|^2\lambda}}{\lambda} - \frac{h(W, v)}{\lambda}.$$

It remains to deduce $F(y)$ for an arbitrary $y \in TM$. Note that every nonzero y is expressible as a positive multiple c of some v with $F(v) = 1$. For $c > 0$, traversing $y = cv$ under the windy conditions should take c seconds. Consequently, F is positively homogeneous: $F(y) = cF(v)$. Using this homogeneity and the formula derived for $F(v)$, we find that

$$F(y) = \frac{\sqrt{[h(W, y)]^2 + |y|^2\lambda}}{\lambda} - \frac{h(W, y)}{\lambda}.$$

Here, $F(y)$ abbreviates $F(x, y)$; the basepoint x has been suppressed temporarily.

As promised, F is a Randers metric. Namely, it has the form $F(x, y) = \sqrt{a_{ij}(x)y^i y^j} + b_i(x)y^i$, where a is a Riemannian metric and b a differential 1-form. Explicitly,

$$a_{ij} = \frac{h_{ij}}{\lambda} + \frac{W_i}{\lambda}\frac{W_j}{\lambda}, \qquad b_i = \frac{-W_i}{\lambda}.$$

Here $W_i := h_{ij}W^j$ and $\lambda = 1 - W^i W_i$. In particular, there is a canonical Randers metric associated to each Zermelo navigation problem with data (h, W). Incidentally, the inverse of a is given by

$$a^{ij} = \lambda(h^{ij} - W^i W^j), \quad \text{and} \quad b^i := a^{ij}b_j = -\lambda W^i.$$

Under the influence of W, the most efficient navigational paths are no longer the geodesics of the Riemannian metric h; instead, they are the geodesics of the Finsler metric F. To see this, let $x(t)$, for $t \in [0, \tau]$, be a curve in M from point p to point q. Return to our imaginary ship sailing about M, with velocity vector u but not necessarily with constant speed. If the ship is to travel along the curve $x(t)$ while the wind blows, the captain must continually adjust the ship's direction $u/|u|$ so that the resultant $u + W$ is tangent to $x(t)$. The travel time along any infinitesimal segment $\dot{x}\,dt$ of the curve is $F(x, \dot{x}\,dt) = F(x, \dot{x})\,dt$, because as explained above it is the positively homogeneous F (not h) that keeps track of travel times. The captain's task is to select a path $x(t)$ from p to q that minimises the total travel time

$$\int_0^\tau F(x, \dot{x})\, dt.$$

This quantity is independent of orientation-preserving parametrisations due to the positive homogeneity of F and the change-of-variables theorem. Such an efficient path is precisely a geodesic of the Finsler metric F, which is said to have solved Zermelo's problem of navigation under the external influence W.

Let's look at some 2-dimensional examples. Being in two dimensions, we revert to the common notation of denoting position coordinates by x, y rather than x^1, x^2, and components of tangent vectors by u, v rather than y^1, y^2.

EXAMPLE (MINKOWSKI SPACE). Consider $\mathbb{R}^2$ equipped with the Euclidean metric $h_{ij} = \delta_{ij}$. Let $W = (p, q)$ be a constant vector field. The resulting Randers metric is of Minkowski type, and is given by

$$F(x, y; u, v) = \frac{\sqrt{(pu+qv)^2 + (u^2+v^2)\big(1 - (p^2+q^2)\big)}}{1 - (p^2+q^2)} + \frac{-(pu+qv)}{1 - (p^2+q^2)}.$$

The condition $|W|^2 = p^2 + q^2 < 1$ ensures that F is strongly convex. The geodesics of both h and F are straight lines. Thus, when navigating on flat-land under the influence of a constant wind, the correct strategy is to steer the ship so that it travels along a straight line. This means the captain should aim the ship, not straight toward the desired destination, but slightly off course with a velocity V, selected so that $V + W$ points at the destination.   ◇

EXAMPLE (SHEN'S FISH POND). Fish are kept in a pond with a rotational current. The pond occupies the unit disc in $\mathbb{R}^2$, and h_{ij} is again the Euclidean metric δ_{ij}. The current's velocity field is $W := -y\partial_x + x\partial_y$, which describes a counterclockwise circulation of angular speed 1. These navigation data give rise to the Randers metric [Shen 2002]

$$F(x, y; u, v) = \frac{\sqrt{(-yu+xv)^2 + (u^2 + v^2)(1 - x^2 - y^2)}}{1 - x^2 - y^2} + \frac{-yu + xv}{1 - x^2 - y^2},$$

with $|W|^2 = x^2 + y^2 < 1$. A feeding station is placed at a fixed location along the perimeter of the pond. The fish, eager to get to the food, swim along geodesics of F. As observed by a stationary viewer at the pond's edge, these geodesics are spirals because it is the fish's instinct to approach the perimeter quickly, where it can obtain the most help from the current's linear velocity (which equals angular speed times radial distance). An experienced fish will aim itself with a velocity V, selected so that $V + W$ is tangent to the spiral.   ◇

For the pond described above, Shen has found numerically that the geodesics of F are spirals as expected. Now suppose that instead of fish we have an excited octopus which secretes ink as it swims toward the feeding station. The resulting ink trail comoves with the swirling water. Its exact shape can be deduced by reexpressing the above spiral with respect to a frame which is comoving with the water. Shen agrees with this trend of thought and finds, somewhat surprisingly, that the trail in question is a straight ray.

fish pond	ink trail	geodesic in space
counterclockwise current	food	food
no current	food	food

3.1.2. Inverse problem.

A question naturally arises: Can every strongly convex Randers metric be realised through the perturbation of some Riemannian metric h by some vector field W satisfying $|W| < 1$?

The answer is yes. Indeed, let us be given an arbitrary Randers metric F with Riemannian metric a and differential 1-form b, satisfying $\|b\|^2 := a^{ij} b_i b_j < 1$. Set $b^i := a^{ij} b_j$ and $\varepsilon := 1 - \|b\|^2$. Construct h and W as follows:

$$h_{ij} := \varepsilon (a_{ij} - b_i b_j), \quad W^i := -b^i / \varepsilon.$$

Note that F is Riemannian if and only if $W = 0$, in which case $h = a$. Also,

$$W_i := h_{ij} W^j = -\varepsilon b_i.$$

Using this, it can be directly checked that perturbing h by the stipulated W gives back the Randers metric we started with. Furthermore,

$$|W|^2 := h_{ij} W^i W^j = a^{ij} b_i b_j =: \|b\|^2 < 1.$$

Incidentally, the inverse of h_{ij} is

$$h^{ij} = \varepsilon^{-1} a^{ij} + \varepsilon^{-2} b^i b^j.$$

This h^{ij}, together with W^i, defines a Cartan metric F^* of Randers type on the cotangent bundle T^*M. A comparison with [Hrimiuc–Shimada 1996] shows that F^* is the Legendre dual of the Finsler–Randers metric F on TM. It is remarkable that the Zermelo navigation data of any strongly convex Randers metric F is so simply related to its Legendre dual. See also [Ziller 1982; Shen 2002; 2004].

We summarise:

PROPOSITION 8. *A strongly convex Finsler metric F is of Randers type if and only if it solves the Zermelo navigation problem on some Riemannian manifold (M, h), under the influence of a wind W with $h(W, W) < 1$. Also, F is Riemannian if and only if $W = 0$.*

EXAMPLE (FINSLERIAN POINCARÉ DISC). As an illustration of the inverse procedure, we apply it to the Randers metric F that describes the Finslerian Poincaré disc (Section 2.3.2). With r and θ denoting polar coordinates on the open disc of radius 2 in $\mathbb{R}^2$, the Randers metric in question is defined by

$$a = \frac{dr \otimes dr + r^2 d\theta \otimes d\theta}{(1 - \frac{1}{4}r^2)^2}, \quad b = \frac{r \, dr}{(1 + \frac{1}{4}r^2)(1 - \frac{1}{4}r^2)}.$$

The underlying Zermelo navigation data is

$$h = \frac{(1 - \frac{1}{4}r^2)^2 \, dr \otimes dr + r^2(1 + \frac{1}{4}r^2)^2 \, d\theta \otimes d\theta}{(1 + \frac{1}{4}r^2)^4}, \quad W = \frac{-r(1 + \frac{1}{4}r^2)}{1 - \frac{1}{4}r^2} \partial_r.$$

It turns out that this Riemannian landscape h on which the navigation takes place is flat (!), and the associated 1-form

$$W^\flat = -r \frac{1 - \frac{1}{4}r^2}{1 + \frac{1}{4}r^2} \, dr = -8 \, d\frac{r^2}{(4 + r^2)^2}$$

is exact. The wind here is blowing radially toward the center of the disc, and its strength decreases to zero there.

The geodesics of h are straight lines. Those of F have been analysed in detail in [Bao et al. 2000]. Their *trajectories* coincide with those of the Riemannian Poincaré model: straight rays to and from the origin, and circular arcs intersecting the rim of the disc at Euclidean right angles. ◇

3.1.3. General relation between two covariant derivatives.

Our goal in Section 3.2 will be to reexpress the Einstein characterisation of Theorem 5 in terms of the navigation data (h, W). To that end, it is helpful to first relate the covariant derivative $b_{i|j}$ of b (with respect to a) to the covariant derivative $W_{i:j}$ of W (with respect to h). Let ${}^a\gamma^i{}_{ij}$ and ${}^h\gamma^i{}_{jk}$ denote the respective Christoffel symbols of the Riemannian metrics a and h. We have

$$b_{i|j} = b_{i,x^j} - b_s \, {}^a\gamma^s{}_{ij} \quad \text{and} \quad W_{i:j} = W_{i,x^j} - W_s \, {}^h\gamma^s{}_{ij}.$$

Since

$${}^a\gamma^i{}_{ij} = ({}^aG^i)_{y^j y^k}, \quad {}^h\gamma^i{}_{ij} = ({}^hG^i)_{y^j y^k},$$

it suffices to compare the geodesic spray coefficients

$${}^aG^i := \tfrac{1}{2} {}^a\gamma^i{}_{jk} y^j y^k =: \tfrac{1}{2} {}^a\gamma^i{}_{00} \quad \text{and} \quad {}^hG^i := \tfrac{1}{2} {}^h\gamma^i{}_{jk} y^j y^k =: \tfrac{1}{2} {}^h\gamma^i{}_{00}.$$

The tool that effects this comparison is Rapcsák's identity (Section 1.2.2).

The symbol α denotes the Finsler norm associated to the Riemannian metric a: $\alpha^2(x,y) := a_{ij}(x)y^i y^j = a(y,y)$. According to Section 1.2.2, the user-friendly form of Rapcsák's identity gives

$$^a G^i = {}^h G^i + \tfrac{1}{4} a^{ij}\big((\alpha^2{}_{:0})_{y^j} - 2\alpha^2{}_{:j}\big).$$

Using the formula $a_{ij} = \tfrac{1}{\lambda} h_{ij} + \tfrac{1}{\lambda^2} W_i W_j$ derived in Section 1.3.1, we find that

$$a^{ij} = \lambda(h^{ij} - W^i W^j) \quad\text{and}\quad \alpha^2 = \frac{1}{\lambda} h_{00} + \frac{1}{\lambda^2} W_0{}^2.$$

Here, $W_i := h_{ij} W^j$, $\lambda := 1 - W^i W_i$, and the formula for the inverse a^{ij} is ascertained by inspection. A straightforward but tedious computation of the right-hand side of Rapcsák's identity yields

$$^a G^i = {}^h G^i + \zeta^i,$$

where (using $W^{s:i}$ to abbreviate $W^s{}_{:r} h^{ri}$)

$$\zeta^i := \frac{1}{\lambda} y^i W^s W_{s:0} + \frac{1}{2} W^i W_{0:0} + \left(\frac{1}{2\lambda} h_{00} + \frac{1}{\lambda^2} W_0{}^2\right)(W^i W^s W^t W_{s:t} - W_s W^{s:i})$$
$$+ \frac{1}{2\lambda} W_0\big(W^i W^s(W_{s:0} + W_{0:s}) + (W^i{}_{:0} - W_0{}^{:i})\big).$$

Differentiating with respect to y^j, y^k gives $^a\gamma^i{}_{jk} = {}^h\gamma^i{}_{jk} + (\zeta^i)_{y^j y^k}$, whence $b_{j|k} = b_{j:k} - b_i(\zeta^i)_{y^j y^k}$. Into the right-hand side we substitute $b_s = -(1/\lambda)W_s$ (Section 1.3.1), resulting in a formula for $b_{j|k}$ in terms of the covariant derivatives of W with respect to h. For later purposes, it is best to split the answer into its symmetric and anti-symmetric parts,

$$b_{j|k} = \tfrac{1}{2}(b_{j|k} + b_{k|j}) + \tfrac{1}{2}(b_{j|k} - b_{k|j}) =: \tfrac{1}{2}\,\text{lie}_{jk} + \tfrac{1}{2}\,\text{curl}_{jk},$$

and to introduce the abbreviations

$$\mathcal{L}_{jk} := W_{j:k} + W_{k:j}, \quad \mathcal{C}_{jk} := W_{j:k} - W_{k:j}.$$

Then the said computation gives $b_{j|k} = \tfrac{1}{2}\,\text{lie}_{jk} + \tfrac{1}{2}\,\text{curl}_{jk}$, with

$$\text{lie}_{jk} = -\mathcal{L}_{jk} - \left(\frac{1}{\lambda} h_{jk} + \frac{2}{\lambda^2} W_j W_k\right) W^s W^t \mathcal{L}_{st}$$
$$+ \frac{1 + |W|^2}{\lambda^2} W^i(W_{i:j} W_k + W_{i:k} W_j) - \frac{1}{\lambda}(W_j W_{k:i} + W_k W_{j:i}) W^i,$$

$$\text{curl}_{jk} = -\frac{1}{\lambda}\mathcal{C}_{jk} + \frac{2}{\lambda^2} W^i(W_{i:j} W_k - W_{i:k} W_j).$$

Observe that since $\text{curl}_{jk} = \partial_{x^k} b_j - \partial_{x^j} b_k$ and $\text{lie}_{jk} = b^i \partial_{x^i} a_{jk} + a_{ik}\partial_{x^j} b^i + a_{ji}\partial_{x^k} b^i$ (where $b^i = -\lambda W^i$), the last two conclusions could have been obtained without relying on the explicit formula of $^a\gamma$. In any case, the relation above between $b_{j|k}$ and the covariant derivatives of W is valid *without* any assumption on b.

3.2. Navigation description of curvature conditions. Theorem 5 (p. 227) characterises Einstein Randers metrics via the defining Riemannian metric a and 1-form b. It says that a Randers metric F is Einstein if and only if both the Basic and Ricci Curvature Equations hold with constant σ (or S-curvature). Though this is a substantial improvement over the Einstein criterion $K^i{}_i = \mathcal{R}ic(x)F^2$, most notably in the realm of computation, the characterisation does little to describe the *geometry* of these metrics. Surprisingly, the breakthrough lies in a change of *dependent* variables. We find that replacing the defining data (a, b) by the navigation data (h, W) (discussed in Section 3.1.2) yields a breviloquent geometric description of Einstein Randers metrics. Explicitly, this change of variables reveals that the Riemannian metric h must be an Einstein metric itself, and the vector field W an infinitesimal homothety of h. The next two subsections are devoted to developing this "navigation description".

3.2.1. Consequences of the Basic Equation. Our first step is to derive the navigation version of the Basic Equation $\text{lie}_{ik} + b_i \theta_k + b_k \theta_i = \sigma(a_{ik} - b_i b_k)$ of page 226. For that we replace a_{ik} by $(1/\lambda)h_{ik} + (1/\lambda^2)W_i W_k$, b_i by $-(1/\lambda)W_i$, and lie_{ik} by the expression derived on the previous page. We also use the formula for curl_{jk} there to compute $\theta_k := b^j \,\text{curl}_{jk}$. Since $b^j := a^{js}b_s = -\lambda W^j$, we get

$$\theta_k = \mathcal{T}_k - \frac{1}{\lambda}(W^i W^j \mathcal{L}_{ij})W_k + \frac{2}{\lambda}|W|^2 W^i W_{i:k}.$$

Here, an abbreviation has been introduced for the ubiquitous quantity

$$\mathcal{T}_k := W^j(W_{j:k} - W_{k:j}) = W^j \mathcal{C}_{jk}.$$

These manoeuvres, followed by some rearranging, convert the Basic Equation to $\lambda \mathcal{L}_{ik} + \mathcal{L}(W, W)h_{ik} = -\sigma h_{ik}$, where $\mathcal{L}(W, W)$ stands for $\mathcal{L}_{st}W^s W^t$. Contracting with W^i, W^k and using $\lambda := 1 - |W|^2$ shows that $\mathcal{L}(W, W) = -\sigma|W|^2$. Consequently, the navigation version of the Basic Equation is

$$\mathcal{L}_{ik} := W_{i:k} + W_{k:i} = -\sigma h_{ik}, \quad \text{that is,} \quad \mathcal{L}_W h = -\sigma h.$$

We name this the $\mathcal{L}_W$ *Equation*. It says that W is an *infinitesimal homothety* of h; see [Kobayashi–Nomizu 1996, p. 309]. In this equation,

$$\sigma \text{ must be zero whenever } h \text{ is not flat.}$$

(In particular, σ must vanish whenever h is not Ricci-flat.) Indeed, let φ_t denote the time t flow of the vector field W. The $\mathcal{L}_W$ Equation tells us that $\varphi_t^* h = e^{-\sigma t}h$. Since φ_t is a diffeomorphism, $e^{-\sigma t}h$ and h must be isometric; therefore they have the same sectional curvatures. If h is not flat, this condition on sectional curvatures mandates that $e^{-\sigma t} = 1$, hence $\sigma = 0$. The argument we presented was pointed out to us by Bryant.

Incidentally, the use of $h = \varepsilon(a - b \otimes b)$ and $W = -b^\sharp/\varepsilon$ (Section 3.1.2) allows us to recover the Basic Equation from the $\mathcal{L}_W$ Equation. Thus the two equations are equivalent. This remains so even if σ were to be a function of x, because neither equation contains any derivative of σ.

We now turn to the derivation of the navigation versions of the Ricci Curvature Equation (Theorem 5, page 227) and the Curvature Equation (Theorem 6, page 228). The $\mathcal{L}_W$ Equation affords simplified expressions for many quantities that enter into the curvature equations of Theorems 5 and 6. The key in all such simplifications can invariably be traced back to the statement

$$W_{i:j} = \tfrac{1}{2}(\mathcal{L}_{ij} + \mathcal{C}_{ij}) = -\tfrac{1}{2}\sigma h_{ij} + \tfrac{1}{2}\mathcal{C}_{ij}.$$

We first address all but one of the terms on the right-hand sides of the curvature equations. Keep in mind that indices on curl, θ, and $^a y_i := a_{ij} y^j$ are manipulated by the Riemannian metric a, while those on $\mathcal{C}$, $\mathcal{T}$, and $^h y_i := h_{ij} y^j$ are manipulated by the Riemannian metric h. The relevant formulae are

$$\text{curl}_{ij} = -\frac{1}{\lambda}\mathcal{C}_{ij} + \frac{1}{\lambda^2}(\mathcal{T}_i W_j - \mathcal{T}_j W_i), \quad {}^a y_i = \frac{1}{\lambda}{}^h y_i + \frac{1}{\lambda^2} W_i W_0,$$

$$\text{curl}^i{}_j = -\mathcal{C}^i{}_j + \frac{1}{\lambda}\mathcal{T}^i W_j, \qquad\qquad \theta_i = \frac{1}{\lambda}\mathcal{T}_i,$$

$$\text{curl}^{ij} = -\lambda\mathcal{C}^{ij}; \qquad\qquad\qquad\qquad \theta^i = \mathcal{T}^i.$$

This does not address the term $\theta_{0|0}$ (equivalently $\theta_{i|k} + \theta_{k|i}$), which appears on the right-hand side of the Ricci Curvature Equation (Section 2.2.3). To tackle this, as well as the left-hand sides of those curvature equations, we shall need the relation between the geodesic spray coefficients $^a G^i$ and $^h G^i$. That relation, first derived in Section 3.1.3, undergoes a dramatic simplification in the presence of the $\mathcal{L}_W$ Equation. The result is

$$^a G^i = {}^h G^i + \zeta^i,$$

where

$$\zeta^i = y^i \frac{1}{2\lambda}(\mathcal{T}_0 - \sigma W_0) - \mathcal{T}^i \left(\frac{1}{4\lambda} h_{00} + \frac{1}{2\lambda^2} W_0{}^2\right) + \frac{1}{2\lambda}\mathcal{C}^i{}_0 W_0.$$

Hence

$$\theta_{0|0} = \theta_{0:0} - 2\theta_i \zeta^i = \frac{1}{\lambda}\mathcal{T}_{0:0} - \frac{1}{\lambda^2} W_0 \mathcal{T}^i \mathcal{C}_{i0} + \left(\frac{1}{2\lambda^2} h_{00} + \frac{1}{\lambda^3} W_0{}^2\right)\mathcal{T}^i \mathcal{T}_i.$$

Differentiating these two statements with respect to y^j, y^k gives an explicit relation between the Christoffel symbols $^a\gamma^i{}_{jk}$ and $^h\gamma^i{}_{jk}$, as well as a formula for $\theta_{j|k} + \theta_{k|j}$ in terms of the navigation data (h, W). However, we refrain from doing so. In the following two subsections, we shall determine the navigation version of the Ricci Curvature Equation in Theorem 5 and of the Curvature Equation in Theorem 6. It is found, in retrospect, that the computational tedium is significantly lessened by working with $^a\text{Ric}_{00}$, $^a K^i{}_k$ rather than $^a\text{Ric}_{ij}$, $^a R_{hijk}$. Consequently, the relation $^a G^i = {}^h G^i + \zeta^i$ and the formula for $\theta_{0|0}$ should suffice.

3.2.2. Einstein–Randers metrics. The contracted form of the Ricci Curvature Equation (Section 2.2.3) reads

$$^a\mathrm{Ric}_{00} = (\alpha^2 + \beta^2)\,\mathcal{R}ic(x) - \tfrac{1}{4}\alpha^2\,\mathrm{curl}^{hj}\,\mathrm{curl}_{hj} - \tfrac{1}{2}\,\mathrm{curl}^j{}_0\,\mathrm{curl}_{j0}$$
$$-(n-1)\big(\tfrac{1}{16}\sigma^2(3\alpha^2 - \beta^2) + \tfrac{1}{4}(\theta_0)^2 + \tfrac{1}{2}\theta_{0|0}\big).$$

With $\alpha^2 = a_{00} = (1/\lambda)\,h_{00} + (1/\lambda^2)W_0{}^2$ and $\beta = b_0 = -(1/\lambda)W_0$, the simplified formulae in Section 3.2.1, and $\mathcal{R}ic(x) =: (n-1)K(x)$, all the terms on the right-hand side are accounted for.

For the left-hand side, note first that $^a\mathrm{Ric}_{00} = {}^aK^i{}_i$. Specialising the split and covariantised form of Berwald's formula (Section 1.2.3) to $F = a$ and $\mathcal{F} = h$, and taking the natural trace, we have

$$^a\mathrm{Ric}_{00} = {}^h\mathrm{Ric}_{00} + (2\zeta^i)_{:i} - (\zeta^i)_{y^j}(\zeta^j)_{y^i} - y^j(\zeta^i{}_{:j})_{y^i} + 2\zeta^j(\zeta^i)_{y^j y^i}.$$

Though ζ^i has a simplified formula (Section 3.2.1), computing the four terms dependent on ζ is still tedious. That task is helped by the $\mathcal{L}_W$ Equation and the navigation description [Robles 2003] of the E_{23} Equation:

$$W^i{}_{:0:i} = (n-1)\big(K(x) + \tfrac{1}{16}\sigma^2\big)W_0.$$

The result is unexpectedly elegant:

$$^h\mathrm{Ric}_{00} = (n-1)\big(K(x) + \tfrac{1}{16}\sigma^2\big)h_{00}.$$

Differentiating away the contracted y^i, y^k gives $^h\mathrm{Ric}_{ik}$.

Conversely, it has been checked that, via $h = \varepsilon(a - b \otimes b)$ and $W = -b^\sharp/\varepsilon$ (Section 3.1.2), the above navigation description reproduces the characterisation in Theorem 5. Thus the characterisation (in terms of a, b) is equivalent to the navigation description (in terms of h, W). In particular, Theorem 5 implies:

THEOREM 9 (EINSTEIN NAVIGATION DESCRIPTION). *Suppose the Randers metric F solves Zermelo's problem of navigation on the Riemannian manifold (M, h) under the external influence W, $|W| < 1$. Then (M, F) is Einstein with Ricci scalar $\mathcal{R}ic(x) =: (n-1)K(x)$ if and only if there exists a constant σ such that*

(i) *h is Einstein with Ricci scalar $(n-1)\big(K(x) + \tfrac{1}{16}\sigma^2\big)$, that is,*

$$^h\mathrm{Ric}_{ik} = (n-1)\big(K(x) + \tfrac{1}{16}\sigma^2\big)h_{ik},$$

   and

(ii) *W is an infinitesimal homothety of h, namely,*

$$(\mathcal{L}_W h)_{ik} = W_{i:k} + W_{k:i} = -\sigma h_{ik}.$$

Furthermore, σ must vanish whenever h is not Ricci-flat.

We call this a 'description' rather than a 'characterisation' because, in contrast with Theorem 5, it makes explicit the underlying geometry of Einstein–Randers metrics. Section 4 will illustrate Theorem 9 with a plethora of examples.

3.2.3. Constant flag curvature Randers metrics. For Randers metrics of constant flag curvature, the equation destined to be recast into navigational form is given in Theorem 6. Contracting it with y^h, y^k, raising the index i with a, and relabelling j as k, we obtain the following expression for the spray curvature $^aK^i{}_k \; (= y^j \, ^aR_j{}^i{}_{kl} y^l$; Section 1.2.3):

$$^aK^i{}_k = \left((K - \tfrac{3}{16}\sigma^2) + (K + \tfrac{1}{16}\sigma^2) b^s b_s - \tfrac{1}{4}\theta^s \theta_s \right) (\delta^i{}_k \alpha^2 - y^i \, ^a y_k)$$
$$+ \tfrac{1}{4}\operatorname{curl}^s{}_0 (\operatorname{curl}_s{}^i \, ^a y_k + y^i \operatorname{curl}_{sk} - \operatorname{curl}_{s0} \delta^i{}_k) - \tfrac{1}{4}\alpha^2 \operatorname{curl}^{si} \operatorname{curl}_{sk} - \tfrac{3}{4}\operatorname{curl}^i{}_0 \operatorname{curl}_{k0}.$$

Here $^a y_k := a_{ik} y^i$. This is equivalent to the Curvature Equation because

$$^aR_{hijk} = \tfrac{1}{3}\left((^aK_{ij})_{y^k y^h} - (^aK_{ik})_{y^j y^h} \right). \tag{$*$}$$

We now recast the equality just given for $^aK^i{}_k$. All the terms on the right-hand side are routinely computed, using $\alpha^2 = a_{00} = (1/\lambda) h_{00} + (1/\lambda^2) W_0{}^2$, $b_s = -(1/\lambda) W_s$, $b^s = -\lambda W^s$, and the simplified formulae in Section 3.2.1.

For the left-hand side, we first specialise the split and covariantised form of Berwald's formula (Section 1.2.3) to $F = a$ and $\mathcal{F} = h$, getting

$$^aK^i{}_k = \,^hK^i{}_k + (2\zeta^i)_{:k} - (\zeta^i)_{y^s}(\zeta^s)_{y^k} - y^s(\zeta^i{}_{:s})_{y^k} + 2\zeta^s(\zeta^i)_{y^s y^k}.$$

Into this we substitute the simplified formula for ζ (Section 3.2.1). The ensuing computation is assisted by the prodigious use of the $\mathcal{L}_W$ Equation and the navigation description of the CC(23) Equation:

$$W_{i:j:k} = \left(K + \tfrac{1}{16}\sigma^2\right)(h_{ik} W_j - h_{jk} W_i).$$

The result is as elegant as the Einstein case:

$$^hK^i{}_k = \left(K + \tfrac{1}{16}\sigma^2\right)(\delta^i{}_k h_{00} - y^i \, ^h y_k),$$

where $^h y_k := h_{ik} y^i$. Lowering the index i with the Riemannian metric h and differentiating in the same fashion as formula $(*)$ gives $^hR_{hijk}$.

We have verified that the use of $h = \varepsilon(a - b \otimes b)$ and $W = -b^\sharp/\varepsilon$ (Section 3.1.2) converts the above navigation description in terms of h, W back to the characterisation in terms of a, b presented by Theorem 6. So the two pictures are indeed equivalent, and Theorem 6 implies:

THEOREM 10 (CONSTANT FLAG CURVATURE NAVIGATION DESCRIPTION). *Suppose the Randers metric F solves Zermelo's problem of navigation on the Riemannian manifold (M, h) under the external influence W, $|W| < 1$. Then (M, F) is of constant flag curvature K if and only if there exists a constant σ such that*

(i) *h is of constant sectional curvature $(K + \tfrac{1}{16}\sigma^2)$, that is,*

$$^hR_{hijk} = \left(K + \tfrac{1}{16}\sigma^2\right)(h_{ij} h_{hk} - h_{ik} h_{hj}),$$

and

(ii) *W is an infinitesimal homothety of h, namely,*

$$(\mathcal{L}_W h)_{ik} = W_{i:k} + W_{k:i} = -\sigma h_{ik}.$$

Furthermore, σ must vanish whenever h is not flat.

EXAMPLES. We present three favorite examples to illustrate the use of Theorem 10. In the examples below: position coordinates are denoted x, y, z rather than x^1, x^2, x^3; components of tangent vectors are u, v, w instead of y^1, y^2, y^3.

FUNK DISC.   The Finslerian Poincaré example was discussed in Section 2.3.2 and Section 3.1.2. Here we revisit it a third time, for the sake of those who prefer to work with simple navigation data.

Fix the angle θ and contract the radius via $r \mapsto r/(1 + \frac{1}{4}r^2)$. This map is an isometry of the Finslerian Poincaré model onto the Funk metric of the unit disc [Funk 1929; Okada 1983; Shen 2001]. The navigation data of the Funk metric is simple: h is the Euclidean metric and the radial $W = -r\partial_r$ is an infinitesimal homothety with $\sigma = 2$. Writing tangent vectors at (r, θ) as $u\partial_r + v\partial_\theta$, we have

$$F = \frac{\sqrt{u^2 + r^2(1 - r^2)v^2}}{1 - r^2} + \frac{ru}{1 - r^2}, \quad \text{with } r^2 = x^2 + y^2.$$

By Theorem 10, $K + \frac{1}{16}\sigma^2 = 0$. Hence the Funk metric on the unit disc has constant flag curvature $K = -\frac{1}{4}$.

The isometry above is a global change of coordinates which transforms the navigation data in the Section 3.1.2 example into a more computationally friendly format.   ◇

A 3-SPHERE THAT IS NOT PROJECTIVELY FLAT. We start with the unit sphere S^3 in $\mathbb{R}^4$, parametrised by its tangent spaces at the poles, as in [Bao–Shen 2002]. For each constant $K > 1$, let h be $1/K$ times the standard Riemannian metric induced on S^3. The rescaled metric has sectional curvature K. Perturb h by the Killing vector field

$$W = \sqrt{K-1}\left(-s(1+x^2), z - sxy, -y - sxz\right),$$

with $s = \pm 1$ depending on the hemisphere. Then $|W| = \sqrt{(K-1)/K}$ and W is tangent to the S^1 fibers in the Hopf fibration of S^3. By Theorem 10, the resulting Randers metric F has constant flag curvature K. Explicitly, $F = \alpha + \beta$, where

$$\alpha = \frac{\sqrt{K(su - zv + yw)^2 + (zu + sv - xw)^2 + (-yu + xv + sw)^2}}{1 + x^2 + y^2 + z^2},$$

$$\beta = \frac{\sqrt{K-1}(su - zv + yw)}{1 + x^2 + y^2 + z^2}.$$

This Randers metric is not projectively flat [Bao–Shen 2002]. This is in stark contrast with the Riemannian case because, according to Beltrami's theorem, a Riemannian metric is locally projectively flat if and only if it is of constant sectional curvature.   ◇

SHEN'S FISH TANK. This example, first presented in [Shen 2002], is a three-dimensional variant of Shen's fish pond (Section 3.1.1). Consider a cylindrical fish tank $x^2 + y^2 < 1$ in $\mathbb{R}^3$, equipped with the standard Euclidean metric h. Suppose the tank has a rotational current with velocity vector $W = y\partial_x - x\partial_y + 0\partial_z$, and a big inquisitive mosquito hovers just above the water. Wishing to reach the bug as soon as possible, the hungry fish swim along a path of shortest time — that is, along a geodesic of the Randers metric $F = \alpha + \beta$ with Zermelo navigation data h and the infinitesimal rotation W. Explicitly,

$$\alpha = \frac{\sqrt{(-yu + xv)^2 + (u^2 + v^2 + w^2)(1 - x^2 - y^2)}}{1 - x^2 - y^2},$$

$$\beta = \frac{-yu + xv}{1 - x^2 - y^2}, \quad \text{with} \quad |W|^2 = x^2 + y^2.$$

Since W is a Killing field of Euclidean space we have $\sigma = 0$, and Theorem 10 tells us that F is of constant flag curvature $K = 0$. The same conclusion holds for the fish pond.   ◇

Theorem 10 implies that every constant flag curvature Randers metric is locally isometric to a "standard model" with navigation data (h, W), where h is a standard Riemannian space form (sphere, Euclidean space, or hyperbolic space), and W is one of its infinitesimal homotheties. It remains to sort these standard models into Finslerian isometry classes.

Let (M_1, F_1), (M_2, F_2) be any two Randers spaces, with navigation data (h_1, W_1) and (h_2, W_2). It is a fact that F_1, F_2 are isometric as Finsler metrics if and only if there exists a Riemannian isometry $\varphi : (M_1, h_1) \to (M_2, h_2)$ such that $\varphi_* W_1 = W_2$. For each standard Riemannian space form (M, h), the isometry group G of h leaves h invariant, but acts on its infinitesimal homotheties W via push-forward. By the cited fact, all (h, W) which lie on the same G-orbit generate mutually isometric standard models. This redundancy can be suppressed by collapsing each G-orbit to a point. For any fixed K, the resulting collection of such "points" constitutes the *moduli space* $\mathcal{M}_K$ for strongly convex Randers metrics of constant flag curvature K. Lie theory effects (a parametrisation and hence) a dimension count of $\mathcal{M}_K$; see [Bao et al. 2003] for details. The table on the next page includes, for comparison, similar information about the Riemannian setting and the case $\theta := \text{curl}(b^\sharp, \cdot) = 0$ (Section 2.3.2).

3.3. Issues resolved by the navigation description

3.3.1. Schur lemma for the Ricci scalar. In essence, this lemma constrains the geometry of Einstein metrics in dimension ≥ 3 by forcing the Ricci scalar to be constant. Historically, this is the second Schur lemma in (non-Riemannian) Finsler geometry. The first Finslerian Schur lemma concerns the flag curvature; see [del Riego 1973; Matsumoto 1986; Berwald 1947]. An exposition can be found in [Bao et al. 2000].

CFC metrics	dim M	Dimension of moduli space			
		$K > 0$	$K = 0$	$K < 0$	
				$\sigma = 0$	$\sigma \neq 0$
Riemannian b equiv. $W = 0$	$n \geqslant 2$	0			empty
Yasuda–Shimada $\theta = 0$	even n	0^*	1	0^*	$0^\dagger$
	odd n	1			
Unrestricted Randers	even n	$n/2$			
	odd n	$(n+1)/2$			$(n-1)/2$

The moduli spaces of dimension 0 consist of a single point.
* The single isometry class is Riemannian.
† The single isometry class is non-Riemannian, of Funk type.

Table 1. The dimension of the moduli space for several families of constant flag curvature (CFC) Randers metrics.

In two dimensions, the Ricci scalar of a Riemannian metric is the Gaussian curvature of the surface. Hence all Riemannian surfaces are Einstein in the sense of Section 1.3.2. Since the Gaussian curvature is not constant in general, the Schur lemma fails for Riemannian (and therefore Randers) metrics in two dimensions. It is natural to ask whether the Schur lemma also fails for non-Riemannian ($W \neq 0$) Randers *surfaces*. The answer is yes. Section 4.1 develops a class of non-Riemannian Randers surfaces whose Ricci scalars are nonconstant functions of x alone. In particular, these non-Riemannian surfaces are Einstein, but fail the Schur lemma.

In dimension $n \geqslant 3$, every Riemannian Einstein metric h is Ricci-constant. This follows readily from tracing the second Bianchi identity and realising that, for such metrics, ${}^h\text{Ric}_{ij} = (S/n)h_{ij}$, where S denotes the scalar curvature of h.

LEMMA 11 (SCHUR LEMMA). *The Ricci scalar of any Einstein Randers metric in dimension greater than two is necessarily constant.*

PROOF. Suppose F is an Einstein metric of Randers type, with navigation data (h, W) and Ricci scalar $\mathcal{R}ic(x) = (n-1)K(x)$. Theorem 9 says that h must be Einstein with Ricci scalar $(n-1)\left(K + \frac{1}{16}\sigma^2\right)$, for some constant σ. Since $n > 2$ here, the Riemannian Schur lemma forces $K + \frac{1}{16}\sigma^2$ to be constant. The same must then hold for K and $\mathcal{R}ic = (n-1)K$. □

Another proof of the Schur lemma, based on the Einstein characterisation of Section 2.2.3 (Theorem 5 on page 227), is given in [Robles 2003].

3.3.2. Three dimensional Einstein–Randers metrics. For Riemannian metrics in three dimensions, being Einstein and having constant sectional curvature are equivalent conditions because the conformal Weyl curvature tensor automatically vanishes. It is not known whether this rigidity holds for Einstein–Finsler metrics in general. However, the said rigidity does hold for Randers metrics. The proof rests on a comparison between the navigation descriptions for Einstein (Section 3.2.2) and constant flag curvature (Section 3.2.3) Randers metrics.

PROPOSITION 12 (THREE-DIMENSIONAL RIGIDITY). *Let F be a Randers metric in three dimensions. Then F is Einstein if and only if it has constant flag curvature.*

PROOF. Metrics of constant flag curvature are always Einstein. As for the converse, let F be an Einstein Randers metric with navigation data (h, W). The Ricci scalar of F is $(n-1)K = 2K$; in view of the Schur lemma above, K has to be constant. According to Theorem 9, h is Einstein with Ricci scalar $2(K + \frac{1}{16}\sigma^2)$, for some constant σ. By Riemannian three-dimensional rigidity, h must have constant sectional curvature $K + \frac{1}{16}\sigma^2$. The navigation description in Theorem 10 then forces F to be of constant flag curvature K.   □

Interestingly, the two navigation descriptions also tell us that *any Einstein Randers metric that arises as a solution to Zermelo's problem of navigation on a Riemannian space form must be of constant flag curvature.*

3.3.3. The Matsumoto identity. This identity first came to light in a letter from Matsumoto to the first author. It says that any Randers metric of *constant* flag curvature K satisfies

$$\sigma\left(K + \tfrac{1}{16}\sigma^2\right) = 0.$$

Since metrics of constant flag curvature are Einstein, it is natural to wonder whether this identity can be extended to Einstein Randers metrics. The answer is yes, by the following result:

PROPOSITION 13. *Let F be an Einstein Randers metric whose Ricci scalar $\mathcal{R}ic(x)$ we reexpress as $(n-1)K(x)$. Then*

$$\sigma\left(K + \tfrac{1}{16}\sigma^2\right) = \begin{cases} W^i K_{;i} & \text{when } n = 2, \\ 0 & \text{when } n > 2. \end{cases}$$

Here, σ is the constant supplied by the navigation data (h, W) of F. According to Theorem 9, h must be Einstein with Ricci scalar $(n-1)(K + \frac{1}{16}\sigma^2)$, and W satisfies the $\mathcal{L}_W$ Equation $W_{i:j} + W_{j:i} = -\sigma h_{ij}$.

PROOF. We begin with the Ricci identity for the tensor $C_{ij} := W_{i:j} - W_{j:i}$, namely $C_{ij:k:h} - C_{ij:h:k} = C_{sj}\,{}^h R_i{}^s{}_{kh} + C_{is}\,{}^h R_j{}^s{}_{kh}$, where ${}^h R_h{}^i{}_{jk}$ is the curvature tensor of h. Trace this identity on (i, k) and (h, j) to obtain

$$\left(W^{r\cdot i}{}_{:i} - W^{r\cdot i}{}_{:i}\right)_{:j} - \left(W^{r\cdot i} - W^{r\cdot i}\right){}^h \mathrm{Ric}_{ij} = 0,$$

where the second equality follows because ${}^h \mathrm{Ric}_{ij}$ is symmetric.

Next, we compute $W^{i:j}{}_{:i} - W^{j:i}{}_{:i}$. To that end, differentiating the $\mathcal{L}_W$ Equation gives $W_{p:q:r} + W_{q:p:r} = 0$. This and the Ricci identity for W imply that

$$W_{i:j:k} - W_{j:i:k} = (W_{i:j:k} - W_{i:k:j}) - (W_{k:i:j} - W_{k:j:i}) + (W_{j:k:i} - W_{j:i:k})$$
$$= W^{s\,h}R_{isjk} - W^{s\,h}R_{ksij} + W^{s\,h}R_{jski}.$$

Using h to trace on (i,k) and raise j, we get $W^{i:j}{}_{:i} - W^{j:i}{}_{:i} = 2W_s{}^h\mathrm{Ric}^{sj}$. Since $^h\mathrm{Ric} = (n-1)(K + \frac{1}{16}\sigma^2)\,h$, we are led to

$$W^{i:j}{}_{:i} - W^{j:i}{}_{:i} = 2(n-1)\left(K + \tfrac{1}{16}\sigma^2\right)W^j. \qquad (*)$$

Finally, tracing the $\mathcal{L}_W$ Equation gives $2W^j{}_{:j} = -n\sigma$, whence

$$0 = (W^{i:j}{}_{:i} - W^{j:i}{}_{:i})_{:j} = 2(n-1)\left(W^j K_{:j} - \tfrac{1}{2}n\sigma(K + \tfrac{1}{16}\sigma^2)\right).$$

The identity now follows from the Schur lemma (Section 3.3.1). $\square$

REMARK. If we assume that the $\mathcal{L}_W$ Equation (or the Basic Equation, which amounts to the same) holds, the E_{23} Equation (Section 2.2.3) can be reexpressed as $(*)$. Thus $(*)$ is the *navigation version* of the E_{23} Equation. It can be further refined, using $W_{i:j:k} + W_{j:i:k} = 0$, to read

$$W^i{}_{:j:i} = (n-1)\left(K + \tfrac{1}{16}\sigma^2\right)W_j.$$

A second derivation of the Matsumoto Identity, based on Theorem 5 (page 227), may be found in [Robles 2003].

4. Einstein Metrics of Nonconstant Flag Curvature

We now present a variety of non-Riemannian Randers metrics that are either Einstein or Ricci-constant. Apart from the 2-sphere, which is included merely because of its simplicity, all examples have nonconstant flag curvatures. Section 3.1.1 will be used without mention.

4.1. Examples with Riemannian–Einstein navigation data

4.1.1. Surfaces of revolution. Our first class of examples comprises surfaces of rotation in $\mathbb{R}^3$. We shall see that solutions to Zermelo's problem of navigation under infinitesimal rotations are Einstein, with Ricci scalar $\mathcal{R}ic(x)$ equal to the Gaussian curvature of the original Riemannian surface. Among the examples below, two (the elliptic paraboloid and the torus) have nonconstant Ricci scalar. These solutions of Zermelo navigation are non-Riemannian counterexamples to Schur's lemma in dimension 2.

To begin, take any surface of revolution M, obtained by revolving a profile curve

$$\varphi \mapsto \left(0, f(\varphi), g(\varphi)\right)$$

in the right half of the yz-plane around the z axis. The ambient Euclidean space induces a Riemannian metric h on M. Parametrise M by

$$(\theta, \varphi) \mapsto \left(f(\varphi) \cos \theta, \; f(\varphi) \sin \theta, \; g(\varphi)\right), \quad 0 \leqslant \theta \leqslant 2\pi.$$

Now consider the infinitesimal isometry $W := \varepsilon \partial_\theta$, where ε is a constant. By limiting the size of our profile curve if necessary, there is no loss of generality in assuming that f is bounded. Choose ε so that $\varepsilon |f| < 1$ for all φ. Expressing h in the given parametrisation, we find that the solution to Zermelo's problem is the Randers metric $F = \alpha + \beta$ on M, with

$$\alpha = \frac{\sqrt{u^2 f^2 + v^2 (1 - \varepsilon^2 f^2)(\dot{f}^2 + \dot{g}^2)}}{1 - \varepsilon^2 f^2}, \quad \beta = \frac{-\varepsilon u f^2}{1 - \varepsilon^2 f^2},$$

and $\|b\|^2 = \varepsilon^2 f^2 = |W|^2$. Here, $u \partial_\theta + v \partial_\varphi$ represents an arbitrary tangent vector on M, and $\dot{f}$, $\dot{g}$ are the derivatives of f, g with respect to φ. Note that the hypothesis $\varepsilon |f| < 1$ ensures strong convexity.

Because W is a Killing vector field, σ vanishes. The Einstein navigation description (Theorem 9 on page 238) then says that the Ricci scalar $\mathcal{R}ic$ of F is identical to that of the Riemannian metric h. The latter is none other than the Gaussian curvature K of h. Hence

$$\mathcal{R}ic(x) = K(x) = \frac{\dot{g}(\dot{f}\ddot{g} - \ddot{f}\dot{g})}{f(\dot{f}^2 + \dot{g}^2)^2},$$

where the dots indicate derivatives with respect to φ.

We examine three special cases of surfaces of revolution:

SPHERE. The unit sphere is given as a surface of revolution by $f(\varphi) = \cos \varphi$ and $g(\varphi) = \sin \varphi$. We will consider the infinitesimal rotation $W = \varepsilon \partial_\theta$, with $\varepsilon < 1$ to effect the necessary $\varepsilon |f| < 1$. The Randers metric solving Zermelo's problem of navigation on the sphere under the influence of W is of constant flag curvature $K = 1$, with

$$\alpha = \frac{\sqrt{u^2 \cos^2 \varphi + v^2 (1 - \varepsilon^2 \cos^2 \varphi)}}{1 - \varepsilon^2 \cos^2 \varphi}, \quad \beta = \frac{-\varepsilon u \cos^2 \varphi}{1 - \varepsilon^2 \cos^2 \varphi}. \qquad \diamondsuit$$

ELLIPTIC PARABOLOID. This is the surface $z = x^2 + y^2$ in $\mathbb{R}^3$. Set the multiple ε in W to be 1. The resulting Randers metric lives on the $x^2 + y^2 < 1$ portion of the elliptic paraboloid, and has Ricci scalar $4/(1 + 4x^2 + 4y^2)^2$. It reads

$$\alpha = \frac{\sqrt{(-yu + xv)^2 + \left((1 + 4x^2)u^2 + 8xy\,uv + (1 + 4y^2)v^2\right)(1 - x^2 - y^2)}}{1 - x^2 - y^2},$$

$$\beta = \frac{yu - xv}{1 - x^2 - y^2}, \quad \text{with } \|b\|^2 = x^2 + y^2. \qquad \diamondsuit$$

TORUS. Specialize to a torus of revolution with parametrisation

$$(\theta, \varphi) \mapsto \left((2 + \cos \varphi) \cos \theta, \; (2 + \cos \varphi) \sin \theta, \; \sin \varphi\right).$$

Set the multiple ε in W to be $\frac{1}{4}$. The resulting Randers metric on the torus has Ricci scalar $\cos\varphi/(2+\cos\varphi)$. It is given by

$$\alpha = \frac{4\sqrt{16\,(2+\cos\varphi)^2\,u^2 + \left(16 - (2+\cos\varphi)^2\right)v^2}}{16 - (2+\cos\varphi)^2},$$

$$\beta = \frac{-4\,(2+\cos\varphi)^2\,u}{16 - (2+\cos\varphi)^2}, \quad \text{with } \|b\|^2 = \tfrac{1}{16}\,(2+\cos\varphi)^2. \qquad \diamond$$

4.1.2. Certain Cartesian products. Recall from Section 1.3.2 the geometrical definition of the Ricci scalar and of Einstein metrics. When specialised to Riemannian n-manifolds, it says that the Ricci scalar $\mathcal{R}ic$ is obtained by summing the sectional curvatures of $n-1$ appropriately chosen sections that share a common flagpole. A Ricci-constant metric is remarkable because this sum is a constant. A moment's thought convinces us of the following:

The Cartesian product of two Riemannian Einstein metrics with the same constant Ricci scalar ρ is again Ricci-constant, and has $\mathcal{R}ic = \rho$.

As we will illustrate, this allows us to construct a wealth of Ricci-constant Randers metrics with nonconstant flag curvature.

Fix ρ. For $i = 1, 2$, let M_i be an n_i-dimensional Riemannian manifold with constant sectional curvature $\rho/(n_i - 1)$. Therefore M_i is Einstein with Ricci scalar ρ. Let W_i be a Killing field on M_i. Let h denote the product Riemannian metric on the Cartesian product $M = M_1 \times M_2$. Then h has constant Ricci scalar ρ, and admits $W = (W_1, W_2)$ as a Killing field.

By Theorem 9 (page 238), the Randers metric F generated by the navigation data (h, W) on M is Einstein, with constant Ricci scalar ρ. When ρ is nonzero, the Einstein metric h on M is not of constant sectional curvature. Hence Theorem 10 (page 239) assures us that F will not be of constant flag curvature. Proposition 8 (page 234) then says that the Randers metric F is non-Riemannian if and only if the wind W is nonzero. So it suffices to select a nonzero W_1.

To that end, let $\widetilde{M}_1$ be the n_1-dimensional, complete, simply connected standard model of constant sectional curvature $\rho/(n_1 - 1)$. The space of globally defined Killing fields on $\widetilde{M}_1$ is a Lie algebra $\mathfrak{g}$ of dimension $\frac{1}{2}n_1(n_1 + 1)$. Select a nonzero $\widetilde{W}_1$ from $\mathfrak{g}$. The isometry group G of $\widetilde{M}$ acts on $\mathfrak{g}$ via push-forwards. Let H be any finite subgroup of the isotropy group of $\widetilde{W}_1$. Then we have a natural projection $\pi : \widetilde{M}_1 \to \widetilde{M}_1/H$. The quotient space $M_1 := \widetilde{M}_1/H$ is of constant sectional curvature $\rho/(n_1 - 1)$, and has a nonzero Killing field $W_1 := \pi_* \widetilde{W}_1$.

As a concrete illustration, we specialise the discussion to spheres. For simplicity, we specify the finite subgroup H to be trivial.

EXAMPLE. Let M_i $(i = 1, 2)$ be the n_i-sphere of radius $\sqrt{n_i - 1}$, $n_i \geqslant 2$. Then M_i has constant sectional curvature $1/(n_i - 1)$, and is therefore Einstein with Ricci scalar 1. The Cartesian product $M = M_1 \times M_2$, equipped with the product metric h, is an $(n_1 + n_2)$-dimensional Riemannian Einstein manifold with Ricci scalar 1, and it is *not* of constant sectional curvature.

The Lie algebra of Killing fields on the n-sphere $S^n(r)$ with radius r is isomorphic to $\mathfrak{so}(n+1)$, regardless of the size of r. The following description accounts for all such vector fields. View points $p \in S^n(r)$ as row vectors in $\mathbb{R}^{n+1}$. For each $\Omega \in \mathfrak{so}(n+1)$, the assignment $p \mapsto p\Omega \in T_p(S^n(r))$ is a globally defined Killing vector field on $S^n(r)$.

Now, for $i = 1, 2$, take $\Omega_i \in \mathfrak{so}(n_i + 1)$. Denote points of M_1, M_2 by p and q, respectively. The map $(p, q) \mapsto (p\Omega_1, q\Omega_2) \in T_{(p,q)}M$ defines a Killing field W of h. This W is nonzero as long as Ω_1, Ω_2 are not both zero. Zermelo navigation on (M, h) under the influence of W generates a non-Riemannian Randers metric with constant Ricci scalar 1, and which is not of constant flag curvature.   ◇

4.2. Examples with Kähler–Einstein navigation data.

In this section, we construct Einstein metrics of Randers type, with navigation data h, which is a Kähler–Einstein metric, and W, which is a Killing field of h. We choose h from among Kähler metrics of constant *holomorphic* sectional curvature, because the formula can be explicitly written down. There exist much more general Kähler–Einstein metrics, for instance those with positive sectional curvature; see [Tian 1997].

4.2.1. Kähler manifolds of constant holomorphic sectional curvature

Suppose (M, h) is a Kähler manifold of complex dimension m (real dimension $n = 2m$) with complex structure J. Let $(z^1, \ldots, z^m)$, where $z^\alpha := x^\alpha + ix^{\bar\alpha}$, denote local complex coordinates. Here, $x^1, \ldots, x^m; x^{\bar 1}, \ldots, x^{\bar m}$ are the $2m$ real coordinates. In our notation, lowercase Greek indices run from 1 to m. The complex coordinate vector fields are

$$Z_\alpha := \partial_{z^\alpha} = \tfrac{1}{2}(\partial_{x^\alpha} - i\partial_{x^{\bar\alpha}}),$$

$$Z_{\bar\alpha} := \partial_{\bar z^\alpha} = \tfrac{1}{2}(\partial_{x^\alpha} + i\partial_{x^{\bar\alpha}}),$$

respectively eigenvectors of J with eigenvalues $+i$ and $-i$. In what follows, uppercase Latin indices run through $1, \ldots, m, \bar 1, \ldots, \bar m$. Also, $\bar z^\alpha := x^\alpha - ix^{\bar\alpha}$ abbreviates the complex conjugate of z^α.

The Kähler metric h is a Riemannian (real) metric with the J-invariance property $h(JX, JY) = h(X, Y)$, and such that the 2-form $(X, Y) \mapsto h(X, JY)$, known as the *Kähler form*, is closed.

Let $h_{AB} := h(Z_A, Z_B)$. The J-invariance of h implies $h_{\alpha\beta} = 0 = h_{\bar\alpha\bar\beta}$. Expanding in terms of the complex basis gives

$$h = h_{\alpha\bar\beta}(dz^\alpha \otimes d\bar z^\beta + d\bar z^\beta \otimes dz^\alpha),$$

where $(h_{\alpha\bar\beta})$ is an $m \times m$ complex Hermitian matrix. By contrast, using the real basis and setting $H_{AB} := h(\partial_{x^A}, \partial_{x^B})$, we have $h = H_{AB}\, dx^A \otimes dx^B$, with $H_{\bar\alpha\bar\beta} = H_{\alpha\beta}$ and $H_{\bar\alpha\beta} = -H_{\alpha\bar\beta}$: the two diagonal blocks are symmetric and identical, while the off-diagonal blocks are skew-symmetric and negatives of each other.

The Kähler form being closed is equivalent to either $Z_\gamma(h_{\alpha\bar\beta}) = Z_\alpha(h_{\gamma\bar\beta})$ or $Z_{\bar\gamma}(h_{\alpha\bar\beta}) = Z_{\bar\beta}(h_{\alpha\bar\gamma})$, which puts severe restrictions on the usual Riemannian connection of h. Consequently, if we expand the curvature operator $^hR(Z_C, Z_D)Z_B$ as $^hR_B{}^A{}_{CD}Z_A$ in the complex basis, it is not surprising to find considerable economy among the coefficients:

$$^hR_{\bar\beta}{}^\alpha{}_{CD} = {}^hR_\beta{}^{\bar\alpha}{}_{CD} = 0 = {}^hR_B{}^A{}_{\gamma\delta} = {}^hR_B{}^A{}_{\bar\gamma\bar\delta}.$$

For each Y in the real tangent space T_xM, the 2-plane spanned by $\{Y, JY\}$ is known as a *holomorphic section* because it is invariant under J, and the corresponding sectional curvature (in the usual Riemannian sense) is called a *holomorphic sectional curvature*. If all such curvatures are equal to the same constant $\mathfrak{c}$, the Kähler metric is said to be of constant holomorphic sectional curvature $\mathfrak{c}$. Such metrics are characterised by their curvature tensor having the following form in the complex basis $\{Z_A\}$:

$$^hR_{\beta\bar\alpha\gamma\bar\delta} = \frac{\mathfrak{c}}{2}(h_{\beta\bar\alpha}h_{\gamma\bar\delta} + h_{\gamma\bar\alpha}h_{\beta\bar\delta}).$$

Equivalently,

$$^hR_\beta{}^\alpha{}_{\gamma\bar\delta} = \frac{\mathfrak{c}}{2}(\delta_\beta{}^\alpha h_{\gamma\bar\delta} + \delta^\alpha{}_\gamma h_{\beta\bar\delta}).$$

Discussion of all this may be found in [Kobayashi–Nomizu 1996]; but one must adjust for the fact that their K_{ABCD} is our R_{BACD}, and that their definition of the curvature operator agrees with ours.

REMARKS. 1. Suppose h is a Kähler metric of constant *holomorphic* sectional curvature $\mathfrak{c}$. Return to the expression for $^hR_\beta{}^\alpha{}_{\gamma\bar\delta}$ above. Tracing on the indices α and γ, we see that $^h\text{Ric} = (m+1)(\mathfrak{c}/2)h$. That is, h must be an Einstein metric with constant Ricci scalar $(m+1)\mathfrak{c}/2$.

2. If a Kähler metric h were to satisfy the stronger condition of constant sectional curvature $\mathfrak{c}$, then it would necessarily be Einstein with Ricci scalar $(n-1)\mathfrak{c} = (2m-1)\mathfrak{c}$; see Section 1.3.1. At the same time, (1) implies that the Ricci scalar is $(m+1)\mathfrak{c}/2$. Hence we would have to have either $\mathfrak{c} = 0$ or $m = 1$.

Thus a Kähler manifold (M, h) can have constant sectional curvature in only two ways: either h is flat, or the real dimension of M is 2. This rigidity indicates that in the Kähler category, the weaker concept of constant holomorphic sectional curvature is more appropriate.

In analogy with the constant sectional curvature case we have a classification theorem for Kähler metrics of constant holomorphic sectional curvature [Kobayashi–Nomizu 1996]:

Any simply connected complete Kähler manifold of constant holomorphic sectional curvature $\mathfrak{c}$ is holomorphically isometric to one of three standard models:

$\mathfrak{c} > 0$: *the Fubini–Study metric on* $\mathbb{C}P^n$ (*see below*),

$\mathfrak{c} = 0$: *the standard Euclidean metric on* $\mathbb{C}^n$,

$\mathfrak{c} < 0$: *the Bergmann metric on the unit ball in* $\mathbb{C}^n$.

4.2.2. Killing fields of the Fubini–Study metric. The Fubini–Study metric is a Kähler metric on $\mathbb{C}P^m$ of constant holomorphic sectional curvature $\mathfrak{c} > 0$. Complex projective space is obtained from $\mathbb{C}^{m+1} \smallsetminus 0$ by quotienting out the equivalence relation $\zeta \sim \lambda\zeta$, where $0 \neq \lambda \in \mathbb{C}$. Denote the equivalence class of ζ by $[\zeta]$. $\mathbb{C}P^m$ is covered by the charts $U^j := \{[\zeta] \in \mathbb{C}P^m : \zeta^j \neq 0\}$, $j = 0, 1, \ldots, m$, with holomorphic coordinate mapping

$$[\zeta] \mapsto \frac{1}{\zeta^j}(\zeta^0, \ldots, \widehat{\zeta^j}, \ldots, \zeta^m) =: (z^1, \ldots, z^m) = z \in \mathbb{C}^m.$$

In these coordinates, the Fubini–Study metric of constant holomorphic sectional curvature $\mathfrak{c} > 0$ has components

$$h_{\alpha\bar{\beta}} := h(Z_\alpha, Z_{\bar{\beta}}) = \frac{2}{\mathfrak{c}}\left(\frac{1}{\rho}\delta_{\alpha\beta} - \frac{1}{\rho^2}\bar{z}_\alpha z_\beta\right),$$

where $\bar{z}_\alpha := \delta_{\alpha\beta}\bar{z}^\beta$, $z_\beta := \delta_{\beta\tau}z^\tau$, and $\rho := 1 + z^\alpha \bar{z}_\alpha$.

Conventional wisdom in complex manifold theory prompts us to construct some explicit Killing vector field of h by considering $\xi := \pi_*(P^i{}_j \zeta^j \partial_{\zeta^i})$. Here, $\pi : \mathbb{C}^{m+1} \smallsetminus 0 \to \mathbb{C}P^m$ is the natural projection and $P \in \mathfrak{u}(m+1)$, the Lie algebra of the unitary group $U(m+1)$. Note that $U(m+1)$ is the group of holomorphic isometries of Euclidean $\mathbb{C}^{m+1} \smallsetminus 0$.

For concreteness, we compute ξ in the local coordinates of the chart U^0. We have $\pi(\zeta^0, \zeta^1, \ldots, \zeta^m) := (1/\zeta^0)(\zeta^1, \ldots, \zeta^m) =: (z^1, \ldots, z^m)$, from which it follows that

$$\pi_*(\partial_{\zeta^0}) = -\frac{1}{\zeta^0}z^\alpha \partial_{z^\alpha} \quad \text{and} \quad \pi_*(\partial_{\zeta^\alpha}) = \frac{1}{\zeta^0}\partial_{z^\alpha}$$

because the differential of π is simply the matrix

$$\pi_* = \frac{1}{\zeta^0}\begin{pmatrix} -z^1 & 1 & & & \\ & & & & 0 \\ \vdots & & & \ddots & \\ & & 0 & & \\ -z^m & & & & 1 \end{pmatrix}.$$

Thus

$$\pi_*(P^i{}_j \zeta^j \partial_{\zeta^i}) = P^0{}_j \zeta^j\left(-\frac{1}{\zeta^0}z^\alpha \partial_{z^\alpha}\right) + P^\alpha{}_j \zeta^j\left(\frac{1}{\zeta^0}\partial_{z^\alpha}\right)$$

$$= (-P^0{}_0 - P^0{}_\beta z^\beta)z^\alpha \partial_{z^\alpha} + (P^\alpha{}_0 + P^\alpha{}_\beta z^\beta)\partial_{z^\alpha}.$$

With the skew-Hermitian property $P^t = -\bar{P}$, and introducing the decomposition

$$P = \begin{pmatrix} E & \bar{C}^t \\ C & Q \end{pmatrix}, \quad \text{where} \quad \begin{cases} E := P^0{}_0 \text{ is pure imaginary,} \\ C = (C^\alpha) := (P^\alpha{}_0) \in \mathbb{C}^m, \\ Q = (Q^\alpha{}_\beta) := (P^\alpha{}_\beta) \in \mathfrak{u}(m), \end{cases}$$

we see that

$$\xi := \pi_*(P^i{}_j \zeta^j \partial_{\zeta^i}) = \left(Q^\alpha{}_\beta z^\beta + C^\alpha + (\bar{C} \cdot z + \bar{E})z^\alpha\right)\partial_{z^\alpha}.$$

The real and imaginary parts of ξ give two real vector fields. A straightforward calculation shows that only $\operatorname{Re}\xi$ is Killing. The failure of $\operatorname{Im}\xi$ to be Killing persists even for $\mathbb{C}P^1$. In that case, the skew-Hermitian Q is simply a pure imaginary number, C is a single complex constant, and

$$\mathcal{L}_{\operatorname{Im}\xi}h = \left(\frac{2}{\rho^2}\left(1 - \frac{2}{\rho}\right)\operatorname{Im}(E + \bar{Q}) + \frac{8}{\rho^3}\operatorname{Im}(E\bar{C})\right)(dx^1 \otimes dx^1 + dx^{\bar{1}} \otimes dx^{\bar{1}}),$$

where $\rho := 1 + |z|^2$. (We hasten to add that for $\mathbb{C}P^m$ with $m > 1$, the $dx^\alpha dx^{\bar{\beta}}$ and $dx^{\bar{\alpha}} dx^\beta$ components of $\mathcal{L}_{\operatorname{Im}\xi}h$ do not vanish.)

Thus, our construction of ξ gives rise to the real Killing fields

$$W := \operatorname{Re}\xi = \tfrac{1}{2}(\xi^\alpha \partial_{z^\alpha} + \overline{\xi^\alpha} \partial_{\bar{z}^\alpha}) = \tfrac{1}{2}\big((\operatorname{Re}\xi^\alpha)\partial_{x^\alpha} + (\operatorname{Im}\xi^\alpha)\partial_{x^{\bar{\alpha}}}\big),$$

where

$$\xi^\alpha := Q^\alpha{}_\beta z^\beta + C^\alpha + (\bar{C} \cdot z + \bar{E})z^\alpha.$$

4.2.3. Non-Riemannian Einstein metrics on $\mathbb{C}P^m$.

Theorem 9 (page 238) assures us that the Randers metric $F = \alpha + \beta$ generated by the navigation data h, W is a globally defined Einstein metric on $\mathbb{C}P^m$, provided that the Killing field W satisfies $|W| < 1$. Since, with $\rho := 1 + z^\gamma \bar{z}_\gamma$, the Fubini–Study metric

$$h := \frac{2}{\mathfrak{c}}\left(\frac{1}{\rho}\delta_{\alpha\beta} - \frac{1}{\rho^2}\bar{z}_\alpha z_\beta\right)(dz^\alpha \otimes d\bar{z}^\beta + d\bar{z}^\beta \otimes dz^\alpha)$$

does *not* have constant sectional curvature for $m > 1$, Theorem 10 (page 239) ensures that F will not be of constant flag curvature. Moreover, the Ricci scalar of F equals that of h, which has the constant value $(m + 1)\mathfrak{c}/2$ (Section 4.2.1).

EXAMPLE. To make explicit the Riemannian metric a underlying α and the 1-form b that gives β, it is necessary to have available the covariant description $W^\flat$ of the Killing field W. If we write $W = W^B \partial_{x^B}$ and $H_{AB} := h(\partial_{x^A}, \partial_{x^B})$, then $W^\flat := (H_{AB}W^B)\,dx^A$. Using $H_{\alpha\beta} = 2\operatorname{Re}h_{\alpha\bar{\beta}} = H_{\bar{\alpha}\bar{\beta}}$ and $H_{\alpha\bar{\beta}} = 2\operatorname{Im}h_{\alpha\bar{\beta}} = -H_{\bar{\alpha}\beta}$, a computation tells us that

$$W^\flat = (\operatorname{Re}\xi_{\bar{\alpha}})\,dx^\alpha + (\operatorname{Im}\xi_{\bar{\alpha}})\,dx^{\bar{\alpha}} = \tfrac{1}{2}(\overline{\xi_{\bar{\alpha}}}\,dz^\alpha + \xi_{\bar{\alpha}}\,d\bar{z}^\alpha).$$

Here, $\xi_{\bar{\alpha}} := h_{\bar{\alpha}B}\xi^B = h_{\bar{\alpha}\beta}\xi^\beta$ has the formula

$$\xi_{\bar{\alpha}} = \frac{2}{\mathfrak{c}\rho^2}\big(\rho(Q_{\alpha\beta}z^\beta + C_\alpha) + (\bar{C} \cdot z - C \cdot \bar{z} + \bar{E} - Q_{\beta\gamma}\bar{z}^\beta z^\gamma)z_\alpha\big),$$

where indices on Q, C, z are raised and lowered by the Kronecker delta.

As long as

$$|W|^2 := h(W, W) = \tfrac{1}{2}\operatorname{Re}(\xi^\alpha\overline{\xi_{\bar{\alpha}}}) < 1,$$

the Randers metric F with defining data

$$a := \frac{1}{\lambda}h + \frac{1}{\lambda^2}W^\flat \otimes W^\flat, \qquad b := -\frac{1}{\lambda}W^\flat,$$

where $\lambda := 1 - |W|^2$, will be strongly convex. Note that $W^\flat \otimes W^\flat$ does have $dz^\alpha dz^\beta$ and $d\bar{z}^\alpha d\bar{z}^\beta$ components, whereas h doesn't; thus the Riemannian metric a is not Hermitian unless $W = 0$.

For any choice of the constant quantities Q, C, E, the resulting function $|W|^2$ is continuous on the compact $\mathbb{C}P^m$, and is therefore bounded. Normalising these quantities by a common positive number if necessary, the strong convexity criterion $|W|^2 < 1$ can always be met. $\Diamond$

4.3. Rigidity and a Ricci-flat example.

In this final set of examples, we consider Einstein–Randers metrics on compact boundaryless manifolds M, with an eye toward those with nonpositive *constant* Ricci scalar. The information obtained will complement the Ricci-positive example presented in Section 4.2.3.

We begin by observing that any infinitesimal homothety of (M, h) is necessarily Killing; that is, $\sigma = 0$. The proof follows from a divergence lemma (for compact boundaryless manifolds) and the trace of the $\mathcal{L}_W$ Equation:

$$0 = \int_M W^i{}_{:i} \, dV_h \qquad \text{by the divergence lemma [Bao et al. 2000]}$$

$$= \int_M -\tfrac{1}{2} n\sigma \, dV_h \qquad \text{by tracing } W_{i:j} + W_{j:i} = -\sigma h_{ij},$$

where $dV_h := \sqrt{h}\, dx$. In conjunction with Theorem 9 (page 238), we have:

LEMMA 14. *Let (M, h) be a compact boundaryless Riemannian manifold. Every infinitesimal homothety W must be a Killing field; equivalently, $\sigma = 0$. In particular, if h is Einstein with Ricci scalar $\mathcal{R}ic$, then the navigation data (h, W) generates an Einstein Randers metric F with Ricci scalar $\mathcal{R}ic$.*

4.3.1. Killing fields versus eigenforms.

Let W be any Killing vector field on a Riemannian Einstein manifold (M, h) with constant Ricci scalar $\mathcal{R}ic$. Let $W^\flat := W_i \, dx^i$ denote the 1-form dual to W. The action of the Laplace–Beltrami operator $\Delta := d\delta + \delta d$ on $W^\flat$ is given by the Weitzenböck formula

$$\Delta W^\flat = (-W_i{}^{:j}{}_{:j} + {}^h\mathrm{Ric}_i{}^j W_j) \, dx^i. \qquad (\dagger)$$

See, for example, [Bao et al. 2000]. Given that ${}^h\mathrm{Ric}_{ij} = \mathcal{R}ic \, h_{ij}$, we have

$${}^h\mathrm{Ric}_i{}^j W_j = \mathcal{R}ic \, W_i.$$

By a Ricci identity, $W^j{}_{:j:i} - W^j{}_{:i:j} = -{}^h\mathrm{Ric}_i{}^s W_s$. Since W is Killing, $W_{i:j} + W_{j:i}$ vanishes. Thus $W^j{}_{:j} = 0$ and $-W^j{}_{:i} = W_i{}^{:j}$, which reduce that Ricci identity to

$$-W_i{}^{:j}{}_{:j} = {}^h\mathrm{Ric}_i{}^s W_s = \mathcal{R}ic \, W_i.$$

(Note for later use that if W is parallel but $\not\equiv 0$, then $\mathcal{R}ic = 0$.)

The Weitzenböck formula ($\dagger$) now becomes

$$\Delta W^\flat = 2\,\mathcal{R}ic \, W^\flat. \qquad (*)$$

Therefore, whenever h is Einstein with *constant* Ricci scalar $\mathcal{R}ic$, all its Killing 1-forms must be eigenforms of Δ, with eigenvalue $2\mathcal{R}ic$.

Now suppose M is compact and boundaryless, so that integration by parts can be carried out freely without generating any boundary terms. Let $\langle\ ,\ \rangle$ denote the L_2 inner product on k-forms; namely,

$$\langle\omega,\eta\rangle := \frac{1}{k!}\int_M h^{i_1 j_1}\cdots h^{i_k j_k}\,\omega_{i_1\cdots i_k}\eta_{j_1\cdots j_k}\,\sqrt{h}\,dx.$$

Since the codifferential δ is the L_2 adjoint of d, we have

$$2\,\mathcal{R}ic\,\langle W^\flat, W^\flat\rangle = \langle\Delta W^\flat, W^\flat\rangle = \langle\delta W^\flat, \delta W^\flat\rangle + \langle dW^\flat, dW^\flat\rangle \geqslant 0. \qquad (\ddagger)$$

In particular, if $\mathcal{R}ic$ is negative, then W must be zero.

On the other hand, using (†) and $^h\mathrm{Ric}_{ij} = \mathcal{R}ic\, h_{ij}$, we get

$$\langle\Delta W^\flat, W^\flat\rangle = \int_M (W_{i:j}\,W^{i:j} + \mathcal{R}ic\,|W|^2)\,\sqrt{h}\,dx.$$

This enables us to make a Bochner type argument: if $\mathcal{R}ic = 0$, so that $W^\flat$ is harmonic by (∗), then $W_{i:j}$ must vanish identically.

4.3.2. Digression on Berwald spaces.
In anticipation of our discussion of non-positive Ricci curvature, we review Berwald spaces. (See [Szabó 1981; 2003] for a complete classification of such spaces.)

Let (M, F) be an arbitrary Finsler space. M need not be compact boundary-less, and F need not be Einstein or of Randers type. Let G^i denote the geodesic spray coefficients of F. Then (M, F) is a *Berwald space* if the Berwald connection coefficients $(G^i)_{y^j y^k}$ do not depend on y. In particular, all Riemannian and locally Minkowski spaces are Berwald; for explicit examples belonging to neither of these two camps, see [Bao et al. 2000].

Now suppose $F = \alpha + \beta$ is Randers but not necessarily Einstein, and has navigation data (h, W). It is known that F is Berwald if and only if the defining 1-form b is parallel. This elegant theorem is due to the efforts of the Japanese school. See [Bao et al. 2000] for an account of the history and references therein; see also the errata for a proof by Mike Crampin.

Decompose $b_{i|j} = \frac{1}{2}\,\mathrm{lie}_{ij} + \frac{1}{2}\,\mathrm{curl}_{ij}$ into its symmetric and skew-symmetric parts. Look back at the expression for lie and curl at the end of Section 3.1.3. (We reiterate here that they were derived under no assumptions on F.) Observe that $W_{i:j} = 0$ implies $b_{i|j} = 0$. We prove the converse: suppose $b_{i|j} = 0$.

o We have $0 = 2b_{j|k}W^j W^k = \mathrm{lie}_{jk}W^j W^k = -(1/\lambda)W_{j:k}W^j W^k$ by referring to Section 3.1.3. Hence $W_{j:k}W^j W^k = 0$.

o Using this, a similar calculation gives $0 = 2b_{j|k}W^j = -(2/\lambda)W^j W_{j:k}$ and $0 = 2b_{j|k}W^k = -(2/\lambda)W_{j:k}W^k$. That is, $W^j W_{j:k} = 0 = W_{j:k}W^k$.

o The formulae for lie_{jk} and curl_{jk} now simplify to $0 = \mathrm{lie}_{jk} = -\mathcal{L}_{jk}$ and $0 = \mathrm{curl}_{jk} = -(1/\lambda)\mathcal{C}_{jk}$. Hence $W_{j:k} = \frac{1}{2}(\mathcal{L}_{jk} + \mathcal{C}_{jk}) = 0$.

LEMMA 15. *Let F be any Randers metric, with defining data (a, b) and naviga-tion data (h, W). The following three conditions are equivalent*:

- *F is Berwald.*
- *b is parallel with respect to a.*
- *W is parallel with respect to h.*

4.3.3. A rigidity theorem. We now focus on Einstein Randers metrics of non-positive *constant* Ricci scalar, and show that there is considerable rigidity.

We begin by addressing the Ricci-flat case. We saw at the end of 4.3.1 that $\mathcal{R}ic = 0$ implies that W is parallel. Conversely, if W is parallel and not identi-cally zero, then $\mathcal{R}ic = 0$ (see parenthetic remark just before $(*)$ on page 251). Whence, in conjunction with Lemma 15 and Proposition 8 (page 234), we obtain:

PROPOSITION 16. *Let F be an Einstein Randers metric on a compact bound-aryless manifold M.*

- *If $\mathcal{R}ic = 0$, then F must be Berwald.*
- *If F is non-Riemannian and Berwald, then $\mathcal{R}ic = 0$.*

The second conclusion is false if we remove the stipulation that F be non-Riemannian. A Riemannian metric is always Berwald and Randers, and being Einstein certainly does not mandate it to be Ricci-flat.

Next, we turn our attention to compact boundaryless Einstein Randers spaces of constant negative Ricci scalar. In this case, by $(\ddagger)$ on page 252, W must be identically zero. Equivalently, $F = h$ is Riemannian.

PROPOSITION 17. *Let F be an Einstein Randers metric with constant negative Ricci scalar on a compact boundaryless manifold M. Then F is Riemannian.*

Together, these two propositions imply the following rigidity theorem.

THEOREM 18 (RICCI RIGIDITY). *Suppose (M, F) is a connected compact bound-aryless Einstein Randers manifold with constant Ricci scalar $\mathcal{R}ic$.*

- *If $\mathcal{R}ic < 0$, then (M, F) is Riemannian.*
- *If $\mathcal{R}ic = 0$, then (M, F) is Berwald.*

Note that locally Minkowskian spaces, being Berwald and of zero flag curvature, are obvious examples of the second camp. The following arguments show that there exist Ricci-flat non-Riemannian Berwald–Randers metrics which are not locally Minkowskian.

EXAMPLE. Take any $K3$ surface[†], namely a complex surface with zero first Chern class and no nontrivial global holomorphic 1-forms. All $K3$ surfaces admit Kähler

---

[†]According to [Weil 1979, v. 2, p. 546], $K3$ surfaces are named after Kummer, Kodaira, Kähler, and "the beautiful mountain K2 in Kashmir" —the second tallest peak in the world. One may conjecture that with this last reference Weil was implying that such surfaces are as hard to conquer as the K2...

metrics (a result due to Todorov and to Siu), and hence Ricci-flat Kähler metrics, by Yau's proof of the Calabi conjecture. Since $\chi(K3) = 24$ by Riemann–Roch, these metrics are not flat by virtue of the Gauss–Bonnet–Chern theorem. See [Besse 1987] for details and references therein. It is futile to consider the Killing fields of such Ricci-flat metrics because, by an argument involving Serre duality, the isometry groups in question are all discrete. To circumvent this difficulty, set $M := K3 \times S^1$ (a compact boundaryless real 5-manifold) and consider the product metric h on M; it can be checked that h is also Ricci-flat but not flat. The vector field $W := 0 \oplus \partial/\partial t$ on M is parallel, hence Killing, with respect to h. Theorem 9 (page 238) tells us that the Randers metric F on M with navigation data (h, W) is Ricci-flat, while Proposition 8 (page 234) guarantees that it is not Riemannian. Theorem 10 (page 239) ensures that F is not of constant (zero) flag curvature; hence it could not be locally Minkowskian.   ◇

Theorem 18 generalises a result of Akbar-Zadeh's for Finsler metrics of constant flag curvature [Akbar-Zadeh 1988]:

Suppose (M, F) is a connected compact boundaryless Finsler manifold of constant flag curvature λ.

- *If $\lambda < 0$, then (M, F) is Riemannian.*
- *If $\lambda = 0$, then (M, F) is locally Minkowski.*

The Ricci rigidity theorem is a straightforward extension of Akbar-Zadeh's result when $\mathcal{R}ic < 0$. To appreciate the generalisation when $\mathcal{R}ic = 0$, it is helpful to note that locally Minkowski spaces are precisely Berwald spaces of constant flag curvature $K = 0$; see, for instance, [Bao et al. 2000].

Akbar-Zadeh's theorem holds for all compact boundaryless Finsler spaces of constant flag curvature, while the Ricci rigidity theorem above is restricted to the Randers setting. So, towards a complete generalisation of Akbar-Zadeh's result: What should the conclusions be if we replace 'Randers' by 'Finsler' in the Ricci rigidity theorem?

5. Open Problems

5.1. Randers and beyond. Table 2 summarises some key information about Randers metrics, which are the simplest members in the much larger family of strongly convex (α, β) metrics. Matusmoto's slope-of-a-mountain metric (Section 1.1.1) is a prime example from this family. Formal discussions of (α, β) metrics are given in [Matsumoto 1986; Antonelli et al. 1993]; see also [Shen 2004] in this volume.

For strongly convex (α, β) metrics, how should the entries of the table be modified?

Also, Randers metrics exhibit three-dimensional rigidity (Section 3.3.2), and their Ricci scalars obey a Schur lemma (Section 3.3.1).

Property	Characterisation with (a, b)	Description with (h, W)
Strong convexity	$\|b\| < 1$	$\|W\| < 1$
Berwald	$^a\nabla b = 0$	$^h\nabla W = 0$
Constant flag curvature	$\mathcal{L}_{b^\sharp} a = \sigma(a - bb) - (b\theta + \theta b)$ $^aR = \text{poly}(K, \sigma, a, b, \text{curl})$ (Theorem 5)	h is a space form, $\mathcal{L}_W h = -\sigma h$
Einstein	$\mathcal{L}_{b^\sharp} a = \sigma(a - bb) - (b\theta + \theta b)$ $^a\text{Ric} = \text{poly}(\mathcal{R}ic, \sigma, a, b, \text{curl}, {}^a\nabla\theta)$ (Theorem 6)	h is Einstein, $\mathcal{L}_W h = -\sigma h$

Table 2. Summary of information about Randers metrics

- Does the passage from Randers metrics to (α, β) metrics allow us to construct, in three dimensions, a Ricci-constant metric which is not of constant flag curvature?

- Does every Einstein (α, β) metric ($\mathcal{R}ic$ a function of x only) in dimension $\geqslant 3$ have to be Ricci-constant?

Finally, fans of Randers metrics can aim to append an extra row to the above table, characterising Randers metrics of scalar curvature (Section 1.2.1).

5.2. Chern's question. Professor S.-S. Chern has openly asked the following question on many occasions:

Does every smooth manifold admit a Finsler Einstein metric?

Topological obstructions prevent some manifolds, such as $S^2 \times S^1$, from admitting Riemannian Einstein metrics; see Section 1.3.3 and references therein. By the navigation description of Theorem 9 (page 238), any manifold that admits an Einstein Randers metric must also admit a Riemannian Einstein metric. Thus the same topological obstructions confront Einstein metrics of Randers type.

As a prelude to answering Chern's question, it would be prudent to first settle the issue for concrete examples such as $S^2 \times S^1$. The discussion above shows that in searching for a Finsler Einstein metric on this 3-manifold, we must look beyond those of Randers type.

5.3. Geometric flows. On the slit tangent bundle $TM \smallsetminus 0$, there are two interesting curvature invariants: the Ricci scalar and the S-curvature. They open the door to evolution equations which may be used to deform Finsler metrics. In this frame of mind, we wonder:

- Would a flow driven by the Ricci scalar, such as[†] $\partial_t \log F = -\mathcal{R}ic$, enable us to prove the existence of Finsler metrics with coveted curvature properties?
- Can deformations tailored to the S-curvature be used to ascertain the existence of Landsberg metrics ($\dot{A} = 0$) which are not of Berwald type ($P = 0$)?

Acknowledgments

We thank G. Beil and Z. Shen for discussions about the physics and the geodesics, respectively, in Zermelo's navigation problem. We thank A. Todorov for information about $K3$ surfaces, and P.-M. Wong for guidance on the Killing fields of $\mathbb{C}P^m$. Sections 3.2.1, 3.2.2, 3.3, 4.1.2, and 4.3 are based on Robles' doctoral dissertation.

References

[Akbar-Zadeh 1988] H. Akbar-Zadeh, "Sur les espaces de Finsler á courbures section-nelles constantes", *Acad. Roy. Belg. Bull. Cl. Sci.* **74** (1988), 281–322.

[Antonelli et al. 1993] P. L. Antonelli, R. S. Ingarden and M. Matsumoto, *The theory of sprays and Finsler spaces with applications in physics and biology*, Fundamental theories of physics **58**, Kluwer, Dordrecht, 1993.

[Aubin 1998] T. Aubin, *Some nonlinear problems in Riemannian geometry*, Springer monographs in mathematics, Springer, Berlin, 1998.

[Auslander 1955] L. Auslander, "On curvature in Finsler geometry", *Trans. Amer. Math. Soc.* **79** (1955), 378–388.

[Bao et al. 2000] D. Bao, S.-S. Chern and Z. Shen, *An introduction to Riemann–Finsler geometry*, Graduate Texts in Mathematics **200**, Springer, New York, 2000.

[Bao et al. 2003] D. Bao, C. Robles and Z. Shen, "Zermelo navigation on Riemannian manifolds", to appear in *J. Diff. Geom.*

[Bao–Robles 2003] D. Bao and C. Robles, "On Randers spaces of constant flag curvature", *Rep. on Math. Phys.* **51** (2003), 9–42.

[Bao–Shen 2002] D. Bao and Z. Shen, "Finsler metrics of constant positive curvature on the Lie group S^3", *J. London Math. Soc.* **66** (2002), 453–467.

[Bejancu–Farran 2002] A. Bejancu and H. Farran, "Finsler metrics of positive constant flag curvature on Sasakian space forms", *Hokkaido Math. J.* **31**:2 (2002), 459–468.

[†]The equation for $\partial_t \log F$ has its genesis in $\partial_t g_{ij} = -2\operatorname{Ric}_{ij}$, which is formally identical to the unnormalised Ricci flow of Richard Hamilton for Riemannian metrics (see [Cao–Chow 1999] and references therein). In the Finsler setting, this equation makes sense on $TM \smallsetminus 0$. Contracting with y^i, y^j (and using Sections 1.1.1 and 1.3.1) gives $\partial_t F^2 = -2F^2\mathcal{R}ic$. Since F is strictly positive on the slit tangent bundle, the stated equation follows readily.

[Bejancu–Farran 2003] A. Bejancu and H. Farran, "Randers manifolds of positive constant curvature", *Int. J. Math. Mathematical Sci.* **18** (2003), 1155–1165.

[Berger 1971] M. Berger, "On Riemannian structures of prescribed Gaussian curvature for compact 2-manifolds", *J. Diff. Geom.* **5** (1971), 325–332.

[Berwald 1929] L. Berwald, "Über die n-dimensionalen Geometrien konstanter Krümmung, in denen die Geraden die kürzesten sind", *Math. Z.* **30** (1929), 449–469.

[Berwald 1947] L. Berwald, "Ueber Finslersche und Cartansche Geometrie IV. Projektivkrümmung allgemeiner affiner Räume und Finslersche Räume skalarer Krümmung", *Ann. Math.* (2) **48** (1947), 755–781.

[Besse 1987] A. L. Besse, *Einstein manifolds*, Ergebnisse der Math. (3) **10**, Springer, Berlin, 1987.

[Bryant 1997] R. Bryant, "Projectively flat Finsler 2-spheres of constant curvature", *Selecta Math. (N.S.)* **3** (1997), 161–203.

[Bryant 2002] R. Bryant, "Some remarks on Finsler manifolds with constant flag curvature", *Houston J. Math.* **28**:2 (2002), 221–262.

[Cao–Chow 1999] H.-D. Cao and Bennett Chow, "Recent developments on the Ricci flow", *Bull. Amer. Math. Soc.* **36** (1999), 59–74.

[Carathéodory 1999] C. Carathéodory, *Calculus of variations and partial differential equations of the first order*, 2 v., Holden-Day, San Francisco, 1965–1967; reprinted by Chelsea, 1999.

[Cheeger–Ebin 1975] J. Cheeger and D. Ebin, *Comparison theorems in Riemannian geometry*, North Holland, Amsterdam, 1975.

[Chen–Shen 2003] X. Chen and Z. Shen, "Randers metrics with special curvature properties", *Osaka J. Math.* **40** (2003), 87-101.

[Foulon 2002] P. Foulon, "Curvature and global rigidity in Finsler manifolds", *Houston J. Math.* **28** (2002), 263–292.

[Funk 1929] P. Funk, "Über Geometrien, bei denen die Geraden die Kürzesten sind", *Math. Ann.* **101** (1929), 226–237.

[Hitchin 1974] N. Hitchin, "On compact four-dimensional Einstein manifolds", *J. Diff. Geom.* **9** (1974), 435–442.

[Hrimiuc–Shimada 1996] D. Hrimiuc and H. Shimada, "On the $\mathcal{L}$-duality between Lagrange and Hamilton manifolds", *Nonlinear World* **3** (1996), 613–641.

[Ingarden 1957] R. S. Ingarden, "On the geometrically absolute optical representation in the electron microscope", *Trav. Soc. Sci. Lettr. Wroclaw* B**45** (1957), 3–60.

[Kobayashi–Nomizu 1996] S. Kobayashi and K. Nomizu, *Foundations of differential geometry*, v. 2, Interscience, New York, 1969; reprinted by Wiley, New York, 1996.

[LeBrun 1999] C. LeBrun, "Four-dimensional Einstein manifolds, and beyond", pp. 247–285 in [LeBrun–Wang 1999].

[LeBrun–Wang 1999] C. LeBrun and M. Wang (editors), *Surveys in differential geometry: essays on Einstein maniolds*, v. 6, International Press, Boston, 1999.

[Matsumoto 1986] M. Matsumoto, *Foundations of Finsler geometry and special Finsler spaces*, Kaiseisha Press, Japan, 1986.

[Matsumoto 1989] M. Matsumoto, "A slope of a mountain is a Finsler surface with respect to a time measure", *J. Math. Kyoto Univ.* **29**:1 (1989), 17–25.

[Matsumoto–Shimada 2002] M. Matsumoto and H. Shimada, "The corrected fundamental theorem on Randers spaces of constant curvature", *Tensor (N.S.)* **63** (2002), 43–47.

[Numata 1978] S. Numata, "On the torsion tensors R_{jhk} and P_{hjk} of Finsler spaces with a metric $ds = (g_{ij}(dx)dx^i dx^j)^{1/2} + b_i(x)dx^{i}$", *Tensor (N.S.)* **32** (1978), 27–32.

[Okada 1983] T. Okada, "On models of projectively flat Finsler spaces of constant negative curvature", *Tensor (N.S.)* **40** (1983), 117–124.

[Rademacher 2004] H.-B. Rademacher, "Nonreversible Finsler metrics of positive flag curvature", pp. 263–304 *A sampler of Riemann–Finsler geometry*, edited by D. Bao et al., Math. Sci. Res. Inst. Publ. **50**, Cambridge Univ. Press, Cambridge, 2004.

[Randers 1941] G. Randers, "On an asymmetrical metric in the four-space of general relativity", *Phys. Rev.* **59** (1941), 195–199.

[Rapcsák 1961] A. Rapcsák, "Über die bahntreuen Abbidungen metrischer Räume", *Publ. Math., Debrecen* **8** (1961), 285–290.

[del Riego 1973] L. del Riego, *Tenseurs de Weyl d'un spray de directions*, doctoral dissertation, Université Scientifique et Médicale de Grenoble, 1973.

[Robles 2003] C. Robles, *Einstein metrics of Randers type*, doctoral dissertation, University of British Columbia, 2003.

[Rund 1959] H. Rund, *The differential geometry of Finsler spaces*, Grundlehren der math. Wissenschaften **101**, Springer, Berlin, 1959.

[Sabau 2003] S. V. Sabau, "On projectively flat Finsler spheres (remarks on a theorem of R. L. Bryant)", pp. 181–191 in *Finsler and Lagrange Geometries* (Iaşi, 2001), edited by M. Anastasiei and P. L. Antonelli, Kluwer, Dordrecht, 2003.

[Shen 2001] Z. Shen, *Differential geometry of sprays and Finsler spaces*, Kluwer, Dordrecht 2001.

[Shen 2002] Z. Shen, "Finsler metrics with $K = 0$ and $S = 0$", *Canadian J. Math.* **55** (2003), 112–132.

[Shen 2003] Z. Shen, "Projectively flat Finsler metrics of constant flag curvature", *Trans. Amer. Math. Soc.* **355** (2003), 1713–1728.

[Shen 2004] Z. Shen, "Landsberg curvature, S-curvature and Riemann curvature", pp. 305–356 in *A sampler of Riemann–Finsler geometry*, edited by D. Bao et al., Math. Sci. Res. Inst. Publ. **50**, Cambridge Univ. Press, Cambridge, 2004.

[Singer–Thorpe 1969] I. M. Singer and J. A. Thorpe, "The curvature of 4-dimensional Einstein spaces", pp. 355–365 in *Global analysis: papers in honour of K. Kodaira*, edited by D. C. Spencer and S. Iyanaga, Princeton mathematical series **29**, Princeton Univ. Press, Princeton, 1969.

[Szabó 1981] Z. I. Szabó, "Positive definite Berwald spaces (structure theorems on Berwald spaces)", *Tensor (N.S.)* **35** (1981), 25–39.

[Szabó 2003] Z. I. Szabó, "Classification of Berwald and symmetric Finsler manifolds via explicit constructions", preprint, 2003.

[Thorpe 1969] J. Thorpe, "Some remarks on the Gauss–Bonnet formula", *M. Math. Mech.* **18** (1969), 779–786.

[Tian 1997] G. Tian, "Kähler–Einstein metrics with positive scalar curvature", *Invent. Math.* **130** (1997), 1–39.

[Weil 1979] André Weil, *Œuvres scientifiques*, Springer, Berlin, 1979.

[Zermelo 1931] E. Zermelo, "Über das Navigationsproblem bei ruhender oder veränderlicher Windverteilung", *Z. Angew. Math. Mech.* **11** (1931), 114–124.

[Ziller 1982] W. Ziller, "Geometry of the Katok examples, *Ergodic Theory Dynam. Systems* **3** (1982), 135–157.

DAVID BAO
DEPARTMENT OF MATHEMATICS
UNIVERSITY OF HOUSTON
HOUSTON, TX 77204-3008
UNITED STATES
bao@math.uh.edu

COLLEEN ROBLES
DEPARTMENT OF MATHEMATICS
UNIVERSITY OF ROCHESTER
ROCHESTER, NY 14627-0001
UNITED STATES
robles@math.rochester.edu

Riemann–Finsler Geometry
MSRI Publications
Volume **50**, 2004

Nonreversible Finsler Metrics
of Positive Flag Curvature

HANS-BERT RADEMACHER

CONTENTS

1. Introduction

Finsler geometry is an essential extension of Riemannian geometry. Instead of an inner product on every tangent space one considers Minkowski norms on every tangent space. For a Finsler metric the unit sphere in each tangent space is a strictly convex hypersurface. One obtains for every nonzero tangent vector an inner product, arising from Minkowski norm; in the Riemannian case these inner products all coincide on a fixed tangent space. The length of a smooth curve is well-defined. Geodesics — locally length-minimizing curves parametrized with constant speed — are uniquely defined for a given initial direction. From the viewpoint of the calculus of variations Finsler metrics are a suitable generalization of Riemannian metrics such that the variational problem for the length of curves between two fixed points is positive and positive regular. In terms of physics a Finsler metric describes a Lagrangian system without a potential; a Riemannian metric can be viewed as the special case of quadratic kinetic energy.

But in contrast to the Riemannian case there is no canonical connection, so several connections have been used in Finsler geometry. We use here the one introduced by S.-S. Chern [Bao et al. 2000, Chapter 2], transposed to vector fields on the manifold for a fixed direction field: Given a nowhere vanishing vector field V in an open nonempty subset U, there is a uniquely determined torsionfree connection ∇^V that is almost metric. Using this connection one can define the *flag curvature*, which generalizes the sectional curvature in Riemannian geometry and controls the infinitesimal behavior of geodesics. Given a geodesic c and a nowhere vanishing geodesic vector field V in an open neighborhood of c extending the velocity field c' of the geodesic, there is a Riemannian metric g_V on U such that c is also a geodesic of the Riemannian manifold (U, g_V) and the flag curvature $K(c'; \sigma)$ for any plane σ containing c' coincides with the sectional curvature $K(\sigma)$ of g_V. In particular the Jacobi fields of the Finsler metric and of g_V coincide along the geodesic c. Thus the flag curvature can be introduced without selecting a connection. The flag curvature does not completely determine the metric. For example, in contrast to the Riemannian case, there are Finsler metrics of constant positive flag curvature on spheres that are not isometric to the standard Riemannian metric; see [Bao et al. 2003; Bryant 2002; Shen 2002] for their characterization.

We call a Finsler metric F *reversible* (or *symmetric*) if opposite vectors have the same length: $F(X) = F(-X)$ for all tangent vectors X. In this case the unit sphere $T_p^1 M = \{X \in T_p M \mid F(X) = 1\}$ is symmetric under reflection through the origin. But this assumption excludes many interesting examples, for example *Randers metrics*, which are Finsler metrics defined by adding a one-form to the norm induced by a Riemannian metric: $F(X) = \sqrt{g(X,X)} + \alpha(X)$, where g is a Riemannian metric and α is a one-form.

In Riemannian geometry a metric with constant positive sectional curvature on a compact simply connected manifold is isometric to the standard sphere of the same curvature. The now classical *Sphere Theorem* states that a compact, simply connected manifold of dimension n with sectional curvature K such that $\frac{1}{4} < K \leq 1$ everywhere is homeomorphic to the n-sphere [Klingenberg 1995, § 2.8; Abresch and Meyer 1997]. In this form the result is contained in [Klingenberg 1961]; earlier contributions are due to M. Berger, H. Rauch and V. A. Toponogov [Berger 1998, I A 2]. The proof uses an estimate for the injectivity radius and the Toponogov comparison theorem for geodesic triangles. In [Klingenberg 1963] it is shown that one can prove the Sphere Theorem without making use of Toponogov's comparison result for geodesic triangles. Instead one uses Morse theory of the energy functional on the space of curves between two fixed points and on the space of loops. The injectivity radius is bounded from below by π, so geodesic loops have length at least 2π and their Morse index is bounded below by $n-1$. This implies that the loop space is $(n-2)$-connected and therefore the manifold is homotopy equivalent to the n-sphere.

Though in Finsler geometry the condition of constant positive flag curvature no longer determines the metric up to isometry, one can show using the exponential map that a simply connected Finsler manifold of constant positive flag curvature is homeomorphic to the n-sphere. P. Dazord [1968a; 1968b] remarked that in the case of a *reversible* Finsler metric one can carry over the Morse-theoretic proof of the Sphere Theorem found in [Klingenberg 1963]. The original proof of the Sphere Theorem does not carry over, since the triangle comparison result cannot be extended to the Finsler case.

The main topic of this article is to show in detail how the estimates for the injectivity radius, the length of a nonminimal geodesic between two fixed points, the length of a nonconstant geodesic loop and the length of a nonconstant closed geodesic can be extended to the case of a *nonreversible* Finsler metric by introducing the notion of *reversibility* $\lambda := \sup\{F(-X) \mid F(X) = 1\} \geq 1$. We will derive from a length estimate (Proposition 9.9) and from Theorem 9.10 the following *Sphere Theorem for nonreversible Finsler metrics*:

THEOREM 9.11. *A simply connected and compact Finsler manifold of dimension $n \geq 3$ with reversibility λ and with flag curvature K satisfying $\left(1 - \frac{1}{1+\lambda}\right)^2 < K \leq 1$ is homotopy equivalent to the n-sphere.*

A proof appears in [Rademacher 2004]. In this article we will present this result in detail, adopt a slightly different approach at places. The examples due to A. Katok of nonreversible Finsler metrics on S^2 with only two geometrically distinct closed geodesics are of great importance in the theory of closed geodesic as a test case for several statements [Rademacher 1992, §5.3]. It was pointed out in [Rademacher 2004, Chapter 5] that the Finsler metric of Katok's example on S^2 coincides with the Finsler metric of constant flag curvature constructed in [Shen 2002]. These examples show that the length estimate for a shortest geodesic given in Theorem 9.10 is sharp. Using the Legendre transformation, we see that Katok's examples describe Finsler metrics of *Randers type*.

It remains an open problem whether one can improve the Sphere Theorem in the nonreversible case by choosing the lower curvature bound $\frac{1}{4}$ as in the reversible case.

2. Conventions

We consider metric structures on a differentiable manifold $M = M^n$ of dimension n. If not otherwise stated, differentiable means C^∞-differentiable. The *tangent bundle* of M is denoted by TM, with projection $\tau : TM \to M$, and $T_x M := \tau^{-1}(x)$ for $x \in M$. We denote by $\mathcal{V}M$ the *vector space of smooth vector fields* on M, that is, the space $\Gamma(TM)$ of smooth sections of the tangent bundle. The zero section $T^0 M$ of TM is the union of the zero vectors $0_x \in T_x M$; it can be identified with M. The *cotangent bundle* of M is denoted by $T^* M$.

If $(x^1, \ldots, x^n)$ are coordinates on M, the coordinate vector fields $(\partial_1, \ldots, \partial_n)$ defined by $\partial_i(x) = (\partial/\partial x^i)(x)$ form a basis for the tangent space $T_x M$. For this set of coordinates, the tangent bundle can be given *canonical coordinates* by associating $(x, y) = (x^1, \ldots, x^n, y^1, \ldots, y^n)$ to the tangent vector $\sum_{i=1}^n y^i \partial_i(x) \in T_x M$. A vector field V can be written as $V(x) = \sum_{i=1}^n v^i(x) \partial_i(x)$.

The real vector space of differentiable functions $f : M \to \mathbb{R}$ is denoted by $\mathcal{F}M = C^\infty(M)$. A multilinear map

$$A : \underbrace{\mathcal{V}M \times \cdots \times \mathcal{V}M}_{k} \to \mathcal{F}M$$

is called a $(0, k)$-*tensor field* on M if it is linear in each argument with respect to the vector space $\mathcal{F}M$. A multilinear map $A : \mathcal{V}M \times \cdots \times \mathcal{V}M \mapsto \mathcal{V}M$ is called a $(1, k)$-*tensor field* on M if it satisfies the same condition. A $(0, k)$-tensor field A on M is *symmetric* if for any $x \in M$ the induced k-linear map $A_x : T_x M \times \ldots \times T_x M \to \mathbb{R}$ is symmetric, that is, satisfies $A_x(X_{\sigma(1)}, \ldots, X_{\sigma(k)}) = A_x(X_1, \ldots, X_k)$ for all $X_1, \ldots, X_k \in T_x M$ and all permutations $\sigma \in \mathfrak{S}_k$. Symmetric $(1, k)$-tensor fields are defined similarly.

Let X be a vector field on M and let $f \in \mathcal{F}M$. For $p \in M$ and $\gamma : (-\varepsilon, \varepsilon) \to M$ a smooth curve with $\gamma(0) = p$ and $\gamma'(0) = X(p)$, the quantity $(d/dt)|_{t=0} f(\gamma(t))$ does not depend on the choice of γ. As p varies, this defines a new function on M, called the *Lie derivative* of f in the direction of X, and written Xf or $X(f)$. We also write $df(p)X$ for $Xf(p)$.

The projection $\tau : TM \to M$ induces by differentiation the *double tangent bundle* $\tau_* : TTM \to TM$; for any $X \in T_x M$, the space $T_X(TM)$ is called the *double tangent space* at $X \in TM$ and has dimension $2n$, where n is the dimension of M. The tangent vectors $Y'(0)$ of *vertical curves* $Y : (-\varepsilon, \varepsilon) \to T_x M \subset TM$ with $Y(0) = X$ span a distinguished n-dimensional subspace of $T_X(TM)$, called the *vertical tangent space* and denoted by $T_X^v(TM) = T_X(T_x M)$. Hence $T_X^v(TM) = \ker(d\tau : T_X(TM) \to T_x M)$. Together the vertical subspaces form the *vertical subbundle* $T^v TM \subset TTM$.

Given a tangent vector $Y \in T_x M$, we define a map $\bar{Y} : T_x M \to T(T_x M) = T_x M \times T_x M$ by setting $\bar{Y}(X) = (X, Y)$. Then from any vector field on M we obtain an associated vertical vector field on TM (that is, a section of $T^v TM$): its value at $X \in T_x M$ is $\bar{Y}(X)$, where Y is the value at x of the given vector field on M. All of this is independent of coordinates.

If Y is a vector field on M with associated vertical vector field $\bar{Y}$ on TM, the Lie derivative of a function $F : U \subset TM \to \mathbb{R}$ with respect to $\bar{Y}$ is given by

$$\bar{Y}F(V) = (d/dt)|_{t=0} F(V + tY).$$

In terms of a set of canonical coordinates $(x, y) = (x^1, \ldots, x^n, y^1, \ldots, y^n)$ on TM, we have $\overline{\partial/\partial x^i} F = \partial/\partial y^i F$ for F equal to each coordinate function; hence

$$\overline{\partial/\partial x^i} = \partial/\partial y^i.$$

3. Finsler Metrics

DEFINITION 3.1. A *Finsler manifold* (M, F) is a differentiable manifold M equipped with a Finsler metric F. A *Finsler metric* on M is a continuous map, $F : TM \to \mathbb{R}$ differentiable outside the zero section $T^0 M$ and satisfying three conditions:

(1) F is *positively homogeneous*, that is, $F(\mu X) = \mu F(X)$ for all positive $\mu \in \mathbb{R}$ and all tangent vectors $X \in TM$.
(2) If $F(X) = 0$ then $X = 0$.
(3) The *Legendre condition* or *strong convexity condition*: for any nonzero $V \in T_x M$, the symmetric bilinear form $g_V : T_x M \times T_x M \to \mathbb{R}$ given by

$$g_V(X, Y) = \langle X, Y \rangle_V := \tfrac{1}{2} \overline{X}\overline{Y} F^2(V) = \frac{1}{2} \frac{\partial^2}{\partial s \partial t} \Big|_{\substack{s=0 \\ t=0}} F^2(V + sX + tY)$$

is positive definite.

REMARK 3.2. (a) In terms of a set of canonical coordinates $(x, y) = (x^1, \ldots, x^n, y^1, \ldots, y^n)$ on TM, and setting

$$g_{ij}(x, y) := g_{(x,y)} \left(\frac{\partial}{\partial x^i}, \frac{\partial}{\partial x^j} \right) = \frac{1}{2} \frac{\partial^2 F^2}{\partial y^i \partial y^j}(x, y),$$

the Legendre condition states that the symmetric matrix $\big(g_{ij}(x, y) \big)_{1 \le i, j \le n}$ is positive definite whenever $y \ne 0$.

(b) Since $F(\mu X) = \mu F(X)$ for all $\mu > 0$, we have

$$\langle V, V \rangle_V = \frac{1}{2} \frac{d^2}{dt^2} \Big|_{t=0} F^2(V + tV) = \frac{1}{2} \frac{d^2}{dt^2} \Big|_{t=0} (1 + t)^2 F^2(V) = F^2(V)$$

In coordinates,

$$\frac{1}{2} y^i y^j \frac{\partial^2}{\partial y^i \partial y^j} F^2(x, y) = F^2(x, y).$$

(c) *Euler's Theorem* states that for a positively homogeneous function $f : V \to \mathbb{R}$ of order k on a vector space V (meaning that $f(\mu X) = \mu^k f(X)$ for $X \in V$ and $\mu > 0$), the radial derivative coincides up to the factor k with f itself:

$$\sum_{i=1}^{n} y^i \frac{\partial}{\partial y^i} f(y) = k f(y).$$

See [Bao et al. 2000, Theorem 1.2.1].

DEFINITION 3.3. The *Legendre transformation* on a Finsler manifold (M, F) is the map $\mathcal{L}_F : TM \to T^* M$ defined by

$$\mathcal{L}_F(V)(W) = g_V(V, W).$$

One can view $\mathcal{L}_F(V)$ as the 1-form dual to V with respect to the metric g_V.

LEMMA 3.4. *If F is a Finsler metric, $F(X + Y) \leq F(X) + F(Y)$ for all $X, Y \in T_x M$. Equality holds only if $Y = \mu X$ for some $\mu \geq 0$.*

This implies that $\langle X, Y \rangle_Y \leq F(X)F(Y)$ for all $X, Y \in T_x M$, with equality if and only if $Y = \mu X$ for some $\mu \geq 0$ [Shen 2001a, § 1.2; Bao et al. 2000, p. 10].

LEMMA 3.5 [Bao et al. 2000, Proposition 14.8.1; Shen 2001a, Lemma 1.2.4]. *Let $V, W \in T_x M$ be nonzero. If $\langle X, V \rangle_V = \langle X, W \rangle_W$ for all $X \in T_x M$, then $V = W$.*

Consequently, the Legendre transformation is an isomorphism.

A *Riemannian metric* g on a manifold M is symmetric $(2, 0)$-tensor field $g : \mathcal{V}M \times \mathcal{V}M \to \mathcal{F}M$ such that for every $x \in M$ the bilinear map $g_x : T_x M \times T_x M \to \mathbb{R}$ is positive definite. The associated Finsler metric is defined by $F(X) = \sqrt{g(X, X)}$; forming the metric g_V for any nonzero V we obtain g for every nonzero V.

EXAMPLE 3.6 (RANDERS METRICS). Suppose given a Riemannian metric α and a differential 1-form β. There is a vector field ζ satisfying $\beta(X) = \alpha(X, \zeta)$ for all X; we say ζ is *dual* to β with respect to α. Define $\|\beta\| := \|\zeta\| = \sqrt{\alpha(\zeta, \zeta)}$. If $\|\beta\| < 1$ everywhere,

$$F(X) := \sqrt{\alpha(X, X)} + \beta(X) = \sqrt{\alpha(X, X)} + \alpha(X, \zeta)$$

defines a Finsler metric. This type of Finsler metric is called a *Randers metric*.

Since $X \neq 0$ implies $\|X\| = \sqrt{\alpha(X, X)} > 0$, we have

$$F(X) = \|X\| \left(1 + \alpha \left(\frac{X}{\|X\|}, \zeta \right) \right) \geq \|X\| \left(1 - \|\zeta\| \right) > 0,$$

showing that F satisfies condition (2) of Definition 3.1. Condition (1) is obvious. For the proof of (3) we refer to [Bao et al. 2000, Chapter 11].

A Randers metric $F(X) = \sqrt{\alpha(X, X)} + \alpha(X, \zeta)$ is only positively homogeneous. If $\alpha(X, \zeta) \neq 0$ then $F(-X) \neq F(X)$. This motivates the following notion:

DEFINITION 3.7. On a Finsler manifold (M, F) the *reversibility function* $\lambda : M \to \mathbb{R}^+$ is defined by

$$\lambda(x) := \sup \left\{ F(-X) \mid X \in T_x M, \, F(X) = 1 \right\}.$$

The number $\lambda = \lambda(M, F) = \sup \left\{ \lambda(x) \mid x \in M \right\}$ is called the *reversibility* of the Finsler manifold (M, F), if it exists—for example, if M is compact.

One has to show that the function λ is continuous, which is done using a standard argument: The subspace $T_x^1 M = \{ X \in T_x M \mid F(X) = 1 \}$ is called the *unit sphere* or *indicatrix* at the point x. It is a compact space diffeomorphic to the sphere S^{n-1}. The subspaces $T_x^1 M$, $x \in M$, form a sphere bundle over M. The function $X \in T_x^1 M \mapsto F(-X) \in \mathbb{R}$ is continuous, therefore the supremum is actually the maximum of this function. The sphere bundle $T^1 M \to M$ is locally trivial, that is, for small open sets $U \subset M$ the restriction $T^1 U \to U$ can be identified via a fiber-preserving diffeomorphism with the canonical projection $U \times S^{n-1} \to U$ on the first factor. The function $T^1 U \cong U \times S^{n-1} \mapsto F(-X) \in \mathbb{R}$ is continuous. A proof by contradiction, using the sequential compactness of S^{n-1}, then shows that the reversibility function $\lambda : M \to \mathbb{R}$ is continuous. If M is compact, the function is bounded and the supremum is actually a maximum. Since F is positively homogeneous we could also write

$$\lambda(x) = \max\left\{ \frac{F(-X)}{F(X)} \;\middle|\; X \in T_x M,\, X \neq 0 \right\}.$$

We call a Finsler metric *reversible* if $F(-X) = F(X)$ for all $X \in TM$. Then obviously $\lambda = 1$. If there is a tangent vector X such that $F(-X) \neq F(X)$ then

$$\lambda \geq \max\left\{ \frac{F(-X)}{F(X)},\, \frac{F(X)}{F(-X)} \right\} > 1.$$

Hence a Finsler metric is reversible if and only if $\lambda = 1$. In this case the *indicatrix* $T_p^1 M$ is symmetric with respect to reflection $X \mapsto -X$. Sometimes a reversible metric is also called *symmetric*, but this terminology conflicts with other notions such as symmetric quadratic forms and symmetric spaces.

For an arbitrary Finsler metric on a compact manifold M we obtain

$$\lambda^{-1} F(X) \leq F(-X) \leq \lambda F(X).$$

If $\gamma : [0,1] \to M$ is a smooth curve on M, we define the length of γ as $L(\gamma) = \int_0^1 F(\gamma'(t))\, dt$. We also introduce $\gamma^{-1} : [0,1] \to M$, the curve γ run in reverse: $\gamma^{-1}(t) = \gamma(1-t)$. The lengths of γ and γ^{-1} satisfy

$$\frac{1}{\lambda} L(\gamma) \leq L(\gamma^{-1}) \leq \lambda L(\gamma). \tag{3-1}$$

EXAMPLE 3.8. Let $F(X) = \sqrt{a(X,X)} + \alpha(X, \zeta)$ be a Randers metric, with vector field ζ. For fixed $x \in M$, we will find $\lambda(x)$ by looking at the quotient $F(-X)/F(X)$ on the unit ball $\{X \in T_x M \mid a(X,X) = 1\}$ of a:

$$X \mapsto \frac{F(-X)}{F(X)} = \frac{1 - \alpha(X, \zeta(x))}{1 + \alpha(X, \zeta(x))}.$$

This quotient attains its maximum for $X = -\zeta(x)/\|\zeta(x)\|$, and we obtain

$$\lambda(x) = \frac{1 + \|\zeta(x)\|}{1 - \|\zeta(x)\|}.$$

For a nonzero tangent vector $V \in T_x M$, we define on $T_x M$ the trilinear form

$$\langle X_1, X_2, X_3 \rangle_V := \tfrac{1}{4} \bar{X}_1 \bar{X}_2 \bar{X}_3 F^2(V)$$

$$= \frac{1}{4} \frac{\partial^3}{\partial s_1 \partial s_2 \partial s_3}\bigg|_{(s_1, s_2, s_3) = (0,0,0)} F^2\left(V + \sum_{i=1}^{3} s_i X_i \right).$$

For a given everywhere nonzero vector field V defined on an open subset $U \subset M$, we obtain a symmetric $(0,3)$-tensor, called the *Cartan tensor*; its coefficients are usually denoted C_{ijk} (that is, the functions C_{ijk} at $V \in T_x M$ express the trilinear form $\langle\ ,\ ,\ \rangle_V$ in a given system of canonical coordinates).

The Cartan tensor vanishes if and only if the Finsler metric comes from a Riemannian metric g (meaning that $F^2(X) = g(X, X)$). Euler's theorem implies that

$$\langle V, X, Y \rangle_V = \langle X, V, Y \rangle_V = \langle X, Y, V \rangle_V = 0 \qquad (3\text{-}2)$$

for all vector fields X, Y.

A *distance function* on a differentiable manifold M is a smooth function $\theta : M \times M \to [0, \infty)$ such that $\theta(p, q) = 0$ if and only if $p = q$, and such that the triangle inequality is satisfied:

$$\theta(p, q) \le \theta(p, r) + \theta(r, q) \quad \text{for all } p, q, r \in M.$$

Given a connected Finsler manifold (M, F), the *induced distance* $\theta : M \times M \to \mathbb{R}$ is defined by

$$\theta(p, q) := \inf \{ L(c) \mid c : [0, 1] \to M \text{ piecewise smooth, } c(0) = p,\ c(1) = q \}.$$

(*Piecewise smooth* means that c is continuous and there is a finite partition of the interval $[0, 1]$ such the restriction of c to each closed subinterval is smooth.) It is easy to check that θ is a distance function. If the Finsler metric is nonreversible, the induced metric is not symmetric: there are points p, q with $\theta(p, q) \ne \theta(q, p)$.

LEMMA 3.9. *The induced distance of a Finsler manifold (M, F) with reversibility $\lambda \ge 1$ satisfies*

$$\frac{1}{\lambda}\theta(p, q) \le \theta(q, p) \le \lambda\theta(p, q) \qquad (3\text{-}3)$$

PROOF. For every $k \in \mathbb{N}$, let $\gamma_k : [0, 1] \to M$ be a piecewise smooth curve with $p = \gamma_k(0)$, $q = \gamma_k(1)$ and $L(\gamma_k) \le \theta(p, q) + 1/k$. Then (3–1) gives $L(\gamma_k^{-1}) \le \lambda\theta(p, q) + \lambda/k$ for all k. Thus $\theta(q, p) \le L(\gamma_k) \le \lambda\big(\theta(p, q) + 1/k\big)$ for all k. $\qquad \square$

Given a Finsler manifold (M, F), the *symmetrized distance* $d : M \times M \to \mathbb{R}$ is defined by $d(p, q) = \frac{1}{2}\big(\theta(p, q) + \theta(q, p)\big)$. The distance functions θ and d of a Finsler manifold coincide if and only if the Finsler metric is reversible.

For U an open subset of a manifold M, recall that $\mathcal{V}U$ is the space of smooth vector fields on U, and let $\mathcal{V}U^+ \subset \mathcal{V}U$ be the subset of nowhere vanishing vector fields. For the next theorem we recall the definition of an *affine connection*: a map $\nabla^V \colon (X, Y) \in \mathcal{V}U \times \mathcal{V}U \mapsto \nabla^V_X Y \in \mathcal{V}U$, linear in Y and satisfying

$$\nabla^V_X(fY) = f\nabla^V_X Y + X(f)Y \quad \text{and} \quad \nabla^V_{fX} Y = f\nabla_X Y \quad \text{for all } f \in \mathcal{F}U,\ X, Y \in \mathcal{V}U.$$

THEOREM 3.10. *Let (M, F) be a Finsler manifold and $U \subset X$ an open subset. There is a map*

$$\nabla : (V, X, Y) \in \mathcal{V}U^+ \times \mathcal{V}U \times \mathcal{V}U \mapsto \nabla^V_X Y \in \mathcal{V}U$$

with the following properties:

(a) *for every $V \in \mathcal{V}U^+$, the map $\nabla^V : (X, Y) \in \mathcal{V}U \times \mathcal{V}U \mapsto \nabla^V_X Y \in \mathcal{V}U$ is an affine connection.*

(b) ∇^V *is torsionfree, that is,*

$$\nabla^V_X Y - \nabla^V_Y X = [X, Y] \quad \text{for all } X, Y \in \mathcal{V}U. \tag{3-4}$$

(c) ∇^V *is almost metric, that is,*

$$X(\langle Y, Z\rangle_V) = \langle \nabla^V_X Y, Z\rangle_V + \langle Y, \nabla^V_X Z\rangle_V + 2\langle \nabla^V_X V, Y, Z\rangle_V. \tag{3-5}$$

Moreover we have

$$\begin{aligned}
2\langle \nabla^V_X Y, Z\rangle_V = {}& X(\langle Y, Z\rangle_V) + Y(\langle Z, X\rangle_V) - Z(\langle X, Y\rangle_V) \\
& + \langle [X, Y], Z\rangle_V - \langle [Y, Z], X\rangle_V + \langle [Z, X], Y\rangle_V \\
& - 2\langle \nabla^V_X V, Y, Z\rangle_V - 2\langle \nabla^V_Y V, Z, X\rangle_V + 2\langle \nabla^V_Z V, X, Y\rangle_V \tag{3-6}
\end{aligned}$$

for all vector fields $X, Y, Z \in \mathcal{V}U$, and this equation, called the generalized Koszul formula, uniquely determines ∇.

SKETCH OF PROOF. From the requirements (3–4) and (3–5) we obtain, through straightforward calculations, the generalized Koszul formula (3–6). Equations (3–5), (3–6) and (3–2) then imply that

$$\langle \nabla^V_V V, Z\rangle_V = 2V(\langle V, Z\rangle_V) - Z(\langle V, V\rangle_V) + 2\langle [Z, V], V\rangle_V$$

and

$$\begin{aligned}
2\langle \nabla^V_X V, Z\rangle_V = {}& X(\langle V, Z\rangle_V) + V(\langle Z, X\rangle_V) - Z(\langle X, V\rangle_V) \\
& + \langle [X, V], Z\rangle_V - \langle [V, Z], X\rangle_V + \langle [Z, X], V\rangle_V - 2\langle \nabla^V_V V, Z, X\rangle_V. \tag{3-7}
\end{aligned}$$

Thus the right-hand side of (3–6) can be expanded into an expression devoid of any reference to ∇, showing that $\nabla^V_X Y$ is uniquely determined. Then one has to check that the ∇^V thus defined is in fact an affine connection, torsionfree and almost metric. □

REMARK 3.11. (a) In the Riemannian case the connection $\nabla^V = \nabla$ is independent of $V \in \mathcal{V}U$, it is *metric*, meaning that $X(\langle Y, Z\rangle) = \langle \nabla_X Y, Z\rangle + \langle Y, \nabla_X Z\rangle$, and it is determined by the *Koszul formula*:

$$\begin{aligned}
2\langle \nabla_X Y, Z\rangle = {}& X\langle Y, Z\rangle + Y\langle Z, X\rangle - Z\langle X, Y\rangle \\
& + \langle [X, Y], Z\rangle - \langle [Y, Z], X\rangle + \langle [Z, X], Y\rangle. \tag{3-8}
\end{aligned}$$

∇ is called the *Levi-Civita connection* or *canonical connection*.

(b) We point out the correspondence between the development adopted above (which one can find in [Matthias 1980, Chapter 2]) and the description given in [Bao et al. 2000, Chapter 2] and [Shen 2001a, § 5.2]. In canonical coordinates

$(x, y) = (x^1, \ldots, x^n, y^1, \ldots, y^n)$ of the tangent bundle TU of an open set $U \subset M$, we obtain the functions

$$g_{ij} : (x, y) \in TU \mapsto g_{ij}(x, y) = g(x, y)\left(\frac{\partial}{\partial x^i}, \frac{\partial}{\partial x^j}\right) = \frac{1}{2}\frac{\partial^2}{\partial y^i \partial y^j}F^2(x, y).$$

The coefficients $C_{ijk} = C_{ijk}(x, y)$ of the Cartan tensor are

$$C_{ijk}(x, y) = \frac{1}{4}\frac{\partial^2}{\partial y^i \partial y^j}F^2(x, y) = \left\langle \frac{\partial}{\partial x^i}, \frac{\partial}{\partial x^j}, \frac{\partial}{\partial x^k} \right\rangle.$$

We also define $g^{ij}(x, y)$ as the coefficients of the inverse matrix of $g_{ij}(x, y)$. Then one can define *formal Christoffel symbols* $\gamma^i_{jk} : TU \to \mathbb{R}$:

$$\gamma^i_{jk}(x, y) = \tfrac{1}{2}g^{il}(x, y)\left(\frac{\partial g_{lj}}{\partial x^k}(x, y) - \frac{\partial g_{jk}}{\partial x^l}(x, y) + \frac{\partial g_{kl}}{\partial x^j}(x, y)\right).$$

We can raise and lower indices by contracting with the coefficients g^{ij} and g_{ij}; for example, $C^i_{jk} = g^{il}C_{ljk}$. Here we use the Einstein summation convention. Then we define the quantities

$$N^i_j = N^i_j(x, y) = \gamma^i_{jk}(x, y)y^k - C^i_{jk}(x, y)\gamma^k_{rs}y^r y^s.$$

It turns out that the coefficients of the Chern connection are

$$\Gamma^i_{jk}(x, y) = \gamma^i_{jk} - g^{li}\left(C_{ijr}N^r_k - C_{jkr}N^r_i + C_{kir}N^r_j\right),$$

(see [Bao et al. 2000, (2.4.9)]), the Chern connection is given by

$$\nabla_{\partial/\partial x^i}\frac{\partial}{\partial x^j}(x, y) = \Gamma^k_{ij}(x, y)\frac{\partial}{\partial x^k},$$

and one shows that

$$N^j_i(x, y) = \Gamma^j_{ik}(x, y)y^k.$$

The Chern connection is torsionfree, which implies that $\Gamma^k_{ij} = \Gamma^k_{jk}$. It is also almost metric, which implies (see [Shen 2001a, (5.22), (5.29)])

$$\frac{\partial g_{jl}}{\partial x^m}(x, y) = g_{kl}\Gamma^i_{jm} + g_{kj}\Gamma^k_{lm} + 2C_{jkl}N^k_m$$

$$= g_{kl}\Gamma^i_{jm} + g_{kj}\Gamma^k_{lm} + 2C_{jkl}\Gamma^k_{mr}y^r.$$

In particular:

LEMMA 3.12. *For two vector fields* $V, W \in \mathcal{V}U^+$ *and a point* $p \in U$ *with* $V(p) = W(p)$ *and for arbitrary vector fields* $X, Y \in \mathcal{V}U$ *we have:*

$$\nabla^V_X Y(p) = \nabla^W_X Y(p).$$

Using the connection ∇^V we introduce the *covariant derivative* ∇^V/dt along a curve $c : [a, b] \to M$. For a vector field X along the curve c with tangent vector field c' define $(\nabla^V/dt)X(t) = \nabla^V_{c'}X(t)$, where on the right-hand side one has to take extensions of the vector fields V, X, c' onto an open subset containing the curve. This expression is independent of the chosen extensions. If the vector fields V, c' coincide, we also write simply $(\nabla^V/dt)X = (\nabla/dt)X$.

For a differentiable map $H : [0, 1] \times [0, 1] \to M$ and a vector field $X(s, t)$ along F (meaning that $X(s, t) \in T_{H(s,t)}M$), we define $(\nabla^V/\partial t)X(t)$ as a vector field along the curve $t \mapsto H(s_1, t)$ for a fixed s_1 and $(\nabla^V/\partial s)X(t)$ as a vector field along the curve $s \mapsto H(s, t_1)$ for a fixed t_1. Then we obtain the following rule for exchanging the order of differentiation:

$$\frac{\nabla^V}{\partial t} \frac{\partial H}{\partial s} = \frac{\nabla^V}{\partial s} \frac{\partial H}{\partial t}$$

This rule follows since the connection ∇^V is torsionfree.

4. First Variation of the Energy Functional

In the Morse-theoretic proof of the Sphere Theorem we use the energy functional E on a suitable space of curves as the Morse function. For a smooth curve $c : [0, 1] \to M$, the *energy* is defined as

$$E(c) = \frac{1}{2} \int_0^1 F^2\big(c'(t)\big)\, dt.$$

For a variation c_s in the first variation formula one studies the first derivative $(d/ds)|_{s=0}E(c_s)$:

LEMMA 4.1 (FIRST VARIATIONAL FORMULA). *If $c_s : [a, b] \to M$, for $s \in (-\varepsilon, \varepsilon)$, is a smooth variation of the curve $c = c_0$ with variation vector field $V(t) = (\partial/\partial s)|_{s=0}c_s(t)$, then*

$$\left.\frac{d}{ds}\right|_{s=0} E(c_s) = \big\langle c'(b), V(b) \big\rangle_{c'(b)} - \big\langle c'(a), V(a) \big\rangle_{c'(a)} - \int_a^b \left\langle \frac{\nabla}{dt} c', V \right\rangle_{c'} dt. \quad (4\text{-}1)$$

PROOF.

$$\frac{1}{2} \frac{\partial}{\partial s} \langle c'_s, c'_s \rangle_{c'_s} = \left\langle \frac{\nabla^{c'_s}}{\partial s} c'_s, c'_s \right\rangle_{c'_s} + \underbrace{\left\langle \frac{\nabla^{c'_s}}{\partial s} c'_s, c'_s, c'_s \right\rangle_{c'_s}}_{=0} = \left\langle (\nabla/\partial t) \frac{\partial c_s}{\partial s}, c'_s \right\rangle_{c'_s}$$

$$= \frac{\partial}{\partial t} \left\langle \frac{\partial c_s}{\partial s}, c'_s \right\rangle_{c'_s} - \left\langle \frac{\partial c_s}{\partial s}, (\nabla/\partial t)c'_s \right\rangle_{c'_s} - \underbrace{\left\langle \frac{\nabla^{c'_s}}{\partial s} c'_s, \frac{\partial c_s}{\partial s}, c'_s \right\rangle_{c'_s}}_{=0}$$

Hence we conclude that

$$\left.\frac{d}{ds}\right|_{s=0} \int_a^b \langle c'_s, c'_s \rangle_{c'_s} dt = \big\langle V(t), c'(t) \big\rangle_{c'} \Big|_a^b - \int_a^b \big\langle V(t), (\nabla/dt)c' \big\rangle_{c'} dt. \qquad \square$$

COROLLARY 4.2. *If $c : [0,1] \to M$ is a piecewise smooth curve such that no other piecewise smooth curve joining $p = c(0)$ and $q = c(1)$ is shorter, then c is a geodesic, that is, c is smooth and $(\nabla/dt)c' = 0$.*

PROOF. Let c be smooth when restricted to each subinterval $[t_j, t_{j+1}]$ of a partition $0 = t_0 < t_1 < \cdots < t_k = 1$ of $[0,1]$. We first want to prove that these restrictions are geodesics, so that c is a *broken geodesic* (also known as a *geodesic polygon*). If $(\nabla/dt)c'(s) \neq 0$ for some $s \in (t_j, t_{j+1})$, we choose a vector field $V(t)$ along the image of c as follows:

$$V(t) = \phi(t) \frac{\nabla}{dt} c'(t),$$

where $\phi : [0,1] \to [0,1]$ is a smooth function with $\phi(s) = 1$ and $\phi(t) = 0$ for $t \notin (t_j, t_{j+1})$. Then we take a smooth variation $c_s : [0,1] \to M$ of the piecewise smooth curve c with variation vector field $V(t) = (\partial/\partial s)|_{s=0} c_s(t)$. (Saying that the family c_s is a *smooth variation of* c is saying that, for each $j = 0, \ldots, k-1$, the restrictions $c_s|[t_j, t_{j+1}]$ form a smooth variation of $c|[t_j, t_{j+1}]$.) Then the first variation formula (Lemma 4.1) gives

$$0 = \frac{d}{ds}\Big|_{s=0} E(c_s) = \int_0^1 \left\langle \frac{\nabla}{dt}c', V \right\rangle_{\dot{c}} dt = \int_0^1 \phi(t) \left\| \frac{\nabla}{dt}c' \right\|^2 dt.$$

Since $\phi(t) \geq 0$ for all t and $\phi(s) = 1$, the right-hand side is positive, a contradiction. Hence no such s exists, and c is a broken geodesic.

Now fix $l \in \{1, 2, \ldots, k-1\}$ and choose any tangent vector $V_0 \in T_{c(t_l)}M$ and a variation vector field $V = V(t)$ along the broken geodesic c such that $V(t_l) = V_0$ and $V(t_j) = 0$ for $j \neq l$. Again, Lemma 4.1 shows that

$$0 = \frac{d}{ds}\Big|_{s=0} E(c_s) = \langle c'(t_l^+), V_0 \rangle_{c'(t_l^+)} - \langle c'(t_l^-), V_0 \rangle_{c'(t_l^-)}$$

for all $V_0 \in T_{c(t_l)}M$. (Here, as usual, $c'(t^{\pm}) = \lim_{\varepsilon \to 0,\, \varepsilon > 0} c'(t \pm \varepsilon)$.) We conclude that $c'(t_l^+) = c'(t_l^-)$, since the Legendre transformation is an isomorphism (Lemma 3.5). Hence c is a smooth curve. $\qquad\square$

A vector field $V \in \mathcal{V}U$ is called a *geodesic vector field* if $\nabla_V^V V = 0$, which says that the flow lines of V are geodesics of the Finsler metric. These lines can also be seen as geodesics of an associated Riemannian metric:

LEMMA 4.3. *Let V be a nowhere vanishing geodesic field defined on an open subset $U \subset M$. Denote by $\bar{\bar{\nabla}}$ the Levi-Civita connection of the Riemannian manifold (U, g_V). Then*

$$\nabla_X^V V = \bar{\bar{\nabla}}_X V$$

for all vector fields X; in particular, the vector field V is also geodesic for the Riemannian manifold (U, g_V).

PROOF. $\bar{\nabla}$ is uniquely determined by the Koszul formula (see (3–8), with $\bar{\nabla}$ playing the role of ∇). Since $\nabla_V^V V = 0$ we conclude from Equation 3–7 that

$$\langle \bar{\nabla}_X V, Z \rangle_V = \langle \nabla_X^V V, Z \rangle_V \tag{4-2}$$

for all vector fields X, Y. □

We obtain a similar statement if we restrict to vector fields along a given geodesic:

LEMMA 4.4. *Let* $c : [0, 1] \to M$ *be a non-self-intersecting geodesic of the Finsler manifold* (M, F), *and* $V \in \mathcal{V}U^+$ *an extension of the velocity vector field* c' *onto an open neighborhood* U *of* $c([0, 1])$. *We call the Riemannian manifold* (U, g_V) *an osculating Riemannian manifold, denote its Levi-Civita connection by* $\bar{\nabla}$ *and the covariant derivative along* c *by* $(\bar{\nabla}/dt)$. *Then* c *is also a geodesic of the osculating Riemannian metric* g_V *and*

$$\frac{\nabla}{dt} X(t) = \frac{\bar{\nabla}}{dt} X(t)$$

for any vector field X *along* c.

PROOF. As in the proof of Lemma 4.3 we show that (4–2) holds along the given geodesic:

$$\langle \bar{\nabla}_X V, Z \rangle_V (c(t)) = \langle \nabla_X^V V, Z \rangle_V (c(t)).$$

Then

$$(\nabla^V/dt)_{c'} X(t) = \nabla_{c'}^V X(t) = \nabla_X^V c'(t) + [c', X]$$
$$= \bar{\nabla}_X c' + [c', X] = \nabla_{c'}^V X(t) = (\bar{\nabla}/dt) X(t).$$

For $X = c'$ it follows that c is also a geodesic of the osculating Riemannian metric. □

5. Flag curvature, Jacobi Fields and Conjugate Points

For a Riemannian manifold (M, g) with Levi-Civita connection ∇, the *Riemann curvature tensor* is a $(1,3)$-tensor defined by

$$R(X,Y)Z = \nabla_X \nabla_Y Z - \nabla_Y \nabla_X Z - \nabla_{[X,Y]} Z, \quad \text{for} X, Y, Z \in \mathcal{V}M.$$

It is determined by the *Jacobi operators* or *directional curvature operators* R^X, $X \in T_x M$, given by

$$Y \in T_x M \mapsto R^X(Y) := R(Y, X)X.$$

The *sectional curvature* $K(\sigma) = K(X, Y)$ of a plane $\sigma \subset T_x M$ spanned by the tangent vectors X, Y is defined by

$$K(X,Y) = \frac{\langle R(X,Y)Y, X \rangle}{|X|^2 |Y|^2 - \langle X, Y \rangle^2} = \frac{\langle R^X(Y), Y \rangle}{|X|^2 |Y|^2 - \langle X, Y \rangle^2}.$$

The Finsler geometry counterparts of these entities can be introduced by considering the osculating Riemannian metric. In the next statement we make use of the notion of a *geodesic variation* of a geodesic c; this is simply a variation c_s of c such that each curve $c_s : [0, 1] \to M$ is geodesic.

PROPOSITION 5.1. (a) [Shen 2001b, Lemma 8.1.1] *For every nonzero tangent vector $X \in T_x M$ on a Finsler manifold M with geodesic $c = c_X : [0, 1] \to M$ and $c'_X(0) = X$, the map $R^X : T_x M \to T_x M$ given by*

$$R^X(Y(0)) = -\frac{\nabla^2}{dt^2} Y(0),$$

where $Y(t) = (\partial/\partial s)|_{s=0} c_s(t)$ is the variation vector field of a geodesic variation of c, is well-defined and linear. It is called the Riemann curvature (operator) of the Finsler manifold.

(b) [Shen 2001b, Proposition 8.4.1] *For every nonzero tangent vector $X \in T_x M$ with a nonzero geodesic vector field V extending $X \in T_x M$ in an open neighborhood U of x, the Riemann curvature operator R^X of the Finsler manifold coincides with the Jacobi operator $\bar{R}^X$ of the osculating Riemannian metric $\bar{g} = g^V$ defined on U.*

We can now introduce the Finsler counterpart of the sectional curvature. In contrast with the Riemannian case, the notion depends not only on the choice of a two-dimensional tangent plane but also on a direction in this plane.

DEFINITION 5.2. For a Finsler manifold (M, F) and a *flag* (X, σ) consisting of a nonzero tangent vector $X \in T_x M$ and a plane $\sigma \subset T_x M$ spanned by the tangent vectors X, Y, the *flag curvature* is defined as

$$K(X; \sigma) = K(X; Y) = \frac{\langle R^X(Y), Y \rangle_X}{|X|_X^2 |Y|_X^2 - \langle X, Y \rangle_X^2}.$$

The notation $\delta < K \le 1$, where $\delta \in (0, 1)$, will be used often; it is a shorthand for the condition $\delta < K(X; \sigma) \le 1$ for all flags $(X; \sigma)$ in the tangent bundle. Given a nonzero vector field V on a Finsler manifold (M, F) with connection ∇^V (see Theorem 3.10), one can consider the curvature tensor R^V defined by

$$R^V(X, Y)Z = \nabla_X^V \nabla_Y^V Z - \nabla_Y^V \nabla_X^V Z - \nabla_{[X,Y]}^V Z.$$

If the vector field V is geodesic, it follows from the definition of the Riemannian curvature that

$$R^V(Y) = R^V(Y, V)V = -\nabla_V^V \nabla_Y^V V - \nabla_{[Y,V]}^V V.$$

As in the Riemannian case, the flag curvature geometrically controls the infinitesimal behavior of geodesics, as described by the *Jacobi fields* along a geodesic:

DEFINITION 5.3. On a Finsler manifold (M, F) we call a vector field $Y = Y(t)$ along a geodesic $c : [0, 1] \to M$ a *Jacobi field* if it satisfies the differential equation

$$\frac{\nabla^2}{dt^2} Y(t) + R^{c'}(Y) = 0.$$

It follows from Proposition 5.1 that the Jacobi fields of an osculating Riemannian metric (U, g_V) along the geodesic coincide with the Jacobi fields of the Finsler metric along c. Therefore the following well-known facts of Riemannian geometry (see [Klingenberg 1995, 1.12], for example) carry over to the Finsler case immediately:

LEMMA 5.4. *Let* (M, F) *be a Finsler manifold and* $c : [0, 1] \to M$ *a geodesic.*

(a) *For any* $Y_0, Y_1 \in T_{c(0)}M$, *there is a uniquely determined Jacobi field* Y *along* c *with initial conditions* $Y(0) = Y_0$ *and* $(\nabla/dt)Y(0) = Y_1$.

(b) *If, in addition,* $\langle Y_0, c'(0) \rangle_{c'(0)} = 0 = \langle Y_1, c'(0) \rangle_{c'(0)}$, *the Jacobi field* Y *thus defined satisfies* $\langle Y(t), c'(t) \rangle_{c'(t)} = 0$ *for all* $t \in [0, 1]$.

The Jacobi equation of Definition 5.3 is the linearization of the geodesic equation:

LEMMA 5.5. *Let* (M, F) *be a Finsler manifold.*

(a) *The variation vector field* $V(t) = (\partial/\partial s)|_{s=0}c_s(t)$ *of a geodesic variation* $c_s : [0, 1] \to M$ *on* M *is a Jacobi field.*

(b) *For every Jacobi field* $Y = Y(t)$ *along the geodesic* $c : [0, 1] \to M$ *there is a geodesic variation* $c_s : [0, 1] \to M$ *whose variation vector field coincides with* Y.

The proof from the Riemannian case carries over; see [Klingenberg 1995, 1.12.4].

DEFINITION 5.6. Let (M, F) be a Finsler manifold and X a unit tangent vector in TM. Let the geodesic parametrized by arc length with initial direction X be defined (at least) in the closed interval $[0, a]$, and denote it by $c : [0, a] \to M$, so that $c'(0) = X$. Suppose that for some $s \in (0, a)$ there is a nontrivial Jacobi field $Y = Y(t)$ along c that vanishes for $t = 0$ and $t = s$. (Nontrivial means that $(\nabla/dt)Y(0) \neq 0$.) Then the point $c(s)$ is called *conjugate* to $p = c(0)$ along c. Moreover, Y can be chosen so that $\langle Y, c' \rangle_{c'} = 0$, and the set of all Y satisfying all these conditions is a vector space whose dimension is called the *multiplicity* of the conjugate point $c(s)$. We define $\mathrm{conj}_X \in (0, \infty]$ as the smallest positive number r such that $c(r)$ is conjugate to p along c. The point $c(r)$ is called the *first conjugate point* to p along c. The *conjugate locus* is the set of all first conjugate points to p. The *conjugate radius* conj_p of a point $p \in M$ is the infimum of the set $\{\mathrm{conj}_X \mid X \in T_pM, F(X) = 1\}$.

The conjugate locus of p consists of critical points of the exponential map (see Section 8). The function $X \in T_p^1 M \mapsto \mathrm{conj}_X \in \mathbb{R}^+ \cup \{\infty\}$ is continuous. We denote by $\mathrm{conj} := \inf \{\mathrm{conj}_p \mid p \in M\}$ the *conjugate radius* of M. If M is compact this is a positive real number or ∞.

REMARK 5.7. In the case of constant flag curvature one can describe Jacobi fields explicitly. Let $c : \mathbb{R} \to M$ be a geodesic parametrized by arc length on a Finsler manifold (M, F) and assume that the flag curvature along c is a constant δ, meaning that $K(c'(t); V) = \delta$ for every t and every $V \in T_{c(t)}M$ forming a flag with $c'(t)$. One can choose an orthonormal basis $(e_1, e_2, \ldots, e_n)$ of the tangent space $T_{c(0)}M$ with respect to the metric $\langle \cdot, \cdot \rangle_{c'}$ with $e_n = c'(0)$. Using parallel transport defined by the covariant derivative ∇/dt along c we obtain a frame $(e_1(t), e_2(t), \ldots, e_n(t))$ along $c = c(t)$ orthonormal with respect to $\langle \cdot, \cdot \rangle_{c'}$ and satisfying $e_1(t) = c'(t)$ for all $t \in \mathbb{R}$. Then with $Y(t) = \sum_{i=2}^{n} y_i(t)e_i(t)$ the Jacobi equation for a Jacobi field orthogonal to c' (with respect to $\langle \cdot, \cdot \rangle_{c'}$) decouples because of the identities

$$\frac{\nabla^2}{dt^2}Y(t) = \sum_{i=2}^{n} y_i''(t)e_i(t)$$

and

$$R^{c'}(Y, c')c'(t) = \sum_{i,j=2}^{n} y_i \langle R^{c'}(e_i), e_j \rangle_{c'} e_j(t) = \sum_{i=2}^{n} y_i K(e_1; e_i)e_i(t) = \delta \sum_{i=2}^{n} y_i e_i(t)$$

into $n-1$ ordinary differential equations

$$y_i''(t) + \delta y_i(t) = 0, \quad i = 2, \ldots, n.$$

The solutions $y'' + \delta y = 0$, $y(0) = 0$, $y'(0) = 1$ are

$$y_\delta(t) = \begin{cases} 1/\sqrt{\delta}\sin(\sqrt{\delta}\,t) & \text{if } \delta > 0, \\ t & \text{if } \delta = 0, \\ 1/\sqrt{-\delta}\sinh(\sqrt{-\delta}\,t) & \text{if } \delta < 0. \end{cases}$$

Hence in the case of constant flag curvature $K(c(t)) = \delta$ along a geodesic c, we obtain for $\mathrm{conj}_{c'(0)}$ the value $\pi/\sqrt{\delta}$ if $\delta > 0$, and ∞ if $\delta \le 0$.

6. Second Variation of the Energy Functional

The first variational formula shows that geodesics can be seen as critical points of the energy functional. Therefore it is natural to study the second-order behavior of the energy functional at a geodesic. This leads to the second variational formula.

For a piecewise smooth curve $c : [0, 1] \to M$ denote by W_c the set of piecewise smooth vector fields along c. The *index form* of a geodesic $c : [0, 1] \to M$ is the symmetric bilinear form $I_c : W_c \times W_c \to \mathbb{R}$ defined by

$$I_c(X, Y) := \int_0^1 \left(\left\langle \frac{\nabla}{dt}X, \frac{\nabla}{dt}Y \right\rangle_{c'}(t) - \langle R^{c'}(X), c' \rangle_{c'}(t) \right) dt.$$

LEMMA 6.1 [Shen 2001a, §10.2; Shen 2001b, §8.5]. *Let $c_s : [0, 1] \to M$, for $s \in (-\varepsilon, \varepsilon)$, be a variation of the geodesic $c = c_0$ with fixed end points $c_s(0) = c(0)$, $c_s(1) = c(1)$; or let $c_s : S^1 \to M$ be a variation of the closed geodesic $c = c_0$ by closed curves. Let the variation vector field be $V(t) = (\partial c_s/\partial s)|_{s=0}c_s(t)$. Then*

$$\frac{d^2}{ds^2}\bigg|_{s=0} E(c_s) = I_c(V, V).$$

For a geodesic $c : [0, 1] \to M$ on a Finsler manifold (M, F) with an osculating Riemannian metric $\bar{g} = g_V$ defined in a neighborhood of $c([0, 1])$, the index forms $I_c, \bar{I}_c$ with respect to the Finsler metric and with respect to the osculating Riemannian metric coincide.

We introduce some subspaces of the space W_c of vector fields along c defined at the beginning of this section. W_c^0 denotes the subspace consisting of vector fields that vanish at the end points. If c is a closed geodesic, also known as a *periodic geodesic* (that is, $c(0) = c(1)$ and $c'(0) = c'(1)$), we denote by W_c^1 the subspace of W_c consisting of periodic vector fields $(X(0) = X(1))$. Note that a closed geodesic can be though of as having domain $S^1 = [0, 1]/\{0, 1\}$.

Using Lemma 6.1 we obtain:

COROLLARY 6.2. (a) *For points $p, q \in M$ let $c : [0, 1] \to M$ be a geodesic joining $p = c(0)$ and $q = c(1)$. The restriction of the index form of c to W_c^0 is denoted by I_c^0. If $c_s : [0, 1] \to M$, $s \in (-\varepsilon, \varepsilon)$, is a piecewise smooth variation of c with variation vector field $Y \in W_c^0$ and with fixed end points $p = c_s(0)$, $q = c_s(1)$, we have*

$$\frac{d^2}{ds^2}\bigg|_{s=0} E(c_s) = I_c^0(Y, Y).$$

(b) *Let $c : S^1 \to M$ be a closed geodesic and denote the restriction of the index form I_c to W_c^1 by I_c^1. If $c_s : S^1 \to M$, $s \in (-\varepsilon, \varepsilon)$ is a variation of c by closed curves with variation vector field Y, we have*

$$\frac{d^2}{ds^2}\bigg|_{s=0} E(c_s) = I_c^1(Y, Y).$$

We now define important invariants of geodesics and closed geodesics.

DEFINITION 6.3. (a) The *index* $\operatorname{ind} c$ of a geodesic $c : [0, 1] \to M$ joining points p and q of a Finsler manifold is by definition the same as the index $\operatorname{ind} W_c^0$ of the quadratic form

$$I_c^0 : W_c^0 \times W_c^0 \to \mathbb{R},$$

that is, the maximal dimension of a subspace on which I_c^0 is negative definite. The *nullity* $\operatorname{nul} c$ is the maximal dimension of a subspace $W' \subset W_c^0$ such that $I_c^0(X, Y) = 0$ for all $X \in W'$ and $Y \in W_c^0$.

(b) The Λ-*index* $\operatorname{ind}_\Lambda c$ of a closed geodesic $c : S^1 \to M$ on a Finsler manifold is the maximal dimension of a subspace on which the index form

$$I_c^1 : W_c^1 \times W_c^1 \to \mathbb{R}$$

is negative definite. The Λ-*nullity* $\operatorname{nul} c$ is the maximal dimension of a subspace $W'' \subset W_c^1$ such that $I_c^1(X, Y) = 0$ for all $X \in W''$ and $Y \in W_c^1$.

As in the Riemannian case, one can show that these numbers are finite:

LEMMA 6.4. *The index $\operatorname{ind} c$ and the nullity $\operatorname{nul} c$ of a geodesic c are finite. So are the Λ-index $\operatorname{ind}_\Lambda c$ and the Λ-nullity $\operatorname{nul} c$ of a closed geodesic c.*

PROOF. Let $c : [0,1] \to M$ be a geodesic. Choose a partition $0 = t_0 < t_1 < \cdots < t_k = 1$ of the unit interval such that there is no pair of conjugate points in $[t_i, t_{i+1}]$. Define the following subspaces of the vector space W_c^0 of piecewise smooth vector fields along c vanishing at the endpoints:

$J := \{X \in W_c^0 \mid X | [t_i, t_{i+1}]$ is a Jacobi field for $i = 0, \ldots, k-1\}$,

$H := \{X \in W_c^0 \mid X(t_i) = 0$ for $i = 1, \ldots, k-1\}$,

$W_c^\perp = \{X \in W_c^0 \mid \langle X, c' \rangle_{c'} = 0$ and $X | [t_i, t_{i+1}]$ is smooth for $i = 0, \ldots, k-1\}$.

Then, for $X, Y \in W_c^\perp$,

$$
I_c(X, Y) = \int_0^1 \left(\left\langle \frac{\nabla}{dt} X, \frac{\nabla}{dt} Y \right\rangle_{c'} - \langle R^{c'}(X, c')c', Y \rangle_{c'} \right) dt
$$

$$
= - \int_0^1 \left\langle \frac{\nabla^2}{dt^2} X + R^{c'}(X, c')c', Y \right\rangle_{c'} dt
$$

$$
+ \sum_{i=1}^{k-1} \left\langle \frac{\nabla}{dt} X(t_i^-) - \frac{\nabla}{dt} X(t_i^+), Y(t_i) \right\rangle_{c'}. \quad (6\text{--}1)
$$

It follows that J and $H \cap W_c^\perp$ are orthogonal with respect to I_c, that is, $I_c(X, Y) = 0$ for all $X \in J$ and $Y \in H \cap W_c^\perp$. Therefore we can conclude that W_c^0 is the direct orthogonal sum $J \oplus (H \cap W_c^\perp)$ with the following argument: For every $X_i \in T_{c(t_i)}M$ and $X_{i+1} \in T_{c(t_{i+1})}M$ there is a unique Jacobi field Y along $c | [t_i, t_{i+1}]$ with $Y(t_i) = X_i$ and $Y(t_{i+1}) = X_{i+1}$, since $c | [t_i, t_{i+1}]$ is by assumption free of pairs of conjugate points. On the other hand, the index form I_c^0 is positive definite on H since there is no conjugate point $c(t^*)$, $t^* \in (t_i, t_{i+1})$, to the point $c(t_i)$ along $c | [t_i, t^*]$. This shows that the indices and nullities match:

$$
\operatorname{ind} c = \operatorname{ind} I_c^0 = \operatorname{ind}(I_c^0 | J), \quad \operatorname{nul} c = \operatorname{nul} I_c^0 = \operatorname{nul}(I_c^0 | J). \quad (6\text{--}2)
$$

Since J is finite-dimensional these invariants are finite. An analogous proof shows that the Λ-index and Λ-nullity of a closed geodesic are also finite. $\square$

We call a geodesic c *nondegenerate* if $\operatorname{nul} c = 0$. This implies that the point $q = c(1)$ is not conjugate to $p = c(0)$ along c.

We call a closed geodesic $c : S^1 \to M$ *nondegenerate* if $\operatorname{nul}_\Lambda c = 1$. Since in this case $I_c^1(c', c') = 0$, the nullity is at least 1. (That's why some other authors define the nullity of a closed geodesic as $\operatorname{nul}_\Lambda c - 1$.) A Finsler metric all of whose closed geodesics are nondegenerate is called *bumpy*.

7. Results from Topology

Using the energy functional on the space $\Omega_{pq}M$ of curves on a Finsler manifold M joining two fixed points $p, q \in M$, we obtain a CW-decomposition of the space $\Omega_{pq}M$. The Morse indices of the geodesics in Ω_{pq} are related to the dimensions of the cells of the CW-decomposition. In this chapter we review results from the topology of CW-complexes and the relation between the topology of $\Omega_{pq}M$ and that of M. As general references we cite [Milnor 1969; Bredon 1995; Spanier 1966].

DEFINITION 7.1. Let A, X be topological spaces with $A \subset X$. We say that X is *obtained from A by adjoining k-cells* e_j^k, $j \in J_k$, if the following conditions hold:

(a) For every $j \in J_k$ the set e_j^k is a closed subset of X.
(b) Let $\dot{e}_j^k := e_j^k \cap A$. Then for all $i, j \in J_k$ with $i \neq j$ the subsets $e_i^k - \dot{e}_i^k$ and $e_j^k - \dot{e}_j^k$ are disjoint.
(c) The topology of $X = A \cup \bigcup_{j \in J_k} e_j^k$ is the *weak topology* with respect to the component subsets A and e_j^k, for $j \in J_k$. (This means that $U \subset X$ is open if and only if $U \cap A$ is open in A and $U \cap e_j^k$ is open in e_j^k for each j.)
(d) For every $j \in J_k$ there is a continuous map

$$\phi_j : (D^k, S^{k-1}) \to (e_j^k, \dot{e}_j^k)$$

with $\phi_j(D^k) = e_j^k$ such that the restriction $\phi_j : D^k - S^{k-1} \to e_j^k - \dot{e}_j^k$ is a homeomorphism, and such that a subset $U \subset e_j^k$ is open if and only if $U \cap \dot{e}_j^k \subset \dot{e}_j^k$ and $\phi_j^{-1}(U) \subset D^k$ are open subsets.

For $k = 0$ the space X is the disjoint union of the space A and a discrete space.

DEFINITION 7.2. A *(relative) CW-complex* (X, A) consists of a topological space X, a closed subspace $A \subset X$, and a sequence of closed subspaces $(X, A)^k \subset X$, $k \geq 0$ (where $(X, A)^k$ is called the *k-skeleton*) with $X = A \cup \bigcup_{k \geq 0} (X, A)^k$ such that the following conditions hold:

(a) $(X, A)^0$ is obtained form A by adjoining 0-cells and for every $k \geq 1$ the space $(X, A)^k$ is obtained from $(X, A)^{k-1}$ by adjoining k-cells.
(b) The topology of X is the weak topology of the union $A \cup \bigcup_{k \geq 0} (X, A)^k$.

Hence it is possible to build up a CW-complex recursively. Start with the topological space $(X, A)^{-1} := A$, and recursively assume that $(X, A)^{k-1}$ has been defined. Given continuous maps $\tilde{\phi}_j : S^{k-1} \to (X, A)^{k-1}$, $j \in J_k$ (called *attaching maps*), we form the subset $(X, A)^k$ as follows: Let $\dot{e}_j^k = \tilde{\phi}_j(S^{k-1})$ and $e_j^k := D^k \cup_{\tilde{\phi}_j} \dot{e}_j^k$, meaning that e_j^k is the quotient space of the disjoint union of D^k and $\dot{e}_j^k$ with the equivalence relation identifying each $x \in S^{k-1}$ with $\tilde{\phi}_j(x) \in \dot{e}_j^k$. The sets $D^k \setminus S^{k-1}$ and $e_j^k \setminus \dot{e}_j^k$ are homeomorphic and can be identified. Next define the *characteristic map* $\phi_j : (D^k, S^{k-1}) \to (e_j^k, \dot{e}_j^k)$ by $\phi_j(x) = x$, for $x \in D^k - S^{k-1}$, and $\phi_j(y) = \tilde{\phi}_j(y)$, for $y \in S^{k-1}$. Then set $(X, A)^k = (X, A)^{k-1} \cup \bigcup_{j \in J_k} e_j^k$. A subset $U \subset (X, A)^k$ is open if and only if the intersection $U \cap (X, A)^{k-1}$ is open in $(X, A)^{k-1}$ and for all $j \in J_k$ the preimages $\phi_j^{-1}(U \cap e_j^k)$ are open subsets of D^k. The subsets e_j^k, $j \in J_k$, are called the *closed k-cells* of the CW-complex. Because of possible identifications on the boundary, these cells are in general not homeomorphic to D^k. The subsets $e_j^k - \dot{e}_j^k$ are called *open cells*; they are homeomorphic to $D^k - S^{k-1}$, but in general these sets are not open subsets of the k-skeleton $(X, A)^k$.

A *subcomplex* of a CW-complex is the union of closed cells of the CW-complex with the same attaching resp. characteristic maps which is itself a CW-complex. For example, the k-skeleton $(X, A)^k$ is a subcomplex. For $A = \emptyset$ we simply write $X^k = (X, \emptyset)^k$.

REMARK 7.3. The advantage of using CW-complexes is that one generally needs fewer cells to write a space as a CW-complexes than to triangulate it. A simple but important example: for $n \geq 1$ the n-dimensional sphere $S^n = \{x \in \mathbb{R}^{n+1} \mid \|x\|^2 = 1\}$ has the structure of a CW-complex with one 0-cell e^0 and one n-cell; the attaching map is the constant map $\phi : S^{n-1} \to e^0$.

CW-complexes are very useful if one considers homotopy properties of topological spaces. For example, Morse theory shows that any manifold is homotopy equivalent to a CW-complexes; see the proof of Corollary 7.7.

Two continuous maps $f, g : X \to Y$ are called *homotopic* if there is a continuous map $F : X \times [0,1] \to Y$, $(x,t) \mapsto F_t(x) = F(x,t)$, with $F_0 = f$ and $F_1 = g$. There is also a relative version of this concept: Let (X, A) and (Y, B) be pairs of topological spaces, meaning that $A \subset X$ and $B \subset Y$. A continuous map $f : (X, A) \to (Y, B)$ between these pairs is a continuous map $f : X \to Y$ with the property $f(A) \subset B$. Two continuous maps $f, g : (X, A) \to (Y, B)$ agreeing on A are *homotopic relative to A*, or *homotopic rel A*, if there is a continuous map $F : (X, A) \times [0,1] \mapsto (Y, B)$, $(x,t) \mapsto F_t(x) = F(x,t)$, with $F_0 = f$, $F_1 = g$ and $F_t|A = f|A = g|B$ for all $t \in [0,1]$. Two topological spaces X, Y are called *homotopy equivalent* if there are continuous maps $f : X \to Y$ and $g : Y \to X$ such that the compositions $f \circ g$ and $g \circ f$ are homotopic to the identity maps id_Y and id_X, respectively. Then f is called a *homotopy equivalence*. We call a pathwise connected topological space X n-*connected* for some $n \geq 1$ if for every $j \in \{1, \ldots, n\}$ every continuous map $f : S^j \to X$ is homotopic to a constant map. (This implies that the *homotopy groups* $\pi_j(X, p) = 0$ vanish for all $j = 1, \ldots, n$, but we do not need this concept here.) A 1-connected space is also called *simply connected*. A topological pair (X, A) is called n-*connected* for some $n \geq 1$ if for every $j \in \{1, \ldots, n\}$ every continuous map $f : S^j \to X$ is homotopic to a continuous map whose image lies in A.

A continuous map $f : X \to Y$ between CW-complexes is called *cellular* if it respects the CW-structure, that is, if the image $f(X^k)$ of the k-skeleton lies in the k-skeleton of Y: $f(X^k) \subset Y^k$.

PROPOSITION 7.4 (CELLULAR APPROXIMATION THEOREM). *Every continuous map $f : X \to Y$ between CW-complexes is homotopic to a cellular map.* (See [Spanier 1966, Theorem 7.6.17] for a proof.)

Thus every continuous map $f : S^j \to S^n$ between spheres is homotopic to a cellular map $F : S^j \to S^n$, where the spheres have the CW-structure described in Remark 7.3. Now suppose that $j < n$; since the j-skeleton of S^n equals the 0-skeleton (a single point), S^n is $(n{-}1)$-connected. The same argument shows:

PROPOSITION 7.5. *If a pathwise connected CW-complex X has no j-dimensional cells for any $j \in \{1, 2, \ldots, k\}$, the space X is k-connected.*

We note the following consequences of Whitehead's and Hurewicz's theorems [Bredon 1995, Theorems VII-11.2 and VII-10.7]:

PROPOSITION 7.6 [Bredon 1995, Corollary VII-11.14]. (a) *Let $f : X \to Y$ be a continuous map between two simply connected CW-complexes, such that the induced homomorphism $f_* : H_i(X; \mathbb{Z}) \to H_i(Y; \mathbb{Z})$ of the singular homology groups with integer coefficients is an isomorphism for all i. Then f is a homotopy equivalence.*

(b) *If a topological space X is $(n{-}1)$-connected for some $n \geq 2$, then for every homology class $h \in H_n(X; \mathbb{Z})$ there is a continuous map $f_h : S^n \to X$ and a generator $g \in H_n(S^n; \mathbb{Z})$ such that $(f_h)_*(g) = h$. (Hence every n-dimensional homology class can be represented by a spherical cycle.)*

COROLLARY 7.7. *A compact and $(n{-}1)$-connected differentiable manifold M is homotopy equivalent to the n-sphere.*

PROOF. We use the fact that an n-dimensional manifold has the homotopy type of a finite CW-complex X (meaning there are only finitely many cells) of dimension n (meaning the maximal cell dimension is n). This can be proved using a Morse function on M; one obtains a CW-structure where every critical point of index k corresponds to a k-dimensional cell. Since the critical points are nondegenerate and the manifold is compact, there are only finitely many of them. The index of a critical point on a differentiable manifold is bounded above by the dimension of the manifold.

The manifold is simply connected, therefore orientable. Hence $H_n(M; \mathbb{Z}) \cong \mathbb{Z}$ [Bredon 1995, Corollary VI-7.2]. It follows from Proposition 7.6(b) that there is a continuous map $f : S^n \to M$ inducing an isomorphism $f_* : H_n(S^n; \mathbb{Z}) \to H_n(M; \mathbb{Z})$. Since S^n and M^n are (up to homotopy equivalence) CW-complexes, we conclude from Proposition 7.6(a) that f is a homotopy equivalence.   □

REMARK 7.8. The *Poincaré conjecture* states that for $n \geq 3$, an n-dimensional simply connected and compact manifold homotopy equivalent to the n-sphere is *homeomorphic* to the n-sphere. For $n \geq 5$ the conjecture was proven by S. Smale, for $n = 4$ by M. Freedman. There is an announcement of a proof for $n = 3$, by G. Perelman [Milnor 2003].

Given a topological space X and a point $p \in X$, we define the *loop space*

$$\Omega_p(X) = \{\gamma : [0,1] \to X \mid \gamma \text{ continuous}, \gamma(0) = \gamma(1) = p\}.$$

PROPOSITION 7.9. *If X is a pathwise connected topological space and $\Omega_p X$ is the loop space of X with $p \in X$, the homotopy groups satisfy*

$$\pi_{k-1}(\Omega_p X) \cong \pi_k X \quad \text{for all } k \geq 1.$$

PROOF. Consider the cube $I^k = [0,1]^k$, with boundary

$$\dot{I}^k = \{(x_1, \ldots, x_k) \in I^k \mid \text{there is } j = 1, \ldots, k \text{ with } x_j \in \{0,1\}\}$$

Given a continuous map $f : (I^{k+1}, \dot{I}^{k+1}) \to (X, p)$ for some point $p \in X$, we define $F(f) : (I^k, \dot{I}^k) \to (\Omega_p X, p)$ by

$$F(f)(x_1, x_2, \ldots, x_k)(t) = f(x_1, x_2, \ldots, x_{k+1}).$$

Also, for a continuous map $g : (I^k, \dot{I}^k) \to (\Omega X, p)$ we define a continuous map $G(g) : (I^{k+1}, \dot{I}^{k+1}) \to (X, p)$ by

$$G(g)(x_1, \ldots, x_{k+1}) = g(x_1, \ldots, x_k)(x_{k+1}).$$

This lets us define the isomorphism between the homotopy groups $\pi_k(\Omega X, p)$ and $\pi_{k+1}(X, p)$.   □

COROLLARY 7.10. *Let M be a simply connected, compact, n-dimensional manifold and $p, q \in M$ arbitrary points. Consider the space*

$$\Omega_{pq} M := \{\gamma : [0,1] \to M \mid \gamma \text{ piecewise smooth}, \gamma(0) = p, \gamma(1) = q\}.$$

If this space is homotopy equivalent to a CW-complex with no cells of dimension $j \in \{1, \ldots, k\}$, then M is k-connected. In particular, if $k = n - 1$ the manifold is homotopy equivalent to the n-sphere.

PROOF. The space $\Omega_{pq} M$ of piecewise continuous curves joining p and q is homotopy equivalent to the space $\Omega_{pq}^* M$ of continuous curves joining p and q with the compact-open topology. This can be shown by using the finite-dimensional approximation $\Omega(k, a)$ for $\Omega^a M = \{\gamma \in \Omega_{pq} M \mid E(\gamma) \leq a\}$ with a sufficiently large k [Milnor 1969, Theorem 17.1]. The next step is that the homotopy type of the spaces $\Omega_{pq} M$ does not depend on the chosen points. For two curves $\gamma_1, \gamma_2 : [0,1] \to M$ with $\gamma_2(1) = \gamma_1(0)$, denote by $\gamma_1 * \gamma_2 : [0,1] \to M$ their *composition,* defined by

$$\gamma_1 * \gamma_2(t) = \begin{cases} \gamma_2(2t) & \text{for } 0 \leq t \leq \frac{1}{2}, \\ \gamma_1(2t - 1) & \text{for } \frac{1}{2} \leq t \leq 1. \end{cases}$$

Now fix a curve $\gamma_1 \in \Omega_{qr} M$; the map $\gamma \in \Omega_{pq} M \mapsto \gamma_1 * \gamma \in \Omega_r M$ defines a homotopy equivalence. Hence we can conclude from Proposition 7.5 that the loop space is k-connected. This finally implies by Proposition 7.9 that the manifold is $(k-1)$-connected. If $k = n - 1$ we conclude from Corollary 7.7 that the manifold is homotopy equivalent to the n-sphere.   □

8. Morse Theory of the Energy Functional

For a Finsler manifold (M, F) and two points $p, q \in M$ we consider the energy functional on the space Ω_{pq} of curves joining the points p and q. A critical value κ equals the energy $E(c)$ of a geodesic c joining these points. Morse theory provides a connection between invariants of critical points of a function on a manifold and global topological invariants, in our case a connection between homology or homotopy invariants of the loop space and indices of geodesics.

We introduce the space $\Omega_{pq}M$ of absolutely continuous curves $\gamma : [0, 1] \to M$ satisfying $\gamma(0) = p$, $\gamma(1) = q$, and

$$\int_0^1 F^2(\gamma'(t))dt < \infty.$$

The energy functional $E : \Omega_{pq}M \longrightarrow \mathbb{R}$, given by

$$E(\gamma) = \frac{1}{2} \int_0^1 F^2(\gamma'(t)) \, dt,$$

is $C^{1,1}$-differentiable, that is, it is C^1-differentiable and its derivative is locally Lipschitz continuous [Mercuri 1971].

If $c_s : [0, 1] \to M$, $s \in (-\varepsilon, \varepsilon)$, is a variation of $c = c_0$ with fixed end points $p = c(0)$, $q = c(1)$, we conclude from the first variation formula (Lemma 4.1) that the variation vector field $Y(t) = (d/ds)|_{s=0} c_s(t)$ satisfies

$$dE(c)Y := \frac{\partial}{\partial s}\Big|_{s=0} E(c_s) = \int_0^1 \left\langle \frac{\nabla}{dt}c', Y \right\rangle_{c'} (t) \, dt.$$

The curve c is a *critical point* of the energy functional if $dE(c)Y = 0$ for every vector field $Y \in W_c^0$. Then it follows from Corollary 4.2 that the critical points are the geodesics $c : [0, 1] \to M$, starting at $p = c(0)$ and ending at $q = c(1)$. For nonnegative $\kappa \geq 0$ we define the sublevel sets

$$\Omega_{pq}^\kappa = \Omega^\kappa := \{\sigma \in \Omega_{pq}M \mid E(\sigma) \leq \kappa\};$$

then $\Omega_{pq}^l M$ with $2l = \theta(p, q)^2$ contains the minimal geodesics joining p and q. For l small enough there is a unique minimal geodesic, that is, $\Omega_{pq}^l M$ contains exactly one element. If the case $p = q$ the subset Ω_{pp}^0 consists of the point curve p.

We choose an arbitrary Riemannian metric g on the manifold, which induces a Hilbert space structure on $\Omega_{pq}M$. If $c : [0, 1] \to M$ is a smooth curve with $c(0) = c(1)$ and X, Y are smooth vector fields along c, a Riemannian metric on $\Omega_{pq}M$ is defined by

$$g_1(X, Y) = \int_0^1 g(X(t), Y(t))dt + \int_0^1 g\left(\frac{\nabla}{dt}X(t), \frac{\nabla}{dt}Y(t)\right) dt,$$

where ∇/dt is the covariant derivative along c induced by the Levi-Civita connection of the Riemannian manifold. The energy functional induces a Lipschitz continuous gradient vector field $\operatorname{grad} E$ through the equation

$$g_1\big(\operatorname{grad} E(c), X\big) = dE(c)\,X$$

for all X. The energy functional satisfies the Palais–Smale condition and the *negative gradient flow*, that is, the flow of the vector field $-\operatorname{grad} E$ on λ, is defined for every $t \geq 0$ [Mercuri 1971].

For the Morse theory of the energy functional we have to consider also the second derivatives of the energy functional at the critical points. For a non-Riemannian Finsler metric the square F^2 of the Finsler metric F is not C^2-differentiable at the zero section. Hence E is C^2-differentiable only at the regular curves, those curves c with $c'(t) \neq 0$ for all t. Geodesics of positive length are regular, so we can use the statement in Corollary 6.2. As remarked just before that corollary, the index form I_c^0 of the geodesic defined with the Finsler metric coincides with the index form $\bar{I}_c$ of an osculating Riemannian metric.

We will not go into details in this construction since instead of the space Ω_{pq} one can use a *finite-dimensional approximation*. This allows us to use Morse theory for a finite-dimensional compact manifold instead of the infinite-dimensional Hilbert manifold $\Omega_{pq}M$. The finite-dimensional approximation was introduced by M. Morse and is explained in [Milnor 1969, Chapter 16]. We start with a compact Finsler manifold (M, F) with injectivity radius $\operatorname{inj} > 0$ (see Definition 9.1 below). For every pair of points $p, q \in M$ with distance $\theta(p, q) < \operatorname{inj}$, there is a unique minimal geodesic from p to q. Choose $a > 0$ and $k \in \mathbb{N}$ such that $1/k < (\operatorname{inj})^2/(2a)$, and set $t_i := i/k$ for $i \in \mathbb{N}$. Define

$$\Omega_{pq}(k, a) := \big\{ c \in \Omega_{pq}^a M \mid c|[t_i, t_{i+1}] \text{ is a geodesic} \big\}.$$

Since

$$\theta^2\big(c(t_i), c(t_{i+1})\big) \leq L^2\big(c|[t_i, t_{i+1}]\big) \leq \frac{2}{k} E\big(c|[t_i, t_{i+1}]\big) \leq \frac{2a}{k} < \operatorname{inj}^2,$$

a curve $c \in \Omega_{pq}(k, a)$ is uniquely determined by the points $c(t_1), \ldots, c(t_{k-1}) \in M \times \cdots \times M$. Therefore we can identify $\Omega_{pq}(k, a)$ with the submanifold with boundary

$$\big\{ (x_1, \ldots, x_{k-1}) \in M \times \cdots \times M \mid \theta(x_i, x_{i+1}) \leq \tfrac{1}{2}\operatorname{inj} \text{ for } i = 0, \ldots, k-1 \big\}$$

of the product manifold $M \times \cdots \times M$, where we set $p = x_0$, $q = x_k$. We conclude that $\Omega_{pq}(k, a)$ has the structure of a compact manifold with boundary of dimension $(k-1)\dim M$. Then there is a strong deformation retraction

$$r_u : \Omega_{pq}^a \to \Omega_{pq}^a, \quad u \in [0, 1],$$

with $r_0(c) = c$ for all $c \in \Omega_{pq}^a$, $r_u(c) = c$ for all $c \in \Omega_{pq}(k, a)$, $u \in [0, 1]$, and $r_1(c) \in \Omega_{pq}(k, a)$ for all Ω_{pq}^a. It is defined for $u \in [t_i, t_{i+1}]$, as follows: For $t \leq t_i$, $r_u(c)(t)$ is the broken geodesic with corners $c(0), c(t_1), c(t_2), \ldots, c(t_i)$; $r_u(t)$ for

$t \in [t_i, u]$ is the minimal geodesic between $c(t_i)$ and $c(u)$ and $r_u(c)(t) = c(t)$ for $t \geq t_{i+1}$. Then the restriction $E' : \Omega_{pq}(k, a) \to \mathbb{R}$ of E given by

$$E'(x_1, \ldots, x_k) = \tfrac{1}{2} \sum_{i=1}^{k} \theta^2(x_i, x_{i+1})$$

is a C^1-smooth function. The first variational formula implies that the critical points are the geodesics from p to q with $E(c) \leq a$, and the function is C^∞ differentiable in the neighborhood of critical points of positive length.

For a broken geodesic $c = (x_1, \ldots, x_k)$ (smooth except at $t = t_1, \ldots, t_{k-1}$, where t_i no longer bears the meaning i/k), a tangent vector $y(t) = (\partial/\partial s)|_{s=0} c_s(t)$ is given by a variation $c_s = (x_1^s, \ldots, x_k^s) \in \Omega_{pq}(k, a)$, $s \in (-\varepsilon, \varepsilon)$, that is, a curve in $\Omega_{pq}(k, a)$. Since the variation vector field of a geodesic variation is a Jacobi field (Lemma 5.5), the tangent space $T_x\Omega_{pq}$ consists of *broken Jacobi fields*,

$$T_x\Omega_{pq}(k, a) = \{X \in W_c \mid X|[t_i, t_{i+1}] \text{ is a Jacobi field}\}.$$

In particular, if c is a geodesic (smooth throughout), the tangent space $T_c\Omega_{pq}$ coincides with the space J introduced in the Proof of Lemma 6.4. Therefore the Proof of Lemma 6.4 also shows that restricting the index form to the tangent space $T_c\Omega_{pq}(k, a)$ changes neither the index nor the nullity.

The energy functional $E' : \Omega_{pq}M(k, a) \to \mathbb{R}$ is a differentiable function on the compact manifold $\Omega_{pq}M(k, a)$. It is a Morse function if all critical points are nondegenerate, that is, if all geodesics c joining the points p and q with energy $\leq a$ are nondegenerate — in symbols, nul $c = 0$. Assume $c \in \Omega_{pq}(k, a)$ is degenerate, so there is $X \in W_c^0$ with $X \neq 0$ and $I_c^0(X, Y) = 0$ for all $Y \in W_c^0$. Since X is a piecewise smooth vector field, choose $0 = t_0 < t_1 < \cdots < t_k = 1$ such that X is smooth when restricted to each subinterval $[t_i, t_{i+1}]$, and also such that no subinterval contains a pair of conjugate points. Then we obtain from Equation 6–1:

$$0 = I_c(X, Y) = -\int_0^1 \left\langle \frac{\nabla^2}{dt^2} X + R^{c'}(X, c')c', Y \right\rangle_{c'} dt$$

$$+ \sum_{i=1}^{k-1} \left\langle \frac{\nabla}{dt} X(t_i^-) - \frac{\nabla}{dt} X(t_i^+), Y(t_i) \right\rangle_{c'}. \quad (8\text{–}1)$$

Let $Y \in W_c^0$ be a broken Jacobi field with $Y(t_i) := (\nabla/dt)X(t_i^-) - (\nabla/dt)X(t_i^+)$, so the restrictions $Y|[t_i, t_{i+1}]$ are Jacobi fields along $c|[t_i, t_{i+1}]$. Then Equation 8–1 implies

$$I_c(X, Y) = \sum_{i=1}^{k-1} \left\| \frac{\nabla}{dt} X(t_i^-) - \frac{\nabla}{dt} X(t_i^+) \right\|_{c'}^2 = 0,$$

hence X is a smooth vector field. We have shown:

LEMMA 8.1. *The energy functional $E' : \Omega_{pq}(k, a) \to \mathbb{R}$ is a Morse function if and only if the point q is not conjugate to p along any geodesic c with $E(c) \leq a$ joining p and q.*

The next proposition relates the conjugate points of $p \in M$ to the critical points of the exponential map at p. (Recall that the *exponential map* $\exp_p : T_pM \to M$ is defined by $\exp_p(X) = c_X(1)$, where for $X \in T_pM$ we denote by $c_X : \mathbb{R} \to M$ the geodesic with $c'_X(0) = X$. This assumes that the metric is complete — more precisely, forward geodesically complete — an assumption that is satisfied in particular if the manifold is compact, thanks to the Finsler version of the Hopf–Rinow Theorem [Bao et al. 2000, § 6.6].)

PROPOSITION 8.2. *Let (M, F) be a complete Finsler manifold and let $p \in M$. A point $q = \exp_p(X)$ is a critical point of the exponential map $\exp_p : T_pM \to M$ if and only if q is a conjugate point of p along the geodesic $t \in [0, 1] \mapsto \exp_p(tX) \in M$ from p to q.*

The proof of the Riemannian case [Milnor 1969, Theorem 18.1] carries over. As an application of Sard's Theorem one obtains:

COROLLARY 8.3 [Milnor 1969, Corollary 18.2]. *Let (M, F) be a compact Finsler manifold and let $p \in M$. For almost all points $q \in M$ (that is, up to a set of measure zero) the point q is not a conjugate point to p along any geodesic from p to q. For almost all $q \in M$ the energy functional $E : \Omega_{pq}M \to \mathbb{R}$ is a Morse function.*

It is the chief observation of Morse theory that the topology of the sublevel sets $\Omega_{pq}^\kappa M := \{\sigma \in \Omega_{pq} \mid E(\sigma) \leq \kappa\}$ and $\Omega_{pq}^\kappa(k, b)$ can only change if κ is a critical value. The change in topology can be described by the indices of the corresponding critical points. Applied to the energy functional, this line of argumentation yields (compare [Milnor 1969, Theorem 17.3]):

THEOREM 8.4 (FUNDAMENTAL THEOREM OF MORSE THEORY). *Let (M, F) be a compact Finsler manifold and $p \in M$ an arbitrary point. For almost all $q \in M$ and for all $a > 0$ the function $E' : \Omega_{pq}(k, a) \to \mathbb{R}$ is a Morse function and there are only finitely many geodesics c joining p and q with $E(c) \leq a$.*

*   The spaces $\Omega_{pq}^\kappa M$ and $\Omega_{pq}^\kappa(k, a)$ have the homotopy type of a CW-complex having as many m-cells as there are geodesics c joining p and q with $E(c) \leq a$ and $\operatorname{ind} c = m$.*

SKETCH OF PROOF. As remarked in Proposition 8.2, for almost all $q \in M$ and all $a > 0$ the energy functional $E' : \Omega_{pq}(k, a) \to \mathbb{R}$ is a Morse function. If there is no critical value in $[\alpha, \beta]$, one can use the flow of the negative gradient field $-\operatorname{grad} E'$ on $\Omega_{pq}(k, a)$ and retract $\Omega_{pq}^\beta(k, a)$ onto $\Omega_{pq}^\alpha(k, a)$.

   The behavior of a Morse function near a critical point is described by the Morse Lemma [Milnor 1969, Lemma 2.2]. Applied to E' it states that near a geodesic c one can introduce local coordinates $y = (y_1, \ldots, y_r)$, with c corresponding to $0 = (0, \ldots, 0)$, such that

$$E'(y_1, \ldots, y_r) = E(c) - \sum_{j=1}^{\operatorname{ind} c} y_j^2 + \sum_{j=\operatorname{ind} c+1}^{r} y_j^2.$$

Here $r = (k-1)\dim M = \dim \Omega_{pq}(k, a)$. Hence the index describes the dimension of a subspace on which the energy of nearby curves decreases quadratically, whereas on a complementary subspace the energy grows quadratically. This implies that the geodesics are isolated. Since $\Omega_{pq}(k, a)$ is compact, there are only finitely many geodesics joining p and q with energy $\leq a$.

Assume for simplicity that there is only one geodesic of energy a joining p and q. Let $\operatorname{ind} c = m$. Then one can show that for sufficiently small $\varepsilon > 0$ the set $\Omega_{pq}^{a+\varepsilon}(k, a)$ has the homotopy type of $\Omega_{pq}^{a-\varepsilon}(k, a)$ with an m-dimensional cell attached [Milnor 1969, Theorem 3.2]. This m-cell corresponds to the set $\{(y_1, \ldots, y_k, 0, \ldots, 0) \mid y_1^2 + \ldots + y_m^2 < \varepsilon\}$ in the coordinates used in the Morse Lemma. $\qquad \square$

REMARK 8.5. In Remark 5.7 we discussed Jacobi fields along a geodesic where the flag curvature is constant and positive. Now we consider the index form $I_c(\delta, l)$ of a geodesic $c = c_l(\delta) : [0, 1] \to M$ of length l with constant flag curvature $K(c'(t); \sigma)$. We can use bounds for the flag curvature to estimate the index and the conjugate radius, as in the Riemannian case.

We choose $e_1, e_2, \ldots, e_n$ in $T_{c(0)}M$, orthonormal with respect to $\langle \cdot, \cdot \rangle_{c'}$ and such that $c'(0) = F(c')e_1$. We extend this frame by parallel transport with respect to (∇/dt) along c. We can write vector fields $X = X(t)$ along c as $X(t) = \sum_{i=1}^{n} x_i(t)e_i(t)$, for smooth functions $x_i : [0, 1] \to \mathbb{R}$. Then

$$I_c(\delta, l)(X, X) = \int_0^1 \left(x_i'(t)^2 - \delta x_i^2(t) \right) dt$$

and one shows that

$$\operatorname{ind} c_l(\delta) = \operatorname{ind} I_c(\delta, l) = k(n - 1) \tag{8-2}$$

for $l \in \left(k\pi/\sqrt{\delta}, (k+1)\pi/\sqrt{\delta} \right)$. See [Klingenberg 1995, Example 2.5.7].

Now let $\gamma : [0, 1] \to M$ be a geodesic of a Finsler metric with a lower bound for the flag curvature: $K \geq \delta$. We again choose along γ an orthonormal frame $(e_1, e_2, \ldots, e_n)(t)$ parallel with respect to (∇/dt). We can estimate the indexes $\operatorname{ind} \gamma$ and $\operatorname{ind} I_\gamma$ by comparing them with the index $\operatorname{ind} c_l(\delta)$ of a geodesic $c_l(\delta)$ of the same length on a space form with constant sectional curvature:

$$I_c(X, X) = \int_0^1 \left(x_i'(t)^2 - K(e_1(t); e_2(t))x_i^2(t) \right) dt$$

$$\leq \int_0^1 \left(x_i'(t)^2 - \delta x_i^2(t) \right) dt = I_c(\delta, l)(X, X).$$

This computation and a similar one in the case of an upper bound for the flag curvature lead to the following estimates for the distance of conjugate points and indices of geodesics. Here we use the fact that the index form I_γ of $\gamma = \gamma_X$ is positive definite for $L(\gamma) < \operatorname{conj}_X$ and degenerate for $L(\gamma) = \operatorname{conj}_X$.

LEMMA 8.6. *Let $\gamma = \gamma_X : [0, a] \to M$ be a geodesic parametrized by arc length on a Finsler manifold (M, F), with $\gamma'_X(0) = X$.*

(a) *If the flag curvature $K = K(\gamma'; \sigma)$ satisfies $K \leq \Delta$ (resp. $K < \Delta$) then $\operatorname{conj}_X \geq \pi/\sqrt{\Delta}$ (resp. $\operatorname{conj}_X > \pi/\sqrt{\Delta}$).*
(b) *If the flag curvature $K = K(\gamma'; \sigma)$ satisfies $\delta \leq K$ (resp. $\delta < K$) then $\operatorname{conj}_X \leq \pi/\sqrt{\Delta}$ (resp. $\operatorname{conj}_X < \pi/\sqrt{\Delta}$).*
(c) *If the Ricci curvature $\operatorname{Ric} = \operatorname{Ric}(\gamma') F^2$ satisfies $\operatorname{Ric} \geq \delta(n-1)F^2$ (resp. $\operatorname{Ric} > \delta(n-1)$) then $\operatorname{conj}_X \leq \pi/\sqrt{\delta}$ (resp. $\operatorname{conj}_X < \pi/\sqrt{\delta}$).*

SKETCH OF PROOF. The argument for cases (a) and (b) is given in Remark 8.5.

The argument in case (c) is the same as in the Riemannian case. As in the preceding remark we choose an orthonormal frame $(e_1, \ldots, e_n)(t)$ along γ with $L(\gamma) = a = \pi/\sqrt{\delta}$, parallel with respect to ∇/dt, and such that $\gamma'(0) = X = F(X)e_1(0)$. Then we define the vector fields

$$X_i(t) = \sin(\sqrt{\delta}\, t)e_i(t), \quad i = 2, \ldots, n.$$

We compute for the index form I_γ :

$$\sum_{i=2}^{n} I_\gamma(X_i, X_i) = \sum_{i=2}^{n} \int_0^a \left(\cos^2(\sqrt{\delta}\, t) - K(e_1(t); e_i(t)) \sin^2(\sqrt{\delta}\, t) \right) dt$$

$$= (n-1) \sum_0^a \left(\delta \cos^2(\sqrt{\delta}\, t) - \frac{\sum_{i=2}^n K(e_1; e_i)}{n-1} \sin^2(\sqrt{\delta}\, t) \right) dt$$

$$\leq (n-1)\delta \int_0^a \left(\cos^2(\sqrt{\delta}\, t) - \sin^2(\sqrt{\delta}\, t) \right) dt = 0.$$

We conclude that $\operatorname{conj}_X \leq \pi/\sqrt{\delta}$. $\qquad\square$

The *diameter* of a complete Finsler manifold M is the maximal distance of two points. By the Hopf–Rinow theorem [Bao et al. 2000, § 6.6] there is a minimal geodesic between two points of maximal distance. Since a geodesic is not minimal after the first conjugate point, the diameter is at most $\max\{\operatorname{conj}_X \mid X \in T^1 M\}$. Therefore we obtain as a consequence of Lemma 8.6 the following generalization of the Bonnet–Myers theorem of Riemannian geometry:

COROLLARY 8.7 [Auslander 1955]. *Let (M, F) be a complete Finsler manifold of dimension n with Ricci curvature $\operatorname{Ric}(V) \geq \delta(n-1) F^2(V)$ for some positive δ and for all nonzero tangent vectors V. Then M is compact and its diameter is at most $\pi/\sqrt{\delta}$.*

Since this estimate also holds for the universal covering space, we conclude that the universal covering space is also compact, so the fundamental group of the manifold is finite. In the proof of the Sphere Theorem the following statement is of importance:

PROPOSITION 8.8. *Let (M, F) be a compact and simply connected Finsler manifold of dimension n, and let $p, q \in M$ be such that q is not conjugate to p along any geodesic joining p and q. Assume there is a number $m \geq 2$ such that every nonminimal geodesic c from p to q has index at least m. Then:*

(a) *The manifold is m-connected* (see page 280 for definition).
(b) *If $m = n - 1$, the manifold is homotopy equivalent to the n-sphere.*

PROOF. It follows from the Fundamental Theorem of Morse Theory (Theorem 8.4) that the space Ω_{pq} has the homotopy type of a CW-complex with no cells of dimension $k \in \{1, 2, \ldots, m - 1\}$. By Proposition 7.5 this implies that the space Ω_{pq} is $(m-1)$-connected; Proposition 7.9 then implies that M itself is m-connected. Part (b) follows from Corollary 7.7. $\qquad\square$

9. Shortest Nonminimal Geodesics and the Sphere Theorem

Now we come to the crucial geometric argument in the proof of the Sphere Theorem. We obtain a lower bound for the length of a nonminimal geodesic c joining two points p, q or a nonconstant geodesic loop. In contrast to a minimal geodesic, this geodesic will meet the cut locus, after which the geodesic is not minimal anymore.

The exponential map $\exp_p : T_p M \to M$ is C^∞-smooth on $T_p M \setminus \{0\}$ and C^1-smooth on $T_p M$ [Shen 2001a, §11.1]. The differential at $0 \in T_p M$ is an isomorphism; hence there is an $\varepsilon > 0$ such that the restriction

$$\exp_p : B_\varepsilon(T_p M) = \{X \in T_p M | F(X) < \varepsilon\} \to M$$

is a local diffeomorphism onto its image $B_\varepsilon(p) \subset M$. If a piecewise smooth curve $c : [0, a] \to M$ is minimal, that is, $L(c) = \theta(c(0), c(a))$, it follows from Corollary 4.2 that c is a smooth geodesic.

DEFINITION 9.1. For a unit tangent vector $X \in T_p M$, set

$$t(X) = \sup \{s > 0 \mid \theta(\exp_p(sX), p) = s\}.$$

Then $q = \exp_p(t(X)X)$ is called a *cut point*. The *cut locus*

$$\mathrm{Cut}(p) := \{\exp_p(t(X)X) \mid F(X) = 1, \ t(X) < \infty\}$$

is the union of all cut points on geodesics starting from p. The *injectivity radius* at p is $\mathrm{inj}\, p := \inf \{\theta(p, q) \mid q \in \mathrm{Cut}(p)\}$. If the manifold is compact we define the *injectivity radius* of M as $\mathrm{inj} = \mathrm{inj}(M; F) = \inf \{\mathrm{inj}\, p \mid p \in M\}$. The *symmetrized injectivity radius* at p is $d(p) := \inf \{d(p, q) \mid q \in \mathrm{Cut}(p)\}$. If the manifold is compact, we define the *symmetrized injectivity radius* $d = d(M; F) = \inf \{d(p) \mid p \in M\}$. Finally, given two points p, q we define

$$\vartheta(p, q) := \inf \{\theta(p, r) + \theta(r, q) \mid r \in \mathrm{Cut}\, p\}.$$

Hence $d(p)$ is the symmetrized distance between p and its cut locus, whereas $\operatorname{inj} p$ is the distance $\theta(p, \operatorname{Cut}(p))$ with respect to the distance function θ. For a reversible Finsler metric these functions coincide. In general we have the bounds

$$\tfrac{1}{2}(1 + 1/\lambda) \operatorname{inj} p \le d(p) \le \tfrac{1}{2}(1 + \lambda) \operatorname{inj} p,$$

$$\frac{2}{1 + \lambda} d(p) \le \operatorname{inj} p \le \frac{2\lambda}{1 + \lambda} d(p),$$

which imply the corresponding estimates for global injectivity radii inj and d in case of a compact manifold. Obviously $\vartheta(p, p) = 2d(p)$ and the triangle inequality for θ implies

$$\vartheta(p, q) + \theta(q, p) \ge 2d(p). \qquad (9\text{--}1)$$

If the manifold is compact, the cut locus of a point is also compact, so the infima in the above definitions of the injectivity radius and the symmetrized injectivity radius are actually minima.

DEFINITION 9.2. A *broken geodesic with one corner* joining p and q is a continuous curve $c : [0, b] \to M$ such that $p = c(0)$, $q = c(b)$, and for some $a \in (0, b)$ the restrictions $c_1 = c|[0, a] \to M$ and $c_2 = c|[a, b] \to M$ are minimal geodesics. The point $r = c(a)$ is the *corner* of c. We call c *smooth at* r if $c_1'(a) = c_2'(a)$. The 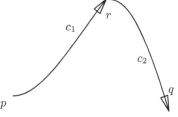 *length* of c is given by $L(c) = L(c_1) + L(c_2) = \theta(p, r) + \theta(r, q)$. If $p = q$, we have a closed broken geodesic, and its length is twice the symmetrized distance between p and r: $L(c) = 2d(p, r)$.

LEMMA 9.3. *Let (M, F) be a compact Finsler manifold with reversibility λ and flag curvature $K \le 1$, and let $p \in M$ be a point on M. If there is a cut point $r \in \operatorname{Cut} p$ with $\theta(p, r) < \pi$, there is a local hypersurface $H \subset M$ with $r \in H$ such that for every smooth curve $\tau : (-1, 1) \to H$ with $\tau(0) = r$ there are two geodesic variations $c_{1,s}, c_{2,s} : [0, \theta(p, q)] \to M$ with $c_{1,s}(\theta(p, r)) = c_{2,s}(\theta(p, r)) = \tau(s)$, $L(c_{1,s}) = L(c_{2,s})$ and such that $c_1 = c_{1,0}, c_2 = c_{2,0}$ are two distinct minimal geodesics joining p and r.*

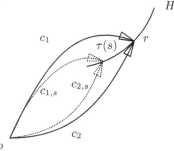

PROOF. We conclude from Lemma 8.6 that, since r is not conjugate to p along a minimal geodesic, there are distinct minimal geodesics $c_1, c_2 : [0, \theta(p, r)] \to M$

parametrized by arc length with $c_1(0) = c_2(0) = p$ and $c_1(\theta(p,r)) = c_2(\theta(p,r)) = r$. Since r is not conjugate to p along c_1 or c_2, we can choose an open neighborhood $U \subset M$ of r and open disjoint neighborhoods $U_j \subset T_pM$ of $c'_j(0)$, for $j = 1, 2$, such that the restrictions of the exponential map $\exp_p : T_pM \to M$ to U_1, U_2 are diffeomorphisms; see Proposition 8.2. We define functions $f_1, f_2 : U \to \mathbb{R}$ by setting

$$f_j(v) = F\big((\exp_p|U_j)^{-1}(v)\big), \quad j = 1, 2.$$

These functions are differentiable and of maximal rank, with $f_j(r) = \theta(p,r)$,

$$\operatorname{grad} f_j(r) = c'_j(\theta(p,r)) \quad \text{and} \quad df_j(r)(X) = \langle c'_j(\theta(p,r)), X \rangle_{c'_j(\theta(p,r))}$$

for $j = 1, 2$ and for all X, as follows from the Gauss lemma (see [Bao et al. 2000, §6.1] and [Shen 2001a, 11.2.1]). One can view f_1, f_2 as distance functions on the Finsler manifolds (U'_1, F) and (U'_2, F), where U'_j is a small open neigborhood of the image of the curve $c([0, \theta(p,r)])$. Since

$$\operatorname{grad} f_1(r) = c'_1(\theta(p,r)) \neq c'_2(\theta(p,r)) = \operatorname{grad} f_2(r),$$

the function $f_1 - f_2$ has maximal rank in an open neighborhood of r, which we again denote by U; thus $H = f^{-1}(0) = \{x \in U \mid f_1(x) = f_2(x)\}$ is a smooth hypersurface with $r \in H$. We finish by setting $c'_{1,s}(0) = f_1^{-1}(\tau(s))$ and $c'_{2,s}(0) = f_2^{-1}(\tau(s))$. $\quad\square$

LEMMA 9.4. *Let (M, F) be a compact Finsler manifold with reversibility λ and with flag curvature $K \leq 1$. Let $p, q \in M$ be two points with $\vartheta(p,q) + \theta(q,p) < \pi(1 + \lambda^{-1})$. Then there is a cut point $r \in \operatorname{Cut} p$ and a geodesic c of length $\vartheta(p,q)$ parametrized by arc length from p to q going through r.*

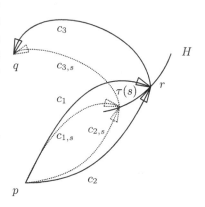

PROOF. Choose $r \in \operatorname{Cut} p$ such that $\vartheta(p,q) = \theta(p,r) + \theta(r,q)$. If $\theta(p,r) \geq \pi$, the definition of the reversibility (Lemma 3.9) implies that $\theta(r,p) \geq \pi/\lambda$, hence $\vartheta(p,q) + \theta(q,p) = \theta(p,r) + \theta(r,q) + \theta(q,p) \geq \theta(p,r) + \theta(r,p) = 2d(p,r) \geq \pi(1 + \lambda^{-1})$, contradicting the assumption. Thus we have proved that $\theta(p,r) < \pi$.

On the other hand, if $\theta(r,q) \geq \pi$, we have $\theta(q,r) \geq \pi/\lambda$ and $\vartheta(p,q) + \theta(q,p) = \theta(p,r) + \theta(r,q) + \theta(q,p) \geq \theta(r,q) + \theta(q,r) = d(r,q) \geq \pi(1 + \lambda^{-1})$. Therefore it follows also that $\theta(r,q) < \pi$.

Since $\theta(p,r) < \pi$, the point r is not conjugate to p along a minimal geodesic joining p. Therefore Lemma 9.3 gives an open hyersurface $H \subset M$ with $r \in H$ such that for every smooth curve $\tau : (-1, 1) \to H$ with $\tau(0) = r$ there are variations $c_{1,s}, c_{2,s} : [0, \theta(p,r)] \to M$ such that $L(\tilde{c}_{1,s}) = L(\tilde{c}_{1,s})$, $p = c_{1,s}(0) =$

$c_{2,s}(0)$, $\tau(s) = c_{1,s}(1) = c_{2,s}(1)$ and $c_1 = c_{1,0}, c_2 = c_{2,0} : [0, \theta(p,r)] \to M$ are distinct minimal geodesics joining p and r.

Now let $c_3 : [\theta(p,r), \vartheta(p,q)] \to M$ be a minimal geodesic parametrized by arc length joining $c_3(\theta(p,r)) = r$ and $c_3(2d(p,q)) = p$. Since $\theta(r,q) < \pi$, the point q is not conjugate to r along a minimal geodesic from r to q. Therefore we can choose a geodesic variation $c_{3,s} : [\theta(p,r), \vartheta(p,q)] \to M$ with $c_{3,0} = c_3$, $c_{3,s}(\theta(p,r)) = \tau(s)$ and $c_{3,s}(\vartheta(p,q)) = q$ for all $s \in (-1,1)$.

Now we can combine the smooth geodesic variations $c_{1,s}$, $c_{2,s}$ and $c_{3,s}$ to obtain two piecewise smooth variations $\tilde{c}_{1,s}, \tilde{c}_{2,s} : [0, \vartheta(p,q)] \to M$ with

$$\tilde{c}_{j,s}(t) = \begin{cases} c_{j,s}(t) & \text{if } t \in [0, \theta(p,r)], \\ c_{3,s}(t) & \text{if } t \in [\theta(p,r), \vartheta(p,q)]. \end{cases}$$

These are variations of the broken geodesics $\tilde{c}_1 = (c_1, c_3)$ and $\tilde{c}_2 = (c_2, c_3)$ by broken geodesics with fixed end points $p = \tilde{c}_j(0)$ and $q = \tilde{c}_j(\vartheta(p,q))$ and with $\tau'(0) = (\partial/\partial s)|_{s=0} \tilde{c}_{j,s}(\theta(p,r))$.

We assume that $c_1'(\theta(p,r)) \neq c_3'(\theta(p,r))$ and $c_2'(\theta(p,r)) \neq c_3'(\theta(p,r))$. Since c_1 and c_2 are distinct, $c_1'(\theta(p,r))$, $c_2'(\theta(p,r))$, $c_3'(\theta(p,r))$ are pairwise disjoint. Recall from Definition 3.3 the Legendre transformation $\mathcal{L}(X)(Y) = \langle Y, X \rangle_X$. Given three pairwise distinct nonzero vectors $X_1, X_2, X_3 \in T_r M$, we have

$$\dim\{Y \in T_r M \mid \mathcal{L}(X_1)(Y) = \mathcal{L}(X_2)(Y) = \mathcal{L}(X_3)(Y)\} \leq n - 2.$$

Applying this $c_1'(\theta(p,r))$, $c_2'(\theta(p,r))$, $c_3'(\theta(p,r))$, we see that there is a tangent vector $V \in T_r H \subset T_r M$ such that $\mathcal{L}(c_3(\theta(p,r)))(V)$ is not equal simultaneously to $\mathcal{L}(c_1(\theta(p,r)))(V)$ and $\mathcal{L}(c_2(\theta(p,r)))(V)$. We assume without loss of generality that

$$\mathcal{L}(c_1(\theta(p,r)))(V) - \mathcal{L}(c_3(\theta(p,r)))(V) \neq 0.$$

The first variational formula for the energy functional (Lemma 4.1), applied to the variation $\tilde{c}_{1,s}$ of the broken geodesic $\tilde{c}_1$, yields

$$\frac{d}{ds}\bigg|_{s=0} E(\tilde{c}_1) = \mathcal{L}(c_1(\theta(p,r))) - \mathcal{L}(c_3(\theta(p,r))) \neq 0.$$

By using $s \mapsto \tau(-s)$ instead of $s \mapsto \tau(s)$, if necessary, we can assume that

$$\frac{d}{ds}\bigg|_{s=0} E(\tilde{c}_{1,s}) < 0.$$

It follows that $\theta(p, \tilde{c}_{1,s}(\theta(p,r))) + \theta(\tilde{c}_{1,s}(\theta(p,r)), q) < \theta(p,r) + \theta(r,q)$ for small $s > 0$. Since for sufficiently small $s > 0$ the geodesics $\tilde{c}_{1,s}, \tilde{c}_{2,s} : [0, \theta(p,r)] \to M$ intersect at $t = \theta(p,r)$, the cut point $\tilde{c}_{1,s}(t_{1,s})$ of $\tilde{c}_{1,s}$ occurs no later than $\theta(p,r)$, that is, $t_{1,s} \leq \theta(p,r)$. Since

$$\theta(p, \tilde{c}_{1,s}(t_{1,s})) + \theta(\tilde{c}_{1,s}(t_{1,s}), q) \leq \theta(p, \tilde{c}_{1,s}(\theta(p,r))) + \theta(\tilde{c}_{1,s}(\theta(p,r)), \tilde{c}_{1,s}(\vartheta(p,q)))$$
$$= L(\tilde{c}_{1,s}) < \vartheta(p,q),$$

we have found for sufficiently small $s > 0$ a cut point $r_{1,s} = \tilde{c}_{1,s}(t_{1,s}) \in \text{Cut}\, p$ satisfying $\theta(p, r_{1,s}) + \theta(r_{1,s}, q) < \theta(p, r) + \theta(r, q) = \vartheta(p, q)$, which contradicts the definition of $\vartheta(p, q)$. Hence $c_1'(\theta(p, r)) = c_3'(\theta(p, r))$, that is, the broken geodesic (c_1, c_3) with break point r is actually smooth.   □

LEMMA 9.5. *Let (M, F) be a compact Finsler manifold with reversibility λ and flag curvature $K \leq 1$.*

(a) *Let $p, q \in M$ with $q \notin \text{Cut}\, p$ and assume that $\vartheta(p, q) + \theta(q, p) < \pi(1 + \lambda^{-1})$. Then $\vartheta(p, q)$ is the length of the shortest nonminimal geodesic from p to q.*
(b) *If the symmetrized injectivity radius d of M satisfies $d < \pi(1 + \lambda^{-1})/2$, there is a shortest geodesic loop c with initial point p and a point $q \in \text{Cut}(p)$ on this loop with $L(c) = 2d = 2d(p, q)$.*

PROOF. (a) The cut locus $\text{Cut}\, p$ is a closed subset; hence there exists $r \in \text{Cut}\, p$ with $\theta(p, r) + \theta(r, q) = \vartheta(p, q)$. It follows from Lemma 9.4 that there is a geodesic c from p to q through r with $L(c) = \vartheta(p, q)$. Since r is a cut point and $r \neq q$, this geodesic is not minimal.

(b) Let $q \in \text{Cut}(p)$ be a point with $d = d(p, q)$. We know from Lemma 9.4 that there is a geodesic loop c with $c(0) = p$ and $L(c) = 2d$. We only have to show that this curve is a shortest geodesic loop. If c_1 is a shortest geodesic loop with $c_1(0) = p$ and with cut point $q = c_1(t_0)$, the restriction $c_1|[0, t_0]$ is minimal and $L(c_1) \geq 2d(p, q)$. But we showed in Lemma 9.4 that there is a geodesic loop $c \in \Omega_p$ with $L(c) = 2d = 2d(p)$. This finally implies that $L(c_1) = 2d$.   □

REMARK 9.6. If the Finsler metric is reversible, the proof of Lemma 9.4 simplifies considerably. The argument for this case was introduced by Klingenberg [1995, 2.1.11] in the Riemannian setting. The minimal geodesic c_3 coincides with one of the minimal geodesics c_1, c_2 (say c_1) up to orientation, that is, $\theta(p, r) = d(p, r)$ and $c_3(t) = c_1(2d(p, r) - t)$. By using the same argument exchanging the roles of p, r one can prove then that there is a closed geodesic c of length $2d$. If c is parametrized by arc length, $c(d)$ is the cut point, and there is no shorter geodesic loop.

If we use the same argument as in the proof of Lemma 9.4, we obtain:

LEMMA 9.7. *Let (M, F) be a compact Finsler manifold with reversibility λ and flag curvature $K \leq 1$. If the symmetrized injectivity radius satisfies $d < \pi\left(1 + \frac{1}{\lambda}\right)$, there is a point $p \in M$ and a cut point $r \in \text{Cut}\, p$ such that either*

(a) *there is a closed geodesic $c : [0, 2d] \to M$ parametrized by arclength with $L(c) = 2d$ and $c(0) = p$, $c(\theta(p, r)) = r$, or*
(b) *there are two distinct geodesic loops $c_1, c_2 : [0, 2d] \to M$ parametrized by arc length (that is, both have the same length $2d$) with $c_1(0) = p = c_2(\theta(r, p))$ and $c_1(\theta(p, r)) = r = c_2(0)$.*

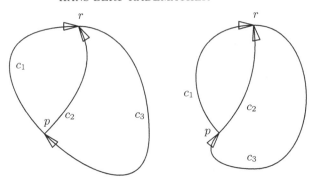

PROOF. If we use the statement of Lemma 9.4 and the argument in the proof of the same lemma, exchanging the roles of p and q, we reach the following statement: There are three minimal geodesics $c_1, c_2 : [0, \theta(p,q)] \to M$ and $c_3 : [\theta(q,p), 2d] \to M$ with $d = d(p,q)$, $p = c_1(0) = c_2(0) = c_3(2d)$, and $q = c_1(\theta(p,q)) = c_2(\theta(p,q)) = c_3(\theta(p,q))$. Without loss of generality we can assume that the broken geodesic (c_1, c_3) with corner q is smooth at q: $c_1'(\theta(p,q)) = c_3'(\theta(p,q))$. In addition there are two cases: Either the broken geodesic formed by c_3, c_1 is smooth at p, so that $c_3'(2d) = c_1'(0)$, which means that the geodesics c_1, c_3 form a closed geodesic, or the broken geodesic formed by c_3, c_2 is smooth at p.   □

In the reversible case we have case (a) of the lemma. It is not clear whether case (b) occurs in the nonreversible case.

LEMMA 9.8. *Let (M, F) be a compact Finsler manifold with reversibility λ and flag curvature $K \leq 1$ and let $p, q \in M$ be two points with distance $\theta(p,q) < \mathrm{inj}\, p$. Let $c_1 : [0,1] \to M$ be a nonminimal geodesic with $c_1(0) = p$ and $c_1(1) = q$, and let $c_0 : [0,1] \to M$ be a curve such that the reversed curve $c_0^{-1} : [0,1] \to M$, $c_0^{-1}(t) = c_0(1-t)$ is a minimal geodesic with $q = c_0(1)$, $p = c_0(0)$ and length $L(c_0) = \theta(q,p)$. If $c_s : [0,1] \to M$, $s \in [0,1]$, is a homotopy of piecewise smooth curves with fixed endpoints $p = c_s(0)$, $q = c_s(1)$ for all $s \in [0,1]$ between the curves c_0, c_1, then*

$$\theta(q,p) + \max_{s \in [0,1]} L(c_s) \geq \pi \left(1 + \frac{1}{\lambda} \right).$$

PROOF. We are assuming that $\theta(q,p) + L(c_s) < \pi(1 + \lambda^{-1})$ for all $s \in (0,1]$; hence there is a $\rho > 0$ such that $\theta(q,p) + L(c_s) \leq (\pi - \rho)(1 + \lambda^{-1})$ for all $s \in [0,1]$. We show by contradiction that

$$\theta(p, c_s(t)) \leq \pi - \rho \quad \text{for all } s, t \in [0,1]. \tag{9-2}$$

If there are $s, t \in [0,1]$ such that $\theta(p, c_s(t)) > \pi - \rho$, then

$$\theta(q,p) + L(c_s) \geq \theta(p, c_s(t)) + \theta(c_s(t), q) + \theta(q,p)$$
$$\geq \theta(p, c_s(t)) + \theta(c_s(t), p) \geq \theta(p, c_s(t))(1 + \lambda^{-1})$$
$$> (\pi - \rho)(1 + \lambda^{-1}),$$

which contradicts our assumption.

Define the closed ball $\bar{B}_a(T_pM) := \{v \in T_pM \mid F(v) \leq a\}$ of radius $a > 0$ in the tangent space T_pM at p. The subset $\bar{B}_a(p) := \{x \in M \mid \theta(p, x) \leq a\} \subset M$ equals the image $\exp_p(\bar{B}_a(T_pM))$ of the exponential map $\exp_p : T_pM \to M$ for an arbitrary $a > 0$. It follows from 9–2 that $c_s(t) \in \bar{B}_{\pi-\rho}(p)$ for all $s, t \in [0, 1]$. The restriction

$$F := \exp_p : \bar{B}_{\pi-\rho}(T_pM) \to \bar{B}_{\pi-\rho}(p)$$

has everywhere maximal rank since the flag curvature satisfies $K \leq 1$; thus it is a local diffeomorphism. The restriction

$$\exp_p : B_{\mathrm{inj}\,p}(T_pM) \to B_{\mathrm{inj}\,p}(p)$$

is a diffeomorphism, since $\theta(q, p) < \mathrm{inj}\,p$ for sufficiently small $\eta > 0$ we have $c_s([0, 1]) \subset B_{\mathrm{inj}\,p}(p)$ for all $s \in [0, \eta)$. Hence there is a uniquely defined lift

$$\tilde{c}_s : t \in [0, 1) \mapsto \tilde{c}_s(t) \in \bar{B}_{\mathrm{inj}\,p}(T_pM)$$

for $s \in [0, \eta)$ with $c_s(t) = \exp_p(\tilde{c}_s(t))$, $0_p = \tilde{c}_s(0)$, and $X = \tilde{c}_0(1) = \tilde{c}_s(1)$ for all $s \in [0, \eta)$. Since the restriction $F = \exp_p|\bar{B}_{\pi-\delta}(T_pM)$. is a local diffeomorphism, there is a uniquely determined extension

$$\tilde{c}_s : t \in [0, 1) \mapsto \tilde{c}_s(t) \in \bar{B}_{\pi-\rho}(T_pM)$$

with $c_s(t) = \exp_p(\tilde{c}_s(t))$ of the lift for all $s \in [0, 1]$. It remains to show that this lift is a homotopy with fixed end points. Define $J_0 := \{s \in [0, 1] | \tilde{c}_s(0) = 0_p\}$, $J_1 := \{s \in [0, 1] | \tilde{c}_s(1) = X\}$; these subsets contain the nonempty interval $[0, \eta)$ and are closed in $[0, 1]$. Since the restriction F of the exponential map is a local diffeomorphism the subsets $J_0, J_1 \subset [0, 1]$ are also open, hence $J_0 = J_1 = [0, 1]$. By assumption $c_1 : [0, 1] \to M$ is a geodesic from p to q; hence $\tilde{c}_1(t) = t c_1'(0)$ for all $t \in [0, 1]$, contradicting $\tilde{c}_1(1) = 0_p$. Therefore we arrive at a contradiction starting from the assumption $\theta(q, p) + L(c_s) < \pi \left(1 + \frac{1}{\lambda}\right)$ for all $s \in [0, 1]$, which finally proves the claim. □

With the long homotopy lemma we are able to gain a lower bound for the length of nonminimal geodesics:

PROPOSITION 9.9. *Let (M, F) be a simply connected, compact Finsler manifold of dimension $n \geq 3$, with reversibility λ and flag curvature $\left(1 - \frac{1}{1+\lambda}\right)^2 < K \leq 1$. If $p \in M$, there exists for every $\varepsilon > 0$ a point q that is a regular point of $\exp_p$ and that satisfies $\theta(q, p) < \varepsilon$ and $\vartheta(p, q) + \theta(q, p) \geq \pi \left(1 + \frac{1}{\lambda}\right)$. (Recall that $\vartheta(p, q)$ is the length of a shortest nonminimal geodesic from p to q.)*

PROOF. Since M is compact we can choose $\delta > \left(1 - \frac{1}{1+\lambda}\right)^2$ such that the flag curvature K of satisfies $\delta < K \leq 1$. For a given $\varepsilon > 0$ we choose a regular point q of $\exp_p$ with $\theta(q, p) < \varepsilon$ and

$$\theta(q, p) < \pi \left(1 + \frac{1}{\lambda} - \frac{1}{\sqrt{\delta}}\right).$$

Now we assume that there is a shortest nonminimal geodesic $c_1 : [0, 1] \to M$ from $p = c_1(0)$ to $q = c_1(1)$ with length $L(c_1)$ and satisfying $L(c_1) + \theta(q, p) < \pi \left(1 + \frac{1}{\lambda}\right)$.

M is simply connected, so there is a path $s \in [0, 1] \mapsto c_s \in \Omega_{pq} M$ going from c_0 (the reverse of the minimal geodesic c_0^{-1} from q to p of length $\theta(q, p)$) to the geodesic c_1. This path describes a map

$$H : ([0, 1], \{0, 1\}) \to \left(\Omega_{pq} M, \Omega_{pq}^{\kappa-} M\right),$$

with $H(s) = c_s$, $\Omega_{pq}^{\kappa-} M := \{\gamma \in \Omega_{pq} M \mid E(\gamma) < \kappa\}$, and κ defined by $2\sqrt{\kappa} = \pi(1 + \lambda^{-1}) - \theta(q, p)$. Let $c^* : [0, 1] \to M$ be a geodesic from p to q with length $L(c^*) \geq \pi\left(1 + \frac{1}{\lambda}\right) - \theta(q, p) \geq \pi/\sqrt{\delta}$. Lemma 8.6 gives the bound $\operatorname{ind} c \geq n - 1$, so $\operatorname{ind} c \geq 2$ in view of the assumption $n \geq 3$. We can conclude from the Fundamental Theorem of Morse Theory (Theorem 8.4) that the pair $(\Omega_{pq} M, \Omega_{pq}^{\kappa-} M)$ has the homotopy type of a CW-complex with no 1-dimensional cells; hence the relative homotopy group

$$\pi_1\left(\Omega_p M, \Omega_p^{\kappa-} M\right) = 0,$$

is 1-connected, as is the pair $\left(\Omega_{pq} M, \Omega_{pq}^{\kappa-} M\right)$; see Proposition 7.5. Therefore there is a map $\Phi : (u, s) \in [0, 1] \times ([0, 1], \{0, 1\}) \mapsto \Phi_u(s) \in (\Omega_{pq} M, \Omega_{pq}^{\kappa-} M)$ with $\Phi_u(0)(t) = c_0(t)$, $\Phi_u(1)(t) = c(t)$ for all $t, u \in [0, 1]$ and $\Phi_0(s) = c_s$ and $\bar{c}_s = \Phi_1(s) \in \Omega_{pq}^{\kappa-}$ for all $s \in [0, 1]$. This implies that $L(\bar{c}_s) < \pi\left(1 + \frac{1}{\lambda}\right) - \theta(q, p)$ for all $s \in [0, 1]$, that and $\bar{c}_0 = c_0$ is up to orientation the minimal geodesic, and that $\bar{c}_1 = c$ is a shortest nonminimal geodesic joining p and q. But we conclude from the Long Homotopy Lemma 9.8 that there is a $s^* \in (0, 1)$ with $L(\bar{c}_{s^*}) \geq \left(1 + \frac{1}{\lambda}\right) - \theta(q, p)$, which is a contradiction. $\square$

The assumption that q is a regular value of the exponential map $\exp_p$ ensures that the energy functional is a Morse function, so all geodesics joining p and q are nondegenerate. If one aims at estimating the length of geodesic loops or closed geodesics it won't be the case in general that p itself is a regular value of the exponential map $\exp_p$. For example, on the standard sphere every point p is conjugate to itself along a great circle; in particular every point p is a critical point of $\exp_p$. But the statement of Proposition 9.9 is also correct if we remove this assumption. In that case either one has to use a version of Morse theory including degenerate critical points [Rademacher 2004, Theorem 3] or one can argue as follows:

THEOREM 9.10. *Let (M, F) be a simply connected, compact Finsler manifold of dimension $n \geq 3$, with reversibility λ and flag curvature $\left(1 - \frac{1}{1+\lambda}\right)^2 < K \leq 1$. Then every nonconstant geodesic loop c has length at least $\pi\left(1 + \frac{1}{\lambda}\right)$ and the injectivity radius satisfies $\operatorname{inj} \geq \pi/\lambda$.*

PROOF. For a point $p \in M$ the function

$$(q, r) \in M \times \operatorname{Cut} p \mapsto \theta(p, r) + \theta(r, q) \in \mathbb{R}$$

is continuous, hence also the map

$$q \in M \mapsto \vartheta(p, q) = \inf \{\theta(p, r) + \theta(r, q) \mid r \in \text{Cut } p\}.$$

Choose a sequence $(q_i)_{i \in \mathbb{N}}$ of regular points of the exponential map $\exp_p$ with $\lim_{i \to \infty} \theta(q_i, p) = 0$. We conclude from Proposition 9.9 that

$$2d(p) = \lim_{i \to \infty} \vartheta(p, q_i) \geq \pi \left(1 + \frac{1}{\lambda}\right).$$

The estimate for the injectivity radius follows from Equation 9–1.   □

Now we can prove the Sphere Theorem:

THEOREM 9.11. *A simply connected and compact Finsler manifold of dimension* $n \geq 3$ *with reversibility* λ *and with flag curvature* K *satisfying* $\left(1 - \frac{1}{1+\lambda}\right)^2 < K \leq 1$ *is homotopy equivalent to the n-sphere.*

For $n = 2$, Synge's Theorem (Theorem 10.2) implies that an orientable compact surface carrying a Finsler metric of positive flag curvature $K > 0$ is diffeomorphic to the 2-sphere.

PROOF OF THE SPHERE THEOREM. Since M is compact we can choose $\delta > \pi \left(1 + \lambda^{-1}\right)$ such that the flag curvature K satisfies $\delta < K \leq 1$. We choose $\varepsilon > 0$ with

$$\varepsilon < \pi \left(1 + \frac{1}{\lambda} - \frac{1}{\sqrt{\delta}}\right). \tag{9–3}$$

We conclude from Proposition 9.9 that there is a regular point $q \in M$ of the exponential map $\exp_p$ with $\theta(q, p) < \varepsilon$ and $\vartheta(p, q) \geq \pi \left(1 + \frac{1}{\lambda}\right) - \varepsilon$; hence $\vartheta(p, q) \geq \pi/\sqrt{\delta}$, by 9–3. We conclude from Lemma 8.6 that the index $\text{ind } c$ of a nonminimal geodesic c joining p and q satisfies $\text{ind } c \geq n - 1$. Then Proposition 8.8 implies that M is homotopy equivalent to the n-sphere.   □

10. Length of Closed Geodesics in Even Dimensions

In even dimensions one obtains a lower bound for the length of closed geodesics for every metric of positive curvature on a simply connected manifold without assuming a lower curvature bound. The crucial point (*Synge's argument*) is that in even dimensions there is a periodic parallel vector field along a closed geodesic. By scaling the metric we can assume that the flag curvature K satisfies $0 < K \leq 1$.

LEMMA 10.1. *Let* (M, F) *be a compact, oriented Finsler manifold of even dimension with positive flag curvature* $0 < K \leq 1$. *For every closed geodesic* c *there is a parallel and periodic vector field* W *with* $\langle W, W \rangle_{c'} = 1$ *and* $\langle W, c' \rangle_{c'} = 0$.

PROOF. The covariant derivative (∇/dt) along the geodesic $c : [0, 1] \to M$ with $\dot{c}(0) = \dot{c}(1)$ defines a parallel transport $P : T_p M \to T_p M$ with $P(X(0)) = X(1)$ and $X = X(t)$ is a parallel vector field along c with respect to (∇/dt).

Since $(d/dt)\langle X(t), X(t)\rangle_{c'} = 2\langle (\nabla/dt)X(t), X(t)\rangle_{c'}$, the parallel transport is an orientation-preserving isometry. Since $P(\dot{c}(0)) = \dot{c}(1)$, the parallel transport defines an isometry of the orthogonal complement

$$T_p^\perp M := \{X \in T_p M \mid \langle X, \dot{c}\rangle_{c'} = 0\}.$$

This vector space has odd dimension, so there exists a nonzero eigenvector to the eigenvalue 1, that is, a vector $v \in T_p^\perp M$ with $Pv = v$. Then the parallel field W along c with $W(0) = v$ is periodic: $W(1) = W(0)$.   □

Now we prove a generalization of Synge's Theorem:

THEOREM 10.2. *Let (M, F) be a compact Finsler manifold of positive flag curvature.*

(a) *If M is orientable, it is simply connected.*
(b) *If M is nonorientable, its fundamental group satisfies $\pi_1(M) = \mathbb{Z}_2$.*

PROOF. Let M be orientable, and assume that $\pi_1(M) \neq 0$. Then there is a nontrivial homotopy class in M and a shortest closed curve in this nontrivial homotopy class is a closed geodesic $c : [0, 1] \to M$. By Lemma 10.1 there is a parallel and periodic vector field W along c; the index form at W satisfies

$$I_c(W; W) = \int_0^1 \langle R^{c'}(W, c')c', W\rangle_{c'} dt < 0;$$

therefore there is a variation by homotopic closed curves c_s with $L(c_s) < L(c)$ for $s > 0$, contradicting the assumption that c is a shortest closed curve in the given homotopy class. Hence $\pi_1(M) = 0$.

If M is nonorientable one passes to the orientable double cover, which by (a) is simply connected, so $\pi_1(M) = \mathbb{Z}_2$.   □

THEOREM 10.3. *Let (M, F) be a simply connected compact Finsler manifold of even dimension $n \geq 2$ with reversibility λ and with flag curvature $0 < K \leq 1$. Then every nonconstant closed geodesic c has length $L(c) \geq \pi\left(1 + \frac{1}{\lambda}\right)$.*

PROOF. Let $c : S^1 \to M$ be a shortest closed geodesic with $0 < L(c) < \pi\left(1 + \frac{1}{\lambda}\right)$. By Lemma 10.1, there exists a parallel unit vector field W along c; it follows that the index form I_c on $V_c^\perp$ satisfies $I_c(W, W) < 0$. Let c_s, $s \in (-\varepsilon, \varepsilon)$, be a variation of $c = c_0$ with variation vector field W. It follows from the second variation formula that $E(c_s) < E(c_0)$ for all $s \in (-\varepsilon, 0) \cup (0, \varepsilon)$. Since there are no critical values of E in the interval $(0, E(c_0))$, there is a map $h_s : S^1 \to M$, $s \in [-1, 1]$, with $c = h_0$, $L(h_1) = L(h_{-1}) = 0$ and $L(h_s) < L(c) = L(h_0)$ for all nonzero $s \in (-1, 1)$. One can generalize the Long Homotopy Lemma 9.8 to the case of homotopies $c_s : S^1 \to M$ of freely homotopic closed curves. This generalization yields a contradiction.   □

11. An Example

Shen [2002] constructed Finsler metrics of constant flag curvature and Randers type. It turns out that in the Hamiltonian description these are the metrics introduced in [Katok 1973] and investigated in [Ziller 1982], as observed in [Rademacher 2004, Chapter 4]. These examples show that the estimates in Proposition 9.9 and Theorem 9.10, for the lengths of nonminimal geodesics between fixed points and of nonconstant geodesic loops, are sharp.

One can describe a Finsler metric using the Legendre transformation with a Hamiltonian function [Ziller 1982, Chapter 1]. The Katok examples on $S^2 = \{(x, y, z) \in \mathbb{R}^3 \mid x^2 + y^2 + z^2 = 1\}$ can be introduced as follows. We start with the standard Riemannian metric g on the 2-sphere S^2, letting g^* be the dual metric on the cotangent bundle T^*M. In the Hamiltonian description the standard metric is determined by the quadratic Hamiltonian function

$$y \in T^*S^2 \mapsto g^*(y, y) \in \mathbb{R},$$

or by the 1-homogeneous Hamiltonian $H_0 : T^*S^2 \to \mathbb{R}$, $H_0(y) = \sqrt{g^*(y, y)}$. Let $\psi_0^t : T^*S^2 \to T^*S^2$ be the corresponding Hamiltonian flow. Then $t \in \mathbb{R} \mapsto \tau^*(\psi_0^t(y))$ is a geodesic of the standard metric; here $\tau^* : T^*S^2 \to S^2$ is the projection of the cotangent bundle.

Let $V(x, y, z) = (y, -x, 0)$ be the Killing field belonging to the 1-parameter subgroup $\phi^t : S^2 \to S^2$ generated by the rotations around the z-axis. A. Katok introduced the following perturbation of Randers type:

$$H_\varepsilon : T^*S^2 \to \mathbb{R}; H_\varepsilon(y) = \sqrt{g^*(y, y)} + \varepsilon y(V).$$

In [Bao et al. 2003] these perturbations are connected to *Zermelo navigation*. For $\varepsilon \in [0, 1)$ this defines a quadratic Hamiltonian $\frac{1}{2}H_\varepsilon^2$ and using the Legendre transformation of this Hamiltonian we obtain a Finsler metric F_ε.

The description of the geodesics appears to be easier in the Hamiltonian picture: Since ϕ^t is a group of isometries leaving H_0 invariant, the Hamiltonian flow ψ_ε^t of the quadratic Hamiltonian $\frac{1}{2}H_\varepsilon^2$ is generated by two commuting flows, $\psi_\varepsilon^t = \psi_0^t \circ (\phi^{\varepsilon t})^*$. Here $(\phi^t)^* : T^*S^2 \to T^*S^2$ is the flow on the cotangent bundle induced by differentiating ϕ^t. The projection of the Hamiltonian flow onto the 2-sphere yields the geodesics of the Finsler metric. As described in [Ziller 1982, Chapter 1] one can visualize the geodesic flow of these Finsler metrics by identifying the cotangent bundle T^*S^2 with the tangent bundle T_*S^2 via the standard Riemannian metric g. Then the geodesic flow can be seen as the geodesic flow of the standard metric observed from a coordinate system rotating around the z-axis with constant speed $2\pi\varepsilon$, as shown in the figure on the next page. For irrational ε the only closed geodesics are $c_\pm(t) = (\cos 2\pi t, \pm \sin 2\pi t, 0)$, $t \in [0, 1]$, i.e., the equator with both orientations. (We consider c_+ and c_- geometrically distinct; for example their lengths $L(c_\pm) = 2\pi/(1 \pm \varepsilon)$ differ.)

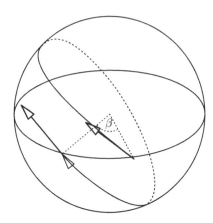

Using the results [Hrimiuc and Shimada 1996, Theorem 5.8], [Shen 2001a, Example 3.1.1] one obtains in geodesic polar coordinates $(r, \phi) \in (0, \pi) \times [0, 2\pi]$ of the standard metric the following formula for F_ε :

$$F_\varepsilon = \frac{\sqrt{(1 - \varepsilon^2 \sin^2 r)\, dr^2 + \sin^2(r)\, d\phi^2} - \varepsilon \sin^2 r \, d\phi}{1 - \varepsilon^2 \sin^2 r}. \tag{11-1}$$

It is shown in [Shen 2002, Remark 3.1] that this metric has constant flag curvature 1. The reversibility of the Finsler metrics F_ε can be computed by

$$\max\left\{ H_\varepsilon(-y) \mid y \in T^* S^2,\, H_\varepsilon(y) = 1 \right\}.$$

Collecting the results from [Ziller 1982] and [Rademacher 2004, Chapter 5] one obtains:

THEOREM 11.1 [Rademacher 2004, Theorem 5]. *There is a one-parameter family F_ε, $\varepsilon \in [0,1)$, of Finsler metrics on the 2-sphere S^2 of constant flag curvature 1. These Finsler metrics are nonreversible for $\varepsilon \in (0,1)$ and F_0 is the standard metric. The reversibility is $\lambda = (1 + \varepsilon)/(1 - \varepsilon)$. If ε is irrational there are exactly two geometrically distinct closed geodesics $c_\pm$ of length $L(c_\pm) = 2\pi(1 \pm \varepsilon)^{-1}$. In particular the shortest closed geodesic c_+ satisfies $L(c_+) = 2\pi(1 + \lambda^{-1}) = \pi/(1 + \varepsilon)$. The injectivity radius and the diameter equal π.*

References

[Abresch and Meyer 1997] U. Abresch and W. Meyer, *Injectivity radius estimates and sphere theorems*, pp. 1–47 in: *Comparison geometry*, MSRI Publications **30** (1997).

[Auslander 1955] L. Auslander, "On curvature in Finsler geometry", *Trans. Amer. Math. Soc.* **79** (1955), 378–388.

[Bao et al. 2000] D. Bao, S.-S. Chern, and Z. Shen, *An introduction to Riemann–Finsler geometry*, Grad. Texts Math. **200**, Springer, New York, 2000.

[Bao et al. 2003] D. Bao, C. Robles, and Z. Shen, "Zermelo navigation on Riemannian manifolds", preprint, 2003.

[Berger 1998] M. Berger, "Riemannian geometry during the second half of the twentieth century", *Jber. d. Dt. Math.-Verein.* **100** (1998), 45–208.

[Bredon 1995] G. E. Bredon, *Topology and geometry*, Grad. Texts Math. **139**, Springer, New York, corr. 2nd printing, 1995.

[Bryant 2002] R. Bryant, "Some remarks on Finsler manifolds with constant flag curvature", *Houston J. Math.* **28** (2002), 221–262.

[Dazord 1968a] P. Dazord, "Variétés finslériennes de dimension paire δ-pincées", *C. R. Acad. Sc. Paris* **266** (1968), 496–498.

[Dazord 1968b] P. Dazord, "Variétés finslériennes en forme des sphères", *C. R. Acad. Sc. Paris* **267** (1968), 353–355.

[Hrimiuc and Shimada 1996] D. Hrimiuc and H. Shimada, "On the $\mathcal{L}$-duality between Lagrange and Hamilton manifolds", *Nonlin. World* **3** (1996), 613–641.

[Katok 1973] A. Katok, "Ergodic properties of degenerate integrable Hamiltonian systems" (Russian), *Izs. Akad. Nauk SSSR Ser. Mat.* **37** (1973), 539–576; translated in *Math. USSR Izv.* **7** (1973), 535–572.

[Klingenberg 1961] W. Klingenberg, "Über Riemannsche Mannigfaltigkeiten mit positiver Krümmung", *Comment. Math. Helv.* **35** (1961), 47–54.

[Klingenberg 1963] W. Klingenberg, "Manifolds with restricted conjugate locus", *Ann. Math.* **78** (1963), 527–547.

[Klingenberg 1995] W. Klingenberg, *Riemannian geometry*, de Gruyter Studies Math. **1**, 2nd rev. ed., de Gruyter, Berlin, New York, 1995.

[Matthias 1980] H. H. Matthias, *Zwei Verallgemeinerungen eines Satzes von Gromoll und Meyer*, Bonner Math. Schr. **126** (1980).

[Mercuri 1971] F. Mercuri, "The critical point theory for the closed geodesic problem", *Math. Zeitschr.* **156** (1971), 231–245.

[Milnor 1969] J. Milnor, *Morse theory*, Annals of Math. Studies **51** Princeton Univ. Press, Princeton 1969.

[Milnor 2003] J. Milnor, "Towards the Poincaré conjecture and the classification of 3-manifolds", *Notices Amer. Math. Soc.* **50**:10 (2003), 1226–1233.

[Rademacher 1992] H.-B. Rademacher, *Morse-Theorie und Geschlossene Geodätische*, Bonner Math. Schr. **229** (1992).

[Rademacher 2004] H.-B. Rademacher, "A sphere theorem for non-reversible Finsler metrics", *Math. Ann.* **328** (2004), 373–387.

[Shen 2001a] Z. Shen, "Lectures on Finsler geometry", World Scientific, Singapore, 2001.

[Shen 2001b] Z. Shen, *Differential geometry of spray and Finsler spaces*, Kluwer, Dordrecht, 2001.

[Shen 2002] Z. Shen, "Two-dimensional Finsler metrics with constant curvature", *Manuscr. Math.* **109** (2002), 349–366.

[Spanier 1966] E. H. Spanier, "Algebraic Topology", McGraw-Hill, New York, 1966; reprinted by Springer, New York, 1989.

[Ziller 1982] W. Ziller, "Geometry of the Katok examples", *Ergod. Th. Dyn. Syst.* **3** (1982), 135–157.

HANS-BERT RADEMACHER
UNIVERSITÄT LEIPZIG
MATHEMATISCHES INSTITUT
AUGUSTUSPLATZ 10/11
04109 LEIPZIG
GERMANY
    rademacher@math.uni-leipzig.de

Riemann–Finsler Geometry
MSRI Publications
Volume **50**, 2004

Landsberg Curvature, S-Curvature and Riemann Curvature

ZHONGMIN SHEN

Contents

1. Introduction

Roughly speaking, Finsler metrics on a manifold are regular, but not necessarily reversible, distance functions. In 1854, B. Riemann attempted to study a special class of Finsler metrics — Riemannian metrics — and introduced what is now called the Riemann curvature. This infinitesimal quantity faithfully reveals the local geometry of a Riemannian manifold and becomes the central concept of Riemannian geometry. It is a natural problem to understand general regular distance functions by introducing suitable infinitesimal quantities. For more than half a century, there had been no essential progress until P. Finsler studied the variational problem in a Finsler manifold. However, it was L. Berwald who first successfully extended the notion of Riemann curvature to Finsler metrics by introducing what is now called the Berwald connection. He also introduced some non-Riemannian quantities via his connection [Berwald 1926; 1928]. Since then, Finsler geometry has been developed gradually.

The Riemann curvature is defined using the induced spray, which is independent of any well-known connection in Finsler geometry. It measures the shape of the space. The Cartan torsion and the distortion are two primary geometric quantities describing the geometric properties of the Minkowski norm in each tangent space. Differentiating them along geodesics gives rise to the Landsberg curvature and the S-curvature. These quantities describe the rates of change of the "color pattern" on the space.

In this article, I am going to discuss the geometric meaning of the Landsberg curvature, the S-curvature, the Riemann curvature, and their relationship. I will give detailed proofs for several important local and global results.

2. Finsler Metrics

By definition, a Finsler metric on a manifold is a family of Minkowski norms on the tangent spaces. A *Minkowski norm* on a vector space V is a nonnegative function $F : V \to [0, \infty)$ with the following properties:

(i) F is positively y-homogeneous of degree one, i.e., $F(\lambda y) = \lambda F(y)$ for any $y \in V$ and any $\lambda > 0$;

(ii) F is C^∞ on $V \setminus \{0\}$ and for any tangent vector $y \in V \setminus \{0\}$, the following bilinear symmetric form $g_y : V \times V \to R$ is positive definite:

$$g_y(u, v) := \frac{1}{2} \frac{\partial^2}{\partial s \partial t} \big(F^2(y + su + tv) \big) |_{s=t=0}.$$

A Minkowski norm is said to be *reversible* if $F(-y) = F(y)$ for $y \in V$. In this article, Minkowski norms are not assumed to be reversible. From (i) and (ii), one can show that $F(y) > 0$ for $y \neq 0$ and $F(u + v) \leq F(u) + F(v)$ for $u, v \in V$. See [Bao et al. 2000] for a proof.

Let $\langle \ , \ \rangle$ denote the standard inner product on R^n, defined by $\langle u, v \rangle := \sum_{i=1}^{n} u^i v^i$. Then $|y| := \sqrt{\langle y, y \rangle}$ is called the standard Euclidean norm on R^n. Let $b \in R^n$ with $|b| < 1$, then $F = |y| + \langle b, y \rangle$ is a Minkowski norm on R^n, which is called a *Randers norm*.

Let M be a connected, n-dimensional, C^∞ manifold. Let $TM = \bigcup_{x \in M} T_x M$ be the tangent bundle of M, where $T_x M$ is the tangent space at $x \in M$. We denote a typical point in TM by (x, y), where $y \in T_x M$, and set $TM_0 := TM \setminus \{0\}$ where $\{0\}$ stands for $\{(x, 0) \mid x \in X, \, 0 \in T_x M\}$. A Finsler metric on M is a function $F : TM \to [0, \infty)$ with the following properties:

(a) F is C^∞ on TM_0;

(b) At each point $x \in M$, the restriction $F_x := F|_{T_x M}$ is a Minkowski norm on $T_x M$.

The pair (M, F) is called a *Finsler manifold*.

Let (M, F) be a Finsler manifold. Let (x^i, y^i) be a standard local coordinate system in TM, i.e., y^i's are determined by $y = y^i(\partial/\partial x^i)|_x$. For a vector $y = y^i(\partial/\partial x^i)|_x \neq 0$, let $g_{ij}(x, y) := \frac{1}{2}[F^2]_{y^i y^j}(x, y)$. The induced inner product $\mathbf{g}_y$ is given by

$$\mathbf{g}_y(u, v) = g_{ij}(x, y)u^i v^j,$$

where $u = u^i(\partial/\partial x^i)|_x$ and $v = v^i(\partial/\partial x^i)|_x$. By the homogeneity of F,

$$F(x, y) = \sqrt{\mathbf{g}_y(y, y)} = \sqrt{g_{ij}(x, y)y^i y^j}.$$

A Finsler metric $F = F(x, y)$ is called a *Riemannian metric* if the $g_{ij} = g_{ij}(x)$ are functions of $x \in M$ only.

There are three special Riemannian metrics.

EXAMPLE 2.1 (EUCLIDEAN METRIC). The simplest metric is the Euclidean metric $\alpha_0 = \alpha_0(x, y)$ on $\mathbf{R}^n$, which is defined by

$$\alpha_0(x, y) := |y|, \quad y = (y^i) \in T_x\mathbf{R}^n \cong \mathbf{R}^n. \tag{2-1}$$

We will simply denote $(\mathbf{R}^n, \alpha_0)$ by $\mathbf{R}^n$, which is called *Euclidean space*.

EXAMPLE 2.2 (SPHERICAL METRIC). Let $\mathbf{S}^n := \{x \in \mathbf{R}^{n+1} \mid |x| = 1\}$ denote the standard unit sphere in $\mathbf{R}^{n+1}$. Every tangent vector $y \in T_x\mathbf{S}^n$ can be identified with a vector in $\mathbf{R}^{n+1}$ in a natural way. The induced metric α_{+1} on $\mathbf{S}^n$ is defined by $\alpha_{+1} = \|y\|_x$, for $y \in T_x\mathbf{S}^n \subset \mathbf{R}^{n+1}$, where $\|\cdot\|_x$ denotes the induced Euclidean norm on $T_x\mathbf{S}^n$. Let $\varphi : \mathbf{R}^n \to \mathbf{S}^n \subset \mathbf{R}^{n+1}$ be defined by

$$\varphi(x) := \left(\frac{x}{\sqrt{1 + |x|^2}}, \frac{1}{\sqrt{1 + |x|^2}}\right). \tag{2-2}$$

Then φ pulls back α_{+1} on the upper hemisphere to a Riemannian metric on $\mathbf{R}^n$, which is given by

$$\alpha_{+1} = \frac{\sqrt{|y|^2 + (|x|^2|y|^2 - \langle x, y\rangle^2)}}{1 + |x|^2}, \quad y \in T_x\mathbf{R}^n \cong \mathbf{R}^n. \tag{2-3}$$

EXAMPLE 2.3 (HYPERBOLIC METRIC). Let $\mathbf{B}^n$ denote the unit ball in $\mathbf{R}^n$. Define

$$\alpha_{-1} := \frac{\sqrt{|y|^2 - (|x|^2|y|^2 - \langle x, y\rangle^2)}}{1 - |x|^2}, \quad y \in T_x\mathbf{B}^n \cong \mathbf{R}^n. \tag{2-4}$$

We call α_{-1} the *Klein metric* and denote $(\mathbf{B}^n, \alpha_{-1})$ by $\mathbf{H}^n$.

The metrics (2–1), (2–3) and (2–4) can be combined into one formula:

$$\alpha_\mu = \frac{\sqrt{(1 + \mu|x|^2)|y|^2 - \mu\langle x, y\rangle^2}}{1 + \mu|x|^2}. \tag{2-5}$$

Of course, there are many non-Riemannian Finsler metrics on $\mathbf{R}^n$ with special geometric properties. We just list some of them below and discuss their geometric properties later.

EXAMPLE 2.4 (FUNK METRIC). Let

$$\Theta := \frac{\sqrt{|y|^2 - (|x|^2|y|^2 - \langle x, y\rangle^2)} + \langle x, y\rangle}{1 - |x|^2}, \quad y \in T_x B^n \cong \mathbb{R}^n. \tag{2-6}$$

$\Theta = \Theta(x, y)$ is a Finsler metric on B^n, called the *Funk metric* on B^n.
For an arbitrary constant vector $a \in \mathbb{R}^n$ with $|a| < 1$, let

$$\Theta_a := \frac{\sqrt{|y|^2 - (|x|^2|y|^2 - \langle x, y\rangle^2)} + \langle x, y\rangle}{1 - |x|^2} + \frac{\langle a, y\rangle}{1 + \langle a, x\rangle}. \tag{2-7}$$

where $y \in T_x B^n \cong \mathbb{R}^n$. $\Theta_a = \Theta_a(x, y)$ is a Finsler metric on B^n. Note that
$\Theta_0 = \Theta$ is the Funk metric on B^n. We call Θ_a the *generalized Funk metric*
on B^n [Shen 2003a].

EXAMPLE 2.5 [Shen 2003b]. Let δ be an arbitrary number with $\delta < 1$. Let

$$F_\delta := \frac{\sqrt{|y|^2 - (|x|^2|y|^2 - \langle x, y\rangle^2)} + \langle x, y\rangle}{2(1 - |x|^2)} - \delta \frac{\sqrt{|y|^2 - \delta^2(|x|^2|y|^2 - \langle x, y\rangle^2)} + \delta\langle x, y\rangle}{2(1 - \delta^2|x|^2)},$$

where $y \in T_x B^n \cong \mathbb{R}^n$. F_δ is a Finsler metric on B^n. Note that $F_{-1} = \alpha_{-1}$ is
the Klein metric on B^n. Let Θ be the Funk metric on B^n defined in (2-6). We
can express F_δ by

$$F_\delta = \tfrac{1}{2}\big(\Theta(x, y) - \delta\Theta(\delta x, y)\big).$$

EXAMPLE 2.6 [Berwald 1929b]. Let

$$B := \frac{\big(\sqrt{|y|^2 - (|x|^2|y|^2 - \langle x, y\rangle^2)} + \langle x, y\rangle\big)^2}{(1 - |x|^2)^2 \sqrt{|y|^2 - (|x|^2|y|^2 - \langle x, y\rangle^2)}}, \tag{2-8}$$

where $y \in T_x B^n \cong \mathbb{R}^n$. Then $B = B(x, y)$ is a Finsler metric on B^n.

EXAMPLE 2.7. Let ε be an arbitrary number with $|\varepsilon| < 1$. Let

$$F_\varepsilon := \frac{\sqrt{\Psi\big(\tfrac{1}{2}(\sqrt{\Phi^2 + (1 - \varepsilon^2)|y|^4} + \Phi)\big) + (1 - \varepsilon^2)\langle x, y\rangle^2} + \sqrt{1 - \varepsilon^2}\langle x, y\rangle}{\Psi}, \tag{2-9}$$

where

$$\Phi := \varepsilon|y|^2 + (|x|^2|y|^2 - \langle x, y\rangle^2), \quad \Psi := 1 + 2\varepsilon|x|^2 + |x|^4.$$

$F_\varepsilon = F_\varepsilon(x, y)$ is a Finsler metric on $\mathbb{R}^n$. Note that if $\varepsilon = 1$, then $F_1 = \alpha_{+1}$ is
the spherical metric on $\mathbb{R}^n$.

R. Bryant [1996; 1997] classified Finsler metrics on the standard unit sphere
S^2 with constant flag curvature equal to $+1$ and geodesics being great circles.
The Finsler metrics F_ε in (2-9) is a special family of Bryant's metrics expressed
in a local coordinate system. See Example 12.7 for further discussion.

The examples of Finsler metrics above all have special geometric properties. They are locally projectively flat with constant flag curvature. Some belong to the class of (α, β)-*metrics*, that is, those of the form

$$F = \alpha\,\phi\left(\frac{\beta}{\alpha}\right), \tag{2-10}$$

where $\alpha = \alpha_x(y) = \sqrt{a_{ij}(x)y^i y^j}$ is a Riemannian metric, $\beta = \beta_x(y) = b_i(x)y^i$ is a 1-form, and ϕ is a C^∞ positive function on some interval $I = [-r, r]$ big enough that $r \geq \beta/\alpha$ for all x and $y \in T_xM$. It is easy to see that any such F is positively homogeneous of degree one. Let $\|\beta\|_x := \sup_{y \in T_xM} \beta_x(y)/\alpha_x(y)$. Using a Maple program, we find that the Hessian $g_{ij} := \frac{1}{2}[F^2]_{y^i y^j}$ is

$$g_{ij} = \rho a_{ij} + \rho_0 b_i b_j + \rho_1(b_i \alpha_j + b_j \alpha_i) + \rho_2 \alpha_i \alpha_j,$$

where $\alpha_i = \alpha_{y^i}$ and

$$\rho = \phi^2 - s\phi\phi', \qquad\qquad \rho_0 = \phi\phi'' + \phi'\phi',$$
$$\rho_1 = -s(\phi\phi'' + \phi'\phi') + \phi\phi', \qquad \rho_2 = s^2(\phi\phi'' + \phi'\phi') - s\phi\phi',$$

where $s := \beta/\alpha$ with $|s| \leq \|\beta\|_x \leq r$. Then

$$\det(g_{ij}) = \phi^{n+1}(\phi - s\phi')^{n-2}\left((\phi - s\phi') + (\|\beta\|_x^2 - s^2)\phi''\right)\det(a_{ij}).$$

If $\phi = \phi(s)$ satisfies

$$\phi(s) > 0, \quad \phi(s) - s\phi'(s) > 0, \quad \left((\phi - s\phi') + (b^2 - s^2)\phi''(s)\right) \geq 0 \tag{2-11}$$

for all s with $|s| \leq b \leq r$, then (g_{ij}) is positive definite; hence F is a Finsler metric.

Sabau and Shimada [2001] have classified (α, β)-metrics into several classes and computed the Hessian g_{ij} for each class. Below are some special (α, β)-metrics.

(a) $\phi(s) = 1 + s$. The defined function $F = \alpha + \beta$ is a Finsler metric if and only if the norm of β with respect to α is less than 1 at any point:

$$\|\beta\|_x := \sqrt{a^{ij}(x)b_i(x)b_j(x)} < 1, \quad x \in M.$$

A Finsler metric in this form is called a *Randers metric*. The Finsler metrics in Example 2.4 are Randers metrics. The Finsler metrics in Example 2.5 is the sum of two Randers metrics.

(b) $\phi(s) = (1 + s)^2$. The defined function $F = (\alpha + \beta)^2/\alpha$ is a Finsler metric if and only if $\|\beta\|_x < 1$ at any point $x \in M$. The Finsler metric in Example 2.6 is in this form.

By a *Finsler structure* on a manifold M we usually mean a Finsler metric. Sometimes, we also define a Finsler structure as a scalar function F^* on T^*M such that F^* is C^∞ on $T^*M \setminus \{0\}$ and $F_x^* := F^*|_{T_x^*M}$ is a Minkowski norm on

$T_x^* M$ for $x \in M$. Such a function is called a *co-Finsler metric*. Given a co-Finsler metric, one can define a Finsler metric via the standard duality defined below.

Let $F^* = F^*(x, \xi)$ be a co-Finsler metric on a manifold M. Define a non-negative scalar function $F = F(x, y)$ on TM by

$$F(x, y) := \sup_{\xi \in T_x^* M} \frac{\xi(y)}{F^*(x, \xi)}.$$

Then $F = F(x, y)$ is a Finsler metric on M, said to be *dual to* F^*. In the same sense, $F^* = F^*(x, \xi)$ is also dual to F:

$$F^*(x, \xi) = \sup_{y \in T_x M} \frac{\xi(y)}{F(x, y)}.$$

Every vector $y \in T_x M \setminus \{0\}$ uniquely determines a covector $\xi \in T_x^* M \setminus \{0\}$ by

$$\xi(w) := \frac{1}{2} \frac{d}{dt} \left(F^2(x, y + tw) \right)|_{t=0}, \quad w \in T_x M.$$

The resulting map $\ell_x : y \in T_x M \to \xi \in T_x^* M$ is called the *Legendre transformation* at x. Similarly, every covector $\xi \in T_x^* M \setminus \{0\}$ uniquely determines a vector $y \in T_x M \setminus \{0\}$ by

$$\eta(y) := \frac{1}{2} \frac{d}{dt} \left(F^{*2}(x, \xi + t\eta) \right)|_{t=0}, \quad \eta \in T_x^* M.$$

The resulting map $\ell_x^* : \xi \in T_x^* M \to y \in T_x M$ is called the *inverse Legendre transformation* at x. Indeed, ℓ_x and ℓ_x^* are inverses of each other. Moreover, they preserve the Minkowski norms:

$$F(x, y) = F^*(x, \ell_x(y)), \quad F^*(x, \xi) = F(x, \ell_x^*(\xi)). \tag{2–12}$$

Let $\Phi = \Phi(x, y)$ be a Finsler metric on a manifold M and let $\Phi^* = \Phi^*(x, \xi)$ be the co-Finsler metric dual to Φ. By the above formulas, one can easily show that if $y \in T_x M \setminus \{0\}$ and $\xi \in T_x^* M \setminus \{0\}$ satisfy

$$\frac{d}{dt} \left(\Phi^*(x, \xi + t\eta) \right)|_{t=0} = \eta(y), \quad \eta \in T_x^* M.$$

Then

$$\Phi(x, y) = 1. \tag{2–13}$$

Let V be a vector field on M with $\Phi(x, -V_x) < 1$ and let $V^* : T^* M \to [0, \infty)$ denote the 1-form dual to V, defined by

$$V_x^*(\xi) = \xi(V_x), \quad \xi \in T_x^* M.$$

We have $\Phi^*(x, -V_x^*) = \Phi(x, -V_x) < 1$. Thus $F^* := \Phi^* + V^*$ is a co-Finsler metric on M. Define $F = F(x, y)$ by

$$F(x, y) := \sup_{\xi \in T_x^* M} \frac{\xi(y)}{F^*(x, \xi)}, \quad y \in T_x M. \tag{2–14}$$

F is a Finsler metric on M, called the *Finsler metric generated from* the pair (Φ, V). One can also define F in a different way without using the duality:

LEMMA 2.8. *Let $\Phi = \Phi(x, y)$ be a Finsler metric on M and let V be a vector field on M with $\Phi(x, -V_x) < 1$ for all $x \in M$. Then $F = F(x, y)$ defined in (2–14) satisfies*

$$\Phi\left(x, \frac{y}{F(x, y)} - V_x\right) = 1, \quad y \in T_x M. \tag{2-15}$$

Conversely, if $F = F(x, y)$ is defined by (2–15), it is dual to the co-Finsler metric $F^* := \Phi^* + V^*$ as defined in (2–14).

PROOF. For the co-Finsler metric $F^* = \Phi^* + V^*$, let $F = F(x, y)$ be defined in (2–14). Fix an arbitrary nonzero vector $y \in T_x M$. There is a covector $\xi \in T_x^* M$ such that

$$F(x, y) = \frac{\xi(y)}{F^*(x, \xi)}. \tag{2-16}$$

Let $\eta \in T_x^* M$ be an arbitrary covector. Consider the function

$$h(t) := \frac{\xi(y) + t\eta(y)}{\Phi^*(x, \xi + t\eta) + \xi(V_x) + t\eta(V_x)}.$$

Then $h(t) \le h(0) = F(x, y)$. Thus $h'(0) = 0$:

$$\eta(y)F^*(x, \xi) - \xi(y)\left(\frac{d}{dt}(\Phi^*(x, \xi + t\eta))|_{t=0} + \eta(V_x)\right) = 0.$$

Dividing by $F^*(x, \xi)$ and using (2–16), one obtains

$$\eta(y) - F(x, y)\left(\frac{d}{dt}(\Phi^*(x, \xi + t\eta))|_{t=0} + \eta(V_x)\right) = 0.$$

From this identity it follows that

$$\frac{d}{dt}(\Phi^*(x, \xi + t\eta))|_{t=0} = \eta\left(\frac{y}{F(x, y)} - V_x\right), \quad \eta \in T_x^* M.$$

Thus $F(x, y)$ satisfies (2–15) as we have explained in (2–13).

Conversely, let $F = F(x, y)$ be defined by (2–15). Then for any $\xi \in T_x^* M$,

$$\Phi^*(x, \xi) = \sup_{y \in T_x M} \eta\left(\frac{y}{F(x, y)} - V_x\right).$$

One obtains

$$\sup_{y \in T_x M} \frac{\xi(y)}{F(x, y)} = \sup_{y \in T_x M} \xi\left(\frac{y}{F(x, y)} - V_x\right) + \xi(V_x) = \Phi^*(x, \xi) + V_x^*(\xi) = F^*(x, \xi).$$

Thus F^* is dual to F and so F is dual to F^*, that is, F is given by (2–14). $\square$

Let $\Phi = \sqrt{\phi_{ij}(x)y^i y^j}$ be a Riemannian metric and let $V = V^i(x)(\partial/\partial x^i)$ be a vector field on a manifold M with

$$\Phi(x, -V_x) = \|V\|_x := \sqrt{\phi_{ij}(x)V^i(x)V^j(x)} < 1, \quad x \in M.$$

Solving (2–15) for $F = F(x, y)$, one obtains

$$F = \frac{\sqrt{(1 - \phi_{ij}V^i V^j)\phi_{ij}y^i y^j + (\phi_{ij}y^i V^j)^2} - \phi_{ij}y^i V^j}{1 - \phi_{ij}V^i V^j}. \tag{2–17}$$

Clearly, F is a Randers metric. It is easy to verify that any Randers metric $F = \alpha + \beta$, where $\alpha = \sqrt{a_{ij}(x)y^i y^j}$ and $\beta = b_i(x)y^i$, can be expressed in the form (2–17). According to Lemma 2.8, any Randers metric $F = \alpha + \beta$ expressed in the form (2–17) can be constructed in the following way. Let $\Phi^* := \sqrt{\phi^{ij}(x)\xi_i \xi_j}$ be the Riemannian metric dual to $\Phi = \sqrt{\phi_{ij}(x)y^i y^j}$ and $V^* := \xi(V_x) = V^i(x)\xi_i$ be the 1-form dual to V. Then $F^* := \Phi^*(x, \xi) + V^*(\xi) = \sqrt{\phi^{ij}(x)\xi_i \xi_j} + V^i(x)\xi_i$ is a co-Finsler metric on M. Moreover, the dual Finsler metric F of F^* is given by (2–17). This is proved in [Hrimiuc and Shimada 1996].

It was discovered in [Shen 2003c; 2002] that if Φ is a Riemannian metric of constant curvature and V is a special vector field, the generated metric F is of constant flag curvature. This discovery opens the door to a classification of Randers metrics of constant flag curvature [Bao et al. 2003]. But Maple programs played an important role in the computations that led to it.

EXAMPLE 2.9. Let $\phi = \phi(y)$ be a Minkowski norm on $\mathbf{R}^n$ and let

$$U_\phi := \{y \in \mathbf{R}^n \mid \phi(y) < 1\}.$$

Define

$$\Phi(x, y) := \phi(y), \quad y \in T_x\mathbf{R}^n \cong \mathbf{R}^n.$$

$\Phi = \Phi(x, y)$ is called a *Minkowski metric* on $\mathbf{R}^n$. Let $V_x := -x$, for $x \in \mathbf{R}^n$. V is a radial vector field pointing toward the origin. For any $x \in U_\phi$,

$$\Phi(x, -V_x) = \phi(x) < 1.$$

The pair (Φ, V) generates a Finsler metric $\Theta = \Theta(x, y)$ on U_ϕ by (2–15):

$$\Theta(x, y) = \phi(y + \Theta(x, y)x). \tag{2–18}$$

Differentiating with respect to x^k and y^k separately, one obtains

$$\left(1 - \phi_{w^l}(w)x^l\right)\Theta_{x^k}(x, y) = \phi_{w^k}(w)\Theta(x, y),$$

$$\left(1 - \phi_{w^l}(w)x^l\right)\Theta_{y^k}(x, y) = \phi_{w^k}(w),$$

where $w := y + \Theta(x, y)x$. It follows that

$$\Theta_{x^k}(x, y) = \Theta(x, y)\Theta_{y^k}(x, y). \tag{2–19}$$

This argument is from [Okada 1983].

A domain U_ϕ in $\mathbf{R}^n$ defined by a Minkowski norm ϕ is called a *strongly convex domain*. A Finsler metric $\Theta = \Theta(x, y)$ defined in (2–18) is called the *Funk metric* on a strongly convex domain in $\mathbf{R}^n$. When $\phi = |y|$ is the standard Euclidean metric on $\mathbf{R}^n$, $U_\phi = \mathbf{B}^n$ is the standard unit ball and $\Theta = \Theta(x, y)$ is given by (2–6). Equation (2–19) is the key property of Θ, from which one can derive other geometric properties of Θ.

DEFINITION 2.10. A Finsler function $\Theta = \Theta(x, y)$ on an open subset in $\mathbf{R}^n$ is called a *Funk metric* if it satisfies (2–19).

EXAMPLE 2.11. Let $\Phi(x, y) := |y|$ be the standard Euclidean metric on $\mathbf{R}^n$ and let $V = V(x)$ be a vector field on $\mathbf{R}^n$ defined by

$$V_x := |x|^2 a - 2\langle a, x\rangle x,$$

where $a \in \mathbf{R}^n$ is a constant vector. Note that

$$\Phi(x, -V_x) = \sqrt{\phi_{ij} V^i V^j} = |V_x| = |a||x|^2 < 1, \quad x \in \mathbf{B}^n(1/\sqrt{|a|}),$$

and that

$$\phi_{ij} y^i V^j = |x|^2 \langle a, y\rangle - 2\langle a, x\rangle \langle x, y\rangle.$$

Given the pair (Φ, V) above, one obtains, by solving (2–15) for F,

$$F = \frac{\sqrt{\left(|x|^2\langle a, y\rangle - 2\langle a, x\rangle\langle x, y\rangle\right)^2 + |y|^2\left(1 - |a|^2|x|^4\right)} - \left(|x|^2\langle a, y\rangle - 2\langle a, x\rangle\langle x, y\rangle\right)}{1 - |a|^2|x|^4}.$$

$$(2–20)$$

This Randers metric F has very important properties. It is of scalar curvature and isotropic S-curvature. But the flag curvature and the S-curvature are not constant. See Example 11.2 below for further discussion.

3. Cartan Torsion and Matsumoto Torsion

To characterize Euclidean norms, E. Cartan [1934] introduced what is now called the Cartan torsion. Let $F = F(y)$ be a Minkowski norm on a vector space V. Fix a basis $\{b_i\}$ for V. Then $F = F(y^i b_i)$ is a function of (y^i). Let

$$g_{ij} := \tfrac{1}{2}[F^2]_{y^i y^j}, \quad C_{ijk} := \tfrac{1}{4}[F^2]_{y^i y^j y^k}(y), \quad I_i := g^{jk}(y)C_{ijk}(y),$$

where $(g^{ij}) := (g_{ij})^{-1}$. It is easy to see that

$$I_i = \frac{\partial}{\partial y^i}\left(\ln\sqrt{\det(g_{jk})}\right). \tag{3–1}$$

For $y \in V \setminus \{0\}$, set

$$C_y(u, v, w) := C_{ijk}(y)u^i v^j w^k, \quad I_y(u) := I_i(y)u^i,$$

where $u := u^i b_i$, $v := v^j b_j$ and $w := w^k b_k$. The family $C := \{C_y \mid y \in V \setminus \{0\}\}$ is called the *Cartan torsion* and the family $I := \{I_y \mid y \in V \setminus \{0\}\}$ is called the *mean Cartan torsion.* They are not tensors in a usual sense. In later sections, we will convert them to tensors on TM_0 and call them the (mean) Cartan tensor.

We view a Minkowski norm F on a vector space V as a *color pattern.* When F is Euclidean, the color pattern is *trivial* or *Euclidean.* The Cartan torsion C_y describes the non-Euclidean features of the color pattern in the direction $y \in V \setminus \{0\}$. And the mean Cartan torsion I_y is the average value of C_y.

A trivial fact is that a Minkowski norm F on a vector space V is Euclidean if and only if $C_y = 0$ for all $y \in V \setminus \{0\}$. This can be improved:

PROPOSITION 3.1 [Deicke 1953]. *A Minkowski norm is Euclidean if and only if* $I = 0$.

To characterize Randers norms, M. Matsumoto introduces the quantity

$$M_{ijk} := C_{ijk} - \frac{1}{n+1}(I_i h_{jk} + I_j h_{ik} + I_k h_{ij}), \qquad (3\text{--}2)$$

where $h_{ij} := FF_{y^i y^j} = g_{ij} - g_{ip}y^p g_{jq}y^q/F^2$. For $y \in V \setminus \{0\}$, set

$$M_y(u, v, w) := M_{ijk}(y)u^i v^j w^k,$$

where $u = u^i b_i$, $v = v^j b_j$ and $w = w^k b_k$. The family $M := \{M_y \mid y \in V \setminus \{0\}\}$ is called the *Matsumoto torsion.* A Minkowski norm is called *C-reducible* if $M = 0$.

LEMMA 3.2 [Matsumoto 1972b]. *Every Randers metric satisfies* $M = 0$.

PROOF. Let $F = \alpha + \beta$ be an arbitrary Randers norm on a vector space V, where $\alpha = \sqrt{a_{ij}y^i y^j}$ and $\beta = b_i y^i$ with $\|\beta\|_\alpha < 1$. By a direct computation, the $g_{ij} := \frac{1}{2}[F^2]_{y^i y^j}$ are given by

$$g_{ij} = \frac{F}{\alpha}\left(a_{ij} - \frac{y_i}{\alpha}\frac{y_j}{\alpha} + \frac{\alpha}{F}\left(b_i + \frac{y_i}{\alpha}\right)\left(b_j + \frac{y_j}{\alpha}\right)\right), \qquad (3\text{--}3)$$

where $y_i := a_{ij}y^j$. The $h_{ij} = FF_{y^i y^j} = g_{ij} - g_{ip}y^p g_{jq}y^q/F^2$ are given by

$$h_{ij} = \frac{\alpha + \beta}{\alpha}\left(a_{ij} - \frac{y_i y_j}{\alpha^2}\right). \qquad (3\text{--}4)$$

The inverse matrix $(g^{ij}) = (g_{ij})^{-1}$ is given by

$$g^{ij} = \frac{\alpha}{F}\left(a^{ij} - (1 - \|\beta\|^2)\frac{y^i}{F}\frac{y^j}{F} + \frac{\alpha}{F}\left(\left(b^i - \frac{y^i}{\alpha}\right)\left(b^j - \frac{y^j}{F}\right) - b^i b^j\right)\right). \qquad (3\text{--}5)$$

The determinant $\det(g_{ij})$ is

$$\det(g_{ij}) = \left(\frac{\alpha + \beta}{\alpha}\right)^{n+1} \det(a_{ij}).$$

From this and (3–1), one obtains

$$I_i = \frac{n+1}{2(\alpha + \beta)}\left(b_i - \frac{y_i}{\alpha}\frac{\beta}{\alpha}\right). \qquad (3\text{--}6)$$

Differentiating (3–3) yields

$$C_{ijk} = \frac{1}{n+1}\left(I_i h_{jk} + I_j h_{ik} + I_k h_{ij}\right), \tag{3-7}$$

implying that $M_{ijk} = 0$.   □

Matsumoto and Hōjō later proved that the converse is true as well, if $\dim V \geq 3$.

PROPOSITION 3.3 [Matsumoto 1972b; Matsumoto and Hōjō 1978]. *If F is a Minkowski norm on a vector space V of dimension at least 3, the Matsumoto torsion of F vanishes if and only if F is a Randers norm.*

Their proof is long and I could not find a shorter proof which fits into this article.

4. Geodesics and Sprays

Every Finsler metric F on a connected manifold M defines a length structure L_F on oriented curves in M. Let $c : [a, b] \to M$ be a piecewise C^∞ curve. The *length* of c is defined by

$$L_F(c) := \int_a^b F\big(c(t), \dot{c}(t)\big)\, dt.$$

For any two points $p, q \in M$, define

$$d_F(p, q) := \inf_c L_F(c),$$

where the infimum is taken over all piecewise C^∞ curves c from p to q. The quantity $d_F = d_F(p, q)$ is a nonnegative function on $M \times M$. It satisfies

(a) $d_F(p, q) \geq 0$, with equality if and only if $p = q$; and
(b) $d_F(p, q) \leq d_F(p, r) + d_F(r, q)$ for any $p, q, r \in M$.

We call d_F the *distance function* induced by F. This is a weaker notion than the distance function of metric spaces, since d_F need not satisfy $d_F(p, q) = d_F(q, p)$ for $p, q \in M$. But if the Finsler metric F is reversible, that is, if $F(x, -y) = F(x, y)$ for all $y \in T_x M$, then d_F is symmetric.

A piecewise C^∞ curve $\sigma : [a, b] \to M$ is *minimizing* if it has least length:

$$L_F(\sigma) = d_F\big(\sigma(a), \sigma(b)\big).$$

It is *locally minimizing* if, for any $t_0 \in I := [a, b]$, there is a small number $\varepsilon > 0$ such that σ is minimizing when restricted to $[t_0 - \varepsilon, t_0 + \varepsilon] \cap I$.

DEFINITION 4.1. A C^∞ curve $\sigma(t)$, $t \in I$, is called a *geodesic* if it is locally minimizing and has constant speed (meaning that $F(\sigma(t), \dot{\sigma}(t))$ is constant).

LEMMA 4.2. *A C^∞ curve $\sigma(t)$ in a Finsler manifold (M, F) is a geodesic if and only if it satisfies the system of second order ordinary differential equations*

$$\ddot{\sigma}^i(t) + 2G^i\big(\sigma(t), \dot{\sigma}(t)\big) = 0, \tag{4-1}$$

where the $G^i = G^i(x, y)$ are local functions on TM defined by

$$G^i := \tfrac{1}{4} g^{il}(x, y) \big([F^2]_{x^k y^l}(x, y) y^k - [F^2]_{x^l}(x, y) \big). \tag{4--2}$$

This is shown by the calculus of variations; see, for example, [Shen 2001a; 2001b].

Let $\{\partial/\partial x^i, \partial/\partial y^i\}$ denote the natural local frame on TM in a standard local coordinate system, and set

$$G := y^i \frac{\partial}{\partial x^i} - 2G^i \frac{\partial}{\partial y^i}, \tag{4--3}$$

where the $G^i = G^i(x, y)$, which are given in (4--2), satisfy the homogeneity property

$$G^i(x, \lambda y) = \lambda^2 G^i(x, y), \quad \lambda > 0. \tag{4--4}$$

G is a well-defined vector field on TM. Any vector field G on TM having the form (4--3) and satisfying the homogeneity condition (4--4) is called a *spray* on M, and the G^i are its *spray coefficients*.

Let

$$N^i_j = \frac{\partial G^i}{\partial y^j}, \qquad \frac{\delta}{\delta x^i} := \frac{\partial}{\partial x^i} - N^i_j \frac{\partial}{\partial y^j}.$$

Then $HTM := \operatorname{span}\{\delta/\delta x^i\}$ and $VTM := \operatorname{span}\{\partial/\partial y^i\}$ are well-defined and $T(TM_0) = HTM \oplus VTM$. That is, every spray naturally determines a decomposition of $T(TM_0)$.

For a Finsler metric on a manifold M and its spray G, a C^∞ curve $\sigma(t)$ in M is a geodesic of F if and only if the canonical lift $\gamma(t) := \dot{\sigma}(t)$ in TM is an integral curve of G. One can use this to define the notion of geodesics for sprays.

It is usually difficult to compute the spray coefficients of a Finsler metric. However, for an (α, β)-metric F, given by equation (2--10), the computation is relatively simple using a Maple program. Let $\bar{G}^i$ be the spray coefficients of the Riemannian metric α, with coefficients $\bar{\Gamma}^i_{jk}$, so that $\bar{G}^i = \tfrac{1}{2}\bar{\Gamma}^i_{jk}(x) y^j y^k$. These coefficients are called the *Christoffel symbols* of α. By (4--2), they are given by

$$\bar{\Gamma}^i_{jk} = \frac{a^{il}}{2} \Big(\frac{\partial a_{jl}}{\partial x^k} + \frac{\partial a_{kl}}{\partial x^j} - \frac{\partial a_{jk}}{\partial x^l} \Big).$$

To find a formula for the spray coefficients $G^i = G^i(x, y)$ of F in terms of α and β, we introduce the *covariant derivatives* of β with respect to α. Let $\theta^i := dx^i$ and $\theta_j{}^i := \bar{\Gamma}^i_{jk} dx^k$. We have

$$d\theta^i = \theta^j \wedge \theta_j{}^i, \qquad da_{ij} = a_{kj}\theta_i{}^k + a_{ik}\theta_j{}^k.$$

Define $b_{i;j}$ by

$$b_{i;j}\theta^j := db_i - b_j\theta_i{}^j.$$

Let

$$r_{ij} := \tfrac{1}{2}\big(b_{i;j} + b_{j;i} \big), \qquad\qquad s_{ij} := \tfrac{1}{2}\big(b_{i;j} - b_{j;i} \big),$$

$$s^i{}_j := a^{ih} s_{hj}, \qquad s_j := b_i s^i{}_j, \qquad e_{ij} := r_{ij} + b_i s_j + b_j s_i.$$

By (4--2) and using a Maple program, one obtains the following relationship:

LEMMA 4.3. *The geodesic coefficients G^i are related to $\bar{G}^i$ by*

$$G^i = \bar{G}^i + \frac{\alpha\phi'}{\phi - s\phi'}s^i{}_0 + \frac{\phi\phi' - s(\phi\phi'' + \phi'\phi')}{2\phi\big((\phi - s\phi') + (b^2 - s^2)\phi''\big)}$$

$$\times \left(\frac{-2\alpha\phi'}{\phi - s\phi'}s_0 + r_{00}\right)\left(\frac{y^i}{\alpha} + \frac{\phi\phi''}{\phi\phi' - s(\phi\phi'' + \phi'\phi')}b^i\right), \quad (4\text{--}5)$$

where $s = \beta/\alpha$, $s^i{}_0 = s^i{}_j y^j$, $s_0 = s_i y^i$, $r_{00} = r_{ij}y^i y^j$ and $b^2 = a^{ij}b_i b_j$.

Consider the metric

$$F = \frac{(\alpha + \beta)^2}{\alpha},$$

where $\alpha = \sqrt{a_{ij}(x)y^i y^j}$ is a Riemannian metric and $\beta = b_i(x)y^i$ is a 1-form with $\|\beta\|_x < 1$ for every $x \in M$. By (4–5), we obtain a formula for the spray coefficients of F:

$$G^i = \bar{G}^i + \frac{2\alpha}{\alpha - \beta}\alpha s^i{}_0 + \frac{\alpha(\alpha - 2\beta)}{2\alpha^2 b^2 + \alpha^2 - 3\beta^2}\left(\frac{-4\alpha}{\alpha - \beta}\alpha s_0 + r_{00}\right)\left(\frac{y^i}{\alpha} + \frac{\alpha}{\alpha - 2\beta}b^i\right),$$

where $b = \|\beta\|_x$.

Given a spray G, we define the covariant derivatives of a vector field $X = X^i(t)(\partial/\partial x^i)|_{c(t)}$ along a curve c by

$$D_{\dot{c}}X(t) := \big(\dot{X}^i(t) + X^j(t)N^i_j(c(t), \dot{c}(t))\big)\frac{\partial}{\partial x^i}\bigg|_{c(t)},$$

$$\nabla_{\dot{c}}X(t) := \big(\dot{X}^i(t) + X^j(t)N^i_j(c(t), X(t))\big)\frac{\partial}{\partial x^i}\bigg|_{c(t)}.$$
$$(4\text{--}6)$$

$D_{\dot{c}}X(t)$ is the *linear covariant derivative* and $\nabla_{\dot{c}}X(t)$ the *covariant derivative* of $X(t)$ along c. The field X is *linearly parallel along c* if $D_{\dot{c}}X(t) = 0$, and *parallel along c* if $\nabla_{\dot{c}}X(t) = 0$. For linearly parallel vector fields $X = X(t)$ and $Y = Y(t)$ along a geodesic c, the expression $g_{\dot{c}(t)}\big(X(t), Y(t)\big)$ is constant, and for a parallel vector field $X = X(t)$ along a curve c, $F\big(c(t), X(t)\big)$ is constant.

5. Berwald Metrics

Consider a Riemannian metric $F = \sqrt{g_{ij}(x)y^i y^j}$ on a manifold M. By (4–2), we obtain $G^i = \frac{1}{2}\Gamma^i_{jk}(x)y^j y^k$, where

$$\Gamma^i_{jk}(x) := \frac{1}{4}g^{il}(x)\left(\frac{\partial g_{lk}}{\partial x^j}(x) + \frac{\partial g_{jl}}{\partial x^k}(x) - \frac{\partial g_{jk}}{\partial x^l}(x)\right). \quad (5\text{--}1)$$

In this case the $G^i = G^i(x, y)$ are quadratic in $y \in T_x M$ at any point $x \in M$. A Finsler metric $F = F(x, y)$ is called a *Berwald metric* if in any standard local coordinate system, the spray coefficients $G^i = G^i(x, y)$ are quadratic in $y \in T_x M$.

There are many non-Riemannian Berwald metrics.

EXAMPLE 5.1. Let $(\bar{M}, \bar{\alpha})$ and $(\underline{M}, \underline{\alpha})$ be Riemannian manifolds and let $M = \bar{M} \times \underline{M}$. Let $f : [0, \infty) \times [0, \infty) \to [0, \infty)$ be a C^∞ function satisfying $f(\lambda s, \lambda t) = \lambda f(s, t)$ for $\lambda > 0$ and $f(s, t) \neq 0$ if $(s, t) \neq 0$. Define

$$F(x, y) := \sqrt{f\big([\bar{\alpha}(\bar{x}, \bar{y})]^2, \, [\underline{\alpha}(\underline{x}, \underline{y})]^2\big)}, \tag{5-2}$$

where $x = (\bar{x}, \underline{x}) \in M$ and $y = \bar{y} \oplus \underline{y} \in T_x M \cong T_{\bar{x}} \bar{M} \oplus T_{\underline{x}} \underline{M}$.

We now find additional conditions on $f(s, t)$ under which the matrix $g_{ij} := \frac{1}{2}[F^2]_{y^i y^j}$ is positive definite. Take standard local coordinate systems $(\bar{x}^a, \bar{y}^a)$ in $T\bar{M}$ and $(\underline{x}^a, \underline{y}^a)$ in $T\underline{M}$, so that $\bar{y} = \bar{y}^a \, \partial/\partial \bar{x}^a$ and $\underline{y} = \underline{y}^a \, \partial/\partial \underline{x}^a$. Then $(x^i, y^j) := (\bar{x}^a, \underline{x}^a, \bar{y}^a, \underline{y}^a)$ is a standard local coordinate system in TM. Express $\bar{\alpha}$ and $\underline{\alpha}$ by

$$\bar{\alpha}(\bar{x}, \bar{y}) = \sqrt{\bar{g}_{\bar{a}\bar{b}}(\bar{x}) \bar{y}^{\bar{a}} \bar{y}^{\bar{b}}}, \quad \underline{\alpha}(\underline{x}, \underline{y}) = \sqrt{\underline{g}_{\underline{a}\underline{b}}(\underline{x}) \underline{y}^{\underline{a}} \underline{y}^{\underline{b}}},$$

We obtain

$$\bar{g}_{ab} = 2 f_{ss} \bar{y}_{\bar{a}} \bar{y}_{\bar{b}} + f_s \bar{g}_{\bar{a}\bar{b}}, \quad g_{ab} = 2 f_{st} \bar{y}_{\bar{a}} \underline{y}_\beta, \quad \underline{g}_{ab} = 2 f_{tt} \underline{y}_{\underline{a}} \underline{y}_{\underline{b}} + f_t \underline{g}_{\underline{a}\underline{b}}, \tag{5-3}$$

where $\bar{y}_{\bar{a}} := \bar{g}_{\bar{a}\bar{b}} \bar{y}^{\bar{b}}$ and $\underline{y}_{\underline{a}} := \underline{g}_{\underline{a}\underline{b}} \underline{y}^{\underline{b}}$. By an elementary argument, one can show that (g_{ij}) is positive definite if and only if $f(s, t)$ satisfies the conditions

$$f_s > 0, \quad f_t > 0, \quad f_s + 2 s f_{ss} > 0, \quad f_t + 2 t f_{tt} > 0, \quad f_s f_t - 2 f f_{st} > 0.$$

In this case,

$$\det(g_{ij}) = h\big([\bar{\alpha}]^2, \, [\underline{\alpha}]^2\big) \det(\bar{g}_{\bar{a}\bar{b}}) \det(\underline{g}_{\underline{a}\underline{b}}), \tag{5-4}$$

where

$$h := (f_s)^{\bar{n}-1} (f_t)^{\underline{n}-1} (f_s f_t - 2 f f_{st}), \tag{5-5}$$

where $\bar{n} := \dim \bar{M}$ and $\underline{n} := \dim \underline{M}$.

By a direct computation, one can show that the spray coefficients of F split as the direct sum of the spray coefficients of $\bar{\alpha}$ and $\underline{\alpha}$:

$$G^{\bar{a}}(x, y) = \bar{G}^{\bar{a}}(\bar{x}, \bar{y}), \quad G^{\underline{a}}(x, y) = \underline{G}^{\underline{a}}(\underline{x}, \underline{y}), \tag{5-6}$$

where $\bar{G}^{\bar{a}}$ and $\underline{G}^{\underline{a}}$ are the spray coefficients of $\bar{\alpha}$ and $\underline{\alpha}$. From (5–6), one can see that the spray of F is independent of the choice of a particular function $f(s, t)$. In particular, the $G^i(x, y)$ are quadratic in $y \in T_x M$. Thus F is a Berwald metric.

A typical example of f is

$$f = s + t + \varepsilon \sqrt[k]{s^k + t^k},$$

where ε is a nonnegative real number and k is a positive integer. The Berwald metric obtained with this choice of f is discussed in [Szabó 1981].

Let (M, F) be a Berwald manifold and $p, q \in M$ be an arbitrary pair of points in M. Let $c : [0, 1] \to M$ be a geodesic going from $p = c(0)$ to $q = c(1)$. Define a linear isomorphism $T : T_p M \to T_q M$ by $T(X(0)) := X(1)$, where $X(t)$ is a linearly parallel vector field along c, so $D_{\dot{c}} X(t) = 0$. Since F is a Berwald metric, the linear covariant derivative $\nabla_{\dot{c}}$ coincides with the covariant derivative $D_{\dot{c}}$ along c, by (4–6). Thus $X(t)$ is also parallel along c, that is, $\nabla_{\dot{c}} X(t) = 0$. Therefore, $F(c(t), X(t))$ is constant. This implies that $T : (T_p M, F_p) \to (T_q M, F_q)$ preserves the Minkowski norms. We have proved the following well-known result:

PROPOSITION 5.2 [Ichijyō 1976]. *On a Berwald manifold (M, F), all tangent spaces $(T_x M, F_x)$ are linearly isometric to each other.*

On a Finsler manifold (M, F), we view the Minkowski norm F_x on $T_x M$ as an *infinitesimal color pattern* at x. As we mentioned early in Section 3, the Cartan torsion C_y describes the non-Euclidean features of the pattern in the direction $y \in T_x M \setminus \{0\}$. In the case when F is a Berwald metric on a manifold M, by Proposition 5.2, all tangent spaces $(T_x M, F_x)$ are linearly isometric, and (M, F) is modeled on a single Minkowski space. More precisely, for any pair points $x, x' \in M$ and a geodesic from x to x', (linearly) parallel translation defines a linear isometry $T : (T_x M, F_x) \to (T_{x'} M, F_{x'})$. This linear isometry T maps the infinitesimal color pattern at x to that at x'. Thus the infinitesimal color patterns do not change over the manifold. If one looks at a Berwald manifold on a large scale, with the infinitesimal color pattern at each point shrunken to a single spot of color, one can only see a space with uniform color. The color depends on the Minkowski model.

A Finsler metric F on a manifold M is said to be *affinely equivalent* to another Finsler metric $\bar{F}$ on M if F and $\bar{F}$ induce the same sprays. By (5–6), one can see that the family of Berwald metrics in (5–2) are affinely equivalent.

PROPOSITION 5.3 [Szabó 1981]. *Every Berwald metric on a manifold is affinely equivalent to a Riemannian metric.*

Based on this observation, Z. I. Szabó [1981] determined the local structure of Berwald metrics.

6. Gradient, Divergence and Laplacian

Let $F = F(x, y)$ be a Finsler metric on a manifold M and let $F^* = F^*(x, \xi)$ be dual to F. Let f be a C^1 function on M. At a point $x \in M$, the differential $df_x \in T_x^* M$ is a 1-form. Define the dual vector $\nabla f_x \in T_x M$ by

$$\nabla f_x := \ell_x^*(df_x), \tag{6-1}$$

where $\ell_x^* : T_x^* M \to T_x M$ is the inverse Legendre transformation. By definition, ∇f_x is uniquely determined by

$$\eta(\nabla f_x) := \frac{1}{2} \frac{d}{dt} \left(F^{*2}(x, df_x + t\eta) \right)|_{t=0}, \quad \eta \in T_x^* M.$$

∇f_x is called the *gradient* of f at x. We have

$$F(x, \nabla f_x) = F^*(x, df_x).$$

If f is C^k ($k \geq 1$), then ∇f is C^{k-1} on $\{df_x \neq 0\}$ and C^0 at any point $x \in M$ with $df_x = 0$.

Given a closed subset $A \subset M$ and a point $x \in M$, let $d(A, x) := \sup_{z \in A} d(z, x)$ and $d(x, A) := \sup_{z \in A} d(x, z)$. The function ρ defined by $\rho(x) := d(x, A)$ is locally Lipschitz, hence differentiable almost everywhere. It is easy to verify [Shen 2001b] that

$$F(x, \nabla \rho_x) = F^*(x, d\rho_x) = 1 \quad \text{almost everywhere.} \tag{6--2}$$

Any function ρ satisfying satisfies (6--2) is called a *distance function* of the Finsler metric F; another example is the function ρ defined by $\rho(x) := -d(x, A)$. In general, a distance function on a Finsler manifold (M, F) is C^∞ on an open dense subset of M. For example, when $A = \{p\}$ is a point, the distance function ρ defined by $\rho(x) := d(x, p)$ or $rho(x) := d(p, x)$ is C^∞ on an open dense subset of M.

Let ρ be a distance function of a Finsler metric F on a manifold M. Assume that it is C^∞ on an open subset $U \subset M$. Then $\nabla \rho$ induces a Riemannian metric $\hat{F} := \sqrt{g_{\nabla \rho}(v, v)}$ on U. Moreover, ρ is a distance function of $\hat{F}$ and $\nabla \rho = \hat{\nabla} \rho$ is the gradient of ρ with respect to $\hat{F}$. See [Shen 2001b].

Every Finsler metric F defines a volume form

$$dV_F := \sigma_F(x) \, dx^1 \cdots dx^n,$$

where

$$\sigma_F := \frac{\text{Vol}\, B^n}{\text{Vol}\left\{(y^i) \in \mathbb{R}^n \mid F(x, y^i(\partial/\partial x^i)|_x) < 1\right\}}. \tag{6--3}$$

Here Vol denotes Euclidean volume in $\mathbb{R}^n$. It was proved by H. Busemann [1947] that if F is reversible, the Hausdorff measure of the induced distance function d_F is represented by dV_F. When $F = \sqrt{g_{ij}(x)y^i y^j}$ is Riemannian,

$$\sigma_F = \sqrt{\det(g_{ij}(x))} \quad \text{and} \quad dV_F = \sqrt{\det(g_{ij}(x))} \, dx^1 \cdots dx^n.$$

For a vector field $X = X^i(x)(\partial/\partial x^i)|_x$ on M, the *divergence* of X is

$$\text{div}\, X := \frac{1}{\sigma_F(x)} \frac{\partial}{\partial x^i}\left(\sigma_F(x) X^i(x)\right). \tag{6--4}$$

The *Laplacian* of a C^2 function f with $df \neq 0$ is

$$\Delta f := \text{div}\, \nabla f.$$

Δ is a nonlinear elliptic operator. If there are points x at which $df_x = 0$, then ∇f is only C^0 at these points. In this case, Δf is only defined weakly in the sense of distributions.

For a C^∞ distance function ρ on an open subset $U \subset M$, $d\rho_x \neq 0$ at every point $x \in U$ and the level set $N_r := \rho^{-1}(r) \subset U$ is a C^∞ hypersurface in U. Thus $\Delta\rho$ can be defined by the above formula and its restriction to N_r, $H := \Delta\rho|_{N_r}$, is called the *mean curvature* of N_r with respect to the normal vector $\boldsymbol{n} = \nabla\rho|_{N_r}$.

7. S-Curvature

Consider an n-dimensional Finsler manifold (M, F). As mentioned in Section 5, we view the Minkowski norm F_x on T_xM as an *infinitesimal color pattern* at x. The Cartan torsion $\boldsymbol{C}_y$ describes the non-Euclidean features of the pattern in the direction $y \in T_xM \setminus \{0\}$. The mean Cartan torsion $\boldsymbol{I}_y$ is the average value of $\boldsymbol{C}_y$. Besides the (mean) Cartan torsion, there is another geometric quantity associated with F_x. Take a standard local coordinate system (x^i, y^i) and let

$$\tau := \ln \frac{\sqrt{\det\left(g_{ij}(x,y)\right)}}{\sigma_F(x)}, \tag{7-1}$$

where σ_F is defined in (6–3). τ is called the *distortion* [Shen 1997; 2001b]. Intuitively, the distortion $\tau(x, y)$ is the directional *twisting number* of the infinitesimal color pattern at x. Observe that

$$\tau_{y^i} = \frac{\partial}{\partial y^i} \ln \sqrt{\det\left(g_{jk}(x,y)\right)} = \frac{1}{2} g^{jk} \frac{\partial g_{jk}}{\partial y^i} = g^{jk} C_{ijk} =: I_i. \tag{7-2}$$

Here σ_F does not occur in the first equality, because it is independent of y at each point x. If the distortion is isotropic at x, that is, if $\tau(x)$ is independent of the direction $y \in T_xM$, then $\tau(x) = 0$ and F_x is Euclidean (Proposition 3.1). In this case, the infinitesimal color pattern is in the simplest form at every point.

It is natural to study the rate of change of the distortion along geodesics. For $y \in T_xM \setminus \{0\}$, let $\sigma(t)$ be the geodesic with $\sigma(0) = x$ and $\dot\sigma(0) = y$. Let

$$\boldsymbol{S} := \frac{d}{dt} \tau\left(\sigma(t), \dot\sigma(t)\right)\Big|_{t=0}. \tag{7-3}$$

$\boldsymbol{S}$ is called the *S-curvature*. It is positively homogeneous of degree one in y:

$$\boldsymbol{S}(x, \lambda y) = \lambda \boldsymbol{S}(x, y), \quad \lambda > 0.$$

In a standard local coordinate system (x^i, y^i), let $G^i = G^i(x, y)$ denote the spray coefficients of F. Contracting (8–2) with g^{ij} yields

$$\frac{\partial G^m}{\partial y^m} = \frac{1}{2} g^{ml} \frac{\partial g_{ml}}{\partial x^i} y^i - 2I_i G^i,$$

so

$$\boldsymbol{S} = y^i \frac{\partial \tau}{\partial x^i} - 2\frac{\partial \tau}{\partial y^i} G^i = \frac{1}{2} g^{ml} \frac{\partial g_{ml}}{\partial x^i} y^i - 2I_i G^i - y^m \frac{\partial}{\partial x^m} \ln \sigma_F$$

and finally

$$S = \frac{\partial G^m}{\partial y^m} - y^m \frac{\partial}{\partial x^m} \ln \sigma_F. \qquad (7\text{–}4)$$

PROPOSITION 7.1 [Shen 1997]. *For any Berwald metric,* $S = 0$.

However, many metrics of zero S-curvature are non-Berwaldian.

Some comparison theorems in Riemannian geometry are still valid for Finsler metrics of zero S-curvature; see [Shen 1997; 2001b].

By definition, the S-curvature is the covariant derivative of the distortion along geodesics. Let $\sigma(t)$ be a geodesic and set

$$\tau(t) := \tau\big(\sigma(t), \dot{\sigma}(t)\big), \quad S(t) := S\big(\sigma(t), \dot{\sigma}(t)\big).$$

By (7–3), $S(t) = \tau'(t)$, so if S vanishes, $\tau(t)$ is a constant. Intuitively, the distortion (twisting number) of the infinitesimal color pattern in the direction $\dot{\sigma}(t)$ does not change along any geodesic. However, the distortion might take different values along different geodesics.

A Finsler metric F is said to have *isotropic S-curvature* if

$$S = (n+1)cF.$$

More generally, F is said to have *almost isotropic S-curvature* if

$$S = (n+1)(cF + \eta),$$

where $c = c(x)$ is a scalar function and $\eta = \eta_i(x)y^i$ is a *closed* 1-form.

Differentiating the S-curvature gives rise to another quantity. Let

$$E_{ij} := \tfrac{1}{2} S_{y^i y^j}(x, y). \qquad (7\text{–}5)$$

For $y \in T_x M \setminus \{0\}$, the form $E_y = E_{ij}(x, y)\, dx^i \otimes dx^j$ is a symmetric bilinear form on $T_x M$. We call the family $E := \{E_y \mid y \in TM \setminus \{0\}\}$ the *mean Berwald curvature*, or simply the *E-curvature* [Shen 2001a]. Let $h_y := h_{ij}(x, y)\, dx^i \otimes dx^j$, where $h_{ij} := FF_{y^i y^j}$. We say that F has *isotropic E-curvature* if

$$E = \tfrac{1}{2}(n+1)cF^{-1}h,$$

where $c = c(x)$ is a scalar function on M. Clearly, if the S-curvature is almost isotropic, the E-curvature is isotropic. Conversely, if the E-curvature is isotropic, there is a 1-form $\eta = \eta_i(x)\, dx^i$ such that $S = (n+1)(cF + \eta)$. However, this η is not closed in general.

Finally, we mention another geometric meaning of the S-curvature. Let $\rho = \rho(x)$ be a C^∞ distance function on an open subset $U \subset M$ (see (6–2)). The gradient $\nabla\rho$ induces a Riemannian metric $\hat{F} = \hat{F}(z, v)$ on U by

$$\hat{F}(z, v) := \sqrt{g_{\nabla\rho}(v, v)}, \quad v \in T_z U.$$

Let Δ and $\hat{\Delta}$ denote the Laplacians on functions with respect to F and $\hat{F}$. Then $H = \Delta\rho|_{N_r}$ and $\hat{H} = \hat{\Delta}\rho|_{N_r}$ are the *mean curvature* of $N_r := \rho^{-1}(r)$ with respect to F and $\hat{F}$, respectively. The S-curvature can be expressed by

$$S(\nabla\rho) = \hat{\Delta}\rho - \Delta\rho = \hat{H} - H.$$

From these identities, one can estimate $\hat{\Delta}$ and obtain an estimate on $\Delta\rho$ under a Ricci curvature bound and an S-curvature bound. Then one can establish a volume comparison on the metric balls. See [Shen 2001b] for more details.

8. Landsberg Curvature

The (mean) Cartan torsion is a geometric quantity that characterizes the Euclidean norms among Minkowski norms on a vector space. On a Finsler manifold (M, F), one can view the Minkowski norm F_x on $T_x M$ as an *infinitesimal color pattern* at x. The Cartan torsion $\boldsymbol{C}_y$ describes the non-Euclidean features of the pattern in the direction $y \in T_x M \setminus \{0\}$. The mean Cartan torsion $\boldsymbol{I}_y$ is the average value of $\boldsymbol{C}_y$. They reveal the non-Euclidean features which are different from that revealed by the distortion. Therefore, it is natural to study the rate of change of the (mean) Cartan torsion along geodesics.

Let (M, F) be a Finsler manifold. To differentiate the (mean) Cartan torsion along geodesics, we need linearly parallel vector fields along a geodesic. Recall that a vector field $U(t) := U^i(T)(\partial/\partial x^i)|_{\sigma(t)}$ along a geodesic $\sigma(t)$ is linearly parallel along σ if $D_{\dot{\sigma}} U(t) = 0$:

$$\dot{U}^i(t) + U^j(t) N_j^i(\sigma(t), \dot{\sigma}(t)) = 0. \tag{8-1}$$

On the other hand, by a direct computation using (4–2), one can verify that g_{ij} satisfy the following identity:

$$y^m \frac{\partial g_{ij}}{\partial x^m} - 2G^m \frac{\partial g_{ij}}{\partial y^m} = g_{im}N_j^m + g_{mj}N_i^m \tag{8-2}$$

Using (8–1) and (8–2), one can verify that for two linearly parallel vector fields $U(t), V(t)$ along σ,

$$\frac{d}{dt} \boldsymbol{g}_{\dot{\sigma}(t)}\big(U(t), V(t)\big) = 0.$$

In this sense, the family of inner products $\boldsymbol{g}_y$ does not change along geodesics. However, for linearly parallel vector fields $U(t), V(t)$ and $W(t)$ along σ, the functions $\boldsymbol{C}_{\dot{\sigma}(t)}\big(U(t), V(t), W(t)\big)$ and $\boldsymbol{I}_{\dot{\sigma}(t)}\big(U(t)\big)$ do change, in general. Set

$$\boldsymbol{L}_y(u, v, w) := \frac{d}{dt}\big(\boldsymbol{C}_{\dot{\sigma}(t)}(U(t), V(t), W(t))\big)|_{t=0} \tag{8-3}$$

and

$$\boldsymbol{J}_y(u) := \frac{d}{dt}\big(\boldsymbol{I}_{\dot{\sigma}(t)}(U(t))\big)|_{t=0},$$

where $u = U(0), v = V(0), w = W(0)$ and $y = \dot{\sigma}(0) \in T_x M$. The family $\boldsymbol{L} = \{\boldsymbol{L}_y \mid y \in TM \setminus \{0\}\}$ is called the *Landsberg curvature*, or *L-curvature*, and the family $\boldsymbol{J} = \{\boldsymbol{J}_y \mid y \in TM \setminus \{0\}\}$ is called the *mean Landsberg curvature*, or *J-curvature*. A Finsler metric is called a *Landsberg metric* if $\boldsymbol{L} = 0$, and a *weakly Landsberg metric* if $\boldsymbol{J} = 0$.

Let (x^i, y^i) be a standard local coordinate system in TM and set $C_{ijk} := \frac{1}{4}[F^2]_{y^i y^j y^k}$. From the definition, $\boldsymbol{L}_y = L_{ijk}\, dx^i \otimes dx^j \otimes dx^k$ is given by

$$L_{ijk} = y^m \frac{\partial C_{ijk}}{\partial x^m} - 2G^m \frac{\partial C_{ijk}}{\partial y^m} - C_{mjk}N_i^m - C_{imk}N_j^m - C_{ijm}N_k^m, \qquad (8\text{-}4)$$

and $\boldsymbol{J} = J_i dx^i$ is given by

$$J_i = y^m \frac{\partial I_i}{\partial x^m} - 2G^m \frac{\partial I_i}{\partial y^m} - I_m N_i^m. \qquad (8\text{-}5)$$

We have

$$J_i = g^{jk} L_{ijk}.$$

It follows from (4–2) that

$$g_{sm} G^m = \frac{1}{4}\left(2\frac{\partial g_{sk}}{\partial x^m} - \frac{\partial g_{km}}{\partial x^s}\right) y^k y^m.$$

Differentiating with respect to y^i, y^j, y^k and contracting the resulting identity by $\frac{1}{2}y^s$, one obtains

$$L_{ijk} = -\tfrac{1}{2} y^s g_{sm} \frac{\partial^3 G^m}{\partial y^i \partial y^j \partial y^k}. \qquad (8\text{-}6)$$

Thus if $G^m = G^m(x, y)$ are quadratic in $y \in T_x M$, then $L_{ijk} = 0$. This proves the following well-known result.

PROPOSITION 8.1. *Every Berwald metric is a Landsberg metric.*

By definition, the (mean) Landsberg curvature is the covariant derivative of the (mean) Cartan torsion along a geodesic. Let $\sigma = \sigma(t)$ be a geodesic and $U = U(t), V = V(t), W = W(t)$ be parallel vector fields along σ. Let

$$\boldsymbol{L}(t) := \boldsymbol{L}_{\dot\sigma(t)}\big(U(t), V(t), W(t)\big), \quad \boldsymbol{C}(t) := \boldsymbol{C}_{\dot\sigma(t)}\big(U(t), V(t), W(t)\big).$$

By (8–3),

$$\boldsymbol{L}(t) = \boldsymbol{C}'(t).$$

If F is Landsbergian, the Cartan torsion $\boldsymbol{C}_{\dot\sigma}$ in the direction $\dot\sigma(t)$ is constant along σ. Intuitively, the infinitesimal color pattern in the direction $\dot\sigma(t)$ does not change along σ. But the patterns might look different at neighboring points.

It is easy to see that in dimension two, a Finsler metric is Berwaldian if and only if $\boldsymbol{E} = 0$ (or $\boldsymbol{S} = 0$) and $\boldsymbol{J} = 0$. It seems that $\boldsymbol{E}$ and $\boldsymbol{L}$ are complementary to each other. So we may ask: *Is a Finsler metric Berwaldian if $\boldsymbol{E} = 0$ and $\boldsymbol{L} = 0$?* A more difficult problem is: *Is a Finsler metric Berwaldian if $\boldsymbol{L} = 0$?* So far, we do not know.

Finsler metrics with $\boldsymbol{L} = 0$ can be generalized as follows. Let F be a Finsler metric on an n-dimensional manifold M. We say that F has *relatively isotropic*

L-curvature $\boldsymbol{L} + c\boldsymbol{F}\boldsymbol{C} = 0$, where $c = c(x)$ is a scalar function on M. We say that F has *relatively isotropic J-curvature* if $\boldsymbol{J} + c\boldsymbol{F}\boldsymbol{I} = 0$.

There are many interesting Finsler metrics having isotropic L-curvature or (almost) isotropic S-curvature. We will discuss them in the following two sections.

9. Randers Metrics with Isotropic S-Curvature

We now discuss Randers metrics of isotropic S-curvature. Let $F = \alpha + \beta$ be a Randers metric on an n-dimensional manifold M, with $\alpha = \sqrt{a_{ij}(x)y^i y^j}$ and $\beta = b_i(x)y^i$. Recall from page 307 that this is a special case of an (α, β)-metric, with $\phi(s) = 1 + s$. By (4–5), the spray coefficients G^i of F and $\bar{G}^i$ of α are related via

$$G^i = \bar{G}^i + Py^i + Q^i, \qquad (9\text{–}1)$$

where

$$P := \frac{e_{00}}{2F} - s_0, \quad Q^i = \alpha s^i{}_0, \qquad (9\text{–}2)$$

and $e_{00} := e_{ij}y^i y^j$, $s_0 := s_i y^i$, $s^i{}_0 := s^i{}_j y^j$. The formula above can be found in [Antonelli et al. 1993].

Let

$$\rho := \ln \sqrt{1 - \|\beta\|_x^2}.$$

The volume forms dV_F and dV_α are related by

$$dV_F = e^{(n+1)\rho(x)}dV_\alpha.$$

Since $s_{ij} = s_{ji}$, $s_{00} := s_{ij}y^i y^j = 0$ and $s^i{}_i = a^{ij}s_{ij} = 0$. Observe that

$$\frac{\partial(Py^m)}{\partial y^m} = \frac{\partial P}{\partial y^m}y^m + nP = (n+1)P, \qquad \frac{\partial Q^m}{\partial y^m} = \alpha^{-1}s_{00} + \alpha s^m{}_m = 0.$$

Since α is Riemannian, we have

$$\frac{\partial \bar{G}^m}{\partial y^m} = y^m \frac{\partial}{\partial x^m} \ln \sigma_\alpha.$$

Thus one obtains

$$\begin{aligned}
\boldsymbol{S} &= \frac{\partial G^m}{\partial y^m} - y^m \frac{\partial}{\partial x^m} \ln \sigma_F \\
&= \frac{\partial \bar{G}^m}{\partial y^m} + \frac{\partial(Py^m)}{\partial y^m} + \frac{\partial Q^m}{\partial y^m} - (n+1)y^m \frac{\partial \rho}{\partial x^m} - y^m \frac{\partial}{\partial x^m} \ln \sigma_\alpha \\
&= (n+1)(P - \rho_0) = (n+1)\left(\frac{e_{00}}{2F} - (s_0 + \rho_0)\right), \qquad (9\text{–}3)
\end{aligned}$$

where $\rho_0 := \rho_{x^i}(x)y^i$.

LEMMA 9.1 [Chen and Shen 2003a]. *For a Randers metric $F = \alpha + \beta$ on an n-dimensional manifold M, the following conditions are equivalent:*

(a) *The S-curvature is isotropic, i.e.,* $\boldsymbol{S}=(n{+}1)cF$ *for a scalar function c on M.*

(b) *The S-curvature is almost isotropic, i.e.,* $\boldsymbol{S} = (n{+}1)(cF + \eta)$ *for a scalar function c and a closed 1-form η on M.*

(c) *The E-curvature is isotropic, i.e.,* $\boldsymbol{E} = \frac{1}{2}(n{+}1)cF^{-1}\boldsymbol{h}$ *for a scalar function c on M.*

(d) $e_{00} = 2c(\alpha^2 - \beta^2)$ *for a scalar function c on M.*

PROOF. (a) $\Longrightarrow$ (b) and (b) $\Longrightarrow$ (c) are obvious.

   (c) $\Longrightarrow$ (d). First, $\boldsymbol{S} = (n+1)(cF+\eta)$, where η is a 1-form on M. By (9–3), (c) is equivalent to the following $e_{00} = 2cF^2 + 2\theta F$, where $\theta := s_0 + \rho_0 + \eta$. This implies that

$$e_{00} = 2c(\alpha^2 + \beta^2) + 2\theta\beta, \quad 0 = 4c\beta + 2\theta.$$

Solving for θ in the second of these equations and plugging the result into the first, one obtains (d).

   (d) $\Longrightarrow$ (a). Plugging $e_{00} = 2c(\alpha^2 - \beta^2)$ into (9–3) yields

$$\boldsymbol{S} = (n + 1)\big(c(\alpha - \beta) - (s_0 + \rho_0)\big). \tag{9-4}$$

On the other hand, contracting $e_{ij} = 2c(a_{ij} - b_ib_j)$ with b^j gives $s_i + \rho_i + 2cb_i = 0$. Thus $s_0 + \rho_0 = -2c\beta$. Plugging this into (9–4) yields (a). $\qquad\square$

EXAMPLE 9.2. Let $V = (A, B, C)$ be a vector field on a domain $U \subset \mathbf{R}^3$, where $A = A(r, s, t)$, $B = B(r, s, t)$ and $C = C(r, s, t)$ are C^∞ functions on U with

$$|V(x)| = \sqrt{A(x)^2 + B(x)^2 + C(x)^2} < 1, \quad \forall x = (r, s, t) \in U.$$

Let $\Phi := |y|$ be the standard Euclidean metric on $\mathbf{R}^3$. Define $F = \alpha + \beta$ by (2–15) for the pair (Φ, V). α and β are given by

$$\alpha = \frac{\sqrt{\langle V(x), y\rangle^2 + |y|^2(1 - |V(x)|^2)}}{1 - |V(x)|^2}, \qquad \beta = -\frac{\langle V(x), y\rangle}{1 - |V(x)|^2},$$

where $y = (u, v, w) \in T_xU \cong \mathbf{R}^3$. One can easily verify that $\|\beta\|_x < 1$ for $x \in U$. By a direct computation, one obtains

$$e_{11} = \frac{B^2(A_r - B_s) + C^2(A_r - C_t) - A_r + H}{1 - A^2 - B^2 - C^2},$$

$$e_{22} = \frac{A^2(B_s - A_r) + C^2(B_s - C_t) - B_s + H}{1 - A^2 - B^2 - C^2},$$

$$e_{33} = \frac{A^2(C_t - A_r) + B^2(C_t - B_s) - C_t + H}{1 - A^2 - B^2 - C^2},$$

$$e_{12} = -\tfrac{1}{2}(A_s + B_r), \quad e_{13} = -\tfrac{1}{2}(A_t + C_r), \quad e_{23} = -\tfrac{1}{2}(B_t + C_s),$$

where $H := 2ABe_{12} + 2ACe_{13} + 2BCe_{23}$. Here as usual we write $A_r = \partial A/\partial r$, etc. On the other hand,

$$a_{ij} - b_ib_j = \frac{\delta_{ij}}{1 - A^2 - B^2 - C^2}.$$

It is easy to verify that $e_{ij} = 2c(a_{ij} - b_i b_j)$ if and only if A, B, and C satisfy

$$A_r = B_s = C_t, \quad A_t + C_r = 0, \quad A_s + B_r = 0, \quad B_t + C_s = 0.$$

In this case,

$$c = -\tfrac{1}{2} A_r = -\tfrac{1}{2} B_s = -\tfrac{1}{2} C_t.$$

By Lemma 9.1, we know that $\boldsymbol{S} = 4cF$.

If $F = \alpha + \beta$ on an n-dimensional manifold M is generated from the pair (Φ, V), where $\Phi = \sqrt{\phi_{ij} y^i y^j}$ is a Riemannian metric and $V = V^i(\partial/\partial x^i)$ is a vector field on M with $\phi_{ij}(x) V^i(x) v^j(x) < 1$ for any $x \in M$, then F has isotropic S-curvature, $\boldsymbol{S} = (n+1)c(x)F$, if and only if

$$V_{i;j} + V_{j;i} = -4c\phi_{ij},$$

where $V_i = \phi_{ij} V^j$ and $V_{i;j}$ are the covariant derivatives of V with respect to Φ. This observation is made by Xing [2003]. It also follows from [Bao and Robles 2003b], although it is not proved there directly.

10. Randers Metrics with Relatively Isotropic L-Curvature

We now study Randers metrics with relatively isotropic (mean) Landsberg curvature. From its definition, the mean Landsberg curvature is the mean value of the Landsberg curvature. Thus if a Finsler metric has isotropic Landsberg curvature, it must have isotropic mean Landsberg curvature. I don't know whether the converse is true as well; no counterexample has been found. Nevertheless, for Randers metrics, "having isotropic mean Landsberg curvature" implies "having isotropic Landsberg curvature". According to Lemma 3.2, the Cartan torsion is given by (3–7). Differentiating (3–7) along a geodesic and using (8–4) and (8–5), we obtain

$$L_{ijk} = \frac{1}{n+1}(J_i h_{jk} + J_j h_{ik} + J_k h_{ij}). \tag{10–1}$$

Here we have used the fact that the angular form $\boldsymbol{h}_y$ is constant along geodesics. By (3–7) and (10–1), one can easily show that $J_i + cFI_i = 0$ if and only if $L_{ijk} + cFC_{ijk} = 0$. This proves the claim.

LEMMA 10.1 [Chen and Shen 2003a]. *For a non-Riemannian Randers metric* $F = \alpha + \beta$ *on an n-dimensional manifold M, these conditions are equivalent:*

(a) $\boldsymbol{J} + cF\boldsymbol{I} = 0$ (*or* $\boldsymbol{L} + cF\boldsymbol{C} = 0$).
(b) $\boldsymbol{S} = (n+1)cF$ *and* β *is closed.*
(c) $\boldsymbol{E} = \tfrac{1}{2} cF^{-1}\boldsymbol{h}$ *and* β *is closed.*
(d) $e_{00} = 2c(\alpha^2 - \beta^2)$ *and* β *is closed.*

Here $c = c(x)$ is a scalar function on M.

PROOF. By (10–1), to compute L_{ijk}, it suffices to compute J_i. First, the mean Cartan torsion is

$$I_i = \frac{1}{2}(n+1)F^{-1}\alpha^{-2}(\alpha^2 b_i - \beta y_i), \qquad (10\text{–}2)$$

where $y_i := a_{ij}y^j$. By a direct computation using (8–5), one obtains

$$J_i = \tfrac{1}{4}(n+1)F^{-2}\alpha^{-2}\Big(2\alpha\big((e_{i0}\alpha^2 - y_i e_{00}) - 2\beta(s_i\alpha^2 - y_i s_0) + s_{i0}(\alpha^2 + \beta^2)\big)$$
$$+\alpha^2(e_{i0}\beta - b_i e_{00}) + \beta(e_{i0}\alpha^2 - y_i e_{00}) - 2(s_i\alpha^2 - y_i s_0)(\alpha^2 + \beta^2) + 4s_{i0}\alpha^2\beta\Big).$$

Using this and (10–2), one can easily prove the lemma. □

Thus, for any Randers metric $F = \alpha + \beta$, the J-curvature vanishes if and only if $e_{00} = 0$ and $d\beta = 0$. This is equivalent to $b_{i;j} = 0$, in which case, the spray coefficients of F coincide with that of α. This observation leads to the following result, first established by the collective efforts found in [Matsumoto 1974; Hashiguchi and Ichijyō 1975; Kikuchi 1979; Shibata et al. 1977].

PROPOSITION 10.2. *For a Randers metric $F = \alpha + \beta$, the following conditions are equivalent*:

(a) *F is a weakly Landsberg metric, $\boldsymbol{J} = 0$.*
(b) *F is a Landsberg metric, $\boldsymbol{L} = 0$.*
(c) *F is a Berwald metric.*
(d) *β is parallel with respect to α.*

EXAMPLE 10.3. Consider the Randers metric $F = \alpha + \beta$ on $\mathbf{R}^n$ defined by

$$\alpha := \frac{\sqrt{(1 - \varepsilon^2)\langle x, y\rangle^2 + \varepsilon|y|^2(1 + \varepsilon|x|^2)}}{1 + \varepsilon|x|^2}, \qquad \beta := \frac{\sqrt{1 - \varepsilon^2}\,\langle x, y\rangle}{1 + \varepsilon|x|^2},$$

where ε is an arbitrary constant with $0 < \varepsilon \le 1$. Since β is closed, $s_{ij} = 0$ and $s_i = 0$. After computing $b_{i;j}$, one obtains

$$e_{ij} = \frac{\varepsilon\sqrt{1 - \varepsilon^2}}{(1 + \varepsilon|x|^2)(\varepsilon + |x|^2)}\delta_{ij}.$$

On the other hand, $a_{ij} - b_i b_j = \dfrac{\varepsilon}{1 + \varepsilon|x|^2}\delta_{ij}$. Thus $e_{ij} = 2c(a_{ij} - b_i b_j)$ with

$$c := \frac{\sqrt{1 - \varepsilon^2}}{2(\varepsilon + |x|^2)}.$$

By Lemma 10.1, F satisfies $\boldsymbol{L} + cF\boldsymbol{C} = 0$, $\boldsymbol{S} = (n+1)cF$, and $\boldsymbol{E} = \tfrac{1}{2}cF^{-1}\boldsymbol{h}$. See [Mo and Yang 2003] for a family of more general Randers metrics with nonconstant isotropic S-curvature.

11. Riemann Curvature

The Riemann curvature is an important quantity in Finsler geometry. It was first introduced by Riemann for Riemannian metrics in 1854. Berwald [1926; 1928] extended it to Finsler metrics using the Berwald connection. His extension of the Riemann curvature is a milestone in Finsler geometry.

Let (M, F) be a Finsler manifold and let $G = y^i(\partial/\partial x^i) - 2G^i(\partial/\partial y^i)$ be the induced spray. For a vector $y \in T_x M \setminus \{0\}$, set

$$R^i{}_k := 2\frac{\partial G^i}{\partial x^k} - y^j\frac{\partial^2 G^i}{\partial x^j \partial y^k} + 2G^j\frac{\partial^2 G^i}{\partial y^j \partial y^k} - \frac{\partial G^i}{\partial y^j}\frac{\partial G^j}{\partial y^k}. \qquad (11\text{--}1)$$

The local curvature functions $R^i{}_k$ and $R_{jk} := g_{ij}R^i{}_k$ satisfy

$$R^i{}_k y^k = 0, \quad R_{jk} = R_{kj}. \qquad (11\text{--}2)$$

$\boldsymbol{R}_y = R^i{}_k(\partial/\partial x^i) \otimes dx^k : T_x M \to T_x M$ is a well-defined linear map. We call the family $\boldsymbol{R} = \{\boldsymbol{R}_y \mid y \in TM \setminus \{0\}\}$ the *Riemann curvature*. The Riemann curvature is actually defined for sprays, as shown in [Kosambi 1933; 1935]. When the Finsler metric is Riemannian, then

$$R^i{}_k(x, y) = R_j{}^i{}_{kl}(x)y^j y^l,$$

where $R(u, v)w = R_j{}^i{}_{kl}(x)w^j u^i v^j (\partial/\partial x^i)|_i$ denotes the Riemannian curvature tensor. Namely, $\boldsymbol{R}_y(u) = R(u, y)y$.

The geometric meaning of the Riemann curvature lies in the second variation of geodesics. Let $\sigma(t)$, for $a \le t \le b$, be a geodesic in M. Take a geodesic variation $H(t, s)$ of $\sigma(t)$, that is, a family of curves $\sigma_s(t) := H(t, s)$, $a \le t \le b$, each of which is a geodesic, with $\sigma_0 = \sigma$. Let

$$J(t) := \frac{\partial H}{\partial s}(t, 0).$$

Then $J(t)$ satisfies the *Jacobi equation*

$$D_{\dot\sigma}D_{\dot\sigma}J(t) + \boldsymbol{R}_{\dot\sigma(t)}(J(t)) = 0, \qquad (11\text{--}3)$$

where $D_{\dot\sigma}$ is defined in (4–6). See [Kosambi 1933; 1935].

There is another way to define the Riemann curvature. Any vector $y \in T_x M$ can be extended to a nonzero C^∞ geodesic field Y in an open neighborhood U of x; a *geodesic field* is one for which every integral curve is a geodesic. Define

$$\hat{F}(z, v) := \sqrt{\boldsymbol{g}_{Y_z}(v, v)}, \quad v \in T_z U, \ z \in U.$$

Then $\hat{F} = \hat{F}(z, v)$ is a Riemannian metric on U. Let $\hat{g} = \boldsymbol{g}_Y$ be the inner product induced by $\hat{F}$ and let $\hat{\boldsymbol{R}}$ be the Riemann curvature of $\hat{F}$. It is well-known in Riemannian geometry that

$$\hat{\boldsymbol{R}}_y(u) = 0, \quad \hat{g}(\hat{\boldsymbol{R}}_y(u), v) = \hat{g}(u, \hat{\boldsymbol{R}}_y(v)), \qquad (11\text{--}4)$$

where $u, v \in T_x U$. An important fact is

$$\boldsymbol{R}_y(u) = \hat{\boldsymbol{R}}_y(u), \quad u \in T_x M. \tag{11-5}$$

See [Shen 2001b, Proposition 6.2.2] for a proof of (11–5). Note that $\hat{\boldsymbol{g}}_x = \boldsymbol{g}_y$. It follows from (11–4) and (11–5) that

$$\boldsymbol{R}_y(y) = 0, \quad \boldsymbol{g}_y(\boldsymbol{R}_y(u), v) = \boldsymbol{g}_y(u, \boldsymbol{R}_y(v)), \tag{11-6}$$

where $u, v \in T_x M$. In local coordinates, this equation is just (11–2). See [Shen 2001b] for the application of (11–5) in comparison theorems in conjunction with the S-curvature.

For a two-dimensional subspace $\Pi \subset T_x M$ and a nonzero vector $y \in \Pi$, define

$$\boldsymbol{K}(\Pi, y) := \frac{\boldsymbol{g}_y(\boldsymbol{R}_y(u), u)}{\boldsymbol{g}_y(y, y)\boldsymbol{g}_y(u, u) - \boldsymbol{g}_y(y, u)^2}, \tag{11-7}$$

where $u \in \Pi$ such that $\Pi = \text{span}\{y, u\}$. One can use (11–6) to show that $\boldsymbol{K}(\Pi, y)$ is independent of the choice of a vector u, but it is usually dependent on y. We call $\boldsymbol{K}(\Pi, y)$ the *flag curvature* of the *flag* (Π, y). When $F = \sqrt{g_{ij}(x)y^i y^j}$ is a Riemannian metric, $\boldsymbol{K}(\Pi, y) = \boldsymbol{K}(\Pi)$ is independent of $y \in \Pi$, in which case $\boldsymbol{K}(\Pi)$ is usually called the *sectional curvature* of the *section* $\Pi \subset T_x M$.

A Finsler metric F on a manifold M is said to be *of scalar curvature* $\boldsymbol{K} = \boldsymbol{K}(x, y)$ if for any $y \in T_x M \setminus \{0\}$ the flag curvature $\boldsymbol{K}(\Pi, y) = \boldsymbol{K}(x, y)$ is independent of the tangent planes Π containing y. From the definition, the flag curvature is a scalar function $\boldsymbol{K} = \boldsymbol{K}(x, y)$ if and only if in a standard local coordinate system,

$$R^i{}_k = \boldsymbol{K} F^2 h^i_k, \tag{11-8}$$

where $h^i_k := g^{ij} h_{jk} = g^{ij} F F_{y^j y^k}$. F is *of constant flag curvature* if this $\boldsymbol{K}$ is a constant. For a Riemannian metric, if the flag curvature $\boldsymbol{K}(\Pi, y) = \boldsymbol{K}(x, y)$ is a scalar function on TM, then $\boldsymbol{K}(x, y) = \boldsymbol{K}(x)$ is independent of $y \in T_x M$ and it is a constant when $n \geq 3$ by the Schur Lemma. In the next section we show that any locally projectively flat Finsler metric is of scalar curvature. Such metrics are for us a rich source of Finsler metrics of scalar curvature.

Classifying Finsler metrics of scalar curvature, in particular those of constant flag curvature, is one of the important problems in Finsler geometry. The local structures of projectively flat Finsler metrics of constant flag curvature were characterized in [Shen 2003b]. R. Bryant [1996; 1997; 2002] had earlier classified the global structures of projectively flat Finsler metrics of $\boldsymbol{K} = 1$ on S^n, and given some ideas for constructing non–projectively flat metrics of $\boldsymbol{K} = 1$ on S^n.

Very recently, some non–projectively flat metrics of constant flag curvature have been explicitly constructed; see [Bao–Shen 2002; Bejancu–Farran 2002; Shen 2002; 2003a; 2003b; 2003c; Bao–Robles 2003], for example. These are all Randers metrics. Therefore the classification of Randers metrics of constant flag curvature is a natural problem, first tackled in [Yasuda and Shimada 1977; Matsumoto 1989]. These authors obtained conditions they believed were necessary and sufficient for a Randers metric to be of constant flag curvature. Using their result strictly as inspiration, Bao and Shen [2002] constructed a family of Randers metrics on S^3 with $\mathbf{K} = 1$; these metrics do satisfy Yasuda and Shimada's conditions. Later, however, examples were found of Randers metrics of constant flag curvature not satisfying those conditions [Shen 2002; 2003c], showing that the earlier characterization was incorrect. Shortly thereafter, Randers metrics of constant flag curvature were characterized in [Bao–Robles 2003] using a system of PDEs, a result also obtained in [Matsumoto and Shimada 2002] by a different method. This subsequently led to a corrected version of the Yasuda–Shimada theorem. Finally, using the characterization in [Bao–Robles 2003], and motivated by some constructions in [Shen 2002; 2003c], Bao, Robles and Shen have classified Randers metrics of constant flag curvature with the help of formula (2–17):

THEOREM 11.1 [Bao et al. 2003]. *Let* $\Phi = \sqrt{\phi_{ij}y^iy^j}$ *be a Riemannian metric and let* $V = V^i(\partial/\partial x^i)$ *be a vector field on a manifold* M *with* $\Phi(x, V_x) < 1$ *for all* $x \in M$. *Let* F *be the Randers metric defined by* (2–17). F *is of constant flag curvature* $\mathbf{K} = \lambda$ *if and only if*

(a) *there is a constant* c *such that* V *satisfies* $V_{i|j} + V_{j|i} = -4c\phi_{ij}$, *where* $V_i := \phi_{ij}V^j$, *and*

(b) Φ *has constant sectional curvature* $\tilde{\mathbf{K}} = \lambda + c^2$,

where $_|$ *denotes the covariant derivative with respect to* Φ *and* c *is a constant.*

The equation $V_{i|j} + V_{j|i} = -4c\phi_{ij}$ of part (a) is by itself always equivalent to $\mathbf{S} = (n+1)cF$, for c a scalar function on M [Xing 2003].

An analogue of Theorem 11.1 still holds for Randers metrics of isotropic Ricci curvature, i.e., $\mathbf{Ric} = (n-1)\lambda F^2$, where $\lambda = \lambda(x)$ is a scalar function on M. See [Bao and Robles 2003b] in this volume.

We have not extended the result above to Randers metrics of scalar curvature. Usually, the isotropic S-curvature condition simplifies the classification problem. It seems possible to classify Randers metrics of scalar curvature and isotropic S-curvature. The following example is our first attempt to understand Randers metrics of scalar curvature and isotropic S-curvature.

EXAMPLE 11.2. Let $F = \alpha + \beta$ be the Randers metric defined in (2–20). Set $\Delta := 1 - |a|^2|x|^4$. We can write $\alpha = \sqrt{a_{ij}(x)y^iy^j}$ and $\beta = b_i(x)y^i$, where

$$a_{ij} = \frac{\delta_{ij}}{\Delta} + \frac{\left(|x|^2a^i - 2\langle a, x\rangle x^i\right)\left(|x|^2a^j - 2\langle a, x\rangle x^j\right)}{\Delta^2}, \quad b_i = -\frac{|x|^2a^i - 2\langle a, x\rangle x^i}{\Delta}.$$

Using Maple for the computation, we obtain, with the notations of Section 9,

$$e_{00} = \frac{2\langle a, x\rangle |y|^2}{\Delta} = 2\langle a, x\rangle (\alpha^2 - \beta^2),$$

$$s_{j0} = 2\frac{\langle a, y\rangle x^j - \langle x, y\rangle a^j}{\Delta^2},$$

$$s_0 = b^i s_{i0} = 2\frac{|a|^2|x|^2\langle x, y\rangle + \langle a, x\rangle\langle a, v\rangle}{\Delta}.$$

By Lemma 9.1, we see that F has isotropic S-curvature:

$$S = (n+1)\langle a, x\rangle F.$$

By (9–1), the spray coefficients $G^i = G^i(x, y)$ of F are

$$G^i = \bar{G}^i + Py^i + \alpha a^{ij} s_{j0},$$

where $P = e_{00}/(2F) - s_0 = \langle a, x\rangle(\alpha - \beta) - s_0$. Using the formulas for G^i and $R^i{}_k$ in (11–1), we can show that F is also of scalar curvature with flag curvature

$$K = 3\frac{\langle a, y\rangle}{F} + 3\langle a, x\rangle^2 - 2|a|^2|x|^2.$$

12. Projectively Flat Metrics

A Finsler metric $F = F(x, y)$ on an open subset $U \subset \mathrm{R}^n$ is *projectively flat* if every geodesic $\sigma(t)$ is straight in U, that is, if

$$\sigma^i(t) = x^i + f(t)y^i,$$

where $f(t)$ is a C^∞ function with $f(0) = 0$, $f'(0) = 1$ and $x = (x^i)$, $y = (y^i)$ are constant vectors. This is equivalent to $G^i = Py^i$, where $P = P(x, y)$ is positively y-homogeneous of degree one. P is called the *projective factor*.

It is generally difficult to compute the Riemann curvature, but for locally projectively flat Finsler metrics, the formula is relatively simple.

Consider a projectively flat Finsler metric $F = F(x, y)$ on an open subset $U \subset \mathrm{R}^n$. By definition, its spray coefficients are in the form $G^i = Py^i$. Plugging them into (11–1), one obtains

$$R^i{}_k = \Xi\delta^i_k + \tau_k y^i, \tag{12–1}$$

where

$$\Xi = P^2 - P_{x^k}y^k, \quad \tau_k = 3(P_{x^k} - PP_{y^k}) + \Xi_{y^k}.$$

Using (11–6), one can show that $\tau_k = -\Xi F^{-1}F_{y^k}$ and

$$R^i{}_k = \Xi\left(\delta^i_k - \frac{F_{y^k}}{F}y^i\right). \tag{12–2}$$

Thus F is of scalar curvature with flag curvature

$$K = \frac{\Xi}{F^2} = \frac{P^2 - P_{x^k}y^k}{F^2}. \tag{12–3}$$

Using (7–4), one obtains

$$S = (n+1)P(x,y) - y^m \frac{\partial}{\partial x^m} \ln \sigma_F(x). \qquad (12\text{–}4)$$

By (12–2), one immediately obtains the following result. (See also [Szabó 1977] and [Matsumoto 1980] for related discussions.)

PROPOSITION 12.1 [Berwald 1929a; 1929b]. *Every locally projectively flat Finsler metric is of scalar curvature.*

There is another way to characterize projectively flat Finsler metrics.

THEOREM 12.2 [Hamel 1903; Rapcsák 1961]. *Let $F = F(x,y)$ be a Finsler metric on an open subset $U \subset \mathbf{R}^n$. F is projectively flat if and only if F satisfies*

$$F_{x^k y^l} y^k - F_{x^l} = 0, \qquad (12\text{–}5)$$

in which case, the spray coefficients are given by $G^i = Py^i$ with $P = \frac{1}{2} F_{x^k} y^k / F$.

PROOF. Let $G^i = G^i(x,y)$ denote the spray coefficients of F in the standard coordinate system in $TU \cong U \times \mathbf{R}^n$. One can rewrite (4–2) as

$$G^i = Py^i + Q^i, \qquad (12\text{–}6)$$

where

$$P = \frac{F_{x^k} y^k}{2F}, \quad Q^i = \frac{1}{2} F g^{il} (F_{x^k y^l} y^k - F_{x^l}).$$

Thus F is projectively flat if and only if there is a scalar function $\tilde{P} = \tilde{P}(x,y)$ such that $G^i = \tilde{P}y^i$, i.e.,

$$Py^i + Q^i = \tilde{P}y^i. \qquad (12\text{–}7)$$

Observe that

$$g_{ij} y^j Q^i = \frac{1}{2} F y^l (F_{x^k y^l} y^k - F_{x^l}) = 0.$$

Assume that (12–7) holds. Contracting with $y_i := g_{ij} y^j$ yields

$$P = \tilde{P}.$$

Then $Q^i = 0$ by (12–7). This implies (12–5). □

Since equation (12–5) is linear, if F_1 and F_2 are projectively flat on an open subset $U \subset \mathbf{R}^n$, so is their sum. If $F = F(x,y)$ is projectively flat on $U \subset \mathbf{R}^n$, so is its reverse $\bar{F} := F(x,-y)$. Thus the symmetrization

$$\tilde{F} := \frac{1}{2}\big(F(x,y) + F(x,-y)\big)$$

is projectively flat.

The Finsler metric $F = \alpha_\mu(x,y)$ in (2–5) satisfies (12–5), so it's projectively flat.

THEOREM 12.3. (Beltrami) *A Riemannian metric $F = F(x,y)$ on a manifold M is locally projectively flat if and only if it is locally isometric to the metric α_μ in (2–5).*

Using the formula (9–1), one can easily prove the following:

THEOREM 12.4. *A Randers metric $F = \alpha+\beta$ on a manifold is locally projectively flat if and only if α is locally projectively flat and β is closed.*

Besides projectively flat Randers metrics, we have the following examples.

EXAMPLE 12.5. Let $\phi = \phi(y)$ be a Minkowski norm on R^n and let U be the strongly convex domain enclosed by the indicatrix of ϕ. Let $\Theta = \Theta(x, y)$ be the Funk metric on U (Example 2.9). By (2–19),

$$\Theta_{x^k y^l} y^k = (\Theta \Theta_{y^k})_{y^l} y^k = \tfrac{1}{2}(\Theta^2)_{y^k y^l} y^k = \tfrac{1}{2}[\Theta^2]_{y^l} = \Theta_{x^l}.$$

Thus Θ is projectively flat with projective factor

$$P = \frac{\Theta_{x^k} y^k}{2\Theta} = \frac{\Theta \Theta_{y^k} y^k}{2\Theta} = \tfrac{1}{2}\Theta.$$

By (12–3), the flag curvature is

$$K = \frac{\Theta^2 - 2\Theta_{x^k} y^k}{4\Theta^2} = \frac{\Theta^2 - 2\Theta^2}{4\Theta^2} = -\frac{1}{4}.$$

EXAMPLE 12.6 [Shen 2003b]. Let $\phi = \phi(y)$ be a Minkowski norm on R^n and let U be the strongly convex domain enclosed by the indicatrix of ϕ. Let $\Theta = \Theta(x, y)$ be the Funk metric on U and define

$$F := \Theta(x, y)\big(1 + \Theta_{y^k}(x, y)x^k\big).$$

Since $F(0, y) = \Theta(0, y) = \phi(y)$ is a Minkowski norm, by continuity, F is a Finsler metric for x nearby the origin. By (2–19), one can verify that

$$F_{x^k y^l} y^k = F_{x^l}, \quad F_{x^k} y^k = 2\Theta F.$$

Thus F is projectively flat with projective factor $P = \Theta(x, y)$. By (2–19) and (12–3), we obtain

$$K = \frac{\Theta^2 - \Theta_{x^k} y^k}{F^2} = \frac{\Theta^2 - \Theta \Theta_{y^k} y^k}{F^2} = 0.$$

Now we take a look at the Finsler metric $F = F_\varepsilon(x, y)$ defined in (2–9).

EXAMPLE 12.7. Let

$$F := \frac{\sqrt{\Psi\big(\tfrac{1}{2}(\sqrt{\Phi^2 + (1 - \varepsilon^2)|y|^4} + \Phi)\big) + (1 - \varepsilon^2)\langle x, y\rangle^2} + \sqrt{1 - \varepsilon^2}\,\langle x, y\rangle}{\Psi},$$

$$(12\text{–}8)$$

where

$$\Phi := \varepsilon|y|^2 + \big(|x|^2|y|^2 - \langle x, y\rangle^2\big), \quad \Psi := 1 + 2\varepsilon|x|^2 + |x|^4.$$

First, one can verify that $F = F_\varepsilon(x, y)$ satisfies (12–5). Thus F is projectively flat with spray coefficients $G^i = Py^i$, where $P = \frac{1}{2}F_{x^k}(x, y)y^k/F(x, y)$. A Maple computation gives

$$P = \frac{\sqrt{\Psi\left(\frac{1}{2}\left(\sqrt{\Phi^2 + (1 - \varepsilon^2)|y|^4} + \Phi\right)\right) - (1 - \varepsilon^2)\langle x, y\rangle^2} - (\varepsilon + |x|^2)\langle x, y\rangle}{\Psi}.$$
$$(12\text{–}9)$$

Further, one can verify that P satisfies

$$P_{x^k}y^k = P^2 - F^2.$$

Thus

$$K = \frac{P^2 - P_{x^k}y^k}{F^2} = \frac{P^2 - (P^2 - F^2)}{F^2} = 1.$$

That is, F has constant flag curvature $K = 1$.

The projectively flat Finsler metrics constructed above are incomplete. They can be pulled back to S^n by (2–2) to form complete irreversible projectively flat Finsler metrics of constant flag curvature $K = 1$. See [Bryant 1996; 1997].

13. The Chern Connection and Some Identities

The previous sections introduced several geometric quantities, such as the Cartan torsion, the Landsberg curvature, the S-curvature and the Riemann curvature. These quantities are not all independent. To reveal their relationships, we use the Chern connection to describe them as tensors on the slit tangent bundle, and use the exterior differentiation method to derive some important identities.

Let M be an n-dimensional manifold and TM its tangent bundle. As usual, a typical element of TM will be denoted by (x, y), with $y \in T_xM$. The natural projection $\pi : TM \to M$ pulls back the tangent bundle TM over M to a vector bundle π^*TM over the slit tangent bundle TM_0. The fiber of π^*TM at each point $(x, y) \in TM_0$ is a copy of T_xM. Thus we write a typical element of π^*TM as (x, y, v), where $y \in T_xM \setminus \{0\}$ and $v \in T_xM$. Let $\partial_{i|(x,y)} := \big(x, y, (\partial/\partial x^i)|_x\big)$. Then $\{\partial_i\}$ is a local frame for π^*TM. Let (x^i, y^i) be a standard local coordinate system in TM_0. Then $HT^*M := \text{span}\{dx^i\}$ is a well-defined subbundle of $T^*(TM_0)$. Let

$$\delta y^i := dy^i - N^i_j dx^j,$$

where $N^i_j := \partial G^i/\partial y^j$. Then $VT^*M := \text{span}\{\delta y^i\}$ is a well-defined subbundle of $T^*(TM_0)$, so that $T^*(TM_0) = HT^*M \oplus VT^*M$. The Chern connection is a linear connection on π^*TM, locally expressed by

$$DX = (dX^i + X^j\omega_j{}^i) \otimes \partial_i, \quad X = X^i\partial_i,$$

where the set of 1-forms $\{\omega_j{}^i\}$ is uniquely determined by

$$dw^i = \omega^j \wedge \omega_j{}^i,$$
$$dg_{ij} = g_{ik}\omega_j{}^k + g_{kj}\omega_i{}^k + 2C_{ijk}\omega^{n+k},$$ (13–1)

where $g_{ij} := \frac{1}{2}[F^2]_{y^i y^j}$, $C_{ijk} := \frac{1}{4}[F^2]_{y^i y^j y^k}$, $\omega^i := dx^i$, and $\omega^{n+i} := \delta y^i$. See [Bao and Chern 1993; Bao et al. 2000; Chern 1943; 1948; 1992]. Each 1-form $\omega_j{}^i$ is horizontal: $\omega_j{}^i = \Gamma_{jk}^i dx^k$. The coefficients $\Gamma_{jk}^i = \Gamma_{jk}^i(x, y)$ are called the *Christoffel symbols*. We have $N_j^i = y^k \Gamma_{jk}^i$. Thus

$$\omega^{n+i} = dy^i + y^j \omega_j{}^i.$$ (13–2)

Put

$$\Omega^i := d\omega^{n+i} - \omega^{n+j} \wedge \omega_j{}^i.$$ (13–3)

One can express Ω^i as

$$\Omega^i = \frac{1}{2} R^i{}_{kl} \omega^k \wedge \omega^l - L^i{}_{kl} \omega^k \wedge \omega^{n+l},$$

where

$$R^i{}_{kl} = \frac{\partial N_l^i}{\partial x^k} - \frac{\partial N_k^i}{\partial x^l} + N_l^s \frac{\partial N_k^i}{\partial y^s} - N_k^s \frac{\partial N_l^i}{\partial y^s},$$

and

$$L^i{}_{kl} := y^j \frac{\partial \Gamma_{jk}^i}{\partial y^l} = \frac{\partial N_k^i}{\partial y^l} - \Gamma_{kl}^i.$$

Let $R^i{}_k$ be defined in (11–1) and L_{ijk} be defined in (8–4). Then

$$R^i{}_k = R^i{}_{kl} y^l, \quad R^i{}_{kl} = \frac{1}{3}\left(\frac{\partial R^i{}_k}{\partial y^l} - \frac{\partial R^i{}_l}{\partial y^k}\right), \quad L^i{}_{kl} = g^{ij} L_{jkl}.$$ (13–4)

Put

$$\Omega_j{}^i := d\omega_j{}^i - \omega_j{}^k \wedge \omega_k{}^i.$$

One can express $\Omega_j{}^i$ as

$$\Omega_j{}^i = \frac{1}{2} R_j{}^i{}_{kl} \omega^k \wedge \omega^l + P_j{}^i{}_{kl} \omega^k \wedge \omega^{n+l}.$$

Differentiating (13–2) yields $\Omega^i = y^j \Omega_j{}^i$. Thus

$$R^i{}_{kl} = y^j R_j{}^i{}_{kl}, \quad L^i{}_{kl} = -y^j P_j{}^i{}_{kl}.$$

There is a canonical way to define the covariant derivatives of a tensor on TM_0 using the Chern connection. For the distortion τ on $TM \setminus \{0\}$, define $\tau_{|m}$ and $\tau_{.m}$ by

$$d\tau = \tau_{|i} \omega^i + \tau_{.i} \omega^{n+i}.$$ (13–5)

It follows from (7–2) that

$$\tau_{.i} = \frac{\partial \tau}{\partial y^i} = I_i.$$ (13–6)

For the induced Riemannian tensor, $g = g_{ij}\omega^i \otimes \omega^j$, define $g_{ij|k}$ and $g_{ij \cdot k}$ by

$$dg_{ij} - g_{kj}\omega_i{}^k - g_{ik}\omega_j{}^k = g_{ij|k}\omega^k + g_{ij \cdot k}\omega^{n+k}.$$

It follows from (13–1) that

$$g_{ij|k} = 0, \quad g_{ij \cdot k} = 2C_{ijk}.$$

Similarly, one can define $C_{ijk|l}$ at $I_{i|l}$. Equations (8–4) and (8–5) become

$$L_{ijk} = C_{ijk|m}y^m, \quad J_i = I_{i|m}y^m. \tag{13-7}$$

Differentiating (13–3) yields the Bianchi identity

$$d\Omega^i = -\Omega^j \wedge \omega_j{}^i + \omega^{n+j} \wedge \Omega_j{}^i. \tag{13-8}$$

It follows from (13–8) that

$$R_j{}^i{}_{kl} = R^i{}_{kl \cdot j} + L^i{}_{kj|l} - L^i{}_{lj|k} + L^i{}_{lm}L^m{}_{kj} - L^i{}_{km}L^m{}_{lj}. \tag{13-9}$$

We are going to find other relationships among curvature tensors. Differentiating (13–1) yields

$$0 = g_{ik}\Omega_j{}^k + g_{kj}\Omega_i{}^k + 2(C_{ijk|l}\omega^l + C_{ijk \cdot l}\omega^{n+l}) \wedge \omega^{n+k} + 2C_{ijk}\Omega^k.$$

It follows that

$$R_{jikl} + R_{ijkl} + 2C_{ijm}R^m{}_{kl} = 0, \tag{13-10}$$

where $R_{jikl} := g_{im}R_j{}^m{}_{kl}$, and

$$P_{jikl} + P_{ijkl} + 2C_{ijl|k} - 2C_{ijm}L^m{}_{kl} = 0,$$

where $P_{jikl} := g_{im}P_j{}^m{}_{kl}$. Then (13–9) can be expressed by

$$R_{jikl} = g_{im}R^m{}_{kl \cdot j} + L_{ikj|l} - L_{ilj|k} + L_{ilm}L^m{}_{kj} - L_{ikm}L^m{}_{lj}.$$

Plugging the formulas for R_{jikl} and R_{ijkl} into (13–10) yields

$$\begin{aligned} L_{ijk|l} - L_{ijl|k} &= -\tfrac{1}{2}(g_{im}R^m{}_{kl \cdot j} + g_{jm}R^m{}_{kl \cdot i}) - C_{ijm}R^m{}_{kl}, \\ I_{k|l} - I_{l|k} &= -R^m{}_{kl \cdot m} - I_m R^m{}_{kl}. \end{aligned} \tag{13-11}$$

The expression for $R^i{}_{kl}$ in (13–4) can be written as

$$R^i{}_{kl} = \tfrac{1}{3}(R^i{}_{k \cdot l} - R^i{}_{l \cdot k}). \tag{13-12}$$

LEMMA 13.1 [Mo 1999]. L_{ijk} and $R^i{}_k$ are related by

$$C_{ijk|p|q}y^p y^q + C_{ijm}R^m{}_k$$
$$= -\tfrac{1}{3}g_{im}R^m{}_{k \cdot j} - \tfrac{1}{3}g_{jm}R^m{}_{k \cdot i} - \tfrac{1}{6}g_{im}R^m{}_{j \cdot k} - \tfrac{1}{6}g_{jm}R^m{}_{i \cdot k}. \tag{13-13}$$

In particular,

$$I_{k|p|q}y^p y^q + I_m R^m{}_k = -\tfrac{1}{3}(2R^m{}_{k \cdot m} + R^m{}_{m \cdot k}). \tag{13-14}$$

PROOF. By (13–7), we have

$$L_{ijk|m}y^m = C_{ijk|p|q}y^p y^q, \quad J_{k|m}y^m = I_{k|p|q}y^p y^q.$$

Contracting the first line of (13–11) with y^l yields (13–13), and contracting (13–13) with g^{ij} yields (13–14). Here we have made use of (13–12). □

The equations above are crucial in the study of Finsler metrics of scalar curvature. Let $F = F(x,y)$ be a Finsler metric of scalar curvature with flag curvature $K = K(x,y)$. Then (11–8) holds. Plugging (11–8) into (13–13) and (13–14) yields

$$C_{ijk|p|q}y^p y^q + KF^2 C_{ijk} = -\tfrac{1}{3}F^2(K_{.i}h_{jk} + K_{.j}h_{ik} + K_{.k}h_{ij}),$$
$$I_{k|p|q}y^p y^q + KF^2 I_k = -\tfrac{1}{3}(n+1)F^2 K_{.k}. \tag{13–15}$$

Using the first of these equations, one shows that any compact Finsler manifold of negative constant flag curvature must be Riemannian [Akbar-Zadeh 1988].

It follows from (13–15) that for any Finsler metric F of scalar curvature with flag curvature K, the Matsumoto torsion satisfies

$$M_{ijk|p|q}y^p y^q + KF^2 M_{ijk} = 0. \tag{13–16}$$

One can use (13–16) to show that any Landsberg metric of scalar curvature with $K \neq 0$ it is Riemannian, in dimension $n \geq 3$ [Numata 1975]. See also Corollary 17.4.

Using (13–16), one can easily prove this:

THEOREM 13.2 [Mo and Shen 2003]. *Let (M, F) be a compact Finsler manifold of dimension $n \geq 3$. If F is of scalar curvature with negative flag curvature, F must be a Randers metric.*

Now we derive some important identities for the S-curvature. Differentiating (13–5) and using (13–3) and (13–6), one obtains

$$0 = d^2\tau = (\tau_{|k|l}\omega^l + \tau_{|k\cdot l}\omega^{n+l}) \wedge \omega^k + (I_{k|l}\omega^l + I_{k\cdot l}\omega^{n+l}) \wedge \omega^k + I_m \Omega^{n+m}.$$

This yields the Ricci identities

$$\tau_{|k|l} = \tau_{|l|k} + I_p R^p{}_{kl}, \tag{13–17}$$
$$\tau_{|k\cdot l} = I_{l|k} - I_p L^p{}_{kl}. \tag{13–18}$$

From the definition (7–3), the S-curvature can be regarded as

$$S = \tau_{|m}y^m. \tag{13–19}$$

Contracting (13–17) with y^k yields

$$S_{\cdot k} = (\tau_{|m}y^m)_{\cdot k} = \tau_{|m\cdot k}y^m + \tau_{|k} = I_{k|m}y^m - I_p L^p{}_{mk}y^m + \tau_{|k} = J_k + \tau_{|k},$$

where we have made use of (13–17) and (13–19). We restate this equation as

$$S_{\cdot k} = \tau_{|k} + J_k. \tag{13–20}$$

LEMMA 13.3 [Mo 2002; Mo and Shen 2003]. *The S-curvature satisfies*

$$S_{.k|m}y^m - S_{|k} = -\tfrac{1}{3}(2R^m_{\ k\cdot m} + R^m_{\ m\cdot k}).\tag{13-21}$$

PROOF. It follows from (13-20) that

$$S_{.k|l} = \tau_{|k|l} + J_{k|l}.\tag{13-22}$$

By (13-17) and (13-22), one obtains

$$
\begin{aligned}
S_{.k|m}y^m - S_{|k} &= (S_{.k|m} - S_{.m|k})y^m = (\tau_{|k|m} - \tau_{|m|k})y^m + (J_{k|m} - J_{m|k})y^m \\
&= I_p R^p_{\ km}y^m + J_{k|m}y^m = I_p R^p_{\ k} - I_p R^p_{\ k} - \tfrac{1}{3}I_m(R^m_{\ k\cdot l} - R^m_{\ l\cdot k}) \\
&= -\tfrac{1}{3}I_m(R^m_{\ k\cdot l} - R^m_{\ l\cdot k}). \qquad\qquad\qquad\qquad\qquad\qquad \square
\end{aligned}
$$

14. Nonpositively Curved Finsler Manifolds

We now use some of the identities derived in the previous section to establish global rigidity theorems.

First, consider the mean Cartan torsion. Let (M, F) be an n-dimensional Finsler manifold. The norm of the mean Cartan torsion $\boldsymbol{I}$ at a point $x \in M$ is defined by

$$\|\boldsymbol{I}\|_x := \sup_{0 \neq y \in T_x M} \sqrt{I_i(x, y)g^{ij}(x, y)I_j(x, y)}.$$

It is known that if $F = \alpha + \beta$ is a Randers metric, then

$$\|\boldsymbol{I}\|_x \leq \frac{n+1}{\sqrt{2}}\sqrt{1 - \sqrt{1 - \|\beta\|_x^2}} < \frac{n+1}{\sqrt{2}}.$$

The bound in dimension two was suggested by B. Lackey. See [Shen 2001b, Proposition 7.1.2] or [Ji and Shen 2002] for a proof. Below is our first global rigidity theorem.

THEOREM 14.1 [Shen 2003d]. *Let (M, F) be an n-dimensional complete Finsler manifold with nonpositive flag curvature. Suppose that F has almost constant S-curvature $\boldsymbol{S} = (n+1)(cF + \eta)$ (with c a constant and η a closed 1-form) and bounded mean Cartan torsion $\sup_{x \in M} \|\boldsymbol{I}\|_x < \infty$. Then $\boldsymbol{J} = 0$ and $\boldsymbol{R} \circ \boldsymbol{I} = 0$. Moreover F is Riemannian at points where the flag curvature is negative.*

PROOF. It follows from (13-14) and (13-21) that

$$I_{k|p|q}y^p y^q + I_m R^m_{\ k} = S_{.k|m}y^m - S_{|k}.\tag{14-1}$$

Assume that the S-curvature is almost isotropic:

$$\boldsymbol{S} = (n+1)(cF + \eta),$$

where $c = c(x)$ is a scalar function on M and $\eta = \eta_i dx^i$ is a closed 1-form on M. Observe that

$$\eta_{.k|m}y^m - \eta_{|k} = (\eta_{k|m} - \eta_{m|k})y^m = \left(\frac{\partial \eta_k}{\partial x^m} - \frac{\partial \eta_m}{\partial x^k}\right)y^m = 0.$$

Thus

$$S_{\cdot k|m}y^m - S_{|k} = (n+1)(c_{x^m}y^m F_{\cdot k} - c_{|k}F + \eta_{\cdot k|m}y^m - \eta_{|k})$$
$$= (n+1)(c_{x^m}y^m F_{\cdot k} - c_{|k}F).$$

In this case, (13–21) becomes

$$2R^m{}_{k\cdot m} + R^m{}_{m\cdot k} = -3(n+1)(c_{x^m}y^m F_{\cdot k} - c_{|k}F) \tag{14-2}$$

and (14–1) becomes

$$I_{k|p|q}y^p y^q + I_m R^m{}_k = (n+1)(c_{x^m}y^m F_{\cdot k} - c_{|k}F).$$

By assumption, c is constant, so this last equation reduces

$$I_{k|p|q}y^p y^q + I_m R^m{}_k = 0. \tag{14-3}$$

Let $y \in T_x M$ be an arbitrary vector and let $\sigma(t)$ be the geodesic with $\sigma(0) = x$ and $\dot{\sigma}(0) = y$. Since the Finsler metric is complete, one may assume that $\sigma(t)$ is defined on $(-\infty, \infty)$. The mean Cartan torsion $\boldsymbol{I}$ and the mean Landsberg curvature $\boldsymbol{J}$ restricted to $\sigma(t)$ are vector fields along $\sigma(t)$:

$$\boldsymbol{I}(t) := I^i(\sigma(t), \dot{\sigma}(t)) \frac{\partial}{\partial x^i}\Big|_{\sigma(t)}, \quad \boldsymbol{J}(t) := J^i(\sigma(t), \dot{\sigma}(t)) \frac{\partial}{\partial x^i}\Big|_{\sigma(t)}.$$

It follows from (8–5) or (13–7) that

$$D_{\dot{\sigma}}\boldsymbol{I}(t) = I^i{}_{|m}(\sigma(t), \dot{\sigma}(t))\dot{\sigma}^m(t) \frac{\partial}{\partial x^i}\Big|_{\sigma(t)} = \boldsymbol{J}(t).$$

It follows from (14–3) that

$$D_{\dot{\sigma}}D_{\dot{\sigma}}\boldsymbol{I}(t) + \boldsymbol{R}_{\dot{\sigma}(t)}(\boldsymbol{I}_{\dot{\sigma}(t)}) = 0.$$

Setting

$$\varphi(t) := \boldsymbol{g}_{\dot{\sigma}(t)}(\boldsymbol{I}(t), \boldsymbol{I}(t)),$$

we obtain

$$\varphi''(t) = 2\boldsymbol{g}_{\dot{\sigma}(t)}(D_{\dot{\sigma}}D_{\dot{\sigma}}\boldsymbol{I}(t), \boldsymbol{I}(t)) + 2\boldsymbol{g}_{\dot{\sigma}(t)}(D_{\dot{\sigma}}\boldsymbol{I}(t), D_{\dot{\sigma}}\boldsymbol{I}(t))$$
$$= -2\boldsymbol{g}_{\dot{\sigma}(t)}(\boldsymbol{R}_{\dot{\sigma}(t)}(\boldsymbol{I}(t)), \boldsymbol{I}(t)) + 2\boldsymbol{g}_{\dot{\sigma}(t)}(\boldsymbol{J}(t), \boldsymbol{J}(t)). \tag{14-4}$$

By assumption, $\boldsymbol{K} \leq 0$. Thus

$$\boldsymbol{g}_{\dot{\sigma}(t)}(\boldsymbol{R}_{\dot{\sigma}(t)}(\boldsymbol{I}(t)), \boldsymbol{I}(t)) \leq 0.$$

It follows from (14–4) that

$$\varphi''(t) \geq 0.$$

Thus $\varphi(t)$ is convex and nonnegative. Suppose that $\varphi'(t_0) \neq 0$ for some t_0. By an elementary argument, $\lim_{t\to+\infty}\varphi(t) = \infty$ or $\lim_{t\to-\infty}\varphi(t) = \infty$. This implies that the mean Cartan torsion is unbounded, which contradicts the assumption. Therefore, $\varphi'(t) = 0$ and hence $\varphi''(t) = 0$. Since each term in (14–4) is nonnegative, one concludes that

$$g_{\dot{\sigma}(t)}\big(R_{\dot{\sigma}(t)}(I(t)), I(t)\big) = 0, \qquad J(t) = 0.$$

Setting $t = 0$ yields

$$g_y\big(R_y(I_y), I_y\big) = 0 \tag{14–5}$$

and $J_y = 0$. By (11–6), $R_y(y) = 0$ and R_y is self-adjoint with respect to g_y, i.e., $g_y\big(R_y(u), v\big) = g_y(u, R_y(v))$, for $u, v \in T_x M$. Thus there is an orthonormal basis $\{e_i\}_{i=1}^n$ with $e_n = y$ such that

$$R_y(e_i) = \lambda_i e_i, \quad i = 1, \ldots, n,$$

with $\lambda_n = 0$. By assumption, the flag curvature is nonpositive. Then

$$g_y(R_y(e_i), e_i) = \lambda_i \leq 0, \quad i = 1, \ldots, n-1.$$

Since I_y is perpendicular to y with respect to g_y, one can express it as $I_y = \mu_1 e_1 + \cdots + \mu_{n-1} e_{n-1}$. By (14–5), one obtains

$$0 = g_y\big(R_y(I_y), I_y\big) = \sum_{i=1}^{n-1} \mu_i^2 \lambda_i.$$

Since each term $\mu_i^2 \lambda_i$ is nonpositive, one concludes that $\mu_i \lambda_i = 0$, or yet

$$R_y(I_y) = \sum_{i=1}^{n-1} \mu_i \lambda_i = 0. \tag{14–6}$$

Now suppose that F has negative flag curvature at a point $x \in M$. Then $\lambda_i < 0$ for $i = 1, \ldots, n-1$. By (14–6), one concludes that $\mu_i = 0$, $i = 1, \ldots, n-1$, namely, $I_y = 0$. By Deicke's theorem [Deicke 1953], F is Riemannian. $\square$

COROLLARY 14.2. *Every complete Berwald manifold with negative flag curvature is Riemannian.*

PROOF. For a Berwald metric F on a manifold M, the Minkowski spaces $(T_x M, F_x)$ are all linearly isometric (Proposition 5.2). Thus the Cartan torsion is bounded from above. Meanwhile, the S-curvature vanishes (Proposition 7.1). Thus F must be Riemannian. $\square$

EXAMPLE 14.3. Let $(\overline{M}, \bar{\alpha})$ and $(\underline{M}, \underline{\alpha})$ be Riemannian manifolds and let $F = F(x, y)$ be the product metric on $M = \overline{M} \times \underline{M}$, defined in Example 5.1. We computed the spray coefficients of F in Example 5.1. Using (11–1), one obtains the Riemann tensor of F:

$$R^{\bar{a}}_{\ \bar{b}} = \overline{R}^{\bar{a}}_{\ \bar{b}}, \quad R^{\bar{a}}_{\ \underline{b}} = 0 = R^{\underline{a}}_{\ \bar{b}}, \quad R^{\underline{a}}_{\ \underline{b}} = \underline{R}^{\underline{a}}_{\ \underline{b}},$$

where $\bar{R}^{\bar{a}}_{\ \bar{b}}$ and $\underline{R}^{\underline{a}}_{\ \underline{b}}$ are the coefficients of the Riemann tensors of $\bar{a}$ and $\underline{a}$. Let $R_{ij} := g_{ik}R^k_{\ j}$ as usual, and define $\bar{R}_{\bar{a}\bar{b}}$ and $\underline{R}_{\underline{a}\underline{b}}$ similarly. Using (5–3), one obtains

$$R_{\bar{a}\bar{b}} = f_s \bar{R}_{\bar{a}\bar{b}}, \quad R_{\bar{a}\underline{b}} = 0 = R_{\underline{a}\bar{b}}, \quad R_{\underline{a}\underline{b}} = f_t \underline{R}_{\underline{a}\underline{b}}.$$

For any vector $v = v^i(\partial/\partial x^i)|_x \in T_x M$,

$$\boldsymbol{g}_y\big(\boldsymbol{R}_y(v), v\big) = f_s \bar{R}_{\bar{a}\bar{b}} v^{\bar{a}} v^{\bar{b}} + f_t \underline{R}_{\underline{a}\underline{b}} v^{\underline{a}} v^{\underline{b}}.$$

Thus if α_1 and α_2 both have nonpositive sectional curvature, F has nonpositive flag curvature.

Using (5–4), one can compute the mean Cartan torsion. First, observe that

$$I_i = \frac{\partial}{\partial y^i} \ln \sqrt{\det(g_{jk})} = \frac{\partial}{\partial y^i} \ln \sqrt{h\big([\alpha_1]^2, [\alpha_2]^2\big)},$$

where $h = h(s,t)$ is defined in (5–5). One obtains

$$I_{\bar{a}} = \frac{h_s}{h} \bar{y}_{\bar{a}}, \quad I_{\underline{a}} = \frac{h_t}{h} \bar{y}_{\underline{a}},$$

where $\bar{y}_{\bar{a}} := \bar{g}_{\bar{a}\bar{b}} y^{\bar{b}}$ and $\bar{y}_{\underline{a}} := \bar{g}_{\underline{a}\underline{b}} y^{\underline{b}}$. Since $\bar{y}_{\bar{a}} \bar{R}^{\bar{a}}_{\ \bar{b}} = 0$ and $\bar{y}_{\underline{a}} \bar{R}^{\underline{a}}_{\ \underline{b}} = 0$, one obtains

$$\boldsymbol{g}_y\big(\boldsymbol{R}_y(\boldsymbol{I}_y), \boldsymbol{I}_y\big) = I_i R^i_{\ j} I^j = \frac{h_s}{h} \bar{y}_{\bar{a}} \bar{R}^{\bar{a}}_{\ \bar{b}} I^{\bar{b}} + \frac{h_t}{h} \bar{y}_{\underline{a}} \bar{R}^{\underline{a}}_{\ \underline{b}} I^{\underline{b}} = 0.$$

Since $\boldsymbol{R}_y$ is self-adjoint and nonpositive definite with respect to $\boldsymbol{g}_y$, $\boldsymbol{R}_y(\boldsymbol{I}_y) = 0$. Therefore F satisfies the conditions and conclusions in Theorem 14.1.

The next example shows that completeness in Theorem 14.1 cannot be replaced by positive completeness.

EXAMPLE 14.4. Let $\phi(y)$ be a Minkowski norm on $\mathbf{R}^n$. Let $\Theta = \Theta(x, y)$ be the Funk metric on $U := \{y \in \mathbf{R}^n \mid \phi(y) < 1\}$ defined in (2–18). Let $a \in \mathbf{R}^n$ be an arbitrary constant vector. Let

$$F := \Theta(x, y) + \frac{\langle a, y \rangle}{1 + \langle a, x \rangle}, \quad y \in TU \cong U \times \mathbf{R}^n.$$

Clearly, F is a Finsler metric near the origin. By (2–19), one sees that the spray coefficients of F are given by $G^i = P y^i$, where

$$P := \frac{1}{2}\left(\Theta(x, y) - \frac{\langle a, y \rangle}{1 + \langle a, x \rangle}\right).$$

Using this and (12–3), one obtains

$$\boldsymbol{K} = \frac{\frac{1}{4}\left(\Theta - \dfrac{\langle a, y \rangle}{1 + \langle a, x \rangle}\right)^2 - \frac{1}{2}\left(\Theta^2 + \left(\dfrac{\langle a, y \rangle}{1 + \langle a, x \rangle}\right)^2\right)}{\left(\Theta(x, y) + \dfrac{\langle a, y \rangle}{1 + \langle a, x \rangle}\right)^2} = -\frac{1}{4}.$$

Thus F has constant flag curvature $\boldsymbol{K} = -\frac{1}{4}$. See also [Shen 2003b, Example 5.3].

Now we compute the S-curvature of F. A direct computation gives

$$\frac{\partial G^m}{\partial y^m} = (n+1)P.$$

Let $dV = \sigma_F(x)\,dx^1\cdots dx^n$ be the Finsler volume form on M. From (12–4), we obtain

$$\boldsymbol{S} = \frac{n+1}{2}F(x,y) - (n+1)\frac{\langle a,y\rangle}{1+\langle a,x\rangle} - y^m\frac{\partial}{\partial x^m}\ln\sigma_F(x)$$

$$= (n+1)\big(\tfrac{1}{2}F(x,y) + d\varphi_x(y)\big),$$

where $\varphi(x) := -\ln\big((1+\langle a,x\rangle)\sigma_F(x)^{1/(n+1)}\big)$. Thus

$$\boldsymbol{E} = \tfrac{1}{4}(n+1)\,F^{-1}\boldsymbol{h},$$

where $\boldsymbol{h}_y = h_{ij}(x,y)\,dx^i \otimes dx^j$ is given by $h_{ij} = F(x,y)F_{y^iy^j}(x,y)$.

When $\phi(y) = |y|$ is the standard Euclidean norm, $U = B^n$ is the standard unit ball in $\mathbf{R}^n$ and

$$\Theta = \frac{\sqrt{|y|^2 - \big(|x|^2|y|^2 - \langle x,y\rangle^2\big)}}{1-|x|^2}.$$

Thus

$$F = \frac{\sqrt{|y|^2 - \big(|x|^2|y|^2 - \langle x,y\rangle^2\big)}}{1-|x|^2} + \frac{\langle a,y\rangle}{1+\langle a,x\rangle}.$$

Assume that $|a| < 1$. It is easy to verify that F is a Randers metric defined on the whole B^n, with constant S-curvature $\boldsymbol{S} = \tfrac{1}{2}(n+1)F(x,y)$. One can show that F is positively complete on B^n, so that every geodesic defined on an interval (λ,μ) can be extended to a geodesic defined on $(\lambda,+\infty)$.

15. Flag Curvature and Isotropic S-Curvature

It is a difficult task to classify Finsler metrics of scalar curvature. All known Randers metrics of scalar curvature have isotropic S-curvature. Thus it is a natural idea to investigate Finsler metrics of scalar curvature which also have isotropic S-curvature.

PROPOSITION 15.1 [Chen et al. 2003]. *Let (M,F) be an n-dimensional Finsler manifold of scalar curvature with flag curvature $\boldsymbol{K} = \boldsymbol{K}(x,y)$. Suppose that the S-curvature is almost isotropic,*

$$\boldsymbol{S} = (n+1)(cF + \eta),$$

where $c = c(x)$ is a scalar function on M and $\eta = \eta_i(x)y^i$ is a closed 1-form. Then there is a scalar function $\sigma = \sigma(x)$ on M such that the flag curvature equals

$$\boldsymbol{K} = 3\frac{c_{x^m}y^m}{F} + \sigma. \tag{15–1}$$

PROOF. By assumption, the flag curvature $\mathbf{K} = \mathbf{K}(x, y)$ is a scalar function on TM_0. Thus (11–8) holds. Plugging (11–8) into (14–2) yields

$$c_{x^m}y^m F_{\cdot k} - c_{x^k}F = -\tfrac{1}{3}\mathbf{K}_{y^k}F^2. \qquad (15\text{–}2)$$

Rewriting (15–2) as

$$\left(\frac{1}{3}\mathbf{K} - \frac{c_{x^m}y^m}{F}\right)_{y^k} = 0,$$

one concludes that the quantity

$$\sigma := \mathbf{K} - \frac{3c_{x^m}y^m}{F}$$

is a scalar function on M. This proves the proposition.   □

COROLLARY 15.2 [Mo 2002]. *Let F be an n-dimensional Finsler metric of scalar curvature. If F has almost constant S-curvature, the flag curvature is a scalar function on M.*

From the definition of flag curvature, one can see that every two-dimensional Finsler metric is of scalar curvature. One immediately obtains the following:

COROLLARY 15.3. *Let F be a two-dimensional Finsler metric with almost isotropic S-curvature. Then the flag curvature is in the form (15–1).*

Let $F = F(x, y)$ be a two-dimensional Berwald metric on a surface M. It follows from Corollaries 15.3 and 15.2 that the Gauss curvature $\mathbf{K} = \mathbf{K}(x)$ is a scalar function of $x \in M$. Since F is a Berwald metric, the $G^i = \tfrac{1}{2}\Gamma^i_{jk}(x)y^j y^k$ are quadratic in $y = y^i(\partial/\partial x^i)|_x \in T_x M$. By (11–1), the Riemann curvature, $R^i{}_k = R^i{}_k(x, y)$, are quadratic in y. This implies that the Ricci scalar $\mathbf{Ric} = R^m{}_m(x, y)$ is quadratic in y. Suppose that $\mathbf{K}(x_0) \neq 0$ at some point $x_0 \in M$. Then

$$F(x_0, y)^2 = \frac{\mathbf{Ric}(x_0, y)}{\mathbf{K}(x_0)}$$

is quadratic in $y \in T_{x_0}M$. Namely, $F_{x_0} = F|_{T_{x_0}M}$ is Euclidean at x_0. By Proposition 5.2, all tangent spaces $(T_x M, F_x)$ are linearly isometric to each other. One concludes that F_x is Euclidean for any $x \in M$ and F is Riemannian. Now we suppose that $\mathbf{K} \equiv 0$. Since F is Berwaldian, F must be locally Minkowskian. See [Szabó 1981] for a different argument.

16. Projectively Flat Metrics with Isotropic S-Curvature

Recall that a Finsler metric F on a manifold M is locally projectively flat if at any point $x \in M$, there is a local coordinate system (x^i) in M such that every geodesic $\sigma(t)$ is straight, i.e., $\sigma^i(t) = f(t)a^i + b^i$. This is equivalent to saying that in the standard local coordinate system (x^i, y^i), the spray coefficients G^i are in the form $G^i = Py^i$ with $P = F_{x^k}y^k/(2F)$. It is well-known that any locally projectively flat Finsler metric F is of scalar curvature, and its flag curvature equals $\boldsymbol{K} = (P^2 - P_{x^k}y^k)/F^2$ (see Proposition 12.1). Our goal is to characterize those with almost isotropic S-curvature.

First, by Beltrami's theorem and the Cartan classification theorem, a Riemannian metric is locally projectively flat if and only if it is of constant sectional curvature. Every Riemannian metric of constant sectional curvature μ is locally isometric to the metric α_μ on a ball in $\mathbf{R}^n$, defined in (2–5). A Randers metric $F = \alpha + \beta$ is locally projectively flat if and only if α is locally projectively flat (hence of constant sectional curvature) and β is closed. This follows directly from a result in [Bácsó and Matsumoto 1997] and the Beltrami theorem on projectively flat Riemannian metrics. If in addition, the S-curvature is almost isotropic, then β can be determined explicitly.

PROPOSITION 16.1 [Chen et al. 2003]. *Let $F = \alpha + \beta$ be a locally projectively flat Randers metric on an n-dimensional manifold M. Suppose that F has almost isotropic S-curvature, $\boldsymbol{S} = (n+1)(cF + \eta)$, where c is a scalar function on M and η is a closed 1-form on M. Then:*

(a) *α is locally isometric to α_μ and β is a closed 1-form satisfying*

$$(\mu + 4c^2)\beta = -c_{x^k}y^k.$$

(b) *The flag curvature is given by*

$$\boldsymbol{K} = \frac{3c_{x^k}y^k}{\alpha + \beta} + 3c^2 + \mu = \tfrac{3}{4}(\mu + 4c^2)\frac{\alpha - \beta}{\alpha + \beta} + \frac{\mu}{4}. \qquad (16\text{--}1)$$

(c) *If $\mu + 4c^2 \equiv 0$, then c is a constant and the flag curvature equals $-c^2$. In this case, $F = \alpha + \beta$ is either locally Minkowskian ($c = 0$) or, up to scaling ($c = \pm\tfrac{1}{2}$), locally isometric to the generalized Funk metric $\Theta_a = \Theta_a(x, y)$ of (2–7) or its reverse $\bar{\Theta}_a = \Theta_a(x, -y)$.*

(d) *If $\mu + 4c^2 \neq 0$, then $F = \alpha + \beta$ must be locally given by*

$$\alpha = \alpha_\mu(x, y), \qquad \beta = -\frac{2c_{x^k}(x)y^k}{\mu + 4c^2} \qquad (16\text{--}2)$$

where $c := c_\mu(x)$ is given by

$$c_\mu = \begin{cases} (\lambda + \langle a, x \rangle)\sqrt{\dfrac{\mu}{\pm(1 + \mu|x|^2) - (\lambda + \langle a, x \rangle)^2}}, & \mu \neq 0, \\[3mm] \dfrac{\pm 1}{2\sqrt{\lambda + 2\langle a, x \rangle + |x|^2}}, & \mu = 0, \end{cases}$$

for $a \in \mathbf{R}^n$ a constant vector and $\lambda \in \mathbf{R}$ a constant number.

PROOF. Let $\alpha_\mu = \sqrt{a_{ij}(x)y^iy^j}$ and $\beta = b_i(x)y^i$. We may assume that $\alpha = \alpha_\mu$ in a local coordinate system

$$a_{ij} = \frac{\delta_{ij}}{1 + \mu|x|^2} - \frac{\mu x^i x^j}{(1 + \mu|x|^2)^2}.$$

The Christoffel symbols of α are given by

$$\bar{\Gamma}^i_{jk} = -\mu \frac{x^j \delta^i_k + x^k \delta^i_j}{1 + \mu |x|^2}.$$

Thus

$$\bar{G}^i = -\frac{\mu \langle x, y \rangle}{1 + \mu |x|^2} \, y^i.$$

The spray coefficients of F are given by $G^i = \bar{G}^i + Py^i + Q^i$, where $P = e_{00}/(2F)$ and $Q^i = \alpha s^i{}_0$ are given by (9–1) and (9–2). Since β is closed, $s_{ij} := \frac{1}{2}(b_{i;j} + b_{j;i}) = 0$ and $s_i := b_j s^j{}_i = 0$. Thus $Q^i = 0$. By assumption, $\boldsymbol{S} = (n+1)(cF + \eta)$ and Lemma 9.1,

$$e_{00} = \beta_{|k} y^k = 2c(\alpha^2 - \beta^2). \qquad (16\text{–}3)$$

Thus $P = e_{00}/(2F) - s_0 = c(\alpha - \beta)$ and

$$\beta_{x^k} y^k = \beta_{|k} y^k + 2\bar{G}^k \beta_{y^k} = 2c(\alpha^2 - \beta^2) - \frac{2\mu \langle x, y \rangle \beta}{1 + \mu |x|^2}.$$

Then $G^i = \tilde{P} y^i$, where $\tilde{P} = -\dfrac{\mu \langle x, y \rangle}{1 + \mu |x|^2} + c(\alpha - \beta)$. By (12–3), we obtain

$$\boldsymbol{K} F^2 = \tilde{P}^2 - \tilde{P}_{x^k} y^k = \mu \alpha^2 + c^2 (3\alpha + \beta)(\alpha - \beta) - c_{x^k} y^k (\alpha - \beta).$$

On the other hand, by Theorem 15.1, the flag curvature is in the following form

$$\boldsymbol{K} = \frac{3 c_{x^k} y^k}{\alpha + \beta} + \sigma,$$

where $\sigma = \sigma(x)$ is a scalar function on M. It follows from the last two displayed equations that

$$2\big(2 c_{x^k} y^k + (\sigma + c^2)\beta\big)\alpha + \big(2 c_{x^k} y^k + (\sigma + c^2)\beta\big)\beta + (\sigma - 3c^2 - \mu)\alpha^2 = 0.$$

This gives

$$2 c_{x^k} y^k + (\sigma + c^2)\beta = 0, \qquad \sigma - 3c^2 - \mu = 0.$$

Solving the second of these equations for σ and substituting into the first we get

$$(\mu + 4c^2)\beta = -2 c_{x^k} y^k. \qquad (16\text{–}4)$$

To prove part (c) of the Proposition, suppose that $\mu + 4c^2 \equiv 0$. Then c is constant. It follows from (16–1) that $\boldsymbol{K} = 3c^2 + \mu = -c^2$. The local structure of F can be easily determined [Shen 2003a].

Now suppose instead that $\mu + 4c^2 \neq 0$ on an open subset $U \subset M$. By (16–4),

$$\beta = -\frac{2 c_{x^k} y^k}{\mu + 4c^2}. \qquad (16\text{–}5)$$

Note that β is exact. It follows from (16–3) and (16–5) that

$$
c_{x^i x^j} + \frac{\mu(x^i c_{x^j} + x^j c_{x^i})}{1 + \mu |x|^2}
$$

$$
= -c(\mu + 4c^2)\left(\frac{\delta_{ij}}{1 + \mu |x|^2} - \frac{\mu x^i x^j}{(1 + \mu |x|^2)^2}\right) + \frac{12 c c_{x^i} c_{x^j}}{\mu + 4c^2}. \quad (16\text{–}6)
$$

We are going to solve for c. Let

$$
f := \begin{cases} \dfrac{2c\sqrt{1 + \mu |x|^2}}{\sqrt{\pm(\mu + 4c^2)}}, & \mu \neq 0, \\[2ex] \dfrac{1}{c^2}, & \mu = 0, \end{cases}
$$

where the sign is chosen so that the radicand $\pm(\mu + 4c^2) > 0$. Then (16–6) reduces to

$$
f_{x^i x^j} = \begin{cases} 0, & \mu \neq 0, \\ 8\delta_{ij}, & \mu = 0. \end{cases}
$$

We obtain

$$
f = \begin{cases} \lambda + \langle a, x \rangle, & \mu \neq 0, \\ 4\left(\lambda + 2\langle a, x \rangle + |x|^2\right), & \mu = 0, \end{cases}
$$

where $a \in \mathbf{R}^n$ is a constant vector and λ is a constant. This gives part (d). $\square$

By Proposition 16.1, one immediately obtains:

COROLLARY 16.2. *Let $F = \alpha + \beta$ be a locally projectively flat Randers metric on an n-dimensional manifold M. Suppose that F has almost constant S-curvature $S = (n + 1)(cF + \eta)$, where c is a constant. Then F is locally Minkowskian, or Riemannian with constant curvature, or up to a scaling, locally isometric to the generalized Funk metric in (2–7).*

PROOF. Let μ be the constant sectional curvature of α. If $\mu + 4c^2 = 0$, by Proposition 16.1(c), $F = \alpha + \beta$ is either locally Minkowskian or, up to a scaling, locally isometric to the generalized Funk metric in (2–7). If $\mu + 4c^2 \neq 0$ instead, $F = \alpha + \beta$ is given by (16–2). Since $c_{x^k} = 0$, we get $\beta = 0$ and $F = \alpha$ is a Riemannian metric. $\square$

Proposition 16.1 completely classifies projectively flat Randers metrics of almost isotropic S-curvature. If a Randers metric has almost isotropic S-curvature, its the E-curvature is isotropic. By Lemma 9.1, the S-curvature is isotropic. Thus a Randers metric is of almost isotropic S-curvature if and only if it is of isotropic S-curvature. This is not true for general Finsler metrics: if $\Theta(x, y)$ is the Funk metric on a strongly convex domain $U \subset \mathbf{R}^n$, the Finsler metric

$$
F = \Theta(x, y) + \frac{\langle a, y \rangle}{1 + \langle a, x \rangle}, \quad y \in T_x U \cong \mathbf{R}^n,
$$

is projectively flat with almost isotropic S-curvature, according to Example 14.4. Thus it has isotropic E-curvature. However, this F is of isotropic S-curvature only for certain U's such as the standard unit ball.

A natural problem is whether there are other types of projectively flat Finsler metrics of almost isotropic S-curvature. Here is the answer:

PROPOSITION 16.3 [Chen and Shen 2003b]. *Let* $F = F(x,y)$ *be a projectively flat Finsler metric on a simply connected open subset* $U \subset R^n$. *Suppose that* F *has almost isotropic* S-*curvature,*

$$S = (n+1)c(F + \eta), \tag{16–7}$$

where c *is a scalar function on* M *and* η *is a closed 1-form on* U.

(a) *If* K *is not of the form* $-c^2 + c_{x^m} y^m / F$ *at every point* $x \in U$, *then* $F = \alpha + \beta$ *is a Randers metric on* U. *Further,* α *is of constant sectional curvature* $\bar{K} = \mu$ *with* $\mu + 4c^2 \neq 0$ *and* α *and* β *are as in Proposition* 16.1(c).
(b) *If* $K \equiv -c^2 + c_{x^m} y^m / F$ *on* U, *then* c *is a constant, and either* F *is locally Minkowskian* ($c = 0$) *or there exist a Funk metric* Θ *and a constant vector* $a \in R^n$ *such that* F *has the form*

$$F = \frac{1}{2c}\left\{ \Psi + \frac{\langle a, y \rangle}{1 + \langle a, x \rangle} \right.,$$

where $\Psi = \Theta(x,y)$ *if* $c = \frac{1}{2}$ *and* $\Psi = -\Theta(x,-y)$ *if* $c = -\frac{1}{2}$.

PROOF. Since F is projectively flat, the spray coefficients are given by $G^i = Py^i$, where

$$P := \frac{F_{x^k} y^k}{2F}.$$

Thus the S-curvature is given by (12–4) and the flag curvature of F is given by (12–3).

By assumption, S is of the form (16–7). Since η is closed on U, it can be written as $\eta(x,y) = dh_x(y)$, where $h = h(x)$ is a scalar function on U. Thus

$$P = cF + d\varphi_x, \tag{16–8}$$

where $\varphi(x) := h(x) + (\ln \sigma_F(x))/(n+1)$. It follows from the last two displayed equations that

$$F_{x^i} y^i = 2FP = 2F(cF + \varphi_{x^i} y^i).$$

Using this together with (16–8) and (12–3), one obtains

$$K = \frac{(cF + \varphi_{x^i} y^i)^2 - (c_{x^i} y^i F + cF_{x^i} y^i + \varphi_{x^i x^j} y^i y^j)}{F^2}$$

$$= \frac{-c^2 F^2 - c_{x^m} y^m F + (\varphi_{x^i} \varphi_{x^j} - \varphi_{x^i x^j}) y^i y^j}{F^2}. \tag{16–9}$$

On the other hand, since F is of scalar curvature, by Proposition 15.1, the flag curvature of F is given by (15–1). Comparing (16–9) with (15–1) yields

$$(\sigma + c^2)F^2 + 4c_{x^m}y^m F + (\varphi_{x^i x^j} - \varphi_{x^i}\varphi_{x^j})y^i y^j = 0. \tag{16–10}$$

Assume that $\boldsymbol{K} \neq -c^2 + c_{x^m}y^m/F$ at every point $x \in U$. Then, by (15–1), for any $x \in U$, there is a nonzero vector $y \in T_x U$ such that

$$\sigma + c^2 + \frac{2c_{x^m}y^m}{F} \neq 0.$$

We claim that $\sigma + c^2 \neq 0$ on U. If not, there is a point $x_0 \in U$ such that $\sigma(x_0) + c(x_0)^2 = 0$. The inequality above implies that $dc \neq 0$ at x_0. Then (16–10) at x_0 reduces

$$4c_{x^m}(x_0)y^m F(x_0, y) + \big(\varphi_{x^i x^j}(x_0) - \varphi_{x^i}(x_0)\varphi_{x^j}(x_0)\big)y^i y^j = 0. \tag{16–11}$$

Differentiating with respect to y^i, then restricting to the hyperplane

$$V := \{y \mid c_{x^m}(x_0)y^m = 0\},$$

one obtains

$$4c_{x^i}(x_0)F(x_0, y) + \big(\varphi_{x^i x^j}(x_0) - \varphi_{x^i}(x_0)\varphi_{x^j}(x_0)\big)y^j = 0.$$

In other words, $F(x_0, y)$ is a homogeneous linear function of $y \in V$. This is impossible, because $F(x_0, y)$ is always positive for $y \in V \setminus \{0\}$.

Now we may assume that $\sigma + c^2 \neq 0$ on U. One can solve the quadratic equation (16–10) for F,

$$F = \frac{\sqrt{(\sigma + c^2)(\varphi_{x^i x^j} - \varphi_{x^i}\varphi_{x^j})y^i y^j + 4(c_{x^m}y^m)^2} - 2c_{x^m}y^m}{\sigma + c^2}.$$

That is, F is expressed in the form $F = \alpha + \beta$, where $\alpha = \sqrt{a_{ij}y^i y^j}$ and $\beta = b_i y^i$ are given by

$$a_{ij} = \frac{(\sigma + c^2)(\varphi_{x^i x^j} - \varphi_{x^i}\varphi_{x^j}) + 4c_{x^i}c_{x^j}}{(\sigma + c^2)^2}, \qquad b_i = -\frac{2c_{x^i}}{\sigma + c^2}.$$

Since F is a Randers metric, by Lemma 9.1, one concludes that $\boldsymbol{S}$ is isotropic, i.e., $\eta = 0$ and

$$\boldsymbol{S} = (n+1)cF.$$

Since F is projectively flat, α is of constant sectional curvature $\bar{\boldsymbol{K}} = \mu$ and β is closed. Moreover, by Proposition 16.1, the flag curvature is given by (16–1). Note that $\sigma + c^2 \neq 0$ is equivalent to the inequality $\mu + 4c^2 \neq 0$. By Proposition 16.1(d), F is given by (16–2).

We now assume that $\boldsymbol{K} \equiv -c^2 + c_{x^i}y^i/F$. It follows from (15–1) that

$$\sigma + c^2 + \frac{2c_{x^m}y^m}{F} \equiv 0.$$

Suppose that $c_{x^m}(x_0)y^m \neq 0$ at some point x_0. From the preceding identity, one sees that $\sigma(x_0) + c(x_0)^2 \neq 0$. Thus

$$F(x_0, y) = -\frac{2c_{x^m}(x_0)y^m}{\sigma(x_0) + c(x_0)^2}$$

is a linear function. This is impossible. One concludes that $c_{x^m}y^m = 0$ on U, and hence c is a constant and $\sigma(x) = -c^2$ is a constant too. In this case, the flag curvature is given by $\boldsymbol{K} = -c^2$. Equation (16–10) reduces to

$$\varphi_{x^i x^j} - \varphi_{x^i}\varphi_{x^j} = 0,$$

which is easily solved to yield

$$\varphi = -\ln(1 + \langle a, x \rangle) + C,$$

where $a \in \mathbb{R}^n$ is a constant vector and C is a constant.

Assume that $c = 0$. Then $\boldsymbol{K} = -c^2 = 0$. It follows from (16–8) that the projective factor $P = d\varphi_x$ is a 1-form, hence the spray coefficients $G^i = Py^i$ are quadratic in $y \in T_xU$. By definition, F is a Berwald metric, and every Berwald metric with $\boldsymbol{K} = 0$ is locally Minkowskian (see [Bao et al. 2000] for a proof).

Assume that $c \neq 0$. By (16–8), $P = cF + d\varphi$. With $\Psi := P + cF = 2cF + d\varphi_x$, we have

$$F = \frac{1}{2c}(\Psi(x, y) - d\varphi_x) = \frac{1}{2c}\left(\Psi(x, y) + \frac{\langle a, y \rangle}{1 + \langle a, x \rangle}\right).$$

Since F is projectively flat and P is the projective factor,

$$F_{x^k} = (PF)_{y^k}, \quad P_{x^k} = PP_{y^k} + c^2 FF_{y^k}.$$

These equations imply that $\Psi_{x^i} = \Psi\Psi_{y^i}$. Let $\Theta := \Psi(x, y)$ if $c > 0$ and $\Theta := -\Psi(x, -y)$ if $c < 0$. Then Θ is a Funk metric and F is of the form stated in part (b) of the theorem. □

17. Flag Curvature and Relatively Isotropic L-Curvature

Although the relatively isotropic J-curvature condition is stronger than the isotropic S-curvature condition for Randers metrics (Lemma 10.1), it seems that there is no direct relationship between these two conditions. Nevertheless, for Finsler metrics of scalar curvature, the relatively isotropic J-curvature condition also implies that the flag curvature takes a special form in certain cases.

PROPOSITION 17.1 [Chen et al. 2003]. *Let F be an n-dimensional Finsler manifold of scalar curvature and of relatively constant J-curvature, so that*

$$\boldsymbol{J} + cF\boldsymbol{I} = 0, \tag{17--1}$$

for some constant c. Then

$$\boldsymbol{K} = -c^2 + \sigma e^{-3\tau/(n+1)},$$

where $\tau = \tau(x, y)$ is the distortion and $\sigma = \sigma(x)$ is a scalar function on M.

PROOF. By assumption, $J_k = -cFI_k$. Using (13–6) and (13–7), one obtains

$$I_{k|p|q}y^p y^q = J_{k|m}y^m = -cFI_{k|m}y^m = c^2 F^2 \tau_{\cdot k}.$$

Plugging this into the second line of (13–15) yields

$$\tfrac{1}{3}(n+1)\boldsymbol{K}_{\cdot k} + (\boldsymbol{K} + c^2)\tau_{\cdot k} = 0.$$

This implies that

$$\left((\boldsymbol{K} + c^2)^{(n+1)/3}e^\tau\right)_{\cdot k} = (\boldsymbol{K} + c^2)^{(n-2)/3}e^\tau\left(\tfrac{1}{3}(n+1)\boldsymbol{K}_{\cdot k} + \boldsymbol{K}\tau_{\cdot k}\right) = 0.$$

Thus the function $(\boldsymbol{K} + c^2)^{(n+1)/3}e^\tau$ is independent of $y \in T_x M$.   □

Proposition 17.1 in the particular case $c = 0$ was essentially achieved in the proof of [Matsumoto 1972a, Proposition 26.2], where it is assumed that F is a Landsberg metric, but what is needed is merely that $\boldsymbol{J} = 0$. Since the notion of distortion is not introduced in [Matsumoto 1972a], the result is stated in a local coordinate system.

COROLLARY 17.2. *Let F be a Finsler metric on a manifold M. Suppose that F has isotropic flag curvature not equal to $-c^2$ and that F has relatively constant J-curvature. Then F is Riemannian.*

PROOF. By Proposition 17.1,

$$\boldsymbol{K}(x) = -c^2 + \sigma(x)e^{-3\tau/(n+1)}.$$

Since $\boldsymbol{K}(x) \neq -c^2$, one concludes that $\sigma(x) \neq 0$, so $\tau = \tau(x)$ is independent of $y \in T_x M$. It follows from (7–2) that $I_i = \tau_{y^i} = 0$. Thus F is Riemannian by Deicke's theorem [Deicke 1953].   □

PROPOSITION 17.3. *Let F be a Finsler metric of scalar curvature on an n-dimensional manifold. Suppose that F has relatively isotropic L-curvature, so*

$$\boldsymbol{L} + cF\boldsymbol{C} = 0, \tag{17–2}$$

where c is a scalar function on M.

(a) *If c is constant, then*

$$\boldsymbol{K} = -c^2 + \sigma e^{-3\tau/(n+1)},$$

   where σ is a scalar function on M.

(b) *If $n \geq 3$ and $\boldsymbol{K} \neq -c^2 + c_{x^m}y^m/F$ for almost all $y \in T_x M \setminus \{0\}$ at any point x in an open domain U of M, then $F = \alpha + \beta$ is a Randers metric in U.*

PROOF. If F has relatively isotropic L-curvature, (17–1) holds by taking the average of (17–2) on both sides. Statement (a) then follows from Proposition 17.1.

Now we assume that $\boldsymbol{K} \neq -c^2 + c_{x^m}(x)y^m/F$ for almost all $y \in T_x M \setminus \{0\}$ at any point x in an open domain $U \subset M$. By assumption, $L_{ijk} = -cFC_{ijk}$, one obtains

$$C_{ijk|p|q}y^p y^q = -c_{x^m}y^m F C_{ijk} - cF L_{ijk} = \left(c^2 - \frac{c_{x^m}y^m}{F}\right)F^2 C_{ijk}.$$

Since $J_k = -cFI_k$ by (17–1), we have

$$I_{k|p|q}y^p y^q = -c_{x^m}y^m F I_k - cF J_k = \left(c^2 - \frac{c_{x^m}y^m}{F}\right)F^2 I_k.$$

By the formula for M_{ijk} in (3–2), one obtains

$$M_{ijk|p|q}y^p y^q = \left(c^2 - \frac{c_{x^m}y^m}{F}\right)F^2 M_{ijk}.$$

Since F is of scalar curvature, equation (13–16) holds. One obtains

$$\left(\boldsymbol{K} + c^2 - \frac{c_{x^m}y^m}{F}\right)F^2 M_{ijk} = 0.$$

It follows that $M_{ijk} = 0$, so the Matsumoto torsion vanishes. By Proposition 3.3, $F = \alpha + \beta$ is a Randers metric on U.   □

Proposition 17.3 was proved by H. Izumi [1976; 1977; 1982], The particular case $c = 0$ is proved by S. Numata [1975].

COROLLARY 17.4 [Numata 1975]. *Let F be a Finsler metric of scalar curvature on an n-dimensional manifold, with $n \geq 3$. Suppose that $\boldsymbol{L} = 0$ and $\boldsymbol{K} \neq 0$. Then F is Riemannian.*

PROOF. By Proposition 17.3, $F = \alpha + \beta$ is a Randers metric with $\boldsymbol{L} = 0$. By Lemma 10.1, $\boldsymbol{S} = 0$ and β is closed. By Proposition 15.1, one concludes that $\boldsymbol{K} = \sigma(x)$ is a scalar function on M. It follows from (13–14) that $0 = -F^2\sigma(x)I_k$. By assumption, $\boldsymbol{K} = \sigma(x) \neq 0$. Thus $I_k = 0$ and F is Riemannian by Deicke's theorem.   □

We may ask again: is there a non-Berwaldian Finsler metric satisfying $\boldsymbol{K} = 0$ and $\boldsymbol{L} = 0$ (or $\boldsymbol{J} = 0$)? If such a metric exists, it cannot be locally projectively flat and it cannot be a Randers metric. (Why?)

EXAMPLE 17.5. Let $F = \alpha + \beta$ be the Randers metric on $\mathbf{R}^n$ defined by

$$F := |y| + \frac{\langle x, y \rangle}{\sqrt{1 + |x|^2}}, \quad y \in T_x\mathbf{R}^n \cong \mathbf{R}^n.$$

Note that

$$\|\beta\|^2 = \frac{|x|^2}{1 + |x|^2} < 1.$$

F is indeed a Randers metric on the whole of $\mathbf{R}^n$. One can verify that F satisfies (12–5). Thus it is a projectively flat Randers metric on $\mathbf{R}^n$. Further, the spray coefficients $G^i = Py^i$ are given by

$$P = c\left(|y| - \frac{\langle x, y \rangle}{\sqrt{1 + |x|^2}}\right),$$

where $c = 1/\left(2\sqrt{1 + |x|^2}\right)$. Let $\rho := \ln\sqrt{1 - \|\beta\|^2} = -\ln\sqrt{1 + |x|^2}$. By (9–3), one obtains $\boldsymbol{S} = (n+1)(P - \rho_0) = (n+1)cF$. Since β is closed, this is equivalent, by Proposition 10.1, to the identity $\boldsymbol{L} + cF\boldsymbol{C} = 0$.

Since F is projectively flat, it is of scalar curvature. Further computation yields the flag curvature:

$$\boldsymbol{K} = \frac{P^2 - P_{x^k}y^k}{F^2} = \frac{3}{4(1 + |x|^2)} \cdot \frac{|y|\sqrt{1 + |x|^2} - \langle x, y \rangle}{|y|\sqrt{1 + |x|^2} + \langle x, y \rangle}.$$

Note that $\boldsymbol{K} \neq -c^2 + c_{x^k}(x)y^k/F(x, y)$ and that F is a Randers metric. This matches the conclusion in Proposition 17.3(b).

The Randers metric in Example 17.5 is locally projectively flat. There are non–projectively flat Randers metrics of scalar curvature and isotropic S-curvature; see Example 11.2. This example is a Randers metric generated by a special vector field on the Euclidean space by (2–15). In fact, we can determine all vector fields V on a Riemannian space form (M, α_μ) of constant curvature μ such that the generated Randers metric $F = \alpha + \beta$ by (α_μ, V) is of scalar curvature and isotropic S-curvature. This work will appear elsewhere.

References

[Akbar-Zadeh 1988] H. Akbar-Zadeh, "Sur les espaces de Finsler á courbures section-nelles constantes", *Acad. Roy. Belg. Bull. Cl. Sci.* **74** (1988), 281–322.

[Antonelli et al. 1993] P. Antonelli, R. Ingarden and M. Matsumoto, *The theory of sprays and Finsler spaces with applications in physics and biology*, Kluwer, Dordrecht, 1993.

[Auslander 1955] L. Auslander, "On curvature in Finsler geometry", *Trans. Amer. Math. Soc.* **79** (1955), 378–388.

[Bácsó and Matsumoto 1997] S. Bácsó and M. Matsumoto, "On Finsler spaces of Dou-glas type: a generalization of the notion of Berwald space", *Publ. Math. Debrecen*, **51** (1997), 385–406.

[Bao and Chern 1993] D. Bao and S. S. Chern, "On a notable connection in Finsler geometry", *Houston J. Math.* **19**:1 (1993), 135–180.

[Bao et al. 2000] D. Bao, S. S. Chern and Z. Shen, *An Introduction to Riemann–Finsler geometry*, Springer, New York, 2000

[Bao et al. 2003] D. Bao, C. Robles and Z. Shen, *Zermelo navigation on Riemannian manifolds*, preprint, 2003.

[Bao–Robles 2003] D. Bao and C. Robles, "On Randers spaces of constant flag curvature", *Rep. on Math. Phys.* **51** (2003), 9–42.

[Bao and Robles 2003b] D. Bao and C. Robles, *Ricci and flag curvatures in Finsler geometry*, pp. 199–261 in *A sampler of Riemann–Finsler Geometry*, edited by D. Bao et al., Math. Sci. Res. Inst. Publ. **50**, Cambridge Univ. Press, Cambridge, 2004.

[Bao–Shen 2002] D. Bao and Z. Shen, "Finsler metrics of constant positive curvature on the Lie group S^3", *J. London Math. Soc.* **66** (2002), 453–467.

[Bejancu–Farran 2002] A. Bejancu and H. Farran, "Finsler metrics of positive constant flag curvature on Sasakian space forms", *Hokkaido Math. J.* **31**:2 (2002), 459–468.

[Bejancu and Farran 2003] A. Bejancu and H. R. Farran, "Randers manifolds of positive constant flag curvature", *Int. J. Math. Sci.* **18** (2003), 1155–1165.

[Berwald 1926] L. Berwald, "Untersuchung der Krümmung allgemeiner metrischer Räume auf Grund des in ihnen herrschenden Parallelismus", *Math. Z.* **25** (1926), 40–73.

[Berwald 1928] L. Berwald, "Parallelübertragung in allgemeinen Räumen", *Atti Congr. Intern. Mat. Bologna* **4** (1928), 263–270.

[Berwald 1929a] L. Berwald, "Über eine characteristic Eigenschaft der allgemeinen Räume konstanter Krümmung mit gradlinigen Extremalen", *Monatsh. Math. Phys.* **36** (1929), 315–330.

[Berwald 1929b] L. Berwald, "Über die n-dimensionalen Geometrien konstanter Krümmung, in denen die Geraden die kürzesten sind", *Math. Z.* **30** (1929), 449–469.

[Bryant 1996] R. Bryant, "Finsler structures on the 2-sphere satisfying $K = 1$", pp. 27–42 in *Finsler Geometry* (Seattle, 1995), edited by D. Bao et al., Contemporary Mathematics **196**, Amer. Math. Soc., Providence, 1996.

[Bryant 1997] R. Bryant, "Projectively flat Finsler 2-spheres of constant curvature", *Selecta Math. (N.S.)* **3** (1997), 161–203.

[Bryant 2002] R. Bryant, "Some remarks on Finsler manifolds with constant flag curvature", *Houston J. Math.* **28**:2 (2002), 221–262.

[Busemann 1947] H. Busemann, "Intrinsic area", *Ann. of Math.* **48** (1947), 234–267.

[Cartan 1934] E. Cartan, *Les espaces de Finsler*, Actualités scientifiques et industrielles **79**, Hermann, Paris, 1934.

[Chen et al. 2003] X. Chen, X. Mo and Z. Shen, "On the flag curvature of Finsler metrics of scalar curvature", *J. London Math. Soc.* (2) **68**:3 (2003), 762–780.

[Chen and Shen 2003a] X. Chen and Z. Shen, "Randers metrics with special curvature properties", *Osaka J. Math.* **40** (2003), 87–101.

[Chen and Shen 2003b] X. Chen and Z. Shen, "Projectively flat Finsler metrics with almost isotropic S-curvature", preprint, 2003.

[Chern 1943] S. S. Chern, "On the Euclidean connections in a Finsler space", *Proc. National Acad. Soc.*, **29** (1943), 33–37; reprinted as pp. 107–111 in *Selected Papers*, v. 2, Springer, New York, 1989.

[Chern 1948] S. S. Chern, "Local equivalence and Euclidean connections in Finsler spaces", *Science Reports Nat. Tsing Hua Univ.* **5** (1948), 95–121.

[Chern 1992] S. S. Chern, "On Finsler geometry", *C. R. Acad. Sc. Paris* **314** (1992), 757–761.

[Deicke 1953] A. Deicke, "Über die Finsler-Räume mit $A_i = 0$", *Arch. Math.* **4** (1953), 45–51.

[Foulon 2002] P. Foulon, "Curvature and global rigidity in Finsler geometry", *Houston J. Math.* **28** (2002), 263–292.

[Funk 1929] P. Funk, "Über Geometrien bei denen die Geraden die Kürzesten sind", *Math. Ann.* **101** (1929), 226–237.

[Funk 1936] P. Funk, "Über zweidimensionale Finslersche Räume, insbesondere über solche mit geradlinigen Extremalen und positiver konstanter Krümmung", *Math. Z.* **40** (1936), 86–93.

[Hamel 1903] G. Hamel, "Über die Geometrien in denen die Geraden die Kürzesten sind", *Math. Ann.* **57** (1903), 231–264.

[Hashiguchi and Ichijyō 1975] M. Hashiguchi and Y. Ichijyō, "On some special (α, β)-metrics", *Rep. Fac. Sci. Kagoshima Univ.* **8** (1975), 39–46.

[Hrimiuc and Shimada 1996] H. Hrimiuc and H. Shimada, "On the L-duality between Finsler and Hamilton manifolds", *Nonlinear World* **3** (1996), 613–641.

[Ichijyō 1976] Y. Ichijyō, "Finsler spaces modeled on a Minkowski space", *J. Math. Kyoto Univ.* **16** (1976), 639–652.

[Izumi 1976] H. Izumi, "On *P-Finsler spaces, I", *Memoirs of the Defense Academy* **16** (1976), 133–138.

[Izumi 1977] H. Izumi, "On *P-Finsler spaces, II", *Memoirs of the Defense Academy* **17** (1977), 1–9.

[Izumi 1982] H. Izumi, "On *P-Finsler spaces of scalar curvature", *Tensor (N.S.)* **38** (1982), 220–222.

[Ji and Shen 2002] M. Ji and Z. Shen, "On strongly convex indicatrices in Minkowski geometry", *Canad. Math. Bull.* **45**:2 (2002), 232–246.

[Kikuchi 1979] S. Kikuchi, "On the condition that a space with (α, β)-metric be locally Minkowskian", *Tensor (N.S.)* **33** (1979), 242–246.

[Kim and Yim 2003] C.-W. Kim and J.-W. Yim, "Finsler manifolds with positive constant flag curvature", *Geom. Dedicata* **98** (2003), 47–56.

[Kosambi 1933] D. Kosambi, "Parallelism and path-spaces", *Math. Z.* **37** (1933), 608–618.

[Kosambi 1935] D. Kosambi, "Systems of differential equations of second order", *Quart. J. Math.* **6** (1935), 1–12.

[Matsumoto 1972a] M. Matsumoto, *Foundations of Finsler geometry and special Finsler spaces*, Kaiseisha Press, Japan, 1986.

[Matsumoto 1972b] M. Matsumoto, "On C-reducible Finsler spaces", *Tensor (N.S.)* **24** (1972), 29–37.

[Matsumoto 1974] M. Matsumoto, "On Finsler spaces with Randers metric and special forms of important tensors", *J. Math. Kyoto Univ.* **14** (1974), 477–498.

[Matsumoto 1980] M. Matsumoto, "Projective changes of Finsler metrics and projectively flat Finsler spaces", *Tensor (N.S.)* **34** (1980), 303–315.

[Matsumoto 1989] M. Matsumoto, "Randers spaces of constant curvature", *Rep. Math. Phys.* **28** (1989), 249–261.

[Matsumoto and Hōjō 1978] M. Matsumoto and S. Hōjō, "A conclusive theorem on C-reducible Finsler spaces", *Tensor (N.S.)* **32** (1978), 225–230.

[Matsumoto and Shimada 2002] M. Matsumoto and H. Shimada, "The corrected fundamental theorem on Randers spaces of constant curvature", *Tensor (N.S.)* **63** (2002), 43–47.

[Mo 1999] X. Mo, "The flag curvature tensor on a closed Finsler space", *Results Math.* **36** (1999), 149–159.

[Mo 2002] X. Mo, *On the flag curvature of a Finsler space with constant S-curvature,* to appear in *Houston J. Math.*

[Mo and Shen 2003] X. Mo and Z. Shen, *On negatively curved Finsler manifolds of scalar curvature,* to appear in *Canadian Math. Bull.*.

[Mo and Yang 2003] X. Mo and C. Yang, "Non-reversible Finsler metrics with non-zero isotropic S-curvature", preprint, 2003.

[Numata 1975] S. Numata, "On Landsberg spaces of scalar curvature", *J. Korea Math. Soc.* **12** (1975), 97–100.

[Okada 1983] T. Okada, "On models of projectively flat Finsler spaces of constant negative curvature", *Tensor (N.S.)* **40** (1983), 117–123.

[Rapcsák 1961] A. Rapcsák, "Über die bahntreuen Abbildungen metrischer Räume", *Publ. Math. Debrecen* **8** (1961), 285–290.

[Sabau and Shimada 2001] V. S. Sabau and H. Shimada, "Classes of Finsler spaces with (α, β)-metrics", *Rep. Math. Phy.* **47** (2001), 31–48.

[Shen 1996] Z. Shen, "Finsler spaces of constant positive curvature", pp. 83–92 in *Finsler Geometry* (Seattle, 1995), edited by D. Bao et al., Contemporary Mathematics **196**, Amer. Math. Soc., Providence, 1996.

[Shen 1997] Z. Shen, "Volume comparison and its applications in Riemann–Finsler geometry", *Advances in Math.* **128** (1997), 306–328.

[Shen 2001a] Z. Shen, *Differential geometry of sprays and Finsler spaces*, Kluwer, Dordrecht 2001.

[Shen 2001b] Z. Shen, *Lectures on Finsler geometry*, World Scientific, Singapore, 2001.

[Shen 2002] Z. Shen, "Two-dimensional Finsler metrics of constant flag curvature", *Manuscripta Mathematica* **109**:3 (2002), 349–366.

[Shen 2003a] Z. Shen, "Projectively flat Randers metrics of constant flag curvature", *Math. Ann.* **325** (2003), 19–30.

[Shen 2003b] Z. Shen, "Projectively flat Finsler metrics of constant flag curvature", *Trans. Amer. Math. Soc.* **355**(4) (2003), 1713–1728.

[Shen 2003c] Z. Shen, "Finsler metrics with $K = 0$ and $S = 0$", *Canadian J. Math.* **55**:1 (2003), 112–132.

[Shen 2003d] Z. Shen, "Nonpositively curved Finsler manifolds with constant S-curvature", Math. Z. (to appear).

[Shibata et al. 1977] C. Shibata, H. Shimada, M. Azuma and H. Yasuda, "On Finsler spaces with Randers' metric", *Tensor (N.S.)* **31** (1977), 219–226.

[Szabó 1977] Z. I. Szabó, "Ein Finslerscher Raum ist gerade dann von skalarer Krümmung, wenn seine Weylsche Projektivkrümmung verschwindet", *Acta Sci. Math.* (Szeged) **39** (1977), 163–168.

[Szabó 1981] Z. I. Szabó, "Positive definite Berwald spaces (structure theorems on Berwald spaces)", *Tensor* (*N.S.*) **35** (1981), 25–39.

[Xing 2003] H. Xing, "The geometric meaning of Randers metrics with isotropic *S*-curvature", preprint.

[Yasuda and Shimada 1977] H. Yasuda and H. Shimada, "On Randers spaces of scalar curvature", Rep. Math. Phys. **11** (1977), 347–360.

ZHONGMIN SHEN
DEPARTMENT OF MATHEMATICAL SCIENCES
INDIANA UNIVERSITY – PURDUE UNIVERSITY INDIANAPOLIS
402 N. BLACKFORD STREET
INDIANAPOLIS, IN 46202-3216
UNITED STATES
zshen@math.iupui.edu

Riemann–Finsler Geometry
MSRI Publications
Volume **50**, 2004

Index

NOTE. This is a partial index and the level of coverage varies from article to article. If you don't find your entry here, it doesn't mean it's not in the book! Searchable PDF files for the articles are available at www.msri.org.